ADVANCES IN CHEMICAL PHYSICS

VOLUME LVI

EDITORIAL BOARD

Advances in
CHEMICAL PHYSICS

EDITED BY

I. PRIGOGINE

University of Brussels
Brussels, Belgium
and
University of Texas
Austin, Texas

AND

STUART A. RICE

Department of Chemistry
and
The James Franck Institute
The University of Chicago
Chicago, Illinois

VOLUME LVI

AN INTERSCIENCE® PUBLICATION
JOHN WILEY & SONS
NEW YORK · CHICHESTER · BRISBANE · TORONTO · SINGAPORE

An Interscience® Publication

Library of Congress Catalog Number: 58-9935

ISBN 0-471-87829-4

Printed in the United States of America

10 9 8 7 6 5 4 3 2 1

CONTRIBUTORS TO VOLUME LVI

STEVEN L. CARNIE, Department of Chemistry, University of Toronto, Toronto, Ontario, Canada

M. R. HOARE, Department of Physics, Bedford College, Regent's Park, London, England

THOMAS KEYES, Department of Chemistry, Boston University, Boston, Massachusetts

DANIEL KIVELSON, Department of Chemistry, University of California, Los Angeles, California

BRANKA M. LADANYI, Department of Chemistry, Colorado State University, Fort Collins, Colorado

PAUL MADDEN, Theoretical Physics Section, Royal Signals and Radar Establishment, Great Malvern, United Kingdom

AKIO MORITA, Department of Chemistry, Akita University, Akita-shi, Japan

GLENN M. TORRIE, Department of Mathematics and Computer Science, Royal Military College, Kingston, Ontario, Canada

HIROSHI WATANABE, Department of Chemistry, College of Arts and Sciences, University of Tokyo, Komaba, Tokyo, Japan

INTRODUCTION

Few of us can any longer keep up with the flood of scientific literature, even in specialized subfields. Any attempt to do more and be broadly educated with respect to a large domain of science has the appearance of tilting at windmills. Yet the synthesis of ideas drawn from different subjects into new, powerful, general concepts is as valuable as ever, and the desire to remain educated persists in all scientists. This series, *Advances in Chemical Physics*, is devoted to helping the reader obtain general information about a wide variety of topics in chemical physics, which field we interpret very broadly. Our intent is to have experts present comprehensive analyses of subjects of interest and to encourage the expression of individual points of view. We hope that this approach to the presentation of an overview of a subject will both stimulate new research and serve as a personalized learning text for beginners in a field.

ILYA PRIGOGINE
STUART A. RICE

CONTENTS

ADVANCES IN CHEMICAL PHYSICS

VOLUME LVI

QUADRATIC TRANSPORT AND SOLUBLE BOLTZMANN EQUATIONS

M. R. HOARE

Department of Physics, Bedford College
Regent's Park, London, England

CONTENTS

I. INTRODUCTION

In 1976 Bobylev in the Soviet Union[1,2] and Krook and Wu in the United States[3,4] almost simultaneously discovered particular exact solutions to a scalar Boltzmann equation derivable from an explicit, physically relevant kinetic model, that of "pseudo-Maxwell molecules." Although there had been hints of such a possibility in earlier work,[5,6] and soluble equations of the Boltzmann type were not altogether unknown, most examples up to this time were either too artificial to generate wide interest or, as in some cases of crucial importance, simply passed unnoticed in the applied probability literature.[7-9]

Whatever its claims to priority, there is no doubt that the Bobylev-Krook-Wu (BKW) model was the first to open a direct avenue to the great wealth of experience surrounding the full Boltzmann equation for a dilute gas and as such generated an immediate response, paralleling the recent upsurge of interest in nonlinear physical systems of many kinds. And just as elsewhere, the realization that for once an intractable-looking nonlinear system could yield improbably simple solutions proved an invaluable stimulus to the search both for more elaborate soluble models and for a better understanding of the underlying structures leading to solubility.

Not all this early research can be said to have led to tangible progress; one of the curious side effects of the success of the BKW model was the emergence of several somewhat ambitious conjectures about the general character of Boltzmann equation solutions—most notably the Krook-Wu conjecture of the "universality" of the similarity solution[4] and the McKean conjecture of the complete monotonicity of the entropy function.[10] Both of these have been decisively proved false, but not before inspiring a plethora of papers, which, to varying degrees have both deepened and confused our understanding of the problems involved.

At the time of writing, the volume of papers more or less directly inspired by the BKW solution has reached proportions which, while not precluding a systematic review, would render one more tedious than instructive. A very thorough and still reasonably up-to-date account from the standpoint of kinetic theory is in any case available and should certainly be studied alongside the present one.[11,12] In these pages a conscious attempt will be made to draw emphasis away from present preoccupations with kinetic theory models toward the consideration of scalar Boltzmann-type equations as formulations of stochastic processes in their own right, with application in the many fields, both inside and outside physics, where quadratic transport is of the essence. We shall thus be concerned here with quite general types of statistical-dynamic systems that share the key characteristic that changes in either population numbers or probability densities are effected through the binary interaction of two participants in an elementary encounter of whatever sort—be they atoms or molecules in collision, polymer molecules connecting, gametes forming a zygote, or players competing pairwise in a collective game or tournament.

While we shall not be able to discuss all these cases in detail, I hope to establish here the main points of connection between the results forthcoming from gas dynamics and those that arise most naturally in polymer science and other areas of physical chemistry. In Section II, quadratic transport equations are considered in general terms, the traditional formulation of the Boltzmann equation being linked with the scalar models introduced by Moran and others. Various simplifying assumptions are considered, and the resulting equations are classified while their general properties are examined. Particular attention is given to discrete variable models and the proper analogies between continuous- and discrete-variable cases. In Section III, we consider specific soluble models for scalar Boltzmann equations, proceeding from the almost trivial two- and three-state models through the cases of increasing difficulty to the recent "persistent-scattering" model of Futcher and Hoare. In Section IV, the connection of these with the original BKW model in velocity space is retraced and the latter is examined in detail. Section V is devoted to various transform properties that bring out the interconnections between apparently separate models as well as providing useful solution techniques. In the final sections we turn to recent advances in aggregation-fragmentation kinetics, some of which are closely connected with Boltzmann equation studies. Particularly important in this connection are the new general solutions to the Blatz-Tobolsky reversible polymerization system and the renewed interest in the treatment of the gelation transition through Smoluchowski-type kinetic equations.

II. QUADRATIC TRANSPORT EQUATIONS

The first study of quadratic transport in a general stochastic context appears to be that of Moran,[7,13] who, though primarily interested in genetic models, consciously used a scalar imitation of the Boltzmann equation as a prototype for more general cases. His work was later extended by Nishimura,[8,9] who gave numerous examples of quadratic interaction models while studying in some generality the behavior of moment equations and the approach to a stationary state. Nishimura's formulation will serve as a useful starting point for the present study.

Before we develop this, however, let us review briefly the more familiar case of a *linear* stochastic ensemble of elements, which can generally be in any of an infinite set of discrete states, indexed $i = 0, 1, 2 \ldots$ at a given time. Adopting for simplicity a discrete time scale indexed $n = 0, 1, 2 \ldots$, and assuming Markovian behavior, we can specify the probability of a state j succeeding a state i by the *transition probability matrix* $\Pi(j, i)$ through which the time evolution of the ensemble is completely determined. If $p(i, n)$ is the probability distribution for state i at time n, then the evolution can be seen to be determined through the equation

$$p(i, n+1) = \sum_j \Pi(i, j) p(j, n) \tag{2.1}$$

By transition to a continuous time scale through the limiting process $n \to \infty$, $h \to 0$, $t = nh$ (see, e.g., Ref. 14 for a proper treatment) we can arrive at a Kolmogorov equation, or "master equation" in the physics tradition. In this way we find

$$\frac{d\bar{p}(i, t)}{dt} = \sum_j \left\{ \bar{k}(i, j) \bar{p}(j, t) - \bar{k}(j, i) \bar{p}(i, t) \right\} \tag{2.2}$$

and its continuous-variable counterpart

$$\frac{dp(x, t)}{dt} = \int dy \left\{ k(x, y) p(y, t) - k(y, x) p(x, t) \right\} \tag{2.3}$$

Note that here the quantities $\bar{k}(.,.)$ and $k(.,.)$ have become *transition rates* with dimension $(\text{time})^{-1}$ in contrast to the pure probabilities $\Pi(.,.)$. The theory of the master equation, in particular its solution in terms of the eigenvalue properties of the kernel k, is well known and has been widely reviewed. (See, for example, Van Kampen.[15])

Returning to the question of quadratic transport, it is natural to enquire at this stage what analogues of the above equations should be written for the case where elements in the system do not change state individually and at random, as implied in (2.2) and (2.3), but instead as a result of binary interactions in pairs. We imagine a pair of elements in states i and j, respectively, scattering to produce a pair in states k and l as a result of the symbolic reaction

$$(i)+(j) \rightarrow (k)+(l)$$

Following Moran and Nishimura, we may assign a probability to this transition in a discrete-time model in the form $\Pi(k, l|i, j)$. If we then write $p(i, n)$ for the probability that a randomly chosen element shall be found in the ith state after the nth successive trial, the evolution equation for $p(i, n)$ will evidently be

$$p(i, n+1) = \sum_j \sum_k \sum_l \Pi(i, j|k, l) P(k, n) P(l, n) \qquad (2.4)$$

In writing this we make the tacit assumptions (1) that all pairs interact at random and in a manner statistically independent of their state at a previous time and (2) that every element shall be paired in the course of each time interval. So far the only mathematical conditions we impose on the above are the *interactional symmetry*

$$\Pi(i, j|k, l) = \Pi(j, i|l, k) \qquad (2.5)$$

and the requirement that $\Pi(i, j|k, l)$ be a probability distribution:

$$\Pi(i, j|k, l) \geqslant 0$$
$$\sum_i \sum_j \Pi(i, j|k, l) = 1 \qquad (2.6)$$

for all k, l.

Moran's equation [(2.4)] points the way to suitable analogues in the continuous-time case. Without going into details of the transition, we can see that, in the same spirit as the derivation of (2.2) and (2.3) from (2.1), the natural equations for quadratic transport in continuous time become

$$\frac{d\bar{P}(i, t)}{dt} = \sum_j \sum_k \sum_l \{ \bar{W}(i, j|k, l) \bar{P}(k) \bar{P}(l) - \bar{W}(k, l|i, j) \bar{P}(i) \bar{P}(j) \}$$

$$(2.7)$$

and

$$\frac{dP(x,t)}{dt} = \int dy \int dv \int dw \left\{ W(x, y|v, w) P(v) P(w) \right.$$
$$\left. - W(v, w|x, y) P(x) P(y) \right\} \tag{2.8}$$

for the discrete and continuous variable, respectively.*

In going over to the continuous-time formulation we have introduced an additional element into the description of the system: the total transition rate to all outcomes may be a function of the initial states. Thus we allow the possibility that

$$\sum_k \sum_l \overline{W}(k, l|i, j) = \bar{z}(i, j) \tag{2.9}$$

with $\bar{z}$ a state-dependent quantity in the nature of a collision number.

Equations (2.7) and (2.8) bear a clear resemblance to the classical Boltzmann equation for a velocity distribution, $f(\mathbf{v}, t)$, in its "transition-probability" form. This we may write

$$\frac{\partial}{\partial t} f(\mathbf{v}, t) = \int \int \int d\mathbf{w} \, d\mathbf{v}' \, d\mathbf{w}' \left\{ W(\mathbf{v}, \mathbf{w}|\mathbf{v}', \mathbf{w}') f(\mathbf{v}') f(\mathbf{w}') \right.$$
$$\left. - W(\mathbf{v}', \mathbf{w}'|\mathbf{v}, \mathbf{w}) f(\mathbf{v}) f(\mathbf{w}) \right\} \tag{2.10}$$

Here $W(\mathbf{v}, \mathbf{w}|\mathbf{v}', \mathbf{w}')$ is the transition rate for scattering from initial velocities $\mathbf{v}, \mathbf{w}$ to final velocities[16] $\mathbf{v}', \mathbf{w}'$ and contains implicit delta functions guaranteeing the conservation of momentum and energy. In fact we recover the more familiar form of the spatially uniform equation in d dimensions on putting

$$W(\mathbf{v}, \mathbf{w}|\mathbf{v}', \mathbf{w}') = 2^d g^{3-d} I(g, \chi) \delta^{(d)}(\mathbf{v} + \mathbf{w} - \mathbf{v}' - \mathbf{w}') \delta(v^2 + w^2 - v'^2 - w'^2) \tag{2.11}$$

to obtain

$$\frac{\partial f(v, t)}{\partial t} = \int d\mathbf{w} \int d\hat{\mathbf{n}} \, g I(g, \chi) \left\{ f(\mathbf{v}') f(\mathbf{w}') - f(\mathbf{v}) f(\mathbf{w}) \right\} \tag{2.12}$$

---

*We shall follow throughout the convention that a bar over a dependent variable, e.g., $\overline{P}, \overline{W}$, distinguishes the analogous function of a discrete argument from the ordinary function of a continuous variable. On occasions, as above, we shall leave out integration and summation limits on the assumption that they can be inferred from the step functions implicit in the integrand. We shall also frequently suppress the time variable in probability distributions appearing on the right of transport equations.

Here $I(g, \chi)$ is the differential scattering cross section, which for central forces depends only on the relative velocity $g = |\mathbf{v} - \mathbf{w}|$ and the scattering angle χ in the collision; and $\mathbf{n}$ is the unit vector for relative velocity after collision. Thus $\hat{\mathbf{n}} = \hat{\mathbf{g}}'$ and $\cos \chi = \hat{\mathbf{g}} \cdot \hat{\mathbf{n}}$. In all the foregoing equations the spatial density has been absorbed into the time scale. We recall that for collisions of particles with unit mass under central forces $\mathbf{g}$ is changed in direction but not in magnitude, while the initial and final velocity vectors are related by

$$\mathbf{v}' = \tfrac{1}{2}(\mathbf{v} + \mathbf{w}) + \tfrac{1}{2}|\mathbf{v} - \mathbf{w}|\hat{\mathbf{n}}$$
$$\mathbf{w}' = \tfrac{1}{2}(\mathbf{v} + \mathbf{w}) - \tfrac{1}{2}|\mathbf{v} - \mathbf{w}|\hat{\mathbf{n}}$$

(2.13)

which we may occasionally write

$$\mathbf{v}' = \tfrac{1}{2}\mathbf{P} + \tfrac{1}{2}\mathbf{g}', \qquad \mathbf{w}' = \tfrac{1}{2}\mathbf{P} - \tfrac{1}{2}\mathbf{g}' \tag{2.14}$$

to emphasize conservation of the total momentum $\mathbf{P}$ and the constancy of g.

Just as in the full Boltzmann equation the transition rate W contains implicit delta functions that reduce the number of independent variables present, so in the Moran equation [(2.4)] it seems reasonable to consider simplifications to the transition matrix $\Pi(i, j|k, l)$ that reduce the number of variables i, j, k, l by one or more. To do this we must first specify i, j, k, l as parametric quantities [which they need not be for Eq. (2.4) to be meaningful] and then see that they are constrained by some form of functional interdependence. The simplest form this might take, and one I shall later justify by numerous examples, is one in which the state quantity i is a scalar conserved in all interactions, i.e.,

$$i + j = k + l \tag{2.15}$$

This would clearly be the case, for example, if i represented the vibrational energy of a molecule in an intermolecular relaxation process or the total money staked in a game between two gamblers. This sort of simplification will be implicit in all the models considered in detail in this article; we cannot really hope to achieve solubility in models that either retain a genuinely vector character or, if scalar, exhibit the full complexity of the Moran equation.

With the conservation condition (2.15), it becomes natural to replace the scattering matrix $\Pi(i, j|k, l)$ by the three-variable matrix:

$$\overline{K}(j, i; k) = \Pi(j, k - j|i, k - i) \tag{2.16}$$

This reduction and its continuous-variable analogue then lead us to the two prototype equations upon which almost the whole of the present treatment is based. Thus,

TYPE A

$$\frac{\partial P(x,t)}{\partial t} = \int_x^\infty du \int_0^u dy \, \{ P(y)P(u-y)K(x,y;u)$$
$$- P(x)P(u-x)K(y,x;u) \} \tag{2.17}$$

and

TYPE A'

$$\frac{\partial \bar{P}(i,t)}{\partial t} = \sum_{k=i}^\infty \sum_{j=0}^k \, \{ \bar{P}(j)\bar{P}(k-j)\bar{K}(i,j;k)$$
$$- \bar{P}(i)\bar{P}(k-i)K(j,i;k) \} \tag{2.18}$$

Here I have put explicit limits to the summations and integrations to reflect the conditions that $P(x) = 0$ for $x < 0$ and $K(y,x;u) = 0$ for x or $y > u$. Note that whereas $P(x)$ is a probability density dimensioned as x^{-1}, $\bar{P}(i)$ is a dimensionless probability. The designations Type A and Type A' are the beginning of a systematic classification of scalar transport equations which shall be developed in some detail throughout this article.

The recognition that Eq. (2.17) is the most natural form for a model scalar Boltzmann equation has proved to be a considerable step toward the generalization of the earlier simple models in the literature and the construction of new ones. Although we shall occasionally refer back to the Moran equation (2.4) in examples and will devote Section IV to further cross connections with the vector equation of kinetic theory, most of our attention will be concentrated on Eqs. (2.17) and (2.18) and their various simplifications. It will be convenient to work with the continuous-variable equation (2.17) wherever possible, since where general properties are concerned the conversion to the discrete variable is only a simple transcription. When detailed models are considered, however, we shall see the emergence of many properties of interest in their own right in the context of difference calculus that do not in all cases have continuous-variable counterparts. In many, however, it will be possible to give a systematic limiting procedure for passing from discrete models governed by Type A' equations to continuous equations of Type A. When this is the case it will invariably be through a limit of the form

$$\lim_{\varepsilon_0 \to 0} \bar{f}\left(\frac{x}{\varepsilon_0}, \frac{y}{\varepsilon_0}, \dots\right)\left(\frac{dx}{\varepsilon_0}\right) = f(x,y,\dots)\,dx \tag{2.19}$$

where ε_0 is a quantum, $f(x, y, z, \ldots)$ is a probability density, and the summations in the discrete case become Riemann integrals. Whenever a limiting relationship of this type occurs between probability distributions and probability densities, between transition matrices and transition kernels, or between complete transport equations, we shall refer to the forms concerned (and the models giving rise to them) as *true analogues* of each other.

In this connection we should note that, on dimensional grounds, the transition kernel $K(y, x; u)$ can be taken to be a homogeneous function of degree zero in the sense that

$$K(\lambda y, \lambda x; \lambda u)\, d(\lambda y) = K(y, x; u)\, dy \qquad (2.20)$$

We cannot speak of homogeneity of the discrete transition matrices $\overline{K}(j, i; k)$ in the same way, but it will occasionally be useful to say that $\overline{K}$ is homogeneous in j, i, k if it possesses a true analogue in the sense of (2.19) and the analogue is homogeneous. [See also the discussion after Eq. (3.63).]

A. Elementary Symmetries

We shall turn now to the examination of the most general properties of Eq. (2.17) and its solutions in relation to the transition kernel $K(y, x; u)$. To emphasize that the results are not exclusive properties of physical systems it will be as well to extend our descriptive terms as follows. Thinking now in terms of the conserved state quantity x, we can portray the symbolic scattering event as the transition from an *initial disposition* $(x) + (u - x)$ to a *final disposition* $(y) + (u - y)$, with the two *elements* concerned simultaneously changing their *state*, one from (x) to (y) and the other from $(u - x)$ to $(u - y)$. The symbol u thus represents the total state quantity available in the scattering event. The italicized terms just introduced are desirable both to remove confusion with the language of linear relaxation systems and to avoid too close an association with any particular physical process. Figure 1 provides a useful visualization.

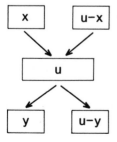

Fig. 1.   Symbolic scattering in energy-space.

As we can effectively see from the diagram, the symmetry (2.5) dictated by the identity of the interacting elements can now be written

$$K(y, x; u) = K(u - y, u - x; u) \tag{2.21}$$

or equally

$$K(u - y, x; u) = K(y, u - x; u) \tag{2.22}$$

We note, however, that while the collision partners are supposed to be indistinguishable in behavior, their identity before and after the interaction is definitely maintained. There is no reason to suppose the "special symmetry" $K(y, x; u) = K(u - y, x; u)$ which several authors have thought to be necessary or of particular interest.

An important characteristic of models, which will prove to be the key to solubility in some cases, concerns the moments of the transition rate. We define these moments of a transition kernel $K(y, x; u)$ as

$$m_n(x, u) = \int_0^u dy\, y^n K(y, x; u) \tag{2.23}$$

and will occasionally use the modified form $\mathcal{M}_n(x, u) = m_n(x, x + u)$. Now we know that by the homogeneity (2.20), the moments $\mathcal{M}_n$ or m_n must be homogeneous of degree n in x and u, that is, that $m_n(\lambda x, \lambda u) = \lambda^n m_n(x, u)$. That they are, in fact, homogeneous polynomials of degree n follows from the symmetries (2.21) and (2.22). Thus, on using (2.21) in the above definition and manipulating by means of the binomial theorem, we find the necessary interrelationship

$$m_n(x, u) = \sum_{k=0}^{n} (-)^k \binom{n}{k} u^{n-k} m_k(u - x, u) \tag{2.24}$$

It is then a simple matter to show by induction on n that $m_n(x, u)$ is homogeneous of degree n given that m_1 (and indeed m_0) is.

These symmetries can be taken over into the discrete variable case in the form

$$\overline{K}(j, i; k) = \overline{K}(k - j, k - i; k) \tag{2.25}$$

$$\overline{K}(k - j, i, k) = \overline{K}(j, k - i; k) \tag{2.26}$$

but the natural definition of moments is now through the *factorial moments*

$$\overline{m}_n(i, k) = \overline{\mathcal{M}}_n(i, k - i) = \sum_{j=n}^{j=k} j^{(n)} \overline{K}(j, i; k) \tag{2.27}$$

Here $i^{(n)}$ is the falling factorial function: $i^{(n)} = i(i-1)(i-2)\cdots(i-n+1)$. [We note that $m_n(x, u)$ and $\overline{m}_n(i, k)$ are proper analogues in the sense of (2.19).] The moment relation analogous to (2.25) must now be obtained through Vandermonde's theorem (Appendix A3.7) and is

$$\overline{m}_n(i, k) = \sum_{\nu=0}^{n} (-)^{\nu} \binom{n}{\nu} k^{(n-\nu)} \overline{m}_\nu(k - i, k) \tag{2.28}$$

B. Conservation Properties

Since (2.17) is a scalar equation, we can hardly expect as full a set of collisional invariants as we associate with the vector Boltzmann equation. In fact, after the conservation of probability

$$\int P(x, t)\, dx = 1, \qquad 0 \leqslant t \leqslant \infty \tag{2.29}$$

we have only the conservation of the mean state quantity $\langle x \rangle$:

$$\int x P(x, t)\, dx = \langle x \rangle, \qquad 0 \leqslant t \leqslant \infty \tag{2.30}$$

Although the former is virtually self-evident in Eq. (2.17), it is instructive to express it as a symmetry property of the right-hand side. Thus, on making the step functions in the transition kernel explicit and integrating left and right, we have [with $H(.)$ the Heaviside function]

$$\frac{\partial}{\partial t} \int P(x, t)\, dx = \int \int \int du\, dy\, dx\, H(x) H(y) H(u-x) H(u-y)$$
$$\times [P(y) P(u-y) K(x, y; u) - P(x) P(u-x) K(y, x; u)]$$
$$= 0$$

by symmetry $x \leftrightarrow y$.

The proof of conservation of state quantity is somewhat more involved. To demonstrate this analytically we multiply left and right by x and integrate to obtain successively

$$\frac{d}{dt} \langle x \rangle = \int \int \int dx\, dy\, du\, x \{ P(y) P(u) K(x, y; u+y) H(u+y-x)$$
$$- P(x) P(u) K(y, x; u+x) H(u+x-y) \}$$

whence, by interchanging $x \leftrightarrow y$ in the second term on the right,

$$\frac{d}{dt} \langle x \rangle = \int_0^\infty dy \int_0^\infty du\, P(y) P(u) \int_0^{u+y} dx\, (x - y) K(x, y; u+y)$$

On further interchange, $u \leftrightarrow y$, it follows that

$$2\frac{d\langle x\rangle}{dt} = \int_0^\infty dy \int_0^\infty du\, P(y)P(u) \int_0^{u+y} dx\, [(x-y)K(x,y;u+y)$$
$$+ (x-u)K(x,u;u+y)]$$

Finally, on transforming $x \to u + y - x$ in the first term and using the symmetry $K(u+y-x, y; u+y) = K(x, u; u+y)$, it follows that

$$2\frac{d\langle x\rangle}{dt} = \int_0^\infty dy \int_0^\infty du\, P(y)P(u) \int_0^{u+y} dx\, (u-x+x-u)K(x,u;u+y)$$
$$= 0$$

In these proofs we have tacitly assumed that the distribution functions P are such that all integrals written are finite for all time. In fact this presumes certain weak conditions on the behaviour of $P(x,t)$ as $x \to \infty$. Discussion of this effect is deferred to Section 6D.

C.   Simplifications of the Transport Equation

Two important simplifications can be applied to the transport equations (2.17) and (2.18) and have been a feature of most published work on soluble models. The first arises when the scattering outcome is in effect independent of the initial state x and the kernel is reduced from $K(y, x; u)$ to $K(y; u)$. The analogy here is with *isotropic* scattering in kinetic theory language, and it is clear that while correlations are still present between initial and final states in a collision (implicitly through u) they are very severely restricted. We might be tempted to describe the more general case as Markovian in character and the special case as "pure random," but this would not be legitimate in the present context. Where physical models are concerned, we shall instead describe them as *persistent* and *diffuse* scattering, respectively. The extreme case of persistent scattering in which x and y differ on average only microscopically is of particular interest; we shall refer to this as the "small-scattering" limit.

The diffuse-scattering equations for continuous and discrete variables thus read:

TYPE B

$$\frac{\partial P(x,t)}{\partial t} = \int_x^\infty du \int_0^u dy\, \{P(y)P(u-y)K(x;u)$$
$$- P(x)P(u-x)K(y;u)\} \tag{2.31}$$

Type B′

$$\frac{\partial \bar{P}(i,t)}{\partial t} = \sum_{k=i}^{\infty} \sum_{j=0}^{k} \left\{ \bar{P}(j)\bar{P}(k-j)\bar{K}(i;k) - \bar{P}(i)\bar{P}(k-i)\bar{K}(j;k) \right\}$$

(2.32)

The second important distinction is between models for which the transition rate for out-scattering from a state x is constant, that is,

$$\int_0^u dy\, K(y, x; u) = \omega$$

(2.33)

and those for which it is a function of state:

$$\int_0^u dy\, K(y, x; u) = z(x; u)$$

(2.34)

with $z(x; u)$ an interaction rate or collision number. In physical contexts this distinction is primarily between systems with relaxation of an *internal* state quantity (e.g., vibrational or rotational energy) and an *external* scalar quantity such as translational kinetic energy or speed. However, the peculiar case of "Maxwell-molecule" scattering falls into the former category by virtue of an accidental lack of dependence of scattering cross section on velocity, a fact crucial to the discovery of the BKW solution. For whichever reason, condition (2.33) leads to an immediate simplification of the transport equation, with the out-scattering term reduced to ωP. Transferring this to the left and absorbing ω into the time scaling, we thus arrive at equations of the form

Type C

$$\left(\frac{\partial}{\partial t} + 1 \right) P(x, t) = \int_x^{\infty} du \int_0^u dy\, P(y) P(u - y) K(x, y; u)$$

(2.35)

Type C′

$$\left(\frac{\partial}{\partial t} + 1 \right) \bar{P}(i, t) = \sum_{k=i}^{\infty} \sum_{j=0}^{k} \bar{P}(j)\bar{P}(k - j)\bar{K}(i, j; k)$$

(2.36)

while for diffuse scattering these reduce further to

Type D

$$\left(\frac{\partial}{\partial t} + 1 \right) P(x, t) = \int_x^{\infty} du \int_0^u dy\, P(y) P(u - y) K(x; u)$$

(2.37)

TYPE D′

$$\left(\frac{\partial}{\partial t}+1\right)\bar{P}(i,t)=\sum_{k=i}^{\infty}\sum_{j=0}^{k}\bar{P}(j)\bar{P}(k-j)\bar{K}(i,k) \qquad (2.38)$$

We have set out these eight categories of transport equation rather carefully in order to refer back later and systematize the various models appearing in the literature. Broadly speaking, the alphabetical order of types corresponds to decreasing order of difficulty in finding soluble models, with Type A equations still, at the time of writing, waiting the discovery of a soluble model of sufficient generality.

Although we are not yet ready to consider particular models in detail, it will be useful at this stage to consider two examples of how these types of transport equations may be applied to simple systems. The first has applications in molecular physics; the second can be shown to have connections with biological and economic equilibria.

D.   Closed System Relaxation of Oscillators

Let the system elements considered above be polyatomic molecules represented by either single or multiple classical harmonic oscillators. (We shall refer to such a system element as a p molecule if it has p internal degrees of freedom in the state variable.) Let the variable x represent the internal vibrational energy per molecule, and consider an ensemble of many such molecules undergoing energy transfer through binary collisions at a fixed collision rate ω. Let the total number of elements and their total energy be N and $N\varepsilon$, respectively, these quantities evidently both conserved. Then, if N is statistically large, we may write a continuous probability density $P(x,t)$ to describe the probability of finding a single molecule with energy dx about x and identify the mean energy per molecule $\langle x \rangle$ with the energy ε. We have only to absorb the collision rate by the time scaling $t \to \omega t$ to see that the time evolution can be represented by Eq. (2.35) of Type C or (2.36) of Type D according to the mechanism involved. In the former case the redistribution of energy upon collision will be explicitly governed by the precise initial disposition of the collision partners; in the latter, relaxation will proceed such that in the statistical outcome all memory of the initial disposition is lost, possibly due to the formation of a long-lived collision complex.

In the case of quantized p molecules (usually assumed to have p degenerate vibrational modes with the same quantum ε_0), similar considerations lead to transport equations of Types C′ or D′. Models of this kind, for $p=1$, were among the earliest to be studied. Shuler[17] obtained an exact solution for the important case of the Landau-Teller transition probabilities (actually a Type

B' equation), and a similar, more elaborate case involving translational energy was solved by Rankin and Light.[18] Other workers, in particular Koura,[19] have published numerical simulations for closed systems of oscillators with similar properties. We shall return to the Shuler model in detail in Section III.

E.   Collective Games

To underline the importance of Boltzmann-type kinetic equations for the description of nonlinear stochastic processes of widely differing origin, we shall briefly examine an example from outside of physics.

Imagine a large collective of N players, each equipped with a given initial capital, who then play a series of two-person zero-sum games with partners selected at random. Clearly, whatever the rules of the individual games, the total money in circulation, the capital available in each game, and the combined stakes committed in a particular game are conserved quantities. Moreover, the outcome of each game may be modeled statistically be assigning kernels $K(y, x; u)$ or $K(y; u)$ as appropriate, with x and y the initial and final capital for one of the two players and u the total capital available for the pair. The former kernel clearly models the outcome of games which consist of a single "throw" for a proportion of the capital staked; the latter models the case where each game is either a long series of contests for small stakes or a single throw upon which both players stake their entire capital. The possibility of delta-function kernels—for winner-takes-all games—and unfair games, in which the advantage lies with the player holding the largest capital, is not ruled out.

It is evident that the distribution of wealth in such a collective can be governed by any of the kinetic equations we have considered, with discrete or continuous time as the case may be and numerous variations according to the discreteness or otherwise of stakes and whether frequency of play is a function of the capital held. While the scope of possible models is limited, such studies of the persistence and distribution of wealth appear to have certain applications in economic theory as well as relevance to the design of scoring systems for leagues and tournaments.[20]

F.   Equilibrium and Microscopic Reversibility

Whether we consider models of physical or nonphysical type, an essential ingredient will be the introduction of *complexity* into the system elements of the model. This complexity may have a mechanical basis, as for example in the case of polyatomic molecules, or it may enter through some form of dimensionality or else purely in statistical terms through an assignment of degrees of freedom. In any case, we shall find it natural to model complexity in terms of a *statistical weight* $g(x)$ that is associated with a state x [or $\bar{g}(i)$

with a state i] and is assumed to be a given at the outset. The function $g(i)$ refers to the density of equivalent *microstates* associated with a *macrostate* x and will in general be a sensitive function of the number of degrees of freedom present.

As is well known, equations of type (2.17) admit a wider choice of kernels than are sanctioned for physical systems, because in the latter, microscopic reversibility arising from time-reversal symmetry puts severe restrictions on the form K can take. Although the question of microscopic reversibility and detailed balance is somewhat commonplace in textbook treatments, it will bear reexamination in the present less-familiar context, where we wish to bring out some of the special features associated with complexity and internal degrees of freedom.

If we consider the reversibility of the symbolic scattering event: $(x)+(u-x) \rightleftarrows (y)+(u-y)$ in terms of the multiplicity of possible microstates consistent with the initial and final disposition of the state variable, we see that the appropriate form of the microscopic reversibility condition on K is

$$g(x)g(u-x)K(y,x;u) = g(y)g(u-y)K(x,y;u) \qquad (2.39)$$

with $g(.)$ the relevant density of states. Only in the case of structureless elements, for example single harmonic oscillators, can we put $g(x)=1$ and state the simple equivalence

$$K(y,x;u) = K(x,y;u) \qquad (2.40)$$

The severity of condition (2.39) restricts considerably our freedom to construct even ad hoc physical models and accounts for their scarcity, particularly in the important case of persistent scattering. We need hardly emphasize that the relation (2.39) embodies our *mechanical* view of the collision process and its symmetry under time reversal. If the interaction takes place, as it were, in a black box with statistically defined outcomes, there is no need at all for the above constraint. As an extreme example of a microscopically irreversible system we might consider the case of two players in a collective game who compete under a winner-takes-all rule. Both the scattering behavior $[K(y,x;u) = \frac{1}{2}\delta(y-u) + \frac{1}{2}\delta(y)]$ and the whole evolution of the system can be seen to be highly anomalous in physical terms.

Finally, if we require our model system to be both of the diffuse scattering type and microscopically reversible, then it becomes clear from (2.39) that the kernel is restricted to be of the form

$$K(y,u) \propto g(y)g(u-y) \qquad (2.41)$$

The closely related question of detailed balance also takes on less familiar aspects when formulated for internal degrees of freedom. In these terms we first recall that a *sufficient* condition for a distribution $P_0(x)$ to be stationary under Eq. (2.17) is that

$$P_0(x)P_0(u-x)K(y,x;u) = P_0(y)P_0(u-y)K(x,y;u) \quad (2.42)$$

This condition is clearly not *necessary*, for other kernels might allow the right hand side of (2.17) to vanish without the uniform vanishing of the integrand. Using microscopic reversibility with the given density of states, we see, taking (2.39) and (2.42) into account, that

$$\frac{P_0(x)P_0(u-x)}{P_0(y)P_0(u-y)} = \frac{g(x)g(u-x)}{g(y)g(u-y)} \quad (2.43)$$

This equation is satisfied by

$$P_0(x) = Cg(x)a^x \quad (2.44)$$

in which C and a are parameters to be fixed by the conservation integrals (2.29) and (2.30).

It is then natural to define a partition function

$$\mathscr{Z}(a) = \int_0^\infty dx\, g(x)a^x \quad (2.45)$$

with a to be determined implicitly through

$$\frac{1}{\mathscr{Z}(a)}\int_0^\infty dx\, xg(x)a^x = \varepsilon \quad (2.46)$$

(We shall use the "energy" language $\langle x \rangle = \varepsilon$ for convenience here.) Now, in a way that parallels the Gibbsian treatment for the canonical ensemble, it is clear that the above implies the formula

$$\varepsilon = \mathscr{Z}(a)^{-1}\left(\frac{d\mathscr{Z}(a)}{da}\right) \quad (2.47)$$

Finally, writing $\mathscr{Z}(\varepsilon)$ to denote the outcome of the computational sequence $\varepsilon \to a(\varepsilon) \to \mathscr{Z}[a(\varepsilon)] \equiv \mathscr{Z}(\varepsilon)$, we may express the equilibrium distribution as a function of x and its mean $\langle x \rangle = \varepsilon$ in the general form

$$P_0(x,\varepsilon) = [\mathscr{Z}(\varepsilon)]^{-1}g(x)[a(\varepsilon)]^x \quad (2.48)$$

The corresponding equation for the case of the discrete state variable requires only a change from integration to summation.

These results are quite general for systems whose scattering kernels obey a condition of the form (2.39) and whose state space is structured by a given density function $g(x)$. Two examples will illustrate these principles while serving as a basis for models to be developed later.

Example 1.   Classical p oscillators.   The case of classical harmonic oscillators (molecules) with p internal degrees of freedom yields the density of states

$$g(\varepsilon) = \Gamma(p)^{-1} x^{p-1} \qquad (2.49)$$

On using the two conditions (2.45) and (2.46) with the definite integral

$$\int_0^\infty x^{p-1} a^x \, dx = \frac{\Gamma(p)}{[\ln(1/a)]^p} \qquad (2.50)$$

we find immediately that

$$\varepsilon = \frac{-p}{\ln a} \qquad a = e^{-p/\varepsilon} \qquad \mathscr{L}(\varepsilon) = \left(\frac{p}{\varepsilon}\right)^p$$

Thus we arrive at the equilibrium distribution

$$P_0(x, \varepsilon) = \Gamma(p)^{-1} \left(\frac{p}{\varepsilon}\right)^p x^{p-1} e^{-px/\varepsilon} \qquad (2.51)$$

This is a gamma distribution in p degrees of freedom with mean ε.

Example 2.   Quantum p oscillators.   The case of quantum harmonic oscillator with p degenerate degrees of freedom is less familiar, though of considerable interest. Let the size of the quantum be $\varepsilon_0 = h\nu$. Then, neglecting for present purposes the zero-point energy, the appropriate density of states is the combinatorial factor

$$g(i) = \Gamma(p)^{-1} (i+1)_{p-1} \qquad (2.52)$$

[We use the Pochhammer, or rising factorial, function defined as $(x)_\alpha = x(x+1)(x+2)\cdots(x+\alpha-1) = \Gamma(x+\alpha)/\Gamma(x)$.* Note that the last form may be used when α is nonintegral.] The previous analysis is essentially un-

---

*See Appendix 1, Eq. (A1.4).

changed, and we seek the discrete equilibrium distribution $P_0(i, \varepsilon)$ analogous to (2.51) with $i \in [0, 1, \ldots, \infty]$. For the "partition-function" $\mathscr{L}$ we have, by the binomial theorem [see Eq. (A3.5)],

$$\mathscr{L}(a) = \sum_{i=0}^{\infty} \Gamma(p)^{-1}(i+1)_{p-1}a^i = (1-a)^{-p} \qquad (2.53)$$

Using the same theorem, or simply referring to the known mean of the negative binomial distribution, we then find that

$$a = \frac{\varepsilon}{\varepsilon_0 p + \varepsilon} \qquad (2.54)$$

Thus the equilibrium distribution is the negative binomial with mean $\varepsilon = \mathscr{L}^{-1}(d\mathscr{L}/da)$. Explicitly,

$$P_0(i, \varepsilon, \varepsilon_0) = \frac{1}{\Gamma(p)}\left(\frac{\varepsilon_0 p}{\varepsilon_0 p + \varepsilon}\right)^p (i+1)_{p-1}\left(\frac{\varepsilon}{\varepsilon_0 p + \varepsilon}\right)^i \qquad (2.55)$$

In our further discussion, systems of any kind that exhibit a density-of-states function $g(x)$ given by (2.49) or $\bar{g}(i)$ given by (2.52) will be said to possess "standard complexity" in p degrees of freedom.

In view of the way the state variable enters expressions (2.51) and (2.55), it will often be convenient, as long as we deal with systems of standard complexity, to work in the rescaled energy variable $x \to px/\varepsilon$ and adjust the discrete variable quantum ε_0 accordingly. If we mark this change by writing $F(x, t)\,dx \equiv P(x\varepsilon/p, t)\,d(x\varepsilon/p)$, the stationary gamma distribution becomes

$$F_0(x) = \Gamma(p)^{-1}x^{p-1}e^{-x} \qquad (2.56)$$

and has the mean $\langle x \rangle = p$.

The corresponding discrete-variable scaling requires that we measure energy such that $\varepsilon_0 \langle i \rangle = p$. This distribution function will then be

$$\bar{F}_0(i) = \frac{1}{\Gamma(p)}\left(\frac{\varepsilon_0}{\varepsilon_0 + 1}\right)^p (i+1)_{p-1}\left(\frac{1}{\varepsilon_0 + 1}\right)^i \qquad (2.57)$$

and connects with (2.56) by virtue of the limit

$$\lim_{\varepsilon_0 \to 0} \bar{F}\left(\frac{x}{\varepsilon_0}, t\right)\frac{dx}{\varepsilon_0} = F(x, t)\,dx$$

However, we do not wish the convenience of this formulation to obscure the fact that the general properties described in the introductory sections of this paper may be interpreted for completely general systems with given density-of-states function $g(x)$, for which only $P(x, t)$ and $\bar{P}(i, t)$ are appropriate.

G. Entropy Production, the H Theorem, and Positivity

Much attention has been paid in the literature to the question of characterizing the approach to equilibrium through laws of entropy production. Of these the Boltzmann H theorem giving $dS/dt \geq 0$ is fundamental, but there has been much speculation as to whether "super-H theorems" might be derived, giving equally universal information about the signs of higher derivatives of the entropy and incidentally justifying the use of the Liapunov function $P \log P$ rather than another acceptable choice of convex function. (See especially Ziff, Merajver, and Stell.[21]) We shall return to these questions after considering particular models, restricting attention here to the ordinary H theorem and the generalities underlying the basic scalar transport equation.

The question of whether Eq. (2.17), with suitable conditions on the scattering kernel, actually gives evolution to the stationary distribution $P_0(x)$ must be decided by examination of a suitable Liapunov function, which can be shown to increase monotonically to a maximum as $t \to \infty$. Before turning to this, however, we shall make some distinctions that, although implicit in some more general work, become of particular interest in the case of a scalar Boltzmann equation.

We observe first that for the case of a scalar macrostate and a multiplicity of microstates with density $g(x)$ there is a natural division into what we shall call the *collective entropy* σ_1 and the *special entropy* σ_2, defined thus:

$$\sigma_1 = -\int dx\, P(x)\ln P(x) \tag{2.58}$$

$$\sigma_2 = \int dx\, P(x)\ln g(x) \tag{2.59}$$

Evidently σ_1 represents the entropy due to partitioning of the state quantity between separate system elements, while σ_2 represents the entropy due to multiplicity of microstates within the complex elements themselves. For simple system $[g(x) = 1]$, σ_2 automatically vanishes. Adding the separate contributions, we obtain the total time-dependent entropy

$$\sigma(t) = -\int P(x)\ln\left[\frac{P(x)}{g(x)}\right] dx \tag{2.60}$$

Alternatively, on expressing $\ln g(x)$ in terms of $P_0(x)$ through (2.48), we have

$$\sigma_2(t) = \ln \mathscr{L}(\varepsilon) - \varepsilon \ln a(\varepsilon) + \int P(x)\ln P_0(x)\, dx$$

by means of which

$$\sigma(t) = \sigma(\infty) - \int P(x)\ln\left[\frac{P(x)}{P_0(x)}\right] dx \tag{2.61}$$

with

$$\sigma(\infty) = \ln \mathscr{L}[a(\varepsilon)] - \varepsilon \ln a(\varepsilon)$$

the relevant value of a being determined throughout by conditions (2.46) and (2.47). For the entropy production at time t, we can now write

$$\dot{\sigma}(t) = - \int \dot{P}(x)\ln\left[\frac{P(x)}{P_0(x)}\right] dx \tag{2.62}$$

this is the ordinary definition of entropy production as used, for example, by Schlögl.[22] However, the above analysis indicates that it should be possible to think of the entropy production as made up of two contributions of physically distinct origin:

$$\dot{\sigma}_1(t) = \int \dot{P}(x)\ln P(x)\, dx \tag{2.63}$$

and

$$\dot{\sigma}_2(t) = - \int \dot{P}(x)\ln P_0(x)\, dx \tag{2.64}$$

We may then raise the question of whether, in addition to the expected positive-definiteness $\dot{\sigma}_{\text{tot}} \geq 0$, at least for certain models, the separate inequalities $\dot{\sigma}_1 \geq 0$ and $\dot{\sigma}_2 \geq 0$ may hold.

With scalar transport equations of the types considered here, it is a relatively simple matter to prove an H theorem. We shall demonstrate this for the general kernel of Eq. (2.17), assuming only that it satisfies conditions (2.21) and (2.42).

Interpreting $\partial P/\partial t$ in (2.62) by the right-hand side of (2.17), we have that

$$\frac{\partial \sigma}{\partial t} = -\int\int\int dx\,dy\,du\,\{\,P(y)P(u-y)K(x,y;u)$$

$$-P(x)P(u-x)K(y,x;u)\}\ln\left[\frac{P(x)}{P_0(x)}\right]$$

We now apply successively the interactional symmetry $K(y,x;u) = K(u-y,u-x;u)$, the detailed-balance condition (2.42), and the general symmetry of the integrand under the interchange of x and y. In this way,

$$\frac{\partial \sigma}{\partial t} = -\frac{1}{2}\int\int\int dx\,dy\,du\,\{\,P(x)P(u-x)K(y,x;u)$$

$$-P(y)P(u-y)K(x,y;u)\}$$

$$\times \ln\left[\frac{P(u-x)}{P_0(u-x)} - \frac{P(u-y)}{P_0(u-y)}\right]$$

$$= \frac{1}{4}\int\int\int dx\,dy\,du\,\{\,P(x)P(u-x)K(y,x;u)$$

$$-P(y)P(u-y)K(x,y;u)\}$$

$$\times \ln\left[\frac{P(x)P(u-x)P_0(y)P_0(u-y)}{P_0(x)P_0(u-x)P(y)P(u-y)}\right] \tag{2.65}$$

Transforming to the new dependent variable $\xi(x) = P(x)/P_0(x)$ and applying the detailed balance condition once more, we see that

$$\frac{\partial \sigma}{\partial t} = \frac{1}{4}\int\int\int dx\,dy\,du\,P_0(x)P_0(u-x)K(y,x;u)$$

$$\times \{\xi(x)\xi(u-x)-\xi(y)\xi(u-y)\}\ln\left[\frac{\xi(x)\xi(u-x)}{\xi(y)\xi(u-y)}\right]$$

$$\geq 0 \tag{2.66}$$

the inequality arising from the usual condition $(x-y)\log(x/y) \geq 0$ when $x, y \geq 0$. We may note that, in the special case $g(x) = 1$ (for example, for independent single harmonic oscillators), it is possible to delete the factors $P_0(x)$ in (2.65) and obtain the final result more directly. The analogous result for discrete systems obeying Eq. (2.18) follows by simple transcription of the above.

It will be noted that our proof gives no indication that the separate entropy productions $\dot{\sigma}_1(t)$ and $\dot{\sigma}_2(t)$ are each positive definite, and indeed it seems unlikely that $\dot{\sigma}_2(t) \geq 0$ holds with any generality.

Finally we must note the question of whether solutions of the transport equation (2.17) are in general positive, or whether guarantees of positivity can be given for reasonable classes of kernels. The answer to the first question is in general no, as later counterexamples will show. Indeed, almost without exception, the soluble models studied so far admit nonpositive solutions for certain initial conditions and time regimes. Fortunately it is possible to prove generally the weaker proposition that if a distribution $P(x, t)$ is positive at any given time t, then it will evolve under Eq. (2.17) such that it remains positive for all times $t' > t$. The proof (see, e.g., Résibois and de Leener[23]) is by a simple contradiction. Suppose that if some positive distribution $P(x, t)$ becomes negative it does so for the first time at some point x_0 and time t_0. Then correspondingly for some time $t - \varepsilon$, $P(x_0, t - \varepsilon) > 0$ and $(\partial P / \partial t)|_{t=t_0} \leq 0$. But, by the transport equation (2.17) we have that $(\partial P / \partial t)|_{t=t_0-\varepsilon} \geq 0$, since by hypothesis $P(y, t_0 - \varepsilon)$ and $P(u - y, t_0 - \varepsilon)$ are either positive or zero and K is certainly positive in the integrand. Thus, by contradiction, $P(x, 0) > 0$ implies $P(x, t) > 0$ for all $t > 0$.

H. Linearization

The possibility of obtaining linear approximations to Boltzmann equations has exercised many workers in the past and led to a considerable literature dealing with the accuracy of such results and their usefulness in numerical computations. While the use of models has been directed primarily to the search for exact solutions, they nevertheless cast light on this problem and have a useful role both in testing the degree to which true behavior may be approximated in the linear case and also in characterizing the nature of the deviations from real behavior produced. We shall therefore discuss the linearization process for scalar Boltzmann equations in some detail, in general terms and particularly in relation to the various moment properties to be treated in Section II.J.

Consider first the full Type A transport equation (2.17) in the continuous variable. Let the time-dependent distribution $P(x, t)$ be represented in the form

$$P(x, t) = P_0(x) + h(x, t) \tag{2.67}$$

with the function $h(x, t)$ assumed small throughout and necessarily subject to the conditions

$$\int_0^\infty dx\, h(x, t) = \int_0^\infty dx\, x h(x, t) = 0 \tag{2.68}$$

for all t.

On substituting (2.67) into (2.17) with neglect of second-order small terms, we find, after use of the equilibrium property (2.42), that

$$\frac{\partial h(x,t)}{\partial t} = \int \int du\, dy\, [h(y)P_0(u)+h(u)P_0(y)]\, K(x,y;u+y)$$

$$- \int \int du\, dy\, [h(x)P_0(u)+h(u)P_0(x)]\, K(y,x;u+x)$$

$$(2.69)$$

By $u \leftrightarrow y$ interchange in the second term of each bracket, we can then cast this into the form

$$\frac{\partial h(x,t)}{\partial t} = \int dy\, K^L(x,y)h(y) - z(x)h(x) \qquad (2.70)$$

where $K^L(x,y)$ is the linear kernel

$$K^L(x,y) = \int du\, P_0(u-y)[K(x,y;u)+K(x,u-y;u)]$$

$$- P_0(x)\int du\, K(u,x;y+x) \qquad (2.71)$$

and

$$z(x) = \int \int du\, dy\, P_0(u-x)K(y,x;u)$$

$$= \int dy\, K^L(y,x) \qquad (2.72)$$

The last line follows from the detailed-balance property.

The solution of the linear integrodifferential equation (2.70) for $h(x,t)$ can be given in terms of the eigenvalue properties of the singular integral operator

$$\mathscr{A}: A(x,y) = K^L(x,y) + z(x)\delta(x-y) \qquad (2.73)$$

By a procedure familiar in linear transport theory (see Hoare[24] and Hoare, Rahman, and Raval[25]), the solution or $P^L(x,t)$ can be constructed as

$$P^L(x,t) = P_0(x) + \sum_{k \neq 0} a_k \phi_k(x)e^{-\mu_k t} + \int_{\mu \in c} d\mu\, a(\mu)\phi(x,\mu)e^{-\mu t}$$

$$(2.74)$$

Here $\phi_k(x)$ and $\phi(x,\mu)$ are the regular and singular eigenfunctions of satisfying

$$\mathscr{A}\phi_k(x) = \mu_k\phi_k(x) \tag{2.75a}$$

$$\mathscr{A}\phi(x,\mu) = \mu\phi(x,\mu) \tag{2.75b}$$

and the integral runs over values of μ in the continuum spectrum which is composed of those values of μ for which a root x_μ exists such that $z(x_\mu) - \mu = 0$. (The spectrum is not, in general, guaranteed to be real.) The expansion coefficients $a_k, a(\mu)$ must be found from the initial condition $h(x,0)$ using the orthogonality properties

$$\int dx\, \phi_n(x)\phi_m(x) = \mathscr{N}_n\delta_{nm} \tag{2.76a}$$

$$\int dx\, \phi(x,\mu)\phi(x,\nu) = \mathscr{N}(\mu)\delta(\mu - \nu) \tag{2.76b}$$

with $\mathscr{N}_n$ and $\mathscr{N}(\mu)$ suitable normalization function. The completeness of the set $\{\phi_k(x), \phi(x,\mu)\}$ for a sufficient class of functions orthogonal to $P_0(x)$ $[= \phi_0(x)]$ cannot be taken for granted and must be proved in each case of interest. Because of the complications of interchanging singular integrals, special theorems such as the Poincaré-Bertrand theorem may have to be invoked. Particular examples of this process are described in Kuščer and Williams[26] and Hoare, Rahman, and Raval.[25]

Various simplifications follow on specializing to Type B, C, and D processes. For Type B [Eq. (2.31)] there is relatively little change, the two contributions to the first integral in (2.71) simply becoming identical. In Type C [Eq. (2.35)], however, the constant collision rate enables us to write (as usual, scaling the collision number to unity)

$$\left(\frac{\partial}{\partial t} + 1\right)h(x,t) = 2\int dy\, k(x,y)h(y,t) \tag{2.77}$$

where $k(x,y)$ is now the simpler kernel

$$k(x,y) = \frac{1}{2}\int du\, P_0(u-y)[K(x,y;u) + K(x,u-y;u)] \tag{2.78}$$

This, we may observe, inherits from (2.42) a detailed-balance condition in the form

$$P_0(y)k(x,y) = P_0(x)k(y,x) \tag{2.79}$$

If $P^L(x, t)$ is the distribution function in the linear approximation corresponding to the solution of (2.77), then we see that it must satisfy

$$\left(\frac{\partial}{\partial t} + 1\right) P^L(x, t) = 2 \int dy\, k(x, y) P^L(y) - P_0(x) \qquad (2.80)$$

The stationarity of $P_0(x)$ is then evident from (2.79). Since there is now no continuum contribution, the solution can be written

$$P^L(x, t) = P_0(x) + \sum_{k=1}^{\infty} a_k \phi_k(x) \exp[-(1 - 2\lambda_k)t] \qquad (2.81)$$

with λ_k and $\phi_k(x)$, respectively, eigenvalues and right eigenfunctions of the kernel $k(x, y)$.

Finally, we note that, in the case of Type D processes, the contributions of the two integrands in (2.78) become identical and $k(x, y)$ is simply

$$k(x, y) = \int_{\max(x, y)}^{\infty} du\, P_0(u - y) K(x; u) \qquad (2.82)$$

This definition is, incidentally, that of the transition kernel for a *distributive process* in the sense recently discussed by Cooper and Hoare[27] (see also Cooper, Hoare, and Rahman[28] and Hoare and Rahman[29]). We shall see the importance of this connection when we treat specific processes in Section III. The factor 2 that appears in (2.77) and (2.80) and causes (2.80) to differ from the usual master equation (2.3) for a linear system is due to the fact that a system with quadratic transport even in its linear regime, must relax twice as fast as an inherently linear system with the same transition kernel, for the simple reason that two rather than one of the elements must change state on each interaction.

Entirely parallel considerations apply to the linearization of Type C' and D' equations in the discrete variable. The solution then depends on the eigenvalue problem for the transition matrix $\bar{k}(i, j)$, defined

$$\bar{k}(i, j) = \frac{1}{2} \sum_{k=0}^{\infty} \bar{P}_0(k) [\bar{K}(i, j; k + j) + \bar{K}(i, k; k + j)] \qquad (2.83)$$

such that, analogously to (2.81), we have

$$\bar{P}^L(i, t) = \bar{P}_0(i) + \sum_{k=1}^{\infty} \bar{\alpha}_k \bar{\phi}_k(i) \exp[-(1 - 2\bar{\lambda}_k)t] \qquad (2.84)$$

in which $\bar{\lambda}_k$ and $\bar{\phi}_k$ are the appropriate eigenvalue and right eigenvector of the infinite matrix $\bar{k}(i, j)$. If the kernels $\bar{k}(i, j)$ and $k(x, y)$ under consideration are proper analogs in the sense of Eq. (2.19), then we may assert that the eigenvalues $\bar{\lambda}_k$ and λ_k are identical. This follows on applying the appropriate limit to the matrix eigenvalue equation for $k(i, j)$.

In the case of type A' and type B' discrete kernels, there is, of course, no complication from a continuous spectrum. The discrete-variable distributive scattering kernel corresponding to (2.82) is

$$\bar{k}(i, j) = \sum_{k = \max(i, j)}^{\infty} \bar{P}_0(k - j)\bar{K}(i; k) \qquad (2.85)$$

and is of the type studied first by Hoare and Rahman.[29]

We shall return to aspects of the linearization approximation in relation to the evolution of moments in Section II.J and in relation to particular models in Section III. A more extended conception of the eigenvalue problem for the linearized equation is introduced later.*

I. Types of Exact Solutions

Part of the charm of the Boltzmann equation field is that, for the simplified scalar equations at least, the type of nonlinearity present is such that some of the standard solution techniques, properly applicable only to linear equations, can nevertheless be applied—although with results that are unpredictable and cannot be taken for granted in the framework of any general theory. Techniques of this kind include the use of integral transforms (Fourier, Hankel, Laplace) and expansions in orthogonal polynomials.

The recent history of soluble Boltzmann equations has been dominated by two types of solution, those of the "moment expansion" type and those of the "similarity" type. Neither of these is all that can be desired in either generality or computational usefulness, yet each to an extent complements the other and is in its way an unexpected gift in a problem of otherwise unyielding difficulty. A brief preview of what is involved will justify the detail into which we must now go.

1. Moment Solutions

The possibility of moment solutions to equations of the Boltzmann type was first demonstrated by Kac,[6] although some of the properties involved

---

*The foregoing "conventional" treatment of the eigenvalue problem for linear relaxation presupposes that we work in a Hilbert space governed by the inner product (2.76a). Although this function space is adequate for many purposes, recent work on Boltzmann models has required its extension to admit solutions decreasing as inverse powers of x. The complete reinterpretation of the eigenvalue problem that results is considered in Section VI.D.

were widely known to workers in the kinetic theory field, possibly even to Maxwell himself.[30] The key discovery is that, under conditions not difficult to satisfy, the moments $\langle x^n(t)\rangle$ for the distribution function, defined

$$m_n(t) = \langle x^n(t)\rangle = \int_0^\infty dx\, x^n P(x, t) \tag{2.86}$$

may be obtained by a sequential algorithm, each $\langle x^n\rangle$ being obtainable from those $\langle x^k\rangle$ with $k < n$. Bearing in mind that necessarily $\langle x^0\rangle = 1$ and $\langle x^1\rangle = \varepsilon$ for all t, the prediction of $P(x, t)$ for given t can proceed in two stages. First the set $\{\langle x^n(t)\rangle\}$ is generated in the sequence $\langle x^2\rangle \to \langle x^3\rangle \to \cdots \to \langle x^n\rangle$, given the initial values $\{\langle x^k(0)\rangle\}$ for sufficiently large n. Then the actual distribution $P(x, t)$ must be recovered to good approximation from the moment set $\{\langle x^k(t)\rangle\}_{k=2}^{k=n}$. The condition that $P(x, 0)$ must possess all the required moments is a first indication of the lack of generality of the method. Given, however, that all, or sufficient, moments exist, it is a standard exercise to recover $P(x, t)$ using the orthogonal polynomial set $\{\phi_k(x)\}$ orthogonal with respect to the stationary distribution $P_0(x)$ as weight. A feature of additional interest, however, is that in certain cases, notably that of "standard complexity," the *polynomial moments* $\langle \phi_k(x)\rangle$ themselves show simplified behavior and can be obtained without recourse to the $\langle x^k(t)\rangle$ by an equally simple sequential procedure: $\langle \phi_2\rangle \to \langle \phi_3\rangle \to \cdots \to \langle \phi_n\rangle$. The details of these algorithms will be considered in the next section.

2. Similarity Solutions

The second of the two recent innovations is the discovery of similarity solutions to equations of Types C, C', D, and D' in our classification. The similarity method, as developed for the study of nonlinear ordinary and partial differential equations, depends on the use of Lie group theory to discover variable transformations that simplify the problem in hand to one of lower order or one in fewer variables—in favorable cases reducing a PDE to an ODE without approximation and, most important, without loss of its essential nonlinearity. The type of simplification relevant in initial-value problems is that in which a *self-similar* solution is discovered that has a structural form invariant with respect to the translation of one variable, in our case the time t. An excellent description of the background and scope of similarity methods for differential equations will be found in Bluman and Cole.[31]

Because the continuous group theory involved is somewhat complex and its adaptation to integrodifferential equations not altogether straightforward, its use with Boltzmann equations has tended to be heuristic rather than systematic. Bobylev[1] and Krook and Wu[3,4] simply guessed the required transformations, the former in the solution for $P(x, t)$ itself and the latter by way of an equivalent generating-function equation. After the two solutions were published, Tenti and Hui provided a rigorous group-theoretical justification of the steps taken.[32]

We shall defer detailed consideration of similarity solutions until they arise for specific models in Section III.

J.   Moment Evolution Equations

With the above considerations in mind, we shall now investigate the time-dependent behavior of the moments $\langle x^n(t) \rangle$ in the light of various properties of the kernels $K(y, x; u)$ and $\overline{K}(j, i; k)$, still independently of particular models. For this purpose it proves convenient to modify the moment definition used and define instead reduced moments $\mu_n(t)$, where

$$\mu_n(t) = \frac{1}{n!} \int_0^\infty dx\, x^n P(x, t) = \frac{\langle x^n \rangle}{n!} \tag{2.87}$$

We note that these must evolve to the equilibrium values

$$\mu_n(\infty) = \frac{1}{\mathscr{Z}(\varepsilon)n!} \int_0^\infty dx\, g(x) x^n [a(\varepsilon)]^x \tag{2.88}$$

with $a(\varepsilon)$ and $\mathscr{Z}(\varepsilon)$ as in Eq. (2.46). For systems of standard complexity these are just the reduced moments of the gamma distribution in p degrees of freedom:

$$\mu_n(\infty) = (p)_n (n!)^{-1} \left(\frac{\varepsilon}{p} \right)^n \tag{2.89}$$

The usefulness of moment equations has so far been exploited primarily in the treatment of Type C and Type D equations with constant collision number. To see how this arises we multiply Eq. (2.35) left and right by $x^n/n!$ and integrate after shifting the variable $u \to u + y$. Thus,

$$\left(\frac{d}{dt} + 1 \right) \mu_n(t) = \frac{1}{n!} \int_0^\infty dy\, P(y) \int_0^\infty du\, P(u) \int_0^{u+y} dx\, x^n K(x, y; u + y)$$

The inner integral will be seen to be in the form of the nth moment of the kernel $m_n(y, u + y)$ introduced earlier (2.23). We know this to be a homogeneous polynomial of degree n [cf. Eq. (2.24) et seq.] so that the modified moment $\mathcal{M}_n(y, u)$ can be written

$$\mathcal{M}_n(y, u) = \int_0^{u+y} dx\, x^n K(x, y; u + y) = \sum_{k=0}^n \binom{n}{k} a_{nk} u^k y^{n-k} \quad (2.90)$$

the binomial coefficient being introduced only for convenience. With the inner integral in the form shown, it is clear that the right-hand side of Eq. (2.90) reduces to a convolution of the moments μ_k, in fact that

$$\left(\frac{d}{dt} + 1\right)\mu_n(t) = \sum_{k=0}^n a_{nk}\mu_k\mu_{n-k} \quad (2.91)$$

We must be careful in using this equation for $n = 0$ and $n = 1$ to remember that $\mu_0(t) = 1$ and $\mu_1(t) = \varepsilon$ have already been imposed and that, necessarily, $a_{00} = 1$ and $a_{10} + a_{11} = 1$. The above is a closed system of equations in which the moments of order n depend only on those of order $k < n$. Since $\mu_0 = 1$, it will be useful to write

$$\left(\frac{d}{dt} + \Lambda_n\right)\mu_n(t) = \sum_{k=1}^{n-1} a_{nk}\mu_k\mu_{n-k} \quad (2.92)$$

in which

$$\Lambda_n = 1 - a_{n0} - a_{nn} \quad (2.93)$$

This quantity will be found to provide a link with both the linearized transport equation and the similarity solutions to be discussed later.

The equilibrium condition for the moments, following (2.91), can be written

$$\mu_n(\infty) = \sum_{k=0}^n a_{nk}\mu_k(\infty)\mu_{n-k}(\infty) \quad (2.94)$$

again with special provisions for $n = 0$ and $n = 1$. Since, for example in the case of standard complexity, the $\mu_n(\infty)$ are fixed [cf. (2.89)], we can see that this imposes a stringent condition on the coefficients a_{nk}.

For the case of Type D models the situation simplifies further. If the inner integral appearing is just the monomial $\mathcal{M}_n(u) = a_n u^n$, then Eq. (2.92) reduces to

$$\left(\frac{d}{dt} + \Lambda_n\right)\mu_n(t) = a_n \sum_{k=1}^{n-1} \mu_k\mu_{n-k} \quad (2.95)$$

with $\Lambda_n = 1 - 2a_n$. Both Eqs. (2.92) and (2.95) provide the type of sequential algorithm for calculation of μ_n to which I have referred.

a. Second Moment Relaxation. The results of the above are particularly simple in the case of the second moment μ_2. Bearing in mind that $\mu_0 = 1$ and $\mu_1 = \varepsilon$ for all time, we can write the relaxation equation

$$\left(\frac{d}{dt} + 1\right)\mu_2(t) = a_{20}\mu_0\mu_2 + 2a_{21}\mu_1^2 + a_{22}\mu_0\mu_2$$
$$= 2a_{21}\varepsilon^2 + (a_{20} + a_{22})\mu_2 \qquad (2.96)$$

This leads immediately to the solution

$$\Delta_2(t) = \Delta_2(0)e^{-\Lambda_2 t} \qquad (2.97)$$

where we have written $\Delta_2(t) = \mu_2(t) - \mu_2(\infty)$ with

$$\Lambda_2 = 1 - a_{20} - a_{22} = \frac{2a_{21}\varepsilon^2}{\mu_2(\infty)}$$

and

$$\mu_2(\infty) = \frac{2a_{21}\varepsilon^2}{1 - a_{20} - a_{22}} \qquad (2.98)$$

When only diffuse scattering is involved, there is further simplification. With $\Lambda_2 = 1 - 2a_2$, we can express the relaxation entirely in terms of $\mu_2(\infty)$ and ε. Thus,

$$\frac{\Delta_2(t)}{\Delta_2(0)} = \exp\left[-\frac{\varepsilon^2 + \mu_2(\infty)}{\varepsilon^2 + 2\mu_2(\infty)}\right] \qquad (2.99)$$

In a similar manner, relatively simple equations can be derived for $\mu_3(t)$ and $\mu_4(t)$ and so on. There is, however, no inductive pattern, and the number of separate exponential transients increases very rapidly with n.

b. Moment Evolution in Discrete Systems. The construction of evolution equations for discrete-variable transport equations of Type C' is broadly parallel to the above, although the character of the moments changes. Defining modified factorial moments of the distribution function,

$$\bar{\mu}_n(t) = \frac{1}{n!} \sum_{i=n}^{\infty} i^{(n)} \bar{P}(i, t) \qquad (2.100)$$

and using the kernel moments $\mathcal{M}_n(j, k)$ defined in (2.27), we arrive at a differential-sum equation analogous to (2.89). The counterpart of the homogeneity condition (2.90) is now seen to be

$$\mathcal{M}_n(j, k) = \sum_{\nu=0}^{n} \binom{n}{\nu} \bar{a}_{n\nu} k^{(\nu)} j^{(n-\nu)} \tag{2.101}$$

Given this, we obtain, by operating on both sides of Eq. (2.36), the moment evolution equation

$$\left(\frac{d}{dt} + \bar{\Lambda}_n\right)\bar{\mu}_n(t) = \sum_{k=1}^{n-1} \bar{a}_{nk}\bar{\mu}_k\bar{\mu}_{n-k} \tag{2.102}$$

with $\bar{\Lambda}_n = 1 - \bar{a}_{n0} - \bar{a}_{nn}$. Not only is this of precisely the same form as Eq. (2.91) for the continuous case, but we can go further and assert that, provided that the original $\bar{K}(j, i; k)$ is a proper analogue of $K(y, x; u)$—and hence $\mathcal{M}_n(j, k)$ is a proper analogue of $\mathcal{M}_n(y, x)$—in the sense of (2.19), then the expansion coefficients a_{nk} and $\bar{a}_{nk}$ are in fact identical and so are Λ_n and $\bar{\Lambda}_n$. For standard complexity the equilibrium moments are those for the negative binomial distribution (2.55), viz.,

$$\bar{\mu}_n(\infty) = \frac{(p)_n}{n!}\left(\frac{\varepsilon}{\varepsilon_0 p}\right)^n \tag{2.103}$$

which must therefore satisfy (2.102) with $(d/dt)\bar{\mu}_n(t) = 0$.

1. Polynomial Moments and Moment Evolution

The possession of moment equations of the forms (2.92) or (2.102) is equivalent to an algorithm for recovering the distribution functions $P(x, t)$ or $\bar{P}(i, t)$ provided always that the conditions for inversion of the associated moment problems are satisfied. Sufficient conditions are that the distribution functions sought should belong to the Hilbert spaces $L_2(0, \infty)$ and $l_2(0, \infty)$, respectively (i.e., that they should be square-integrable or square-summable on the infinite range of the state variable). Given this, the standard method of inversion using the appropriate set of orthogonal polynomials may be applied. To this end we construct polynomials $\phi_n(x)$ or $\bar{\phi}_n(i)$ satisfying the orthogonality relations with respect to $P_0(x)$ and $\bar{P}_0(i)$ as weight function, viz.:

$$\int_0^{\infty} dx\, P_0(x)\phi_n(x)\phi_m(x) = \mathcal{N}_n\delta_{nm} \tag{2.104}$$

and

$$\sum_{i=0}^{\infty} \overline{P}_0(i)\overline{\phi}_n(i)\overline{\phi}_m(i) = \overline{\mathcal{N}}_n\delta_{nm} \tag{2.105}$$

respectively. Given the equilibrium distributions $P_0(x)$ or $P_0(i)$ as functions of the mean quantity $\langle x \rangle = \varepsilon$ via Eq. (2.48) or its analogue, the polynomials $\phi_n(x)$ or $\overline{\phi}_n(i)$ can always, in principle, be found, together with their normalization functions $\mathcal{N}_n$ and $\overline{\mathcal{N}}_n$. Thus, if they are known explicitly as

$$\phi_n(x,\varepsilon) = \sum_{\nu=0}^{n} d_{n\nu}x^{\nu} \tag{2.106}$$

$$\overline{\phi}(i,\varepsilon,\varepsilon_0) = \sum_{\nu=0}^{\min(i,n)} \overline{d}_{n\nu}i^{(\nu)} \tag{2.107}$$

(in which we emphasize dependence on the energy-like parameter ε and the quantum ε_0 connecting the index i with the scale of the continuous variable), we can define *polynomial moments* $\gamma_n(t) = \langle \phi_n(x) \rangle$ $\overline{\gamma}_n(t) = \langle \overline{\phi}_n(i) \rangle$, respectively. Expressing these in terms of the ordinary moments $\mu_n(t) = \langle x^n \rangle / n!$ and $\mu_n(t) = \langle i^{(n)} \rangle / n!$, we thus have that

$$\gamma_n(t) = \langle \phi_n(x) \rangle = \sum_{\nu=0}^{n} d_{n\nu}\nu!\mu_\nu(t) \tag{2.108}$$

and

$$\overline{\gamma}_n(t) = \langle \overline{\phi}_n(i) \rangle = \sum_{\nu=0}^{\min(n,\nu)} \overline{d}_{n\nu}\nu!\overline{\mu}_\nu(t) \tag{2.109}$$

The reconstruction of the distribution functions then takes the form of the Fourier series

$$P(x,t) = P_0(x) \sum_{n=0}^{\infty} \gamma_n(t)\mathcal{N}_n^{-1}\phi_n(x) \tag{2.110}$$

$$\overline{P}(i,t) = \overline{P}_0(i) \sum_{n=0}^{\infty} \overline{\gamma}_n(t)\overline{\mathcal{N}}_n^{-1}\overline{\phi}_n(i) \tag{2.111}$$

[Note that with $P(x,t)$ and $\overline{P}(i,t)$ normalized to unity, we can always take $d_{00} = \mathcal{N}_0 = 1$.]

The polynomial moments are not merely of interest in the algorithm for recovering $P(x, t)$ from the set of moments $\{\mu_n(t)\}$ obtained via Eqs. (2.91); they also prove to have interesting properties in their own right.

In the first place it is evident by the orthogonality relation that, in addition to $\gamma_0(t) = 1 \ (= d_{00})$,

$$\gamma_1(t) = d_{10} + d_{11}\langle x \rangle = 0$$

and

$$\gamma_n(t) \to 0 \qquad \text{for all } n \geq 2 \qquad (2.112)$$

The latter property leads to the conclusion that, unlike the $\mu_k(t)$, each $\gamma_k(t)$ can be expressed as a sum of pure exponential transients, such that the approach to equilibrium is self-evident. All the above considerations apply, with suitable redefinition of the moments, to the discrete-variable case.

When we specialize to systems of standard complexity, further interesting properties appear. For the continuous variable with density of states $g(x) = \Gamma(p)^{-1}x^{p-1}$, the appropriate polynomial set is the Laguerre polynomials proportional to $_1F_1(-n, p; x)$ (see Appendix Eq. A5.1). By the usual definition and notation, we have that

$$\phi_n(x) = L_n^{(p-1)}\left(\frac{px}{\varepsilon}\right) = (n!)^{-1}(p)_n \, _1F_1\left(-n, p; \frac{px}{\varepsilon}\right) \qquad (2.113)$$

which, on interpretation of the confluent hypergeometric function, gives

$$\phi_n(x) = \frac{(p)_n}{n!} \sum_{\nu=0}^{n} \frac{(-n)_\nu}{(p)_\nu \nu!} \left(\frac{px}{\varepsilon}\right)^\nu \qquad (2.114)$$

and hence

$$d_{nk} = \frac{(p)_n(-n)_\nu}{n!(p)_\nu \nu!}\left(\frac{p}{\varepsilon}\right)^\nu, \qquad \mathcal{N}_n = \frac{(p)_n}{n!} \qquad (2.115)$$

Further properties of these polynomials are given in Appendix 5.

In the case of the discrete variable, the corresponding set with $\bar{g}(i) = \Gamma(p)^{-1}(i+1)_{p-1}$ are the Meixner polynomials[33,34] $\{M_n(i)\}$:

$$M_n(i, p; a) = \, _2F_1\left(-n, -i, p; 1 - a^{-1}\right) \qquad (2.116)$$

The set $\{M_n(i, p; a)\}$ are orthogonal in the sense of (2.105) with weight function given by (2.55) provided the parameter a conforms to the value (2.54). However, in order to create a proper analogue of the set

$\{L_n^{(p-1)}(px/\varepsilon)\}$ for the continuous case, we shall alter the normalization constant and take

$$\bar{\phi}_n(i) = (n!)^{-1}(p)_n {}_2F_1(-n, -i, p; 1-a^{-1}) \qquad (2.117)$$

Interpreting this $[i^{(\nu)} = (-1)^{\nu}(i)_{\nu}]$, we find that

$$\bar{\phi}_n(i) = \frac{(p)_n}{n!} \sum_{\nu=0}^{\min(i,n)} \frac{(-n)_{\nu}}{(p)_{\nu}\nu!} \left(\frac{\varepsilon_0 p}{\varepsilon}\right)^{\nu} i^{(\nu)} \qquad (2.118)$$

The limiting process $\bar{\phi}_n(i/\varepsilon_0) \to \phi(x)$ as $\varepsilon_0 \to 0$ is obvious, and we may recognize that $M_n(i, p; a) \to n!(p)_n^{-1}L_n^{(p-1)}(px/\varepsilon)$ [cf. (A5.19)] and thus

$$\bar{d}_{n\nu} = d_{n\nu}\varepsilon_0^{\nu}, \qquad \bar{\mathcal{N}}_n = \left(\frac{\varepsilon}{\varepsilon_0 p + \varepsilon}\right)^{-n}\mathcal{N}_n \qquad (2.119)$$

Further properties of the Meixner polynomials are given in Appendix 5 and will be examined in connection with the special models treated in Section III.

Thus, to summarize, we shall seek sets of expansion coefficients γ_n that enable us to write solutions for systems of standard complexity as series of the form

$$P(x, t) = P_0(x)\left\{1 + \sum_{n=1}^{\infty} \frac{n!}{(p)_n}\gamma_n(t)L_n^{(p-1)}\left(\frac{px}{\varepsilon}\right)\right\} \qquad (2.120)$$

for the continuous variable and

$$\bar{P}(i, t) = \bar{P}_0(i)\left\{1 + \sum_{n=1}^{\infty} \bar{\gamma}_n(t)a^n M_n(i, p; a)\right\} \qquad \text{with} \qquad a = \frac{\varepsilon}{(\varepsilon_0 p + \varepsilon)} \qquad (2.121)$$

in the discrete case. These will be referred to as Laguerre series and Meixner series solutions, respectively. Their different appearance is due only to the historical accident of the way in which Meixner and Laguerre polynomials have been defined. In the limit $\varepsilon_0 \to 0$, the right-hand sides of each become equal term by term by virtue of (A5.19) and $\bar{\gamma}_n(t) \to \gamma_n(t)$ as we would expect for natural analogues. The scaling [(2.56) and (2.57)] is obtained on writing $p = \varepsilon = 1$ in the parameter a.

The procedure just set out is certainly a valid, if somewhat tedious, computational algorithm for obtaining $P(x, t)$ or $\bar{P}(i, t)$ at any time to specified accuracy, given the moments of the kernel, or preferably the expansion coefficients a_{nk} in the homogeneous case—provided always that the initial distribution possesses moments to sufficiently high order. It does not rule out the possibility that solutions may exist whose moments diverge for some order n.

Fortunately, a number of simplifications to the procedure outlined have been discovered that render the task of obtaining distribution functions less forbidding than might be supposed. We have already noted that in favorable cases the expansion coefficients for discrete and continuous models that are true analogues may prove to be the same. There is also the possibility of a much simplified determination of the kernel moments by the use of a transform method, a topic to which we shall return in Section V. Most important, however, is the discovery that, for reasons fundamentally connected with the detailed-balance symmetry of the kernel, the polynomial moments $\gamma_n(t)$ and $\bar{\gamma}_n(t)$ can be shown to satisfy precisely the same transport equation of type (2.91) as the simple moments $\bar{\mu}_n(t)$ and $\bar{\mu}_n(t)$ (although with different initial conditions). The origin of this effect, which seemed accidental when first discovered in particular models, has recently been exposed by Ernst and Hendriks.[35]

The Ernst-Hendriks proof, which is restricted to the continuous variable and standard complexity, proceeds in outline as follows. It is convenient to define alternative polynomial moments

$$c_n = \langle {}_1F_1(-n, p, x) \rangle \tag{2.122}$$

and a set of reduced moments

$$M_n = \frac{\langle x^n \rangle}{\langle x^n \rangle_0} \tag{2.123}$$

with $\langle x^n \rangle_0$ the appropriate equilibrium value. We note that the equilibrium moments are $\langle x^n \rangle_0 = (p)_n(\varepsilon/p)^n$ or $\langle x^n \rangle_0 = (p)_n$ according to whether we are using the scaling (2.51) or (2.56). (It will always be clear from the context which equation applies.) On interpreting the ${}_1F_1$ function, the relation between the polynomial and reduced moments is seen to be

$$c_n = \sum_{k=0}^{n} \binom{n}{k}(-)^k M_k \tag{2.124}$$

Now, using the expansion of the Kronecker delta:

$$\sum_{k=\nu}^{n} \binom{n}{k}\binom{k}{\nu}(-)^k = \sum_{k=0}^{n-\nu} \binom{n-\nu}{k}(-)^k = \delta_{nk} \tag{2.125}$$

it is easy to show that the discrete transform (2.124) can be inverted to give

$$M_n = \sum_{k=0}^{n} \binom{n}{k}(-)^k c_k \tag{2.126}$$

At the same time the scaled moments can be shown to satisfy an equation of the form (2.91)

$$\left(\frac{d}{dt}+1\right)M_n = \sum_{k=0}^{n} b_{nk} M_k M_{n-k} \tag{2.127}$$

but with coefficients b_{nk} related to the original expansion coefficients a_{nk} by

$$b_{nk} = a_{nk}\binom{n}{k}\frac{(p)_k(p)_{n-k}}{(p)_n} \tag{2.128}$$

On substituting for the M_k on the right of Eq. 2.127 by means of Eq. 2.126 and back-transforming via Eq. 2.124, it is a straightforward matter to show that the condition for the right-hand side to be invariant is that

$$b_{nr} = \sum_{k=0}^{n} b_{kk}\binom{n}{k}\binom{k}{r}(-)^{k-r} \tag{2.129}$$

the $n+1$ values of b_{kk} being arbitrary. It remains to show that this is actually a consequence of the detailed-balance condition (2.42) which, for standard complexity in p degrees of freedom, reads

$$x^{p-1}(u-x)^{p-1}K(y,x;u) = y^{p-1}(u-y)^{p-1}K(x,y;u) \tag{2.130}$$

On multiplying the homogeneous moment Eq. (2.90) by y^m, integrating, and using (2.128), it is apparent that the condition on the b_{nk} coefficients equivalent to (2.129) is

$$\sum_{k=0}^{n} b_{nk}\frac{(p+m)k}{(p)_k} = \sum_{k=0}^{m} b_{mk}\frac{(p+n)k}{(p)_k} \tag{2.131}$$

By a somewhat intricate argument, which we shall not reproduce here, Ernst and Hendriks were able to show that, in fact, (2.131) is implied by the symmetry (2.130), the homogeneity (2.20), and the density of states $g(x) \propto x^{p-1}$. This completes the proof that the polynomial moments c_n satisfy the same equation, (2.127), as that satisfied by the ordinary reduced moments $M_n(t)$.

These results, as well as considerably simplifying computations of $P(x, t)$, have important consequences for the construction of alternative model kernels and the actual determination of the moment expansion coefficients a_{nk} in particular cases. These flow from the recognition that, according to (2.90),

$$b_{nn} = a_{nn} = \int_0^1 dx \, x^n K(x, 0; 1) \qquad (2.132)$$

and that, since the sequence $\{b_{nn}\}$ is undoubtedly positive and monotone increasing, we can assert the existence of a function $\bar{g}(\mu)$ that generates b_{nn} through the integrals[36]

$$b_{nn} = \int_{-1}^1 d\mu \, \bar{g}(\mu) \left[\frac{1+\mu}{2} \right]^n \qquad (2.133)$$

Now, on referring to Eq. (2.90), we can identify the function $\bar{g}(\mu)$ with $K(\frac{1}{2}(1+\mu), 0, 1)$.* It then remains only to compute the b_{nk} for $k \neq n$ using (2.129). The final result is

$$b_{nk} = \frac{1}{2} \binom{n}{k} \int_{-1}^1 d\mu \, K\left(\tfrac{1}{2}(1+\mu), 0; 1\right) \left[\frac{1}{2}(1+\mu) \right]^n \left[\frac{1}{2}(1-\mu) \right]^{n-k}$$

$$= \binom{n}{k} \int_0^1 ds \, K(s, 0, 1) s^k (1-s)^{n-k} \qquad (2.134)$$

Ernst and Hendriks have demonstrated that the above route to the kernel moments is considerably less tedious in particular cases than the direct integration of the kernel according to Eq. (2.23). A parallel proof for the discrete variable can be given but is unnecessary in practice in view of the identity between expansion coefficients a_{nk} and $\bar{a}_{nk}$ in systems that are true analogues.

2. Linearized Moment Equations

The linearization of the moment equations yields a number of interesting properties and further insight into the nature of the linearization approximation. We have a choice of either forming moments on the linearized transport equation (2.77) or alternatively making a linearization of the full moment equations (2.91) or (2.102). Taking the second course here, we write the reduced moments $\mu_n(t)$ in the form $\mu_n(t) = \mu_n(\infty) + \zeta_n(t)$, the functions

*In some cases it may be necessary to obtain $K(s, 0; 1)$ as the limit $\lim_{\delta \to 0} K(s, \delta; 1)$. See, e.g., Eq. (3.101).

$\zeta_n(t)$ assumed "small." On substituting into Eq. (2.91), neglecting second-order terms, and using the equilibrium property (2.94), we find that

$$\left(\frac{d}{dt}+1\right)\zeta_n(t) = \sum_{k=0}^{n} a_{nk}[\mu_k(\infty)\zeta_{n-k}+\mu_{n-k}(\infty)\zeta_k]$$

$$= \sum_{k=0}^{n} \zeta_k\mu_{n-k}(\infty)[a_{nk}+a_{nn-k}] \qquad (2.135)$$

Alternatively, denoting the moments thus approximated by $\mu_n^L(t)$, we obtain

$$\left(\frac{d}{dt}+1\right)\mu_n^L(t) = \sum_{k=0}^{n} \mu_k^L\mu_{n-k}(\infty)[a_{nk}+a_{nn-k}]-\mu_n(\infty) \qquad (2.136)$$

Equations (2.136) are easily solved by an eigenvector expansion, if we note that the eigenvalues λ_k of the triangular matrix on the right are just its diagonal elements: $\lambda_k = a_{nn}+a_{n0}$. Thus, somewhat in parallel to the solution (2.81), we find that

$$\mu_n^L(t) = \mu_n(\infty) + \sum_{k=1}^{n} \beta_k v_n^{(k)}e^{-\Lambda_k t} \qquad (2.137)$$

with, as before, $\Lambda_k = 1 - a_{nn} - a_{n0}$.

However, a much simplified alternative exists that obviates the necessity of finding the eigenvectors $v_n^{(k)}$. We consider instead the polynomial moments $\gamma_k(t)$, which, according to the analysis of Section II.J, satisfy the same set of equations as (2.91) but with the property that $\gamma_k(\infty) = 0$ for $k > 0$. Thus the γ_k themselves form the small quantities in the linearization approximation, while at the same time all but the $n = 0$ term must vanish in the summation on the right. The result is that

$$\frac{d}{dt}\gamma_n^L(t) = -\Lambda_n\gamma_n^L$$

with the obvious solution

$$\gamma_n^L(t) = \gamma_n(0)e^{-\Lambda_n t} \qquad (n>1) \qquad (2.138)$$

From the linearized polynomial moments determined in this way, the corresponding distribution function $P^L(x, t)$ can be reconstructed by the Fourier series (2.110), the result being equivalent to (2.81). We shall illustrate this process later with particular soluble models.

Given that $a_{nn} = b_{nn}$ and $a_{n0} = b_{n0}$, we can write an integral expression for Λ_n by means of Eqs. (2.93) and (2.132). This is

$$\Lambda_n = \int_0^1 ds\, K(s,0;1)\{1 - s^n - (1-s)^n\} \qquad (n > 1) \qquad (2.139)$$

It is a curious feature of this expression that whatever the form of the kernel K, the eigenvalues Λ_2 and Λ_3 stand in the simple ratio $\frac{1}{2}\Lambda_2 = \frac{1}{3}\Lambda_3$. We shall see that this has important consequences later. [See Eq. (5.39) et seq.]

K. Diffusion Approximations

A particularly interesting possibility is opened up by models for persistent, as opposed to diffuse, scattering, because, under suitable conditions, an approximation of "small" scattering may be applied with the objective of obtaining a diffusion-type kinetic equation. The condition for diffusionlike evolution in a scalar state space is the informal one that $|y - x| \ll u$ for "almost all" collisions and bears an obvious analogy to the case of small-angle scattering in the vector state space of kinetic theory. Needless to say, the condition just given is simply uninterpreted in diffuse-scattering model with kernels of type $K(y; u)$.

To obtain the diffusion approximation to Eqs. (2.17) or (2.31), we carry out the time-honored procedure of redefining the transition kernel $K(y, x; u)$ in terms of a quantity $\xi = x - y$ assumed to be effectively small. The procedure then follows a natural extension of that used for linear master equation problems (see, e.g., Van Kampen[15]).

With the usual reservations as to mathematical rigor, we may substitute $k(x, \xi, u + x)\, d\xi$ for $K(y, x; u)\, dy$ and expand the integrand in Eq. (2.17) about $\xi = 0$ to obtain formally

$$\left(\frac{\partial P(x,t)}{\partial t} \right) = - P(x) \int du\, P(u) \int d\xi\, k(x, \xi; u + x)$$

$$+ \sum_{\nu = 0}^{\infty} \frac{1}{\nu!} \left(\frac{\partial}{\partial x} \right)^{\nu} P(x) \int du\, P(u) \int d\xi (-\xi) k(x, \xi; u + x)$$

$$+ P(x) \int du\, P(u) \int d\xi\, k(x, -\xi, u + x) \qquad (2.140)$$

The first and last terms canceling, we make a truncation of the Taylor expansion of the second to obtain

$$\frac{\partial P(x,t)}{\partial t} = - \frac{\partial}{\partial x} \{ \mathscr{A}_1[x, P(x,t)] P(x,t) \}$$

$$+ \frac{1}{2} \left(\frac{\partial^2}{\partial x^2} \right) \{ \mathscr{A}_2[x, P(x,t)] P(x,t) \} \qquad (2.141)$$

where $\mathcal{A}_n$ represents the functional

$$\mathcal{A}_n[x, P(x,t)] = \int du\, P(u,t) \int dy\, (y-x)^n K(y,x; u+x)$$

$$= \int du\, P(u,t) A_n(x,u) \qquad (2.142)$$

In writing this we have reverted to the original kernel and the integration variable y. The inner integral is evidently a type of *transfer moment* superficially similar to those occurring in treatment of the linear master equation, though with a somewhat different role here.

The result (2.141) can evidently be referred to as a nonlinear Fokker-Planck equation, the nonlinearity entering through a solution-dependent "drift" term and diffusion coefficient. Notwithstanding the complicated nature of $\mathcal{A}_1$ and $\mathcal{A}_2$, it follows on partial integration of (2.141) that both number and energy are correctly conserved.

The problems posed by the solution of Eq. (2.141) could hardly be said to be less than those for the original integral equation were it not for one remarkable property. Should the kernel satisfy a homogeneous moment condition for at least the first and second orders, that is, if

$$A_1(x,u) = \beta_{10}x + \beta_{11}u \qquad (2.143a)$$

$$A_2(x,u) = \beta_{20}x^2 + \beta_{21}xu + \beta_{22}u^2 \qquad (2.143b)$$

then it is clear that, through number and state-variable conservation, the nonlinearity is effectively removed. We then arrive at a linear Fokker-Planck equation with a known parametric time dependence—that of $\mu_2(t)$ from (2.99)—in the diffusion coefficient. The result can be written

$$\frac{\partial}{\partial t} P(x,t) = \frac{\partial}{\partial x} \{[\beta_{10} + \beta_{11}\varepsilon] P(x,t)\}$$

$$+ \frac{1}{2} \frac{\partial^2}{\partial x^2} \{[\beta_{20}x^2 + \beta_{21}x\varepsilon + \beta_{22}\mu_2(t)] P(x,t)\} \qquad (2.144)$$

Again, little can be said in general terms, although the equation is clearly susceptible to treatment by approximate methods. We may note that the exponential transient in $\mu_2(t)$ leads to a partial differential difference equation on taking the Laplace transform in time. This would seem to rule out the possibility of exact solutions except, perhaps, where the time dependence is negligible, a condition effectively limited to either the case of short-time expansions or that of a sufficiently "aged" system close to equilibrium.

III. SOLUBLE STOCHASTIC MODELS

I shall begin the main part of this review by introducing a number of exactly soluble models, treated roughly in order of increasing complexity. As throughout, the intention will be primarily to give instances of Boltzmann-type equations with definite physical interest rather than to explore the ramifications of models which, however interesting mathematically, have consequences that vary between the unrealistic and the nonsensical. This will mean an emphasis on models whose scalar variable is energy-like.

The first models considered, relating to systems with only two or three states, are almost trivial compared with those that follow but are by no means without interest, for they expose an important connection between Boltzmann-type transport equations and simple differential equations of the type familiar in chemical rate theory. These are followed with the little-known model of Shuler, still in the discrete-variable domain, before we turn to the so-called VHP model of Ernst and Hendriks and the variety of models that have emerged as more elaborate successors to the original Bobylev-Krook-Wu (BKW) case. After a resumé of these models in both continuous and discrete variables, we shall return to more general considerations, to discuss the underlying structures that link them and lead to solution methods of wider application and to recently discovered techniques for manufacturing new soluble models from certain prototypes.

A. Two- and Three-State Models

If all requirements to satisfy conservation laws are waived, it is possible to construct models having as few as two states; the simplest with energy-like conservation requires no less than three. Two types of two-state models were known before the BKW solution, those due to Carleman[37] and McKean.[38,39] When interpreted as velocity models in a one-dimensional space, both of these lead to insoluble PDEs that are somewhat similar to the telegraph equation and reduce to it on linearization. (See, e.g., Ernst,[11] Secs. 2.1 and 2.2.) The spatially uniform cases are less interesting but can be solved, as we indicate below.

1. The Carleman Model

Carleman[37] considered a closed system of particles possessing only two states, which we may label plus $(+)$ and minus $(-)$, with the simple scattering law: Identical particles change sign, nonidentical particles are unchanged on collision. (See Fig. 2a.) Denoting the probability distribution for each by $F_+(t)$ and $F_-(t)$ with $F_+ + F_- = 1$, we can see that the evolution equations are simply

$$\frac{dF_+}{dt} = F_-^2 - F_+^2 = 1 - 2F_+ \qquad \text{and} \qquad \frac{dF_-}{dt} = F_+^2 - F_-^2 = 1 - 2F_-$$

$$(3.1)$$

and are thus linear. Solving, we obtain

$$F_\pm(t) = \tfrac{1}{2}\left[1 - \left[1 - 2F_\pm(0)\right]e^{-2t}\right] = \tfrac{1}{2}\left(1 \pm \Delta_0 e^{-2t}\right) \qquad (3.2)$$

with $\Delta_0 = F_+(0) - F_-(0)$. The stationary state is clearly $F_\pm(\infty) = \tfrac{1}{2}$.

2. The McKean Model

The McKean model[38,39] is somewhat more interesting than Carleman's because it is truly nonlinear. The scattering law, as indicated in Fig. 2b is not symmetrical in the two species and gives for the spatially homogeneous case

$$\frac{dF_+}{dt} = F_-^2 - F_+ F_- \qquad \text{and} \qquad \frac{dF_-}{dt} = F_+ F_- - F_-^2 \qquad (3.3)$$

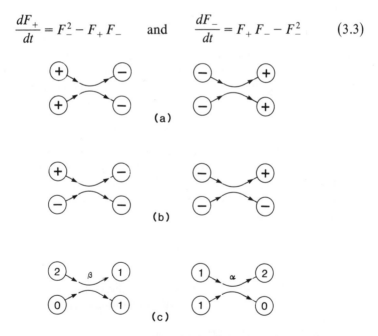

Fig. 2.   Two- and three-state scattering models. (a) The Carleman model. (b) The McKean model. (c) The three-state model.

Using $F_+ + F_- = 1$ in the first equation, it follows that

$$\frac{dF_+}{dt} = (F_+ - 1)(2F_+ - 1) \tag{3.4}$$

This can be solved, after a partial fraction resolution, to give

$$F_+(t) = \frac{1 - Ae^{-t}}{2 - Ae^{-t}}, \qquad F_-(t) = \frac{1}{2 - Ae^{-t}} \tag{3.5}$$

with

$$A = \frac{2F_+(0) - 1}{F_+(0) - 1} \tag{3.6}$$

By further manipulations, both states can be included in the alternative

$$F_\pm(t) = \frac{1}{2}\left(1 \pm \frac{\Delta_0 e^{-t}}{1 - \Delta_0(1 - e^{-t})}\right) \tag{3.7}$$

with Δ_0 as before. $F_\pm = \frac{1}{2}$ is again seen to be stationary.

The McKean model has been used by Henin and Prigogine[40] in discussions of entropy production in Boltzmann-type systems. A much deeper account of the McKean model from the point of view of its underlying algebraic structure, fluctuation theory, and the central limit theorem will be found in Ref. 38.

3. The Three-State Model

The simplest case in which an energy-exchange model can be constructed is where three states occur corresponding to the possession of zero, one, and two quanta of energy, respectively. Simple considerations show that if we index the states $i = 0, 1, 2$, then energy exchange can take place effectively only through the forward and reverse process (see Fig. 2c):

$$(0) + (2) \underset{\alpha}{\overset{\beta}{\rightleftarrows}} (1) + (1)$$

with rate constants α and β as indicated. In these terms we are dealing with a Type C' process on three states with transition matrix elements

$$\begin{aligned}
K(2,1;2) &= K(0,1;2) = \alpha \\
K(1,2;2) &= K(1,0;2) = \beta \\
K(j,i;k) &= 0 \text{ otherwise} \quad (k = 0,1,2; i, j \leqslant k)
\end{aligned} \tag{3.8}$$

With this we can see that the evolution of the system formally described by Eq. (2.36) is in fact reducible to the three coupled differential equations:

$$\frac{dF_0}{dt} = \alpha F_1^2 - \beta F_0 F_2$$

$$\frac{dF_1}{dt} = -2\alpha F_1^2 + 2\beta F_0 F_2 \qquad (3.9)$$

$$\frac{dF_2}{dt} = \alpha F_1^2 - \beta F_0 F_2$$

Here we have written $F_0(t) = P(0, t)$, $F_i(t) = P(i, t)$ for the transient populations and $F_i(\infty) = P_0(\infty) = P(i, \infty)$ for this section alone. The conservation conditions

$$\frac{d}{dt}(F_0 + F_1 + F_2) = 0, \qquad \frac{d}{dt}(F_1 + 2F_2) = 0 \qquad (3.10)$$

are evident and we may take these to hold with the normalization and scaling

$$F_0 + F_1 + F_2 = 1, \qquad F_1 + 2F_2 = \varepsilon \qquad (3.11)$$

The solution of these equations is a straightforward exercise of a kind familiar in chemical kinetics. Noticing first that necessarily

$$F_1 = 2 - \varepsilon - 2F_0, \qquad F_2 = F_0 + \varepsilon - 1 \qquad (3.12)$$

we can concentrate on the solution for $F_0(t)$ alone. Use of conditions (3.12) then leads to the single nonlinear equation

$$\frac{dF_0}{dt} = (4\alpha - \beta)F_0^2 + [4\alpha(\varepsilon - 2) - \beta(\varepsilon - 1)]F_0 + \alpha(\varepsilon - 2)^2 \qquad (3.13)$$

which we can rewrite as

$$\frac{dF_0}{dt} = (F_0 - \lambda_1)(F_0 - \lambda_2)$$

This is immediately integrable if we express the reciprocal of the right-hand side in partial fractions. Supplying the constant of integration from the initial conditions $F_0(0)$, we find

$$\ln\left[\frac{[F_0(t) - \lambda_2]/[F_0(0) - \lambda_2]}{[F_0(t) - \lambda_1]/[F_0(0) - \lambda_1]}\right] = (\lambda_2 - \lambda_1)t$$

which on rearrangement gives

$$F_0(t) = \left[\frac{\lambda_2[F_0(0)-\lambda_1]-\lambda_1(F_0(0)-\lambda_2)\exp[-(\lambda_1+\lambda_2)t]}{F_0(0)-\lambda_1-[F_0(0)-\lambda_2]\exp[-(\lambda_1-\lambda_2)t]}\right] \quad (3.14)$$

The quantities λ_1, λ_2 are identifiable as

$$\lambda_1 = A + B^{1/2} \qquad \lambda_2 = A - B^{1/2} \quad (3.15)$$

with

$$A = \beta(\varepsilon-1)-4\alpha(\varepsilon-2)$$
$$B = \beta^2(\varepsilon-1)^2-4\alpha\beta\varepsilon(\varepsilon-2) \quad (3.16)$$

The inverse "relaxation time" for the process $\Lambda = \lambda_1 - \lambda_2$ is then

$$\Lambda = \lambda_1 - \lambda_2 = \frac{\left[\beta^2(\varepsilon-1)^2-4\alpha\beta\varepsilon(\varepsilon-2)\right]^{1/2}}{\alpha-\beta} \quad (3.17)$$

Finally we note the equilibrium distribution

$$F_0(\infty) = \lambda_2, \qquad F_1(\infty) = 2-\varepsilon-2\lambda_2, \qquad F_2(\infty) = \lambda_2+\varepsilon-1 \quad (3.18)$$

The solutions can be seen to become trivial in the limits $\varepsilon \to 0$ and $\varepsilon \to 2$.

B.   Shuler's Model: Landau-Teller Oscillator Relaxation

A paper by Shuler in 1960[17] gave what appears to be the first example of a soluble Boltzmann equation of Type A'. The system treated is the closed ensemble of single, quantum harmonic oscillators $[\bar{g}(i)=1]$ interacting under the Landau-Teller transition probabilities. These require nearest-neighbor transitions with rate proportional to the state index and in our present notation lead to the transition matrix

$$K(j,i;k) = (i+1)(k-i)\delta_{j,i+1} + i(k-i+1)\delta_{j,i-1} \quad (3.19)$$

We may easily check the interactional symmetry

$$K(k-j,k-i;k) = (k-i+1)i\delta_{k-j,k-i+1} + (k-i)(i+1)\delta_{k-j,k-i-1}$$
$$= K(j,i;k)$$

and the microscopic reversibility condition

$$K(i, j; k) = j(k - j + 1)\delta_{j, i+1} + (j + 1)(k - j)\delta_{j, i-1}$$
$$= (i + 1)(k - i)\delta_{j, i+1} + i(k - i + 1)\delta_{j, i-1}$$
$$= K(j, i; k)$$

Since the density of states is $\bar{g}(i) = 1$, the equilibrium distribution must be just the $p = 1$ case of Eq. (2.55). Taking $\varepsilon_0 = 1$ for the present, this can be written

$$P_0(i) = \frac{1}{1 + \varepsilon}\left(\frac{\varepsilon}{1 + \varepsilon}\right)^i \qquad (3.20)$$

Returning to the transport equation, we find, on substitution of the transition probabilities (3.19) into (2.18) and use of the conservation conditions

$$\sum_{i=0}^{\infty} P(i) = 1, \qquad \sum_{i=1}^{\infty} iP(i) = \varepsilon \qquad (3.21)$$

that

$$\frac{dP(i)}{dt} = (i + 1)(\varepsilon + 1)P(i + 1) - [(2i + 1)\varepsilon + i] P(i) + i\varepsilon P(i - 1) \qquad (3.22)$$

By introducing the quantity $a = \varepsilon/(1 + \varepsilon)$ and scaling the time as $\tau = (1 - a)t$, we then arrive at the simpler alternative

$$\frac{dP(i)}{dt} = (i + 1)P(i + 1) - [i + (i + 1)a] P(i) + aiP(i - 1) \qquad (3.23)$$

Thus, by a fortunate property of the nearest-neighbor structure, the exact transport equation is reduced to a *linear* differential difference form. This equation has previously occurred and been solved in connection with linear transport problems for oscillators in a heat bath. Details of the solution, which need not concern us here, were first given by Montroll and Shuler.[41] Using a generating-function method, they showed that

$$P(i, \tau) = (1 - a)a^i + a^i \sum_{\nu=1}^{\infty} c_\nu a^{-\nu} l_\nu(i; a)e^{-\nu\tau} \qquad (3.24)$$

where

$$c_\nu = (1-a) \sum_{i=0}^{\infty} P(i,0) l_\nu(i;a) \qquad (3.25)$$

and the $l_\nu(i;a)$ are the Gottlieb polynomials defined

$$l_\nu(i,a) = a^\nu \sum_{r=0}^{\nu} (1-a^{-1})^r \binom{\nu}{r}\binom{i}{r} = a^\nu {}_2F_1(-i, -\nu, 1; 1-a^{-1})$$

$$(3.26)$$

These polynomials, although sometimes discussed separately,[42] are evidently just the $p=1$ case of the Meixner polynomials treated earlier [Eq. (2.116)] and discussed in Appendix 5. A generating function for the moments of $P(i,\tau)$ based on the relation (A5.16) can be found in the Montroll and Shuler paper.

The Shuler model is of considerable importance inasmuch as the transition probabilities (3.19) may be derived by a perturbational treatment in the quantum theory of vibration-vibration energy transfer for diatomic molecules.[43] This type of system, when excited by laser or shock-wave techniques, is one of the few that can be studied experimentally under conditions where the internal-state Boltzmann equation may apply. The Shuler model would appear to be so far a unique example of a Type A' Boltzmann equation with a solution that is both exact and general. An extension of the Shuler solution for the case of simultaneous vibrational and translational energy transfer has been given by Rankin and Light.[18]

C. The Scattering $K(y, x; u) = $ const.
(Ernst-Hendriks VHP Model)

This, one of the first soluble models distinct from the BKW case, was introduced by Ernst and Hendriks in 1979 as the "very hard particle" model.[44] Taken out of the kinetic theory context that gives rise to this name (and to which we return in Section IV), it corresponds to a Type B equation with kernel

$$K(x, y; u) = C \qquad 0 < y < u$$
$$= 0 \qquad y > u \qquad (x < u) \qquad (3.27)$$

that is, to a system for which the transition rate to all final states is a constant up to the total state quantity u and zero otherwise. Since the collision-number function is evidently $z(x; u) = u$ in this case, we can simplify the out-scattering term to obtain the transport equation

$$\left(\frac{\partial}{\partial t} + x + \varepsilon\right) P(x,t) = \int_x^\infty du \int_0^u dy\, P(y) P(u - y) \qquad (3.28)$$

the constant C being here absorbed into the time scaling.

The convolution on the right invites the use of a Laplace transform

$$G(z,t) = \int_0^\infty dx\, e^{-zx} P(x,t) \qquad (3.29)$$

through which (3.28) is reduced to a nonlinear partial differential equation

$$\left(\frac{\partial}{\partial t} - \frac{\partial}{\partial z} + \varepsilon\right) G = \frac{1}{z}(1 - G^2) \qquad (3.30)$$

By suitable changes of both dependent and independent variables, Ernst and Hendriks were able to reduce this to a Riccatti equation that could be solved to give

$$G(z,t) = \frac{\phi(z+t) + (z-1)e^{-t}}{(1+z)\phi(z+t) + e^{-t}} \qquad (3.31)$$

with $\phi(.)$ an arbitrary function to be fitted through the initial condition $G(z,0)$. Letting $t = 0$ in (3.31), the result is

$$\phi(z) = \frac{G(z,0) + z - 1}{(z+1)G(z,0) - 1} \qquad (3.32)$$

The solution is thus determined to within a Laplace inversion. Only in special cases can an explicit inversion to the distribution function $P(x,t)$ be given, but in most cases of practical interest standard numerical inversion methods may be used straightforwardly.

Some of the special cases are useful, however, for the insight they provide. Ernst and Hendriks consider the case

$$P(x,0) = e^{-\alpha x}(A + Bx) \qquad (3.33)$$

with $1 < \alpha < 2$ and the conditions $\langle 1 \rangle = 1$ and $\langle x \rangle = 1$ dictating the connections $A = \alpha(2 - \alpha)$ and $B = \alpha^2(\alpha - 1)$. The initial-time Laplace transform becomes

$$G(z,0) = \frac{\alpha^2 + \alpha(2 - \alpha)z}{(z + \alpha)^2} \qquad (3.34)$$

with a double pole at $z = -\alpha$. The function $\phi(z)$ follows in the form

$$\phi(z) = \frac{1 - 2\alpha - z}{(\alpha - 1)^2} \tag{3.35}$$

From this we can deduce that $G(z, t)$ has two time-dependent poles $z_1(t)$ and $z_2(t)$, and takes the form

$$G(z, t) = \frac{az + b}{(z - z_1)(z - z_2)} \tag{3.36}$$

with functions $a(t)$ and $b(t)$ that can be determined. The corresponding distribution function is then

$$P(x, t) = \frac{1}{z_1 - z_2}\left\{ (az_1 + b)e^{z_1 x} - (az_2 + b)e^{z_2 x} \right\} \tag{3.37}$$

This behavior, although special to the case considered, would seem to mimic the general solution, but the latter will as a rule be governed by a quite complex pattern of motion of further poles and branch points in the complex z plane. Ernst and Hendriks show, however, that for sufficiently long time the evolution is always determined by a particular pole $z_1(t)$ whose behavior is necessarily of the form

$$z_1(t) = -1 + \tfrac{1}{2}(\alpha - 1)e^{-t} + \cdots$$

The approach of the distribution function is correspondingly

$$P(x, t) = \left\{ 1 + O(e^{-t}) \right\} e^{-x}\exp\left\{ xO(e^{-t}) \right\} + O(e^{-t})e^{-xt + \cdots} \tag{3.38}$$

This analysis is important in that it provides a counterexample to the conjecture that the approach to the Maxwellian for large times is uniform in energy and described by the appropriate linearized transport equation. This is patently not the case in the above example, where the decay of the high-energy tail is anomalous.

A discrete-variable version of the VHP model has recently assumed new importance through its relationship to the aggregation-fragmentation equations of aerosol and polymer science. (See Section VII.) As the unique example of an exactly and generally soluble model of Type B' and one that has appeared at several points on the literature, this case deserves at least brief mention.

The required analogue of Eq. (3.28) can be written

$$\left(\frac{\partial}{\partial t}+i+\varepsilon\right)\bar{P}(i,t)=\sum_{k=i}^{\infty}\sum_{j=0}^{k}\bar{P}(j)\bar{P}(k-j) \tag{3.39}$$

under the assumption, as before, that $\langle l \rangle =1$, $\langle i \rangle = \varepsilon$, and $\varepsilon_0 =1$.

This equation was, in fact, first introduced by Rouse and Simons,[45] who noticed that its solution may be obtained by a sequential algorithm and analyzed it extensively on this basis. More recently, however, Ernst and Hendriks[44] have found the exact general solution by methods parallel to the continuous-variable solution just treated. On defining a Laplace generating function

$$\bar{G}(x,t)=\sum_{n=0}^{\infty}x^{i}\bar{P}(i,t) \tag{3.40}$$

an evolution equation for this can be obtained and solved. The result is

$$\bar{G}(x,t)=\frac{\phi(xe^{-t})+(\varepsilon-\varepsilon x-x)e^{-(\varepsilon+1)t}}{(1+\varepsilon-\varepsilon x)\phi(xe^{-t})-xe^{-(\varepsilon+1)t}} \tag{3.41}$$

with $\phi(\cdot)$ again an arbitrary function to be determined by initial conditions. The function $\bar{G}(x,t)$ may be used to determine moments or asymptotic behavior directly or submitted to a numerical inversion procedure leading to the distribution function $\bar{P}(i,t)$ itself. In this connection we may note that $\bar{G}(x,t)$ is the "Z transform" of $\bar{P}(i,t)$ in engineering terminology and that a considerable body of information under this heading is available.[46]

An interesting recent development concerning both types of VHP models, is the discovery that for the general solutions given to satisfy energy conservation they must decay more rapidly than x^{-3} or i^{-3}, respectively, as the variable tends to infinity. If this is not the case, a situation arises in which energy is in effect being supplied from infinity in the tail of the distribution function at a rate inconsistent with conservation. The analogous property for polymer kinetics is of particular importance and will be considered again in Section VII. For further details of this effect we refer to Ernst (Ref. 12, Sec. 2.1).

D.   The Tjon-Wu Scattering $K(y,x;u)=u^{-1}$

The first detailed analysis of this model was carried out by Tjon and Wu[47,48] after they had shown that it is the energy-space equivalent of the original BKW model in velocity space. It had earlier been introduced by

Nishimura[9] as a discrete-time model. The Tjon-Wu model marked a crucial step in the understanding of soluble Boltzmann models, and we shall use it here as a prototype through which to demonstrate techniques applicable in more elaborate cases. Again, the kinetic theory background will be left in abeyance, to be taken up more comprehensively in Section IV.

Seen as a scalar model in an energy-like variable, the Tjon-Wu scattering can be taken to represent the behavior of a collision complex of total energy u, which divides such that all dispositions $(y)+(u-y)$ in the outgoing elements are formed with equal probability. This might correspond to single classical oscillators $[g(x)=1]$ in what is occasionally referred to in chemical kinetics as a "sticky collision"—an inelastic collision of such long duration that all details of the initial disposition $(x)+(u-x)$ are lost in the statistical outcome. Since, by the implied convention, $K(y, x; u)=0$ for $x > u$ and $y > u$, we obtain $z(x, u)=1$ on integration, and the transport equation must be of Type D.

$$\left(\frac{\partial}{\partial t}+1\right)P(x, t)=\int_x^\infty \frac{du}{u}\int_0^u dy\, P(y)P(u-y) \qquad (3.42)$$

1. Moment Evolution

Let us consider first the derivation of the ordinary moment equations, following the procedures of Section II.J. From the moments of the uniform distribution we find that $\mathcal{M}_n(y, u)=(u+y)^n/(n+1)$, with obvious homogeneity in y and u. The expansion coefficients being just $a_{nk}=a_n=(n+1)^{-1}$, the required moment evolution equation can be written

$$\left(\frac{d}{dt}+1\right)\mu_n(t)=\frac{1}{n+1}\sum_{k=0}^n \mu_k(t)\mu_{n-k}(t) \qquad (3.43)$$

This is the continuous-time version of the equation first derived by Nishimura.[9]

Now consider the polynomial moments $\gamma_n(t)$ defined as in (2.108) et seq. The appropriate polynomial set being the Laguerre polynomials $L_k^0(x/\varepsilon)$, we can take moments on Eq. (3.42) to obtain

$$\left(\frac{d}{dt}+1\right)\gamma_n(t)=\int_0^\infty dy\, P(y)\int_0^\infty \frac{du\, P(u)}{u+y}\int_0^{u+y} dx\, L_n^0\left(\frac{x}{\varepsilon}\right)$$

The inner integral can now be reduced by the ladder-operator formula (A5.7) in the special case $p = q = 1$ with the appropriate argument. The result is that

$$\left(\frac{d}{dt}+1\right)\gamma_n(t) = \frac{1}{n+1}\int_0^\infty dy\, P(y)\int_0^\infty du\, P(u) L_n^{(1)}\left(\frac{u+y}{\varepsilon}\right)$$

It remains only to use the Laguerre addition formula (A5.6), again for $p = q = 1$, whereupon

$$\left(\frac{d}{dt}+1\right)\gamma_n(t) = \frac{1}{n+1}\sum_{k=0}^n \gamma_k(t)\gamma_{n-k}(t) \tag{3.44}$$

Thus the Laguerre moments $\gamma_n(t)$ are seen to satisfy an equation identical in form to that already derived for the simple moments $\mu_n(t)$ [though of course with necessarily different initial conditions $\gamma_n(0)$]. We know this to be a foregone conclusion from the analysis of Section II.J, but the detailed derivation is nevertheless interesting and will be found useful in constructing the analogous equations for the discrete-variable model.

2. The Similarity Solution

We turn now to the similarity solution. As indicated earlier, the search for self-similar solutions to model Boltzmann equations has in practice been largely heuristic in character, although sometimes justified after the event by more careful analysis. Bobylev himself proceeded in this way[1]; Krook and Wu[3] allowed themselves a degree of more inspired guesswork by working with a more suggestive generating function method. (See Section IV.B).

The solution chosen by Bobylev is easily adapted to Tjon and Wu's equivalent in energy space. We consider the following hypothetical form for the solution under the u^{-1} scattering kernel:

$$P^S(x,t) = [\alpha(t)+\beta(t)x]\exp[-x\psi(t)] \tag{3.45}$$

Here there are three as yet unknown time-dependent functions $\alpha(t)$, $\beta(t)$, and $\psi(t)$, which, it may be hoped, yield the required equilibrium distribution, $P(x,\infty) = \exp(-x)$, as $t \to \infty$.

To test whether such a solution to Eq. (3.42) is possible, we first note that the two conservation conditions (2.29) and (2.30) imply the interrelationships

$$\alpha + \frac{\beta}{\psi} = \psi \quad \text{and} \quad \alpha + \frac{2\beta}{\psi} = \psi^2\varepsilon$$

which may be used to reduce the number of unknown functions to one: ψ,

M. R. HOARE

given by

$$\alpha = \psi(2 - \varepsilon\psi) \qquad \text{and} \qquad \beta = \psi^2(\varepsilon\psi - 1) \qquad (3.46)$$

Substitution of the solution (3.45) directly into the transport equation leads to the postulated equivalence

$$\left[-x\left(\frac{\partial\psi}{\partial x}\right) + \frac{\partial}{\partial t} + 1\right][\alpha + \beta x]$$

$$\overset{?}{=} \psi^{-3}\left\{ \alpha^2\psi^2 + \alpha\beta\psi + \frac{1}{3}\beta^2 + \left(\alpha\beta\psi^2 + \frac{1}{3}\psi\right)x + \frac{1}{6}\beta^2 x^2\psi^2\right\}$$

For this actually to be a solution, we require identity of the coefficients of x^0, x^1, and x^2 on either side. When this is tested, each condition is found separately to give the same requirement in the form of a single differential equation:

$$\frac{d\psi}{dt} = -\tfrac{1}{6}(\varepsilon\psi - 1)\psi$$

This may be solved to give

$$\psi(t) = \varepsilon^{-1}\left[1 - \eta\exp(-\tfrac{1}{6}t)\right]^{-1} \qquad (3.47)$$

with η a constant of integration that now serves to parameterize the family of allowed initial conditions $P^S(x,0)$. The final form of the solution becomes, on using (3.46),

$$P^S(x,t) = \psi[2 - \varepsilon\psi + \psi(\varepsilon\psi - 1)x]e^{-\psi x} \qquad (3.48)$$

Since $\psi \to \psi_0 = \varepsilon^{-1}$ as $t \to \infty$, we can better expose the approach to equilibrium by writing

$$P^S(x,t) = \left(\frac{\psi}{\psi_0}\right)\left[2 - \frac{\psi}{\psi_0} + \frac{\psi}{\psi_0}\left(\frac{\psi}{\psi_0} - 1\right)\frac{x}{\varepsilon}\right]\cdot\frac{1}{\varepsilon}\exp\left[-\frac{x}{\varepsilon}\left(\frac{\psi}{\psi_0}\right)\right] \qquad (3.49)$$

Thus $P^S(x,t) \to P_0(x) = \varepsilon^{-1}\exp(-x/\varepsilon)$ in accordance with (2.51) for $p = 1$.

Now, while it is very satisfying to possess an exact solution to a nonlinear initial-value problem of the kind considered, several factors limit this achievement and even cast doubt on the practical usefulness of (3.49). The similarity solution is very far from general and shows no obvious feature that would lend itself to an extension of the function space parameterized by the quantity η. Worse than this, we have no immediate guarantee that the positivity condition $P^S(x, t) \geq 0$ can be satisfied over any useful range; in fact, $P^S(x, t)$ is manifestly negative for certain values of η and/or t. Investigating this more closely, we see that positivity requires that both α and β must be positive and hence [noting (3.46)] that ψ must satisfy the inequalities

$$1 \leqslant \varepsilon \psi(t) \leqslant 2$$

which in turn imply that η must be restricted by

$$0 \leqslant \eta \leqslant \tfrac{1}{2} \tag{3.50}$$

However, we do know that if ψ satisfies the inequalities given above at a given time, then it will certainly satisfy them at all subsequent times, as we would expect from the more general requirement on the distribution function (Section II.G). And, if a nonpositive initial condition is specified such that $\eta > \frac{1}{2}$ in violation of (3.50), then we can easily calculate that it will become positive at a time t_0, where $t_0 = 6 \ln 2\eta$. We shall return to questions of the range of validity of similarity solutions again in Section IV. Before we leave the present model, however, let us consider briefly the *similarity moments* arising from (3.48). Defining ordinary and polynomial similarity moments as

$$\mu_n^S(t) = \frac{1}{n!} \int_0^\infty dx\, x^n P^S(x, t) \tag{3.51}$$

and

$$\gamma_n^S(t) = \int_0^\infty dx\, L_n^0\left(\frac{x}{\varepsilon}\right) P^S(x, t) \tag{3.52}$$

respectively, we see that both are easily derived from the solution (3.48). For the former,

$$\mu_n^S(t) = \psi^{-n}(1 + n(\varepsilon\psi - 1)) \tag{3.53}$$

To obtain the latter we use the standard integral (A5.8), whence, after some algebra,

$$\gamma_n^S(t) = \eta^n(1 - n)\exp(-\tfrac{1}{6}nt) \tag{3.54}$$

Using this we can then write a Fourier expansion of the similarity solution in the form

$$P^S(x, t) = P_0(x)\left\{1 + \sum_{k=2}^\infty (1 - k)\eta^k L_k^0\left(\frac{x}{\varepsilon}\right)\exp\left(-\frac{1}{6}kt\right)\right\} \tag{3.55}$$

This is completely equivalent to the finite expression (3.48).

3. Linearization

Let us compare this result with the one obtained for the linearized approximation of Section II.H. The linearized Laguerre moments $\gamma_k^L(t)$ relax as $\gamma_k^L(t) = \gamma_k(0)\exp(-\Lambda_k t)$, with Λ_k the eigenvalue

$$\Lambda_k = 1 - 2a_k = \frac{k-1}{k+1} \qquad (3.56)$$

We can thus write the linearized solution as

$$P^L(x,t) = P_0(x)\left\{1 + \sum_{k=2}^{\infty} \gamma_k(0)L_k^0\left(\frac{x}{\varepsilon}\right)\exp\left[-\frac{k-1}{k+1}t\right]\right\} \qquad (3.57)$$

Since, by (3.54), $\gamma_k(0) = (1-k)\eta^k$ for the similarity initial conditions, Eqs. (3.55) and (3.57) are identical except for the exponential factors. These, it will be noticed, are also identical for $k = 2$ and $k = 3$, diverging more or less rapidly thereafter. This is an interesting instance of the generally believed tendency for linear approximations to be most accurate in approximating the lower moments, and hence long-term behavior, and to falsify increasingly the higher moment transients responsible for short-term behavior and more severe deviation from equilibrium.

Finally we may consider the linearized transport equation itself. Since, for one degree of freedom, $P_0(x) = \varepsilon^{-1}e^{-x/\varepsilon}$, the linear kernel $k(x,y)$ according to Eq. (2.82) becomes

$$k(x,y) = \frac{e^{x/\varepsilon}}{\varepsilon}\int_{\max(x,y)}^{\infty}\frac{du\,e^{-u/\varepsilon}}{u} = \varepsilon^{-1}e^{x/\varepsilon}\text{erf}[\max(x,y)] \qquad (3.58)$$

The eigenvalue problem for this kernel was studied some time ago by the author,[49] who showed that it can be solved to give eigenvalues

$$\lambda_k = (k+1)^{-1} \qquad (3.59)$$

and right eigenfunctions

$$\phi_k(x) = P_0(x)L_k^0\left(\frac{x}{\varepsilon}\right) \qquad (3.60)$$

We are thus led to the solution (3.57) by this alternative route.

E. The Discrete Scattering $\bar{K}(j,i;k) = (k+1)^{-1}$

This type of scattering was introduced by Nishimura,[8] who obtained moment equations for the discrete-time case. Futcher, Hoare, Hendriks, and Ernst[50,51] were the first to obtain similarity solutions and carry out a complete analysis of the polynomial moment problem in parallel to the continuous model just described. A closely related model, not susceptible to exact solution, was the subject of a numerical study by Tjon.[48] For purposes of exposition we again concentrate on this simplest case of the more general negative hypergeometric scattering model treated by Futcher et al. and summarized in Section III.G.

The model we consider is based on a discrete-variable uniform distribution:

$$\bar{K}(j,i;k) = (k+1)^{-1} \qquad (j,i \leqslant k) \tag{3.61}$$

leading to the transport equation

$$\left(\frac{\partial}{\partial t} + 1\right)\bar{P}(i,t) = \sum_{k=i}^{\infty} \frac{1}{k+1} \sum_{j=0}^{k} \bar{P}(j)\bar{P}(k-j) \tag{3.62}$$

This might be taken to represent a system of single quantized oscillators that interact such that the total quanta in every collision complex are distributed between the partners with equal probability in all possible dispositions. We shall obtain both moment solutions and similarity solutions by steps which, though less familiar, are strictly parallel to those of Section III.D at every stage.

As a first step we compute the factorial moments of the kernel according to Eq. (2.27), finding that

$$\mathscr{M}_n(j,k) = \frac{(j+k)^{(n)}}{n+1} \tag{3.63}$$

These can be seen to be homogeneous of degree n in the factorial monomials $k^{(\nu)}$, $j^{(\nu)}$ by virtue of Vandermonde's theorem (A3.7), which also confirms that the expansion coefficients in Eq. (2.102) are $\bar{a}_{nk} = (n+1)^{-1}$, unchanged from the continuous-variable case.* Thus, as we expect from the analysis of

---

*The property of homogeneity for functions of a discrete variable is not altogether natural. A sufficient definition for present purposes is to say, by extension of the analogue property (2.19), that a function $\bar{f}(i,j,\dots)$ of the discrete variables $i, j,\dots$ is homogeneous of degree n if there exists a continuous function $f(x, y,\dots)$ in the variables $x, y,\dots$ such that

$$\lim_{\varepsilon_0 \to 0} \varepsilon_0^n \bar{f}\left(\frac{\lambda x}{\varepsilon_0}, \frac{\lambda y}{\varepsilon_0},\dots\right) = \lambda^n f(x, y,\dots)$$

The "true analogue" property [Eq. (2.19)] is possible only when there is homogeneity of degree zero ($n = 0$).

Section II.J, the discrete-moment equation is identical in form, though not in interpretation, to (3.43):

$$\left(\frac{d}{dt}+1\right)\bar{\mu}_n(t) = \frac{1}{n+1}\sum_{k=0}^{n}\bar{\mu}_k(t)\bar{\mu}_{n-k}(t) \tag{3.64}$$

The discrete-time version of the same was first given by Nishimura.[8] We note that in the present scaling, $\bar{\mu}_1 = \varepsilon/\varepsilon_0$ for all time, while asymptotically $\bar{\mu}_n(\infty) = (\varepsilon/\varepsilon_0)^n$ in agreement with (2.103) for the equilibrium distribution

$$\bar{P}_0(i) = (1-a)a^i, \qquad a = \frac{\varepsilon}{\varepsilon_0} + \varepsilon \tag{3.65}$$

Derivation of the polynomial moment equations requires the use of the Meixner set $\{M_n(i)\}$ orthogonal with respect to the above weight. Since there is only a single degree of freedom, these are in fact

$$M_n(i,1,a) = l_n(i;a) = {}_2F_1\left(-n,-i,1;-\frac{\varepsilon_0}{\varepsilon}\right) \tag{3.66}$$

which are just the Gottlieb polynomials of Section III.B. To derive the evolution equation for the moments $\bar{\gamma}_n(t)$ [cf. (3.44)], we operate on Eq. (3.62) to obtain first

$$\left(\frac{d}{dt}+1\right)\bar{\gamma}_n(t) = \sum_{j=0}^{\infty}P(j)\sum_{k=0}^{\infty}\frac{P(k)}{k+j}\sum_{i=0}^{k+j}M_n(i,1,a)$$

$$= \sum_{j=0}^{\infty}\sum_{k=0}^{\infty}P(j)P(k)M_n(k+j,2,a)$$

The Meixner polynomial ladder-operator formula (A5.18) has been used to simplify the triple summation. It remains only to apply the Meixner addition formula (A5.17) to show that

$$\left(\frac{d}{dt}+1\right)\bar{\gamma}_n(t) = \frac{1}{n+1}\sum_{k=2}^{n}\bar{\gamma}_k(t)\bar{\gamma}_{n-k}(t) \tag{3.67}$$

as expected. In writing the summation we recognize that since $M_0(i,1,a)=1$ and $M_1(i,1,a)=1+i(1-a^{-1})$, $\bar{\gamma}_0(t)=1$ and $\bar{\gamma}_1(t)=0$, while, as in the continuous-variable case, $\bar{\gamma}_n(\infty)=0$ for $n\geq 1$.

The linearization of the moment equations also proceeds in straightforward analogy with that for the continuous-variable case. We find that the

Meixner moments relax as $\bar{\gamma}_n(t) = \bar{\gamma}_n(0)\exp(-\Lambda_n t)$. The linear approximation to the solution is thus

$$\bar{P}^L(i, t) = \bar{P}_0(i)\left\{1 + \sum_{k=2}^{\infty} \bar{\gamma}_k(0)M_k(i, 1, a)e^{-\Lambda_k t}\right\} \qquad (3.68)$$

with $\Lambda_n = 1 - 2\bar{a}_n = (n-1)/(n+1)$ as before.

The linearized form of the transport equation is derived from the kernel

$$\bar{k}(i, j) = 2a^{-j} \sum_{k=\max(i, j)}^{\infty} \frac{a^k}{k+1}, \qquad a = \frac{\varepsilon}{\varepsilon_0} + \varepsilon \qquad (3.69)$$

which is a finite-difference analogue of the exponential integral in Eq. (3.58). The eigenvalue problem for this was also studied long ago in the context of linear master equations,[52] with results anticipating the above.

We shall postpone until Section III.G the question of finding a similarity solution for the present model, which would be a proper analogue to the result (3.48). It will be obtained as a special case of the more general result to be derived there.

F.   The Diffuse Scattering $K(y; u) = W_{pp}(y, u)$ (Beta Distribution)

The assumption of a uniform distribution of energy after scattering is clearly inappropriate when the elements concerned have internal degrees of freedom, or when the dimensionality of the system enters otherwise as a variable. In the case of normal complexity, with density of states $g(x) \propto x^{p-1}$, the hypothesis of equal a priori probabilities for all out-scattering dispositions in the microstate space of two p molecules leads naturally to the symmetrical beta distribution[27]

$$K(y; u) = W_{pp}(y; u) = B(p, p)^{-1} \frac{y^{p-1}(u-y)^{p-1}}{u^{2p-1}} \qquad (3.70)$$

Here $B(p, q) = \Gamma(p)\Gamma(q)/\Gamma(p+q)$ is the beta function. This is still a diffuse scattering model, the kernel integrating to unity by the beta integral and hence of Type D. In the special case $p = 1$, the Tjon-Wu model of the previous section is recovered.

The transport equation (2.4) with the above kernel was first considered by Nishimura,[8] who proved somewhat laboriously that evolution occurs to the stationary gamma distribution (2.51). Given the analysis of Section II.F, we can assert such evolution as an immediate consequence of the symmetries (2.39) and (2.42), which are clearly exhibited by the kernel.

The beta-function scattering model was reintroduced by Futcher, Hoare, Hendriks, and Ernst[51,52] and examined in more detail, it having been shown earlier that, apart from its reference to internal-state scattering, it can arise by suitable transformation of a $2p$-dimensional Maxwell model in velocity space.[53,54] (See Section IV.) Since the operations involved are parallel in almost all essentials to those set out in detail in the previous section, we can afford to summarize the main results. As before, these relate mainly to the discovery of moment-evolution equations, similarity solutions, and aspects of the linearized transport problem.

1. Moment Evolution

Using the moment formula for the beta distribution (3.70), the kernel moments $\mathcal{M}_n(u + y)$ are immediately seen to be

$$\mathcal{M}_n(u + y) = \frac{(p)_n}{(2p)_n}(u + y)^n \tag{3.71}$$

These are clearly homogeneous and such that $a_n = (p)_n/(2p)_n$ in the expansion (2.90). Thus we can immediately write the moment evolution equations in the form

$$\left(\frac{d}{dt} + 1\right)\mu_n(t) = \frac{(p)_n}{(2p)_n} \sum_{k=0}^{n} \mu_k(t)\mu_{n-k}(t) \tag{3.72}$$

It is easily checked that the moments of the gamma distribution (2.51) $\mu_n(\infty) = (p)_n(\varepsilon/p)^n n!^{-1}$ satisfy (3.72). The eigenvalues of the linearized equation are

$$\Lambda_n = 1 - \frac{2(p)_n}{(2p)_n} \tag{3.73}$$

and thus the second moment relaxes such that

$$\Delta\mu_2(t) = \Delta\mu_2(0)e^{-\Lambda_2 t} \tag{3.74}$$

with $\Lambda_2 = 1 - (p+1)/(2p+1) = p/(2p+1)$.

To obtain the polynomial moments $\gamma_n(t)$, we use the orthogonal set $L_n^{(p-1)}(x/\varepsilon)$ and the relations (A5.6) and (A5.7). After some algebra we then obtain, as expected, Eq. (3.72) with $\gamma_n(t)$ replacing $\mu_n(t)$ throughout.

2. The Similarity Solution

Hendriks, Ernst, Futcher, and Hoare[51] obtained a similarity solution for beta scattering by a generalization of the method described earlier for the special case $p = 1$. The supposition this time is that the distribution function takes the form

$$P^S(x, t) = (\alpha + \beta x) x^{p-1} e^{-\psi x} \tag{3.75}$$

and the functions $\alpha(t)$, $\beta(t)$, and $\psi(t)$ are determined by an entirely similar route to that which led up to Eq. (3.48). The final result is

$$P^S(x, t) = \Gamma(p)^{-1} x^{p-1} \psi^p e^{-\psi x} \left[1 + p - \varepsilon\psi + x\psi \left(\frac{\varepsilon\psi}{p} - 1 \right) \right] \tag{3.76}$$

with

$$\psi(t) = \frac{p}{\varepsilon} (1 - \eta e^{-\lambda t})^{-1} \tag{3.77}$$

and

$$\lambda = \frac{\frac{1}{2}p}{2p+1} = \frac{1}{2}\Lambda_2 \tag{3.78}$$

The ordinary and polynomial moments are

$$\mu_n^S(t) = \left[1 + (n-1)\eta e^{-\lambda t} \right] (1 - \eta e^{-\lambda t})^{n-1} \left(\frac{\varepsilon}{p} \right)^n \frac{(p)_n}{n!} \tag{3.79}$$

and

$$\gamma_n^S(t) = (1-n) \eta^n e^{-n\lambda t} \frac{(p)_n}{n!} \tag{3.80}$$

respectively. The "similarity eigenvalues" λ_n^S are therefore

$$\lambda_n^S = \frac{\frac{1}{2}np}{2p+1} \tag{3.81}$$

The Fourier representation of the solution $P^S(x, t)$ is then

$$P^S(x, t) = P_0(x) \left\{ 1 + \sum_{n=2}^{\infty} (1-n) \eta^n L_n^{(p-1)} \left(\frac{px}{\varepsilon} \right) \exp\left[-\frac{\frac{1}{2}npt}{2p+1} \right] \right\} \tag{3.82}$$

Again, this is completely equivalent to Eq. (3.76).

3. The Linearized Transport Equation

The linearized version of the transport equation is again of interest. Following the procedure of Section II.H, we find the linear kernel to be

$$k(x, y) = \int_{\max(x, y)}^{\infty} du\, W_{pp}(x, u) P_0(u - y) \tag{3.83}$$

or explicitly

$$k(x, y) = \frac{\Gamma(2p)}{\Gamma(p)^2} \left(\frac{p}{\varepsilon} \right)^p x^{p-1} e^{py/\varepsilon}$$

$$\times \int_{\max(x, y)}^{\infty} du\, \frac{(u - x)^{p-1}(u - y)^{p-1} e^{-pu/\varepsilon}}{u^{2p-1}} \tag{3.84}$$

This is the symmetrical case of the distributive kernel previously introduced by Cooper and Hoare[27] for the study of a linear ensemble of p molecules interacting with a heat bath of similar type. The eigenvalues of $k(x, y)$ were shown to be

$$\lambda_n = \frac{(p)_n}{(2p)_n} \tag{3.85}$$

with eigenfunctions $L_k^{(p-1)}(px/\varepsilon)$. The linear approximation to the beta-distribution model thus yields the solution

$$P^L(x, t) = P_0(x) \left\{ 1 + \sum_{n=2}^{\infty} \gamma_n(0) L_n^{(p-1)}\left(\frac{px}{\varepsilon} \right) e^{-\Lambda_n t} \right\} \tag{3.86}$$

with $\Lambda_n = 1 - 2\lambda_n$ as before [cf. (3.73)]. The same follows by linearization of (3.72) and use of the eigenvalues (3.73). Again there is an opportunity to compare the transient terms of the exact similarity solution and the linearized equation for $P^L(x, t)$. Comparing values from (3.73) and (3.81), it is clear that $\Lambda_2 = \lambda_2^S = p((2p + 1)$ and also that $\Lambda_3 = \lambda_3^S = 3p/2(2p + 1)$. Thereafter, increasingly, $\lambda_n^S < \Lambda_n$ as expected.

G. The Diffuse Scattering $\bar{K}(j; k) = \bar{W}_{pp}(j, k)$ (Negative Hypergeometric Distribution)

The discrete analogue of the beta-distribution diffuse-scattering model can be given in terms of the symmetrical negative hypergeometric (NHG) distribution, defined (A4.7) as

$$\bar{K}(j, k) = \bar{W}_{pp}(i, k) = B(p, p)^{-1}(i + 1)_{p-1} \frac{(k - i + 1)_{p-1}}{(k + 1)_{2p-1}} \tag{3.87}$$

This is normalized over the state space $j \in \{0, 1, \ldots, k\}$ by Vandermonde's theorem (A3.6) and is a true analogue of the beta distribution in the sense that

$$\lim_{\varepsilon_0 \to 0} \overline{W}_{pp}\left(\frac{x}{\varepsilon_0}; \frac{u}{\varepsilon_0}\right)\frac{dx}{\varepsilon_0} = W_{pp}(x, u)\, dx \tag{3.88}$$

The property $\overline{W}_{pp}(j, k) = \overline{W}_{pp}(k - j, j)$ guarantees interactional symmetry.

Recalling the discussion in Section III.F and drawing the proper analogy with the beta-distribution model considered there, we can see that the NHG scattering model represents the interaction of a closed ensemble of quantized p molecules with the property that all combinations of quanta in the p degrees of freedom of the outgoing collision partners have equal a priori probability. Since the structure of the kernel is $\overline{K}(j; k) \propto \overline{g}(j)\overline{g}(k - j)$, with $\overline{g}(j) = \Gamma(p)^{-1}(j + 1)_{p-1}$ the density-of-states function, we can deduce via the argument of Section II.F that the system must evolve to the stationary negative binomial distribution (2.55).

1. Moment Evolution

By the now familiar pattern, we can expect a moment relaxation of precisely the form (3.72) given for the continuous model, provided we replace ordinary moments $\mu_n(t)$ by factorial moments $\overline{\mu}_n(t)$ throughout and reinterpret the kernel moments $\mathcal{M}_n$ in the same manner. This is easily confirmed using the known moments of the NHG distribution (A4.8) to give

$$\overline{\mathcal{M}}_n(i + k) = \frac{(p)_n}{(2p)_n}(i + k)^{(n)} \tag{3.89}$$

Vandermonde's theorem (A3.7) again guarantees that the kernel moments are indeed homogeneous in the factorial monomials $i^{(n)}$ and $k^{(n)}$, and thus that the moments $\overline{\mu}_n(t)$ simply replace the $\overline{\mu}_n$ in Eq. (3.72).

To derive the polynomial moment equations directly, we use the orthogonal set $M_n(i, p, a)$ with $a = \varepsilon/(\varepsilon_0 p + \varepsilon)$ [cf. (2.116)] and $\overline{\gamma}_n(t)$ defined as in Eqs. (2.109), (2.117), and (2.119). Operating on the transport equation (2.38) with kernel (3.87), we need this time the general cases of the relations (A5.17) and (A5.18). Using both of these together with the identity

$$(n - k + 1)_{p-1}(k + 1)_{p-1} = \frac{(p)_{n-k}(p)_n \Gamma(p)^2}{k!(n - k)!}$$

we arrive, by steps parallel to those set out in detail in the case of the continuous variable, at an evolution equation of precisely the form (3.72), but with $\bar{\gamma}_n(t)$ replacing $\mu_n(t)$.

2. Similarity Solutions

We deferred from Section III.E the question of finding a similarity solution in the discrete variable which, like the transition kernel $K(j; k)$ itself, would be a true analogue of the continuous solutions (3.45) and (3.75). With our experience of the passage from the negative binomial (2.55) to the gamma distribution (2.51), we can recognize easily enough that the nonexponential part of the solution (3.75) translates into $(A + Bi)(i+1)_{p+1}$ in the discrete variable; the problem is to render the exponential factor correctly. Closer examination shows that the correct form is likely to contain a factor $(1 - \Psi)^i$, since this will lead to $\exp(-\Psi x)$ in the limit $\varepsilon_0 \to 0$ provided we identify the unknown functions through $\Psi = \varepsilon_0 \Psi$. We thus take as the assumed solution

$$\bar{P}^S(i, t) = (A + Bi)(i+1)_{p-1}(1 + \Psi)^i \tag{3.90}$$

with $A(t)$, $B(t)$, an $\Phi(t)$ yet to be determined.

Hendriks, Ernst, Futcher, and Hoare[51] showed by a procedure essentially parallel to that of Section III.E that (3.90) leads to the exact solution

$$\bar{P}^S(i, t) = \Gamma(p)^{-1}(i+1)_{p-1}\Psi^p(1 - \Psi)^i$$
$$\times \left\{1 + p - \left(\frac{\varepsilon}{\varepsilon_0} + p\right)\Psi + \frac{i\Psi}{1 - \Psi}\left[\left(\frac{\varepsilon}{\varepsilon_0 p} + 1\right)\Psi - 1\right]\right\} \tag{3.91}$$

with

$$\Psi(t) = \left[\frac{\varepsilon_0 p}{\varepsilon + \varepsilon_0 p}\right]\left[1 - \eta \exp\left(-\frac{1}{2}\Lambda_2 t\right)\right]^{-1} \tag{3.92}$$

and $\Lambda_2 = p/(2p+1)$ as before. We note the obvious evolution to the negative binomial distribution (2.55) as $t \to \infty$. The special case for $p = 1$ is clearly the solution for the discrete Tjon-Wu scattering (3.61).

Once again, the limits of validity for positive solutions must be carefully considered. It emerges that the conditions $A(0) \geq 0$ and $B(0) \geq 0$ imply, respectively, that $\eta \leq (p+1)^{-1}$ and $\eta \leq \varepsilon/(\varepsilon + \varepsilon_0 p)$. Thus, with somewhat more complexity than in the continuous-variable case, we have that

$$0 \leq \eta \leq \min\left[\frac{1}{1+p}, \frac{\varepsilon}{\varepsilon + \varepsilon_0 p}\right] \tag{3.93}$$

the second constraint disappearing as $\varepsilon_0 \to 0$. For $p > 1$ this turns out to be a considerable restriction on any practical use of the model. As before, however, we can rephrase our result to the effect that if for some initial condition η is positive but violates the above inequality, then $P^S(i, t)$ will nevertheless become positive after a time:

$$\tau_0 = \lambda_2^{-1} \ln\left[\eta \max\left(1 + p, 1 + \frac{\varepsilon_0 p}{\varepsilon} \right) \right] \qquad (3.94)$$

3.   Linearization

Linearization of the transport equation according to Eq. (2.85) leads us to the linear kernel

$$\bar{k}(i, j) = \sum_{k = \max(i, j)}^{\infty} \bar{W}_{pp}(i, k) \bar{P}_0(k - j)$$

$$= 2 \frac{\Gamma(2p)}{\Gamma(p)^3} (i + 1)_{p-1} (1 - c)^p c^{-i}$$

$$\times \sum_{k = \max(i, j)}^{\infty} \frac{c^k (k - i + 1)_{p-1} (k - j + 1)_{p-1}}{(k + 1)_{2p-1}} \qquad (3.95)$$

The eigenvalue problem for the kernel $\bar{k}(i, j)$ was studied by Hoare and Rahman.[29] They showed the eigenvalues to be $\lambda_n = (p)_n / (2p)_n$ with left eigenfunctions $\bar{\psi}_n(i) = M_n(i, p, c)$. Thus, by this route, or equally by linearization of the previous moment equations, we arrive at the approximate solution

$$\bar{P}^L(i, t) = \bar{P}_0(i) \left\{ 1 + \sum_{n=2}^{\infty} \bar{\gamma}_n(0) M_n(i, p, c) e^{-\Lambda_n t} \right\} \qquad (3.96)$$

with $\Lambda_n = 1 - 2\lambda_n$ as before. This may be compared with the Fourier expansion of the similarity solution (3.91) having the same form as the above, but now $\bar{\gamma}_n(0) = \eta^n (1 - n)(p)_n / n!$ and $\lambda_n^S = np / 2(2p + 1)$ replaces Λ_n. The comparison of exact and linearized similarity solutions is thus precisely as in the continuous-variable model, and similar remarks apply.

H.   Persistent Scattering: The p - q Distributive Model

It has proved extremely difficult to devise Type C Boltzmann models that give positive solutions identifiable with a definite physical process. One of the few such models treated so far is the "persistent-scattering" (or p-q)

model of Futcher and Hoare,[55-58] the first successful attempt to remove the shortcomings of Type D scattering and allow a genuine correlation to occur between initial and final dispositions in a collision. Up to now we have restricted consideration entirely to models in which initial and final states correlate only implicitly through the total state quantity u in $K(y, x; u)$. This severely limits the "smallness" of possible scattering and excludes altogether the interesting special case of the "small-scattering limit" described in Section II.K. The p-q distributive model, as we shall call it here, has been analyzed in both discrete- and continuous-variable versions and is the most elaborate soluble model to have appeared so far, including as it does all the results of Sections III.D to III.G as special cases.

The p-q scattering kernel is an adaption of a more complex kernel described in the context of linear ensembles by Cooper and Hoare[27] and Hoare and Rahman.[29] It enables us to specify, without violating the symmetries of the problem, that only some limited amount of energy or the analogous state variable is effectively available for transfer in each collisional interaction. The underlying principle is shown schematically in Fig. 3 and may be explained verbally as follows. For convenience we continue to use the language of internal energy exchange between molecules. The translation, for example, to the case of collective games, is obvious.

1. Each interacting "molecule" shall possess an internal structure represented by $s = p + q$ degrees of freedom and thus a density of states $g(x) = \Gamma(s)^{-1}x^{s-1}$.

2. The total energy x in such a molecule shall be supposed to reach a microcanonical equilibrium in a time scale short compared to the interval between collisions.

3. Only binary collisions are considered, and in these only the energy present in a subset of q degrees of freedom at the moment of collision shall be deemed to be "exchangeable."

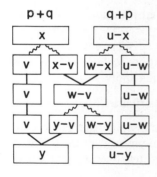

Fig. 3. The p-q persistent-scattering model of Futcher and Hoare. Molecules of $p + q$ degrees of freedom exchange energy such that only energy present at random in the subset of q degrees of freedom is available for transfer. Stochastic steps are indicated by wavy lines. The total energy u is conserved at each stage.

4. The act of energy transfer consists in the microcanonical equilibration of that proportion of energy u that is "pooled" for transfer in the combined $2q$ "active" degrees of freedom in the collision complex. This equilibration is assumed to be complete within the duration of a collision.

5. After collision, complete microcanonical equilibration occurs over all degrees of freedom, both active and inactive, in each separated collision partner and is effectively complete before the next collision takes place.

On working out the implications of these steps it becomes clear that the scattering kernel $K(y, x; u)$ is given by a double convolution of three beta distributions. The result is

$$
K(y, x; u) = \int_0^{\min(x, y)} dv \, W_{pq}(v, x)
$$
$$
\times \int_{\max(x, y)}^u dw \, W_{qp}(w - x, u - x) W_{qq}(y - v, w - v)
$$

$$(3.97)$$

in which $W_{pq}(v, x)$ is now the unsymmetrical beta distribution

$$
W_{pq}(v, x) = B(p, q)^{-1} v^{p-1} \frac{(x - v)^{q-1}}{x^{p+q-1}}
\tag{3.98}
$$

Writing K explicitly, we find

$$
K(y, x; u) = \frac{\Gamma(p + q)\Gamma(2q)}{\Gamma(p)^2 \Gamma(q)^4} \frac{1}{x^{p+q-1}(u - x)^{p+q-1}}
$$
$$
\times \int_0^{\min(x, y)} dv \, v^{p-1}(x - v)^{q-1}(y - v)^{q-1}
$$
$$
\times \int_{\max(x, y)}^u dw \, \frac{(w - x)^{q-1}(w - y)^{q-1}(u - w)^{p-1}}{(w - v)^{2q-1}}
$$

$$(3.99)$$

This is a more symmetrical version of the kernel derived by Cooper and Hoare[27] for the distributive transfer of energy between complex molecules. The properties of a still more general kernel, which includes that of (3.99) but is in general not stochastic, have been studied by Rahman.[59]

As can easily be verified, the normalization (2.33) ($\omega = 1$) is guaranteed by that of the separate beta distributions in the convolution, while the interactional and detailed-balance symmetries are clearly evident. The special case $p = q = 1$ in which the integrals give logarithmic terms is worth noting. This is

$$K(y, x; u) = x^{-1}(u-x)^{-1}\{|x - y|\ln|x - y| + u\ln u$$

$$- [u - \min(x, y)]\ln[u - \min(x, y)]$$

$$- \max(x, y)\ln[\max(x, y)]\} \qquad (3.100)$$

On the other hand, the replacement $p = 0$, $q = s$, making the entire energy subspace available for transfer in the collision, causes K to revert to the simpler beta-distribution kernel (3.70). [Note that formally $W_{0s}(v, x) = \delta(v)$ and $W_{s0}(v, x) = \delta(v - x)$.] By contrast, the limit $q \to 0$ leads to $K(y, x; u) = \delta(y - x)$, and all scattering ceases. Although this is a trivial case, the condition $p \gg q$ is of great interest in that it leads to the diffusionlike small-scattering behavior dealt with in Section II.K. We shall return to this shortly.

1. Moment Solutions

The p-q model gives us our first opportunity to evaluate moments of a kernel with the full functional dependence $\mathcal{M}_n(x, u)$ and so make use of the methods developed in Section II.J. In fact, this case is an excellent illustration of the power of Eqs. (2.129) and (2.134) as compared to the direct approach of evaluating the integrals (2.90).

We first find the function $\mathcal{K}(s) = \lim_{y \to 0} K(s, y; 1)$ on the kernel (3.99), noting that in the integrand $\lim_{y \to 0} W_{pq}(v, y) = \delta(v)$. Thus,

$$\mathcal{K}(s) = \int_0^1 dw\, W_{qq}(s, w) W_{qp}(w, 1) \qquad (3.101)$$

Next we evaluate (2.132) for b_{nn}, obtaining, after some beta-function integrations,

$$b_{nn} = \int_0^1 ds\, s^n \mathcal{K}(s) = \frac{B(n + q, q)B(n + q, p)}{B(q, q)B(q, p)} \qquad (3.102)$$

Finally, on rewriting the beta functions and using (2.129), we have

$$b_{nk} = \sum_{\nu = k}^{\nu = n} \binom{n}{\nu}\binom{\nu}{k}(-1)^{\nu - k}\frac{(q)_\nu(q)_\nu}{(2q)_\nu(p + q)_\nu} \qquad (3.103)$$

and hence, by (2.128),

$$a_{nk} = \frac{(p+q)_n}{(p+q)_k(p+q)_{n-k}} \sum_{\nu=k}^{n} (-1)^{\nu-1} \binom{n-k}{\nu-k} \frac{(q)_\nu(q)_\nu}{(2q)_\nu(p+q)_\nu}$$

(3.104)

At no extra cost we can also write the eigenvalues $\Lambda_n = 1 - a_{nn} - a_{n0}$ of the linearized transport equation. Using the relation $\binom{n}{k} = (-n)_k k!^{-1} (-1)^k$, it follows that

$$\Lambda_n = 1 - \frac{(q)_n(q)_n}{(2q)_n(p+q)_n} \cdot {}_3F_2\left[\begin{matrix} -n, q, q \\ 2q, p+q \end{matrix}; 1\right]$$

(3.105)

Note that in the limiting case $p \to 0$, the ${}_3F_2$ function reduces to ${}_2F_1[-n, q; 2q; 1] = (q)_n/(2q)_n$ and we recover the result (3.73) for the diffuse-scattering model. Alternative forms of the coefficients a_{nk} can be derived by direct integration of the kernel.

The lower order values of a_{nk}, which we shall need shortly, are

$$a_{10} = \frac{2p+q}{2(p+q)}$$

(3.106)

$$a_{11} = \frac{q}{2(p+q)}$$

(3.107)

$$a_{20} = \frac{(p+q+1)(q^2+3pq+p+q)+p(p+1)(q+1)}{2(p+1)(p+q+1)(2q+1)}$$

(3.108)

$$a_{21} = \frac{q(q^2+2pq+p+q)}{2(2q+1)(p+q)^2}$$

(3.109)

$$a_{22} = \frac{q(q+1)^2}{2(p+q)(p+q+1)(2q+1)}$$

(3.110)

The eigenvalue Λ_2 governing the second moment relaxation (2.97) and, as we shall see presently, the similarity solution, is thus

$$\Lambda_2 = 1 - a_{20} - a_{22} = \frac{q(q^2+2pq+p+q)}{(p+q)(p+q+1)(2q+1)}$$

(3.111)

We may also note the alternative extreme case in which the energy transferred per collision becomes extremely small as $p \gg q$, and the relaxation is correspondingly slow. The qualitative character of the scattering in the p-q model is best seen in the first moment of the kernel which, from (3.106) and (3.107), can be written

$$m_1(x, u) = \frac{px}{p+q} + \frac{qu}{2(p+q)} \tag{3.112}$$

This clearly demonstrates the suppression of all scattering as $q = 0$ and $m_1(x, u) \rightarrow x$ and the loss of persistence as $p \rightarrow 0$ $[m_1(X, u) = \frac{1}{2}u]$.

Possession of the expansion coefficients a_{nk} enables us to write the moment evolution equations (2.91) as before and obtain $P(x, t)$ by the sequential algorithm.

2. The Similarity Solution

The similarity solution for p-q distributive scattering was obtained by Futcher and Hoare[55-57] by means of the Ansatz $P^S(x, t) = (\alpha + \beta x)x^{s-1}e^{-\psi t}$, in which now $s = p + q$. A derivation that is algebraically more complex but otherwise entirely parallel to the one in Section III.F leads to the result

$$P^S(x, t) = \Gamma(s)^{-1}x^{s-1}\psi^s e^{-\psi x}\left[1 + s - \varepsilon\psi + x\psi\left(\frac{\varepsilon\psi}{s} - 1\right)\right] \tag{3.113}$$

in which $\Psi(t)$ is the function

$$\psi(t) = \frac{s}{\varepsilon}\left[1 - \eta\exp\left(-\tfrac{1}{2}\Lambda_2 t\right)\right]^{-1} \tag{3.114}$$

and Λ_2 is as already defined in (3.111). Somewhat surprisingly, the structure of this expression is precisely like the one found for diffuse scattering [cf. (3.76)], the more elaborate character of the p-q model being contained only in the quantity Λ_2. The time dependence of passage to equilibrium is clearly a sensitive function of the relative values of the degree-of-freedom parameters p and q and their sum s, the latter having a particularly marked influence through the factor ψ^s. In the limit $p \rightarrow 0$, $q = s$, we have the diffuse-scattering limit $\Lambda_2 \rightarrow s/(2s + 1)$, while at the opposite extreme $p \gg q$ we find $\Lambda_2 \simeq q/p$. The case $p = q = 1$ [cf. (3.100)] leads to $\Lambda_2 = \frac{5}{18}$. Not surprisingly, relaxation in the p-q model is always slower than for a diffuse scattering model with the same number of degrees of freedom.

Analysis of the conditions for positivity now yields the inequalities

$$\frac{s}{\varepsilon} \leqslant \psi(t) \leqslant \frac{1+s}{\varepsilon} \tag{3.115}$$

and

$$0 \leqslant \eta \leqslant (p+1)^{-1} \qquad (3.116)$$

It is likewise obvious that for some "nonphysical" initial condition with $\eta > p+1$, the distribution $P^S(x,t)$ will always become positive after a time $\tau_0 = 2\Lambda_2^{-1}\log[\eta(p+1)]$ and will continue so as $t \to \infty$. The above conditions are once again quite a severe restriction on the use of the similarity solution for more than illustrative purposes. With these reservations about the similarity solution, we can, however, emphasize the appropriateness of the p-q model in representing, for example, the transfer of energy between s molecules, by noting that whereas the relaxation constant $\frac{1}{2}\Lambda_2$ for diffuse scattering can at most be varied between the narrow limits of $\frac{1}{6}$ and $\frac{1}{4}$, in the p-q case the range available is between 0 and $\frac{1}{4}$. These and other interrelationships between the models so far discussed are summarized in Table I.

We shall see later, in considering the transformation properties of our various models, how the underlying structure of the kernels involved implies a unique standard form of similarity solution of the form given, even with systems that appear to differ considerably in complexity.

3.   "Small Scattering" and the Diffusion Equation

The small-scattering limit $p \gg q$ leads us to what is in some respects the most interesting property of the p-q model. We can for the first time arrive at a diffusion equation of the type (2.144) and interpret it in terms of the small parameter q/p. We must first write the transfer moments $A_1(x,u) = \beta_{10}x + \beta_{11}u$ and $A_2(x,u) = \beta_{20}x^2 + \beta_{21}xu + \beta_{22}u^2$ as in Eq. (2.143). Using the values of the a_{nk} coefficients already listed in (3.106) to (3.110), we can

TABLE I
Interrelationship of Relaxation Parameters λ for
Different Scalar Boltzmann Models

Small scattering ($p \gg q$):	$\lambda = \dfrac{q}{2p} \ll 1$
	$\uparrow \quad p \gg q$
Distributive scattering:	$\lambda = \dfrac{q(q^2 + 2pq + p + q)}{2(p+q)(p+q+1)(2q+1)}$
	$\downarrow \quad p \to 0$
Diffuse scattering:	$\lambda = \dfrac{q}{2(2q+1)}$
	$\downarrow \quad q=1$
Bobylev-Krook-Wu model:	$\lambda = \frac{1}{6}$

show straightforwardly that

$$\beta_{10} = a_{10} - 1 = \frac{-q}{2(p+q)} \tag{3.117}$$

$$\beta_{11} = a_{11} = \frac{q}{2(p+q)} \tag{3.118}$$

$$\beta_{20} = 1 + a_{20} - 2a_{10} = \frac{q(q+1)^2}{2(p+q)(p+q+1)(2q+1)} \tag{3.119}$$

$$\beta_{21} = 2(a_{21} - a_{11}) = \frac{-q^3}{(p+q)^3(2q+1)} \tag{3.120}$$

$$\beta_{22} = a_{22} = \frac{q(q+1)^2}{2(p+q)(p+q+1)(2q+1)} \tag{3.121}$$

For simplicity, let us take the case $q = 1$ and examine Eqs. (3.117)–(3.121) in terms of the small parameter $\gamma = 1/p$. Denoting by $\hat{\beta}_{nk}$, etc., the value of β_{nk} to lowest order in γ, we can see that: $\hat{\beta}_{10} = -\gamma/2$, $\hat{\beta}_{11} = \gamma/2$, $\hat{\beta}_{20} = \hat{\beta}_{22} = 2\gamma^2/3$, $\hat{\beta}_{21} = -\gamma^2/3$. With these we see that to the same approximation, $\hat{A}_1(x, u) = -(\gamma/2)(x - u)$ and $\hat{A}_2(x, u) = (\gamma^2/3)(2x^2 - xu + 2u^2)$. Applying these in the diffusion equation (2.144), we obtain the following result for the approximate distribution functions now labeled $P^F(x, t)$:

$$\frac{\partial P^F(x, t)}{\partial t} = \frac{1}{2}\gamma\frac{\partial}{\partial x}\{(x - \varepsilon)P^F(x, t)\}$$
$$+ \frac{1}{6}\gamma^2\frac{\partial^2}{\partial x^2}\{(2x^2 - x\varepsilon + 4\mu_2(t))\}P^F(x, t) \tag{3.122}$$

Here $\mu_2(t)$ is the time-dependent second moment given by Eq. (2.97) with Λ_2 as in (3.111).

We cannot solve this equation in closed form, although a number of properties can be extracted. The second moment is correctly reproduced at all times, and the third moment derived from Eq. (3.122) for the similarity initial conditions is identical with the similarity moment from (3.113). The similarity solution may also be used as a test to show that all moments for $n \geq 3$ decay faster in the exact case than in the diffusion approximation. These and other aspects of the diffusion approximation are described more fully in Reference 56.

4. The Discrete-Variable p-q Model

The whole of the foregoing analysis may be translated into a discrete-variable formulation that is a true analogue in the sense of the limit (2.19). We shall give only the barest outline of results, referring to Reference 57 for further detail.

The scattering kernel analogous to (3.99) is formed using the negative hypergeometric distribution $\overline{W}_{pq}(i, k)$ (A4.7) in place of the beta distribution $W_{pq}(x, u)$ throughout and replacing integrals by summations. The result is

$$
\overline{K}(j, i; k) = \sum_{\mu=0}^{\min(i, j)} \overline{W}_{pq}(\mu, i)
$$

$$
\times \sum_{\nu=\max(i, j)}^{k} \overline{W}_{qp}(\nu - i, k - i) \overline{W}_{qq}(j - \mu, \nu - \mu)
$$

(3.123)

or explicitly

$$
\overline{K}(j, i; k) = \frac{\Gamma(p+q)\Gamma(2q)}{\Gamma(p)^2 \Gamma(q)^4} \left(\frac{1}{(i+1)_{p+q-1}(k-i+1)_{p+q-1}} \right)
$$

$$
\times \sum_{\mu=0}^{\min(i, j)} (\mu+1)_{p-1}(i-\mu+1)_{q-1}(j-\mu+1)_{q-1}
$$

$$
\times \sum_{\nu=\max(i, j)}^{k} \frac{(\nu-i+1)_{q-1}(\nu-j+1)_{q-1}(k-\nu+1)_{p-1}}{(\nu-\mu+1)_{2q-1}}
$$

(3.124)

The matrix thus defined gives the probability that an s molecule with i quanta entering a collision with another having $k - i$ quanta will emerge having j quanta when there is unbiased combinatorical rearrangement of all quanta under the rules set down earlier. $\overline{K}$ is normalized to unity on summation over j, has the interactional symmetry in the form $K(j, i; k) = K(k - j, k - i, k)$, and satisfies microscopic reversibility in the form $(i+1)_{s-1}(k-i+1)_{s-1}K(j, i; k) = (j+1)_{s-1}(k-j+1)_{s-1}K(i, j; k)$. The special case $p = q = 1$ leads to

$$
\overline{K}(j, i; k) = \frac{1}{(i+1)(k-i+1)} \sum_{\mu=0}^{\min(i, j)} \sum_{\nu=\max(i, j)}^{k} \frac{1}{(\nu-\mu+1)}
$$

(3.125)

which, however, in the absence of a finite-difference equivalent of the logarithm, cannot be written so as to resemble (3.100).

By the analysis of Section II.F we know that a closed ensemble evolving through the action of $\overline{K}$ must tend to the stationary distribution (2.55) with s

replacing p. Furthermore, since $\overline{K}(j, i; k)$ in (3.116) is a true analogue of $K(y, x; u)$ in (3.97), we know that the expansion coefficients $\bar{a}_{nk}$ for the kernel moments must be identical with the a_{nk} given in Eqs. (3.106)–(3.110). This may be checked by direct evaluation, Vandermonde's theorem replacing the binomial theorem at the appropriate stages. Moment evolution equations of the form (3.72) and (3.74) can then be written, the expansion set being again the Meixner polynomials $M_k(i, s, a)$ with $a = \varepsilon/(\varepsilon_0 s + \varepsilon)$.

The similarity solution analogous to (3.113) is also readily derived. Starting with the Ansatz (3.90), we arrive again at a result of the form (3.91) but with s replacing p and Λ_2 of (3.111) replacing that of (3.74). Similar parallels apply to the conditions for positivity. Passage to the continuous analogue (3.113) is straightforward once we let $\varepsilon_0 \to 0$ and make the association $\Psi = \varepsilon_0 \psi$. The discrete-variable diffuse scattering model of Section III.G is recovered on letting $q \to s = p$ and the discrete Tjon-Wu scattering model follows for $p = 1$. The interrelationships of Table I hold equally for discrete- and continuous-variable models.

Hendriks and Ernst[58] have recently noticed the curious property that

$$b_{nk} = \overline{K}(k, 0; n)$$

IV. MODELS IN VELOCITY SPACE

As we stressed earlier, it will not be possible to examine in full detail the many recently derived properties of model Boltzmann equations in the setting of gas kinetics and collision theory. We shall, however, be at some pains to establish the interconnections between true vector models, isotropic models, and the simpler scalar models in an energy-like variable upon which we have concentrated so far. In fact, the development of the subject has been marked by a very fruitful interplay not only between collision theory and scalar modeling but also between each of these and the fields of aggregation-fragmentation theory in polymer and aerosol science. Leaving the latter to our final section, we shall return here to a consideration of the original soluble models of Bobylev and Krook and Wu, which were the influence behind most of what we have so far described. As well as deserving prominence on historical grounds, these studies provide an excellent exposure of the many interesting mathematical aspects of the relationship between scalar interodifferential equations of the Boltzmann type and related nonlinear partial differential equations, differential-difference equations, integral transforms, and so on. Before discussing the BKW solution, however, we shall first set out the formal interconnections between models in velocity and energy space and then revisit briefly the earliest soluble model to attract general interest, that invented by Kac in 1955.[6]

A. Velocity-to-Energy Transformations*

So far we have emphasized the construction of models in an energy-like variable x rather than a velocity-like variable appropriate to kinetic theory. We must now examine the relationship between these two types of models in the case of general d-dimensional space, with and without isotropic velocity distributions.

Distinguishing carefully between functions of vector and scalar argument, we define a velocity distribution function $f(\mathbf{v}, t)$ such that

$$\int d\mathbf{v} f(\mathbf{v}, t) = \langle 1 \rangle_{\mathbf{v}} = 1 \tag{4.1}$$

$$\int d\mathbf{v} \mathbf{v} f(\mathbf{v}, t) = \langle \mathbf{v} \rangle_{\mathbf{v}} = 0 \tag{4.2}$$

$$\int d\mathbf{v} v^2 f(\mathbf{v}, t) = \langle v^2 \rangle_{\mathbf{v}} = d \tag{4.3}$$

in d space dimensions. The last condition presupposes a scaling of the velocity in a way that will facilitate the choice of an energy-like quantity x. If $f(\mathbf{v}, t)$ is isotropic, its normalization may be written

$$\Omega_d \int_0^\infty dv\, v^{d-1} f(\mathbf{v}, t) = 1 \tag{4.4}$$

where Ω_d is the surface area of the d-dimensional sphere

$$\Omega_d = \frac{2\pi^{1/2}}{\Gamma(\tfrac{1}{2}d)} \tag{4.5}$$

Thus $d\mathbf{v} \equiv \Omega_d v^{d-1} dv$. [We shall write the ambiguous $f(v, t)$ only when the nature of its normalization is absolutely clear from the context.] With the above scaling, the equilibrium Maxwellian becomes

$$f(\mathbf{v}, \infty) = f_0(v) = (2\pi)^{-(1/2)d} \exp\left(-\tfrac{1}{2}v^2\right) \tag{4.6}$$

---

*In this section and the next we draw liberally on the presentation of Ernst[11] and conform to his notation as far as possible.

To convert to an energy-like variable x and a probability distribution $F(x, t)$, we require that

$$F(x, t)\, dx = f(\mathbf{v}, t)\, d\mathbf{v} = \Omega_d (2x)^{(1/2)d - 1} f(\mathbf{v}, t)\, dv\big|_{v^2 = 2x} \qquad (4.7)$$

which, with (4.5) and (4.6), corresponds to the Maxwellian

$$F(x, \infty) = F_0(x) = \Gamma(p)^{-1} x^{p-1} e^{-x} \qquad (4.8)$$

with $p = \tfrac{1}{2}d$. The function $F(x, t)$ is thus related to $P(x, t)$ by the scaling $x \to px/\varepsilon$ as in (2.56). We recall that in this variable the conserved quantities are

$$\int_0^\infty dx\, F(x, t) = \langle 1 \rangle_x = 1 \qquad (4.9)$$

$$\int_0^\infty dx\, xF(x, t) = \langle x \rangle_x = \tfrac{1}{2}d = p \qquad (4.10)$$

In passing from the velocity to the energy description we shall have to consider angular averages of scalar functions of the velocity and some other variable in the form $\langle h(\mathbf{a} \cdot \mathbf{b}) \rangle_{\hat{\mathbf{b}}}$, where the subscript $\hat{\mathbf{b}}$ denotes averaging over all directions of the unit vector $\hat{\mathbf{b}}$. Using d-dimensional polar coordinates it is not difficult to show that the required averages take the form

$$\langle h(\mathbf{a} \cdot \mathbf{b}) \rangle_{\hat{\mathbf{b}}} = \frac{1}{\Omega_d} \int d\hat{\mathbf{b}}\, h(\mathbf{a} \cdot \mathbf{b})$$

$$= \frac{\displaystyle\int_0^\pi d\theta \sin^{d-2}\theta\, h(ab\cos\theta)}{\displaystyle\int_0^\pi d\theta \sin^{d-2}\theta} \qquad (4.11)$$

The main examples of this kind that we shall require are $h(\mathbf{k} \cdot \mathbf{v}) = (\mathbf{k} \cdot \mathbf{v})^{2n}$ and $h(\mathbf{k} \cdot \mathbf{v}) = e^{i\mathbf{k} \cdot \mathbf{v}}$. These are as follows:

$$\langle (\mathbf{k} \cdot \mathbf{v})^{2n} \rangle_{\hat{\mathbf{v}}} = \frac{\Gamma(n + \tfrac{1}{2})\Gamma(\tfrac{1}{2}d)}{\Gamma(n + \tfrac{1}{2}d)\Gamma(\tfrac{1}{2})} \qquad (4.12)$$

and

$$\langle \exp(-i\mathbf{k} \cdot \mathbf{v}) \rangle_{\hat{\mathbf{v}}} = \Gamma(\tfrac{1}{2}d) \left(\frac{2}{kv}\right)^{\frac{1}{2}d - 1} J_{\frac{1}{2}d - 1}(kv) \qquad (4.13)$$

where $J_n(\cdot)$ is the Bessel function of given order. Using the series form for this, we see that

$$\langle \exp(-i\mathbf{k}\cdot\mathbf{v})\rangle_{\hat{v}} = \sum_{n=0}^{\infty} \frac{(-)^n(kv)^{2n}}{2^{2n}(\frac{1}{2}d)_n n!}$$

$$= {}_0F_1(\tfrac{1}{2}d, -\tfrac{1}{4}k^2v^2) \tag{4.14}$$

In the greater part of what follows we shall confine our attention to the case of ordinary three-dimensional space. For this the interconnection between energy and velocity distributions in the isotropic case is simply

$$F(x,t)\,dx = 4\pi v^2 f(\mathbf{v},t)\,dv\big|_{v^2=2x} \tag{4.15}$$

$$F_0(x)\,dx = \Gamma\big(\tfrac{3}{2}\big)^{-1} x^{1/2} e^{-x}\,dx \tag{4.16}$$

B. The Kac Model

In 1955 Kac[6] introduced what was in effect the first nontrivial model leading to a soluble Boltzmann-type equation. This was based on the equation

$$\frac{\partial}{\partial t} f(v,t) = \int_{-\infty}^{\infty} dw \int_{-\pi}^{\pi} d\theta\,\sigma(\theta)\{f(v')f(w') - f(v)f(w)\} \tag{4.17}$$

with the artificial scattering law

$$\begin{aligned} v' &= v\cos\theta + w\sin\theta \\ w' &= -v\sin\theta + w\cos\theta \end{aligned} \tag{4.18}$$

(Caution: v, w are here full-range one-dimensional velocities $v, w \in (-\infty, +\infty)$, *not* the scalars $v = |\mathbf{v}|$, $w = |\mathbf{w}|$ of our usual context.) Since $v^2 + w^2 = v'^2 + w'^2$, the model conserves energy, although conservation of momentum is violated. The "scattering cross section" $\sigma(\theta)$ must be specified and the total cross section is assumed finite.

$$\int_{-\pi}^{\pi} d\theta\,\sigma(\theta) = \nu < \infty$$

The Kac model is sometimes referred to as a caricature of the hard-sphere gas; but for all its simplicity it leads to a rich variety of properties.

The usual case to be considered is $\sigma(\theta) = 2\pi^{-1}$, for which evidently

$$\left(\frac{\partial}{\partial t} + 1\right) f(v, t) = \frac{1}{2\pi} \int_{-\infty}^{\infty} dw \int_{-\pi}^{\pi} d\theta f(v') f(w') \qquad (4.19)$$

Let us now make two successive transformations of the independent variables $(\theta, w) \to (v', w') \to (x, y)$ and one of the dependent variable $f(v, t) \to F(x, t)$, where $\frac{1}{2}v^2 = x$, $\frac{1}{2}w^2 = u$. Noting the Jacobians

$$\left| \frac{\partial(\theta, w)}{\partial(v', w')} \right| = |w|^{-1} \qquad \left| \frac{\partial(v', \omega')}{\partial(y, u)} \right| = \frac{1}{2 y^{1/2} u^{1/2}}$$

and the ranges of integration $v'^2 + w'^2 \geq v^2$ and $y + v \geq x$, we then have successively

$$\left(\frac{\partial}{\partial t} + 1\right) f(v) = \frac{1}{2\pi} \int \int_{\{v'^2 + w'^2 \geq v^2\}} \frac{dv' \, dw' \, f(v') f(w')}{(v'^2 + w'^2 - v^2)^{1/2}}$$

and

$$\left(\frac{\partial}{\partial t} + 1\right) F(x, t) = \frac{1}{\pi} \int_x^{\infty} du \int_0^u dy \frac{F(y) F(u - y)}{[x(u - x)]^{1/2}} \qquad (4.20)$$

The last is evidently a Type D scalar equation with scattering kernel

$$K(x, u) = W_{\frac{1}{2}\frac{1}{2}}(x, u) = \pi^{-1} [x(u - x)]^{-1/2} \qquad (4.21)$$

Thus K is the symmetrical beta distribution with parameter $\frac{1}{2}$, otherwise known as the arcsine distribution. The association of the Kac model with the beta distribution scattering in energy space was first made by Nishimura.[9] (See also Hendriks and Ernst.[6]) The density of states in energy space is $g(x) = \pi^{-1/2} x^{1/2}$, and the system must tend to the stationary distribution $F_0(x) = \pi^{-1/2} x^{-1/2} e^{-x}$.

The above formulation enables us to write moment evolution equations as a special case of (3.72). Noting that $a_n = \Gamma(n + \frac{1}{2}) \pi^{-1/2} n! =$

*We shall occasionally use the notation

$$(2k + 1)!! = (2k + 1)(2k - 1) \cdots 5 \cdot 3 \cdot 1$$

$$= \frac{(2k + 1)!}{2^k k!} = \frac{2^{k+1} \Gamma\left(k + \frac{3}{2}\right)}{\Gamma\left(\frac{3}{2}\right)} = 2^{k+2} \pi^{-1/2} \Gamma\left(k + \frac{3}{2}\right)$$

$(2n-1)!!2^{-n}n!^{-1}$, it follows that

$$\left(\frac{d}{dt}+1\right)\mu_n(t) = \frac{(2n-1)!!}{2^n n!} \sum_{k=0}^{n} \mu_k(t)\mu_{n-k}(t) \qquad (4.22)$$

An equation of precisely the same form leads to the polynomial moments $\gamma_k(t)$, which in this case are with respect to the set $\{L_k^{(-1/2)}(x)\}$. Thus a series of the form (2.120) can be written for $P(x,t)$ after a sequential solution for the $\gamma_k(t)$. The result of reverting to the velocity-space formulation is the series

$$f(v,t) = f_0(v)\left\{1 + \sum_{n=1}^{\infty} \frac{(-1)^n \gamma_n(t)}{2^n(2n-1)!!} H_n\left(\frac{v}{\sqrt{2}}\right)\right\} \qquad (4.23)$$

in which we have used the equivalence (A5.9) between Laguerre and Hermite polynomials. The stationary distribution is

$$f_0(v) = (2\pi)^{-1/2}\exp\left(-\tfrac{1}{2}v^2\right) \qquad (4.24)$$

Referring back to Eq. (3.76), we can immediately write the similarity solution in the rescaled energy variable as

$$F^S(x,t) = \pi^{-1/2}x^{-1/2}\psi^{1/2}e^{-\psi x}\{\tfrac{3}{2} - \psi + x\psi[2\psi - 1]\} \qquad (4.25a)$$

$$\psi(t) = \tfrac{1}{2}\left[1 - \eta\exp\left(-\tfrac{1}{8}t\right)\right]^{-1} \qquad (4.25b)$$

In his original work, Kac derived the equivalent of (4.23) but did not consider similarity solutions.

C. Maxwell Models and Pseudo-Maxwell Models

We have already referred to the simplification of the vector Boltzmann equation (2.10) that occurs when the collision rate $gI(g,\chi)$ depends only on the scattering angle χ between initial and final trajectories and not on the relative velocity $\mathbf{g} = \mathbf{v} - \mathbf{w}$. The usefulness of this case has been appreciated since Maxwell's original work, and it has been applied to advantage in the calculation of transport coefficients.[61,62] Nevertheless, its key role in the generation of exact solutions came to light only in the nearly simultaneous papers of Bobylev[1] and Krook and Wu.[3,4]

Writing the scattering function $gI(g,\chi) = h(\cos\chi) = h(\hat{\mathbf{g}}\cdot\hat{\mathbf{n}})$ and reformulating (2.12) in the scalar distribution function $f(v,t)$, we can thus confine attention to the homogeneous, isotropic equation

$$\frac{\partial}{\partial t}f(v,t) = \frac{1}{4\pi}\int d\mathbf{w}\int d\hat{\mathbf{n}}\, h(\cos\chi)\{f(\mathbf{v}')f(\mathbf{w}') - f(\mathbf{v})f(\mathbf{w})\} \qquad (4.26)$$

There is some confusion of terminology in the literature in this context. We shall follow authors who refer to all models whose scattering and transport equation can be expressed in this way as "Maxwell models," although the model Maxwell himself derived from the particular scattering law for the potential $V(r) = \kappa/r^4$ yields a particular, and quite complicated, form for $h(\cdot)$ [see Eq. (4.61)]. The term *pseudo-Maxwell model* will here be reserved for the special case $h = $ const., which gives rise to the Krook-Wu solution [but is sometimes applied to all models satisfying the above when $h(\cdot)$ is arbitrarily chosen and not derivable from a known scattering law]. Systems that obey the potential scattering with $V(r) \propto r^{-4}$ will always be referred to as *true Maxwell molecules*. When a true scattering law is not implied, the choice of the function $h(\cos\chi)$ [$= \phi(\chi)$] offers scope for creating models of mathematical interest that may mimic, to a greater or lesser extent, the behavior of real molecules.

1. The Bobylev-Krook-Wu Model

We shall now show how, by suitable definition of reduced even velocity moments of $f(\mathbf{v}, t)$, the BKW model for pseudo-Maxwell molecules can be seen to be governed by moment evolution equations of the familiar form (3.43). The method used is not that of Krook and Wu, but follows a somewhat simpler derivation from unpublished work by Corngold and Mathews[63] generously made available to the author.

Consider first the natural even velocity moments

$$m_i(t) = \int d^3\mathbf{v}\, v^{2k} f(\mathbf{v}, t) \tag{4.27}$$

noting at the same time their reduced form $M_k(t) = m_k(t)/m_k(\infty) = m_k(t)/(2k+1)!!$ in the case of the three-dimensional Maxwellian $f(v, \infty) = (2\pi)^{-3/2}\exp(-\frac{1}{2}v^2)$. By taking moments on Eq. (2.12) we can write $(dm_k/dt) = \mathscr{S}_k^+ - \mathscr{S}_k^-$, where $\mathscr{S}_k^+$, $\mathscr{S}_k^-$ are in- and out-scattering terms. The out-scattering term, being of the form

$$\mathscr{S}_k^- = n\int d\mathbf{v}f(\mathbf{v})\mathbf{v}^{2k}\int d\mathbf{w}f(\mathbf{w})\int d\hat{\mathbf{g}}\, g I(g, \chi)$$

reduces easily to $4\pi n\kappa m_k(t)$ under the assumption $gI(g, \chi) = h(\cos\chi) = \kappa$. The particle number density n and the factor $4\pi\kappa$ will subsequently be absorbed into the time scaling. The main analysis required involves the $\mathscr{S}_k^+$ term. Under the same assumption this can be written

$$\mathscr{S}_k^+ = 4\pi n\int d\mathbf{v}' f(\mathbf{v}')\int d\mathbf{w}f(\mathbf{w})(4\pi)^{-1}\int d\hat{\mathbf{g}}\left[\tfrac{1}{2}(\mathbf{P}+\mathbf{g}\cdot\hat{\mathbf{g}})\right]^{2k}$$

Here $\mathbf{P} = \mathbf{v} + \mathbf{w}$ is a total momentum-like variable and $\mathbf{g} = \mathbf{v} - \mathbf{w}$ is the relative velocity as before. The inner integral is thus the angular average of the function $(\mathbf{P}' + g\hat{\mathbf{g}})^{2k}$. However, since the distributions $f(\mathbf{v})$ and $f(\mathbf{w})$ are isotropic, we can consider them angularly averaged at the same time on introducing the appropriate factors of 4π. Thus the problem can be identified with that of finding a general average of the form we may choose to write

$$A_k(a^2, b^2) = 4^{-k} \langle (a + b + |a - b|\hat{\mathbf{g}})^{2k} \rangle_{\hat{\mathbf{a}}, \hat{\mathbf{b}}, \hat{\mathbf{g}}} \tag{4.28}$$

the brackets implying angular averaging in three dimensions over the three unit vectors $\hat{\mathbf{a}}$, $\hat{\mathbf{b}}$, and $\hat{\mathbf{g}}$. On binomial expansion of the argument, the $\hat{\mathbf{g}}$ average may be taken, giving

$$A_k(a^2, b^2) = \frac{1}{2k} \sum_m \binom{k}{m} \frac{1}{2m+1} (a^2 + b^2)^{k-2m} \langle (a+b)^{2m} (a-b)^{2m} \rangle$$

A further binomial expansion yields, on putting $\alpha = a^2$, $\beta = b^2$,

$$A_k(\alpha, \beta) = \sum_l \frac{(-)^l}{2l+1} (\alpha + \beta)^{k-2l} \alpha^l \beta^l 2^{2l-k} \sum_m \frac{1}{2m+1} \binom{k}{2m} \binom{m}{l}$$

The m summation can be recognized as a $_2F_1$ function of unit argument and summed by Gauss' theorem (A3.8):

$$_2F_1(a, b; c; 1) = \frac{\Gamma(c)\Gamma(c - a - b)}{\Gamma(c - a)\Gamma(c - b)}$$

This yields

$$A_k(\alpha, \beta) = \frac{1}{k+1} \sum_l \frac{(-)^l}{2l+1} \binom{k-l}{l} (\alpha + \beta)^{k-2l} \alpha^l \beta^l \tag{4.29}$$

After a further expansion of the binomial, the l summation can be carried out to give A_k as the single series

$$A_k(\alpha, \beta) = \frac{1}{k+1} \sum_{n=0}^{k} \alpha^{k-n} \beta^n \frac{(2k+1)!!}{[2(k-n)+1]!!(2n+1)!!} \tag{4.30}$$

Thus Corngold and Mathews' identity reads ($k = 0, 1, 2, \ldots$)

$$\frac{n+1}{(2n+1)!!} \frac{1}{4^n} \langle (a + b + |a - b|\hat{\mathbf{g}})^{2n} \rangle_{\hat{\mathbf{a}}, \hat{\mathbf{b}}, \hat{\mathbf{g}}}$$

$$= \sum_{\nu=0}^{n} \frac{a^{2\nu}}{(2\nu+1)!!} \left(\frac{b^{2(n-\nu)}}{[2(n-\nu)+1]!!} \right) \tag{4.31}$$

and can immediately be applied to the integrals for $\mathscr{S}^+$. On identifying $\mathbf{a} \equiv \mathbf{v}$, $\mathbf{b} \equiv \mathbf{w}$ in the above, we then find

$$\left(\frac{d}{dt} + 1\right) m_n(t) = \frac{(2n+1)!!}{n+1} \sum_{k=0}^{n} \frac{m_k}{(2k+1)!!} \left(\frac{m_{n-k}}{[2(n-k)+1]!!}\right)$$

(4.32)

with the factor $4\pi n\kappa$ absorbed into the time scaling. It is immediately clear that the $m_n(t)$ need only be replaced by the reduced moments $M_n = m_n/(2n+1)!!$ for (4.32) to fall into the expected form

$$\left(\frac{d}{dt} + 1\right) M_n(t) = \frac{1}{n+1} \sum_{k=0}^{n} M_k M_{n-k}$$

(4.33)

We have now arrived, by Corngold and Mathews' algebraic route, at the result of Krook and Wu's rather more involved tirgonometric derivation. Nishimura[8] had previously derived the discrete-time version of this equation; Tjon and Wu[47,48] were later to show that it followed from the energy-space equation (3.42) equivalent by integral transformation to the above. We shall see later that the variable changes from (isotropic) velocity space to energy space and consequent rescaling of moments are part of an intricate system of integral transform relationships between different formulations of the same problem (see Section V).

At this point we could use the Tjon-Wu solution in energy space to work back to the velocity-space similarity solution. It will be instructive, however, to follow the original Krook-Wu derivation, since this illustrates various points of interrelationship between Eqs. (3.42) and (4.33) and a number of other nonlinear problems in the field of partial differential equations and differential difference equations.

Krook and Wu introduced the generating function

$$G(\xi, t) = \sum_{n=0}^{\infty} \xi^n M_n$$

(4.34)

through which, on summing (4.23) left and right and using the convolution property, it follows that

$$\left(\frac{\partial}{\partial t} + 1\right) G(\xi, t) = \frac{1}{\xi} \int_0^\xi d\eta \, G(\eta, t)^2$$

(4.35)

and hence that

$$\frac{\partial^2(\xi G)}{\partial \xi \, \partial \tau} + \frac{\partial(\xi G)}{\partial \xi} = G^2(\xi, t)$$

(4.36)

The conditions $M_0 = 1$, $M_1 = 1$ then impose the behavior $G(\xi, \tau) = 1 + \xi + 0(\xi^2)$ as $\xi \to 0$ and $G(\xi, \tau) \to (1 - \xi)^{-1}$ as $t \to \infty$. A variable transformation $x = (1 - \xi)/\xi$, $y(x, t) = G(\xi, t)$ then changes (4.36) into

$$\frac{\partial^2 y}{\partial x \, \partial t} + \frac{\partial y}{\partial x} + y^2 = 0 \tag{4.37}$$

where now $y(x, t) = 1/x + O(x^{-2})$ as $x \to \infty$ and $y(x, t) \to x^{-1}$ as $\tau \to \infty$.

At this point Krook and Wu depart from complete generality and, with the structure of Eqs. (4.36) and (4.37) in mind, seek a similarity solution in the form

$$y(x, t) = x^{-1} z(\eta) \tag{4.38}$$

with $\eta = \ln x + ct$ and c a constant to be determined. Finally, a function $Z = dz/dx$ is introduced that reduces (4.38) to the first-order equation

$$cZ \frac{dZ}{dz} + (1 - c)Z - z(1 - z) = 0 \qquad (0 < z < 1)$$

On reconstructing the boundary conditions in the new variables it is found that $Z(z) \sim 2(1 - z)$ as $z \to 1$; $Z(z) \sim z$ as $z \to 0$. The first can be used to determine that $c = \frac{1}{6}$ so that $\eta = \ln x + \frac{1}{6}$ and the equation becomes

$$Z \frac{dZ}{dz} + 5Z + 6z(1 - z) = 0 \tag{4.39}$$

This has the solution

$$Z(z) = 2(1 - z)\left[1 - (1 - z)^{1/2}\right] \tag{4.40}$$

and thus

$$\eta = \frac{1}{2} \int \frac{dz}{(1 - z)\left[1 - (1 - z)^{1/2}\right]} \tag{4.41}$$

Finally, the special case of the generating function is obtained in the form

$$G^S(\xi, \tau) = \frac{1 + [1 - 2K(\tau)]\xi}{[1 - K(\tau)\xi]^2} \tag{4.42}$$

where

$$K(\tau) = 1 - e^{-(1/6)\tau} \tag{4.43}$$

On expanding the denominator in (4.42), the similarity moments are seen to be

$$M_n^S(t) = K^{n-1}[n - (n-1)K] \qquad (4.44)$$

With hindsight, we may recognize the identity of this formula with the moments (3.53) of the Tjon-Wu model in the energy variable (replacing K by ψ^{-1}). In this way the Krook-Wu model in velocity space is shown to be equivalent to the Tjon-Wu model in energy space, provided the moments are redefined in the manner above. We can omit the direct derivation of the Tjon-Wu transport equation by transformation of the Krook-Wu equation, which is given in detail in Reference 47. Ziff et al.[64] also give further ramifications of the energy-space–velocity-space connection, a subject we shall return to in the next section.

It remains to convert the similarity solution (4.42) into its counterpart for the velocity distribution $f(v, t)$. The necessary connection is given by Krook and Wu via a Fourier transform. Since $f(v, t)$ is even, we write the cosine transform

$$\tilde{h}(p, t) = 2\int_0^\infty dv\, v^2 f(v, t)\cos pv \qquad (4.45)$$

so that, on expanding $\cos pv$,

$$\tilde{h}(p, t) = 2\sum_{n=0}^\infty \frac{(-p^2)^n}{(2n)!}\int_0^\infty dv\, f(v, t) v^{2n+2}$$

Using (4.17), (4.22), and (4.34), it follows that

$$\tilde{h}(p, t) = \frac{1}{\pi}\sum \frac{(2n+1)!!}{(2n)!}(-p)^n[K^n + n(1-K)K^{n-1}]$$

The series may be summed to give

$$\tilde{h}(p, t) = (4\pi)^{-1}[2 + (K-3)p^2 + K(K-1)p^4]\exp(-\tfrac{1}{2}Kp^2) \qquad (4.46)$$

On taking the Fourier inverse by standard integrals, the now celebrated Krook-Wu similarity solution emerges in the form

$$f^S(v, t) = \frac{e^{-v^2/2K}}{2K(2\pi K)^{3/2}}\left[5K - 3 + \frac{1-K}{K}v^2\right] \qquad (4.47)$$

As before, the range of positivity must be carefully considered. In the present time scale it can be shown that positivity is guaranteed only for $t > 6\ln(\frac{5}{2}) = 5.4977$.

D. Nonisotropic Scattering

Krook and Wu also extended their analysis to the case of *true* Maxwell molecules for which $h(\cos\chi)$ is not a constant. This time a solution of the form (4.47) is assumed heuristically and is fitted by consistency with the conservation conditions in the manner used in Section III.D. They were able to prove after some analysis that the only difference in the case of true Maxwell molecules is the replacement of K in Eq. (4.43) by a function $\alpha(t)$, where

$$\alpha(t) = 1 - \exp\left(-\tfrac{1}{6}\phi_1 t\right) \tag{4.48}$$

and ϕ_1 is defined by the integral

$$\phi_1 = \frac{3}{4}\int_0^\pi d\chi\, h(\cos\chi)\sin^3\chi \tag{4.49}$$

The question immediately suggests itself as to whether the solution for general Maxwell molecules can also be found via moment evolution equations in the standard form (2.91), and, if so, by what procedure the expansion coefficients are to be found. This was not settled by Krook and Wu, though several authors soon provided an affirmative answer.

Ernst[53] was the first to show that Eq. (4.33) could be generalized to the form

$$\left(\frac{d}{dt} + \Lambda_n\right)M_n(t) = \sum_{k=1}^{n-1} \alpha_{nk} M_k M_{n-k} \tag{4.50}$$

with $\Lambda_n = 1 - \alpha_{nn} - \alpha_{n0}$ as before and with the coefficients α_{nk} given explicitly in terms of the scattering function $\phi(\chi) = h(\cos\chi)$. The original derivation required a restriction to scattering for which $\int_0^\delta d\chi \sin\chi\,\phi(\chi) < \infty$, this excluding, in fact, the case of true Maxwell molecules [for which $\phi(\chi) \sim \chi^{-5/2}$ as $\chi \to 0$]. Cornille and Gervois[65] later weakened this to $\phi_1 < \infty$ in Eq. (4.49) with proper attention to the cancellation of divergencies. Their formula for the expansion coefficients is

$$\alpha_{nk} = \frac{1}{2}\binom{n}{k}\int_0^\pi d\chi\,\phi(\chi)\sin\chi\cos^{2n-2k}\tfrac{1}{2}\chi\sin^{2k}\tfrac{1}{2}\chi \qquad (k=1,2\ldots n)$$

$$\tag{4.51}$$

The eigenvalues of the linearized transport equation, identified as such by arguments paralleling those of Section II.H, can then be written as

$$\Lambda_n = \frac{1}{2} \int_0^\pi dx \sin \chi \phi(\chi) \left[1 - \sin^{2n}\tfrac{1}{2}\chi - \cos^{2n}\tfrac{1}{2}\chi\right] \qquad (4.52)$$

We may note that in the case of pseudo-Maxwell molecules $[\phi(\chi) = \text{const.}]$ these equations reduce immediately to the Krook-Wu values

$$\alpha_{nk} = (n+1)^{-1} \quad \text{and} \quad \Lambda_k = \frac{k-1}{k+1} \qquad (4.53)$$

Cornille and Gervois went on to obtain polynomial moments $C_n(t)$ in the basis $\{L_n^{(\frac{1}{2})}(\tfrac{1}{2}v^2)\}$, which were shown to satisfy the identical equation (4.50) and to be related to the velocity moments $M_k(t)$ through

$$C_n = \sum_{k=0}^{n} (-1)^{n+1}\binom{n}{k} M_k \qquad (4.54)$$

and

$$M_k = \sum_{k=0}^{n} \binom{n}{k} C_k \qquad (4.55)$$

These equations clearly echo the relationships (2.124) and (2.126), and we shall later see why this is necessarily so. It follows that the distribution function $f(v,t)$ for the general nonisotropic scattering should be expressible in the familiar form

$$f(v,t) = (2\pi)^{-3/2} e^{-v^2/2} \left\{1 + \sum_{n=1}^{\infty} (-1)^n C_n L_n^{(1/2)}(\tfrac{1}{2}v^2)\right\} \qquad (4.56)$$

provided always that convergence of the right-hand side can be established.

A number of authors have pointed to the intricate relationships that exist among the coefficients b_{nk} and between these and the eigenvalues Λ_k, and explicit expressions for the first few moments $M_n(t)$ have been given for the pseudo-Maxwell case.[64-66] The most useful procedure for computational purposes is probably that of Hauge and Prestgaard[66] in which the combinations $b_{nk} + b_{nn-k}$ are expressed in terms of Λ_k in the form

$$b_{nk} + b_{nn-k} = -\binom{n}{k} \sum_{\nu=0}^{k} \binom{k}{n}(-1)^\nu \Lambda_{n-k+\nu} \qquad (4.57)$$

and the eigenvalues are connected by

$$\Lambda_n = \sum_{\nu=0}^{n} \binom{n}{\nu}(-1)^{\nu}\Lambda_{\nu} \qquad (4.58)$$

By repeated application of the latter, the combinations $b_{nk} + b_{nn-k}$ as well as the eigenfunctions of odd index Λ_{2n+1} may be generated entirely in terms of the even eigenvalues Λ_{2n}. We quote the results of this procedure for the first few cases. Letting $\Lambda_{2n} = 2n\bar{\lambda}_{2n}$,

$$
\begin{aligned}
b_{21} &= 2\bar{\lambda}_2 \\
b_{31} + b_{32} &= 3\bar{\lambda}_2, \\
b_{41} + b_{43} &= 4\bar{\lambda}_2 + 16(\bar{\lambda}_4 - \bar{\lambda}_2) \\
b_{42} &= -12(\bar{\lambda}_4 - \bar{\lambda}_2) \\
b_{51} + b_{54} &= 5\bar{\lambda}_2 + 30(\bar{\lambda}_4 - \bar{\lambda}_2) \\
b_{52} + b_{53} &= -20(\bar{\lambda}_4 - \bar{\lambda}_2)
\end{aligned}
\qquad (4.59)
$$

Ziff, Stell, and Cummings[64] have recently computed numerical values for the first twenty Λ_k, thus checking earlier work by Alterman, Frankowski, and Pekeris[67] in a study of the linearized Boltzmann equation. To do this it is necessary to have the scattering function $\phi(\chi)$ for true Maxwell molecules, the derivation of which we shall now outline.

From the classical theory of scattering under two-body central forces, we have that the scattering angle χ as a function of the interaction potential $V(r)$, the relative velocity g, and the impact parameter b takes the form

$$\chi = \pi - 2\int_0^{\eta_0} \frac{d\eta}{\left[1 - \eta^2 - 2/g^2 V(b/\eta)\right]^{1/2}} \qquad (4.60)$$

in which η_0 is the smallest root of the denominator. For the particular case $V(r) = Kr^{-4}$, it follows, on forming the differential scattering cross section $\sigma = (\sin\chi)^{-1}|(db/d\chi)|$, that

$$\phi(\chi) = \frac{1}{4}\frac{1}{\sin\chi}\left(\frac{1-2s}{s(1-s)}\right)^{1/2}\left[(1-s)K(s) - (1-2s)E(s)\right]^{-1} \qquad (4.61)$$

where $K(s)$ and $E(s)$ are complete elliptic integrals of the first and second kinds:

$$
\begin{aligned}
K(s) &= \int_0^{\pi/2} d\theta\,(1 - s^2\sin^2\theta)^{-1/2} = \tfrac{1}{2}\pi\,{}_2F_1\left(\tfrac{1}{2}, \tfrac{1}{2}; 1; s^2\right) \\
E(s) &= \int_0^{\pi/2} d\theta\,(1 - s^2\sin\theta)^{1/2} = \tfrac{1}{2}\pi\,{}_2F_1\left(-\tfrac{1}{2}, \tfrac{1}{2}; 1; s^2\right)
\end{aligned}
\qquad (4.62)
$$

and $s(\chi)$ is to be determined by the relation

$$\tfrac{1}{2}(\pi - \chi) = (1 - 2s)^{1/2} K(s) \tag{4.63}$$

An analysis of the limit $\chi \to 0$ shows that $\phi(\chi) \sim \chi^{-5/2}$ or, more precisely, that

$$\phi(\chi) = \frac{(3\pi)^{1/2}}{8\chi^{5/2}} \left\{ 1 + \frac{35}{24}\left(\frac{\chi}{\pi}\right) + \left(\frac{\pi^2}{6} - \frac{35}{384}\right)\left(\frac{\chi}{\pi}\right)^2 + \cdots \right\} \tag{4.64}$$

a formula sufficient for some integrations over $\phi(\chi)$.[64]

A number of authors have examined the existence of solutions of type (4.56) and their positivity, neither of which properties is to be taken for granted.[65,68,69] Existence in this context presupposes boundedness of the coefficients C_k and convergence of the polynomial series. Of particular importance are the "fundamental solutions" $f_p(v, t)$, which arise when the initial conditions $f_p(v, 0)$ are such that $C_n = 0$ for all $n \neq p$. The BKW similarity solution is, of course, a very special case of a distribution $f_1(v, t)$ and can be derived via algebraic properties of the expansion coefficients.[65] (We shall consider a simpler method in Section V.) On general questions of positivity, results are so far somewhat inconclusive, with useful sufficient conditions difficult to find. Certain classes of positive solutions may, however, be constructed and have been studied both analytically and by numerical computation. We refer particularly to the extensive work of Cornille and Gervois,[65] Barnsley and Cornille,[69] and Ernst (Ref. 11, Sec. 10) for further details. Some of the objectives and conclusions of these studies will be examined here briefly in Section VI.D, particularly in relation to anomalous convergence at high energies. Convergence of the infinite polynomial solutions has been proved by several authors, the most elegant method being that of Bobylev.[70] Useful bounds to expansion coefficients and eigenvalues can be obtained at the same time.

Finally, we should mention the extension of expressions (4.51) and (4.52) to the case of general dimensionality $d > 3$. The fundamental equation of type (4.56) is confirmed, and the coefficients are given explicitly by obvious generalizations of the three-dimensional formulas.[60,65] Refer to the original papers for further details, but we will revert to some aspects of the relationship between dimensionality in scattering models and degrees of freedom in the next section.

V.   INTEGRAL TRANSFORM METHODS

As indicated earlier, a number of techniques normally associated only with linear systems find application in the solution of Boltzmann models, though in a somewhat unsystematic manner and without the benefit of such tools as the superposition principle. We have already discussed the use of orthogonal polynomial methods and seen the usefulness of generating functions in the original Krook and Wu solution. Here we shall conclude with a review of the various integral transform methods that have been prominent in more recent work.

Although Fourier transforms and related methods have been emphasized in some of the most elegant papers to appear on the subject, it is probably true to say that their role has been not so much in obtaining particular solutions as in exposing the interrelationships between apparently distinct models, between continuous- and discrete-variable formulations of a problem and, in the case of vector models, the natural generalizations to dimensions other than three.

A. Fourier Transforms

The significance of the Fourier transform in the solution of Maxwell models was first recognized by Bobylev[71] and extensively exploited by Ernst and coworkers.[54,60,72] We shall begin with a brief outline of Bobylev's original observations.

Defining the Fourier transform pair in d dimensions by

$$\phi(\mathbf{k}, t) = \int d\mathbf{v}\, e^{-i\mathbf{k}\cdot\mathbf{v}} f(\mathbf{v}, t) \tag{5.1}$$

$$f(\mathbf{v}, t) = \frac{1}{2\pi^d} \int d\mathbf{k}\, e^{i\mathbf{k}\cdot\mathbf{v}} \Phi(\mathbf{k}, t) \tag{5.2}$$

and applying to the Maxwell molecule form of the Boltzmann equation in velocity space we see that the right-hand side can be split into in- and out-scattering contributions as follows:

$$\frac{\partial\phi(\mathbf{k}, t)}{\partial t} = \int d\mathbf{v} \int d\mathbf{w} \int d\mathbf{n}\, h(\hat{\mathbf{g}}\cdot\hat{\mathbf{n}}) e^{-i\mathbf{k}\cdot\mathbf{v}} f(\mathbf{v}') f(\mathbf{w}')$$

$$- \int d\mathbf{v} \int d\mathbf{w} \int d\mathbf{n}\, h(\hat{\mathbf{g}}\cdot\hat{\mathbf{n}}) e^{-i\mathbf{k}\cdot\mathbf{v}} f(\mathbf{v}) f(\mathbf{w})$$

with $\mathbf{v}'$ and $\mathbf{w}'$ given by the usual collision equations (2.13) and (2.14). On using the inverse collision symmetry to remove the primes together with $\mathbf{v}' = \frac{1}{2}[\mathbf{v} + \mathbf{w} + g\hat{\mathbf{n}}]$, the in- and out-scattering terms can be recombined in the form

$$\frac{\partial}{\partial t}\phi(\mathbf{k}, t) = \int d\mathbf{v} \int d\mathbf{w} f(\mathbf{v}) f(\mathbf{w}) \exp\left[-\tfrac{1}{2} i\mathbf{k}(\mathbf{v} + \mathbf{w})\right]$$

$$\times \int d\hat{\mathbf{n}}\, h(\hat{\mathbf{g}}\cdot\hat{\mathbf{n}}) \left\{ \exp\left[-\tfrac{1}{2} ikg(\hat{\mathbf{k}}\cdot\hat{\mathbf{n}})\right] - \exp\left[-\tfrac{1}{2} i(\mathbf{k}\cdot\mathbf{g})\right] \right\}$$

the caret always indicating a unit vector.

Bobylev's crucial observation was that the inner integral is actually invariant under simultaneous rotations of $\mathbf{g}$ and $\mathbf{k}$ and can thus depend only on the scalar quantities g, k, and kg. It is likewise clear that the inner integral is symmetric under the interchange of $\mathbf{k}$ and $\mathbf{g}$. Using this and further manipulations like $kg(\hat{\mathbf{g}}\cdot\hat{\mathbf{n}}) = k(\mathbf{g}\cdot\hat{\mathbf{n}}) = k\mathbf{n}\cdot(\mathbf{v}-\mathbf{w})$, we can recast the exponents in the form of combinations of k and n in dot products with $\mathbf{v}$ and $\mathbf{w}$. In this way the right-hand side can be expressed in terms of Fourier transforms with different arguments, the final result being

$$\frac{\partial}{\partial t}\phi(\mathbf{k},t) = \int d\hat{\mathbf{n}}\, h(\hat{\mathbf{k}}\cdot\hat{\mathbf{n}})\left\{\phi\left[\tfrac{1}{2}k(\hat{\mathbf{k}}+\hat{\mathbf{n}})\right]\phi\left[\tfrac{1}{2}k(\hat{\mathbf{k}}-\hat{\mathbf{n}})\right] - \phi(\mathbf{k})\phi(\mathbf{o})\right\}$$

(5.3)

where, incidentally, $\phi(0) = \langle 1 \rangle = 1$. The other boundary conditions are

$$\nabla_{\mathbf{k}}\phi(\mathbf{k},t)|_{k=0} = -i\langle v \rangle$$
$$\nabla_{\mathbf{k}}^2\phi(\mathbf{k},t)|_{k=0} = -\langle v^2 \rangle$$

(5.4)

with $\langle v \rangle = 0$ in the isotropic case. The stationary distribution $f(\mathbf{v},t) = (2\pi)^{-d/2}e^{-v^2/2}$ gives $\phi(\mathbf{k},\infty) = e^{-k^2/2}$.

Equation (5.3) reveals a hidden symmetry present in the Maxwell models. If we define a new function

$$\phi_\rho(\mathbf{k},t) = \phi(\mathbf{k},t)e^{-\rho k^2/2}, \qquad s > 0$$

(5.5)

then it is clear on substitution that this equally well satisfies (5.3). Translated back into the velocity variable, the effect of the "Bobylev symmetry" is that the vector Boltzmann equation for Maxwell molecules is invariant under the one-parameter semigroup of transformations defined by

$$f_\rho(\mathbf{v},t) = \frac{1}{(2\pi s)^{d/2}}\int d\mathbf{w} f(\mathbf{w},t)e^{-(\mathbf{v}-\mathbf{w})^2/2}$$

(5.6)

As we shall show presently, the so-called Bobylev symmetry is the origin of various simplifications in the structure of moment equations, similarity solutions, and related properties.

B. Isotropic Velocity Distributions

So far we have worked within the framework of the full vector equation in general dimensionality. Further simplifications emerge if we specialize to the case of isotropic distributions. To study these we keep the dimensionality general and consider angular averages $\langle(\mathbf{k}\cdot\mathbf{v})^{2n}\rangle_{\hat{v}}$ and $\langle\exp(-i\mathbf{k}\cdot\mathbf{v})\rangle_{\hat{v}}$ given by Eqs. (4.12) and (4.14). Applying these in (5.3), it is clear that the scalar transform $\phi(k,t)$ can be written as

$$\phi(k,t) = \int_0^\infty dv_0\, F_1\left(\tfrac{1}{2}d, -\tfrac{1}{4}k^2v^2\right) f(v,t) \tag{5.7}$$

or, on substitution of $x = \tfrac{1}{2}v^2$, $y = \tfrac{1}{2}k^2$, the energy-variable alternative

$$\Phi\left(\tfrac{1}{2}k^2, t\right) \equiv \Phi(y,t) = \int_0^\infty dx_0\, F_1\left(\tfrac{1}{2}d, -yx\right) F(x,t) \tag{5.8}$$

Thus $\phi(y,t) = \langle_0 F_1(\tfrac{1}{2}d, -\tfrac{1}{4}k^2v^2)\rangle$. The two inverses are accordingly

$$f(v,t) = \frac{\Omega_d}{(2\pi)^d} \int_0^\infty dk\, k^{d-1}\,_0 F_1\left(\tfrac{1}{2}d, -\tfrac{1}{4}v^2k^2\right) \Phi\left(\tfrac{1}{2}k^2, t\right) \tag{5.9}$$

and

$$F(x,t) = \frac{1}{\Gamma\left(\tfrac{1}{2}d\right)^2} \int_0^\infty dy\,(xy)^{d/2-1}\,_0 F_1\left(\tfrac{1}{2}d, -xy\right) \Phi(y,t) \tag{5.10}$$

Although these expressions are technically Hankel transforms, they are often referred to in the literature as Fourier transforms, without distinction.

It is clear that the transforms (5.9) and (5.10) play the role of moment-generating functions such that, for the energy-variable in particular,

$$\begin{aligned}
\Phi(y,t) &= \langle_0 F_1\left(\tfrac{1}{2}d, -yx\right)\rangle \\
&= \sum_{n=0}^\infty \frac{(-y)^n\langle x^n\rangle}{(\tfrac{1}{2}d)_n n!} = \sum_{n=0}^\infty \frac{(-y)^n}{n!} M_n
\end{aligned} \tag{5.11}$$

and

$$M_n(t) = (-1)^n \frac{d}{dy}\Phi(y,t)\big|_{y=0} \tag{5.12}$$

The assumption of isotropic conditions also leads to simplification of the Fourier-transformed transport equation (5.3). Since in the terms of the integrand $\Phi(\tfrac{1}{2}\mathbf{k}\cdot(\hat{\mathbf{k}}\pm\hat{\mathbf{n}})) = \Phi(\tfrac{1}{2}k^2(1\pm\hat{\mathbf{k}}\cdot\hat{\mathbf{n}}))$, we can see that the whole of the right-hand side must reduce to an integral over some function of $\mu =$

$\cos \chi = \hat{\mathbf{k}} \cdot \hat{\mathbf{n}}$ only. Referring to (4.11), it is then a straightforward matter to rewrite the integration in the variable μ. In this way the evolution of the function $\Phi(y, t)$ can be seen to be given by

$$\frac{\partial}{\partial t} \Phi(y, t) = \int_{-1}^{1} d\mu \, \bar{g}(\mu) \{ \Phi[\tfrac{1}{2}(1+\mu)y] \Phi[\tfrac{1}{2}(1-\mu)y] - \Phi(y)\Phi(0) \}$$

$$(5.13)$$

[where actually $\Phi(0) = 1$]. Here $\bar{g}(\mu)$ is a modification of $h(\mu)$ due to geo-metrical factors that arise on forming the average (4.11) in dimensions other than three, viz.:

$$\bar{g}(\mu) = \frac{2\pi^{d/2 - 1/2}}{\Gamma(\tfrac{1}{2}d - \tfrac{1}{2})} (1 - \mu^2)^{(d-3)/2} h(\mu) \tag{5.14}$$

Provided that the average collision rate

$$b_{00} = \int_{-1}^{1} d\mu \, \bar{g}(\mu) \tag{5.15}$$

is finite, Eq. (5.13) can be simplified to

$$\left(\frac{\partial}{\partial t} + b_{00} \right) \Phi(y, t) = \int_{-1}^{1} d\mu \, \bar{g}(\mu) \Phi[\tfrac{1}{2}(1+\mu)y] \Phi[\tfrac{1}{2}(1-\mu)y] \quad (5.16)$$

Otherwise the right-hand side of (5.13) must be interpreted as it stands, the divergences in the two terms canceling in favorable cases, which include that of true Maxwell molecules.

We can now recognize that the "Bobylev symmetry" property given earlier for $\phi(k, t)$ takes an even simpler form in the case of the energy-space trans-forms, namely, that substitutions of the form

$$\Phi_\rho(y, t) = e^{\rho y} \Phi(y, t) \tag{5.17}$$

leave the transport equation (5.13) invariant. This has a number of im-portant consequences to which we shall return presently.

C. Deterministic, Stochastic, and "Maxwell" Models

As we hinted earlier, the Fourier and Hankel transformed representations play a crucial role in unifying not only models of different complexity, but also those arising from deterministic scattering mechanics on the one hand and purely stochastic postulates on the other—the former interpreting both momentum and energy conservation in physical terms, the latter energy conservation alone. Following the lead of Ernst and Hendriks,[11,60] we are now in a position to amplify these insights.

As an example, consider the three-dimensional Maxwell model for which, in suitable time scaling, we can take $\bar{g}(\mu) = \frac{1}{2}$, $b_{00} = 1$ in Eq. (5.13). On making the substitution $\mu = 2s - 1$, this yields

$$\left(\frac{\partial}{\partial t} + 1\right)\Phi(y, t) = \int_0^1 ds\, \Phi(ys)\Phi(y(1-s)) \tag{5.18}$$

or alternatively

$$\left(\frac{\partial}{\partial t} + 1\right)\Phi(y, t) = \frac{1}{y}\int_0^y dz\, \Phi(z)\Phi(y-z) \tag{5.19}$$

If, by contrast, we take the Boltzmann equation for the stochastic Tjon-Wu model in one degree of freedom (3.42) and form the Hankel transform thus

$$\left(\frac{\partial}{\partial t} + 1\right)\Phi(y, t) = \int_0^\infty dx \sum_{\nu=0}^\infty \frac{(-yx)^\nu}{\nu!^2}\int_x^\infty \frac{du}{u}\int_0^u d\eta\, F(\eta)F(u-\eta) \tag{5.20}$$

we arrive, after straightforward partial integrations, at precisely the same equation, (5.19). Thus it is clear that each of the two models, open to quite different physical interpretations, can be solved through a unified mathematical formulation. Moreover, the common transformed equation, which involves only a single integral, is manifestly simpler than either of the originals.

The importance of this correspondence lies not so much in the simplified working just illustrated as in the demonstration of the fact that both cases belong to the much larger class of models whose evolution can be represented in the form (5.13) and whose solutions, either in polynomial series or BKW-type similarity solutions, are then found automatically. To generalize the example just given to other stochastic models, we must return to the standard moment-evolution equations (2.102) and the expression (2.134) for the coefficients appearing there. On multiplying the former left and right by $(-y)^n/n!$ and using (2.128) and (2.134), we obtain the energy-space transport equation

$$\left(\frac{\partial}{\partial t} + b_{00}\right)\Phi(y, t) = \int_0^1 ds\, K(s, 0; 1)\Phi(ys)\Phi(y(1-s)) \tag{5.21}$$

This is clearly the same as (5.16) if we put $s = \frac{1}{2}(\mu + 1)$ and identify $K(s, 0; 1) \equiv \bar{g}(2s - 1) = \mathcal{K}(s)$.

This far-reaching result brings together all models for which the derivation of (2.134) is valid and enables us to write down immediately from the transition kernel both the coefficients b_{nk} in the moment-evolution equations and the form of the transformed Boltzmann equation itself. Since the above can equally be seen as arising in a stochastic or deterministic context, it is reasonable to refer to all such models collectively as Maxwell models. This is now accepted terminology, although authors differ in the precise choice of defining relation. The standard interrelationships through which solutions can be written once a model is identified as of the Maxwell type will be described in detail in the Section VI, where the power and simplicity of the integral transform methods will be evident.

The correspondences between different interpretations of the beta-distribution diffuse-scattering model of Section III.D, first pointed out by Tjon and Wu[47] and Futcher, Hoare, Hendriks, and Ernst,[50] are particularly interesting. As we have seen, the result $\mathcal{K}(s) = W_{pp}(s; 1)$ implies that

$$\bar{g}(\mu) = \frac{1}{2} W_{pp}\left(\frac{1}{2}(1 + \mu); 1\right) = \frac{(1 - \mu^2)^{p-1}}{2^{2p-1} B(p, p)} \tag{5.22}$$

Comparing this with (5.14) it is clear that it can equally be interpreted as a d-dimensional model with scattering function $h(\mu) \propto (1 - \mu^2)^{p - d/2 + 1/2}$, that is, $h(\mu) \propto |\sin \chi|^{2p+d-1}$. Taking $p = 1$, $d = 3$, we have $\bar{g}(\mu) = \frac{1}{2}$, $h(\mu) = \text{const.}$ in the familiar case, but the same is equally true for any value of p in $d = 2p + 1$ dimensions. The choice $p = 1$, d general, for example, leads to the class of nonisotropic scattering models with $h(\cos \chi) = |\sin \chi|^{3-d}$, a case studied in detail by Ziff et al.[64] Thus the Tjon-Wu scattering model, originally conceived as stochastic in one degree of freedom, is open to a whole family of different interpretations, beyond the simple connection with the Krook-Wu model.

Hendriks and Ernst[60] have also solved the difficult problem of obtaining a stochastic kernel $K(y, x; u)$ corresponding to an arbitrary deterministic scattering law, that is to say the reverse of the train of thought above. The result, which is obtained by integration of the transition kernel form of the full Boltzmann equation (2.10), is formidably complex, but it is worth quoting, since it is certainly one of the most intricate and interesting results to come within the scope of this article.

Beginning with the expression

$$K(x, y; u) = \frac{1}{[4y(u - y)]^{d/2-1} \Omega_d^2} \int \int dv\, dw \int d\hat{n}\, \sigma(g, \hat{g} \cdot \hat{n})$$
$$\times \delta\left(x - \tfrac{1}{2}v^2\right) \delta\left(y - \tfrac{1}{2}v'^2\right) \delta\left(u - \tfrac{1}{2}v^2 - \tfrac{1}{2}w^2\right) \tag{5.23}$$

with $\sigma(g, \cos\chi) = gI(g, \chi)$ in Eq. (2.11), they reduce the indicated integrals to obtain

$$K(x, y; u) = \frac{a_d u^{d-3}}{[y(u-y)]^{d/2-1}}$$

$$\times \int_{\mu_-}^{\mu_+} d\mu \int_0^1 ds \frac{(1-s^2)^{d/2-2}\sigma(s, \mu)\Delta^{(d-3)/2}}{[1-\mu^2]^{1/2}} \quad (5.24)$$

in which

$$a_d = \frac{(d-2)\Omega_d}{\pi B\left(\frac{1}{2}d - \frac{1}{2}, \frac{1}{2}\right)}$$

$$\bar{\sigma}(s, \mu) = \frac{1}{2}\left[\sigma(g_+, \mu) + \sigma(g_-, \mu)\right]$$

$$g_\pm(\mu) = (2u)^{1/2}\left\{1 \mp s\left[\frac{4\Delta}{1-\mu^2}\right]^{1/2}\right\}^{1/2} \quad (5.25)$$

$$\mu_\pm = (1-2\sigma)(1-2\tau) \pm 4[\sigma\tau(1-\sigma)(1-\tau)]^{1/2}$$

$$\Delta = \frac{1}{4}(\mu - \mu_-)(\mu_+ - \mu)$$

$$\sigma = \frac{x}{u}, \qquad \tau = \frac{y}{u}$$

The formula as written is valid only in the region $x < y$, $x + y < u$ but is easily extended to the full range of variables by use of the symmetry properties $K(u - x, u - y; u) = K(x, y; u)$ and $K(x, u - y; u) = K(x, y; u)$. Fortunately there is some simplification in special cases such as that of Maxwell molecules. Thus, if $\bar{\sigma}(s, \mu) = h(\mu)$, the s integral may be evaluated with the result

$$K(x, y; u) = \frac{u^{d-3}}{[y(u-y)]^{d/2-1}}\int_{\mu_-}^{\mu_+} d\mu \frac{\bar{g}(\mu)\Delta^{(d-3)/2}}{B(d/2 - \frac{1}{2}, \frac{1}{2})(1-\mu^2)^{d/2-1}}$$

$$(5.26)$$

We may check that both the formulas given for $K(x, y; u)$ satisfy all the symmetry properties set out in Section II.A and vanish correctly outside the region $x, y \leqslant u$. Although the general expression (5.24) seems too complicated to be of much help in solving deterministic models, Hendriks and

Ernst[60] have used it very effectively to systematize a variety of special cases and provide connections between less general results previously known,[73,74] as well as to extend the concept of the "very hard particle" model to systems of arbitrary dimensionality. A particularly interesting class of kernels is the two-parameter set in which one parameter fixes the quantity d in Eq. (5.26) and the other, in general different, is identified with the d in expression (5.14) for $\bar{g}(\mu)$, $h(\mu)$ taken to be constant. Many special cases of the integral (5.26) can then be worked out, for combinations of integer and half-integer values of the parameters, in terms of standard functions. Refer to Hendriks and Ernst[60] for an extensive selection of these.

D. Transform Properties and Soluble Models

The results of the previous sections have been confined to the formalities of Fourier and Hankel transforms in relation to scattering kernels and their transport equations; it remains to demonstrate the insight they give into the properties of actual soluble models. To do this we shall return to the spirit of the earlier sections of this article and work with stochastic models of p degrees of freedom in standard complexity, leaving the relationship to energy-space formulations of deterministic models in $2p$ dimensions to be understood. It will be convenient to continue, in this expanded context, with the use of the distribution function $F(x, t) = P(cx/p, t)$ rather than the $P(x, t)$ of the earlier sections. We are thus concerned with the transform pair

$$\Phi(y, t) = \int_0^\infty dx \, _0F_1(p, - yx) F(x, t) \tag{5.27}$$

$$F(x, t) = \frac{1}{\Gamma(p)^2} \int_0^\infty dy \, (xy)^{p-1} \, _0F_1(p, - xy) \Phi(y, t) \tag{5.28}$$

with $\Phi(y, t)$ satisfying the evolution equation

$$\left(\frac{\partial}{\partial t} + b_{00}\right) \Phi(y, t) = \int_0^1 ds \, K(s) \Phi(ys) \Phi(y(1-s)) \tag{5.29}$$

and $\mathcal{K}(s) = K(s, 0; 1)$ as before.

Given the equilibrium distribution (2.56) we have that

$$\Phi(y, \infty) = e^{-y} \tag{5.30}$$

while the conservations $\langle 1 \rangle$ and $\langle x \rangle = 1$ lead to the boundary conditions

$$\Phi(0, t) = 1 \quad \text{and} \quad \Phi'(0, t) = -1 \tag{5.31}$$

Three main items will be considered in this section: (1) The derivation of standard polynomial moment equations, (2) the universality of the BKW-type similarity solution for "Maxwell" systems, and (3) the general relationship between analogous discrete- and continuous-variable models.

1. Polynomial Moment Expansions

The use of the Hankel-transformed Boltzmann equation in energy space offers a particularly straightforward route to the polynomial moment equations of type (2.120) and one that, incidentally, does not presume any advance knowledge of the appropriate polynomial set. To see this we make use of the Bobylev symmetry property (5.17) to write

$$\Phi_\rho(y,t) = e^{\rho y}\Phi(y,t) = \sum_{n=0}^{\infty} \frac{c_n(t)}{n!}y^n \tag{5.32}$$

considering thus the $c_n(t)$ to be moments corresponding to the left-hand side as generating function. The function $\Phi_\rho(y,t)$ does not need to satisfy particular boundary conditions at this stage. But since its time evolution under Eq. (5.16) must be precisely the same as that of $\Phi(y,t)$, it follows that the moments $c_n(t)$ must satisfy the same equation of type (2.102) as the scaled ordinary moments $M_n(t)$. A choice of the parameter ρ may now be made according to convenience. Examining the structure of the series on the right, we see immediately that [because of the asymptotic requirement $\Phi(y,\infty) = e^{-y}$] the choice $\rho = 1$ yields the simple behavior $c_0(t) = 1$, $c_1(t) = 0$, $c_n(\infty) = \delta_{n0}$. The expansion

$$\Phi_1(y,t) = \sum_{n=0}^{\infty} \frac{c_n(t)}{n!}y^n e^{-y} \tag{5.33}$$

can now be inverted term by term to provide a series expression for $F(x,t)$. Using (5.28) in this way for a system of p degrees of freedom, we obtain

$$F(x,t) = F_0(x)\left\{1 + \sum_{n=1}^{\infty} \frac{c_n(t)(p)_n}{n!}{}_1F_1(-n,p;x)\right\} \tag{5.34}$$

Since $n!^{-1}(p)_n {}_1F_1(-n,p;x) = L_n^{(p-1)}(x)$, the coefficients $c_n(t)$ are clearly the modified Laguerre moments $c_n(t) = \langle {}_1F_1(-n,p;x)\rangle$ of Eq. (2.122), and the series is the appropriate form of (2.120). The result is common to all Maxwell models, whose particular moments $c_n(t)$ must satisfy the identical evolution equation (2.127) with coefficients determined by (2.128) and (2.134).

2. Similarity Solutions

A variation of the above formulation due to Hendriks and Ernst[60] leads neatly to the BKW-type similarity solution by a route that makes clear its universality for all Maxwell models. Consider now Eq. (5.32) with, instead of $\rho = 1$, the choice left as a function $\rho(t)$ to be determined. Substituting this and using the Bobylev symmetry, we see that the transport equation to be satisfied is now

$$\left(\frac{\partial}{\partial t} - \rho y + b_{00}\right)\Phi_\rho(y, t) = \int_0^1 ds\, \mathscr{K}(s)\Phi(ys)\Phi(y(1-s)) \quad (5.35)$$

This time we form the expansion

$$\Phi_{\rho(t)}(y, t) = e^{\rho(t)y}\Phi(y, t) = \sum_{n=0}^{\infty} \frac{\xi_n(t)}{n!} y^n \quad (5.36)$$

whereupon, noting that the boundary conditions $\Phi(0, t) = 1$, $\Phi'(0, t) = -1$ imply that $\Phi_{\rho(t)}(y, t) = 1 + (\rho - 1)y + O(y^2)$, it follows that $\xi_0(t) = 1$ and $\xi_1(t) = \rho(t) - 1$. Now, provided always that $\rho(t) > 1$, we can again invert the series term by term to obtain this time

$$F(x, t) = \frac{x^{p-1}}{\Gamma(p)} \frac{\exp[-x/(1+\xi_1)]}{[1+\xi_1]^p} \sum_{n=0}^{\infty} \frac{\xi_n(t)}{(1+\xi_1)^n} L_n^{(p-1)}\left(\frac{x}{1+\xi_1}\right)$$

$$(5.37)$$

It is now necessary to substitute $\Phi_{\rho(t)}$ from (5.36) back into (5.35) to obtain, using (2.134) and (2.139), the coupled equations

$$\dot{\xi}_n + n\bar{\lambda}_n\xi_n = n\xi_n\xi_{n-1} + (b_{n1} + b_{nn-1})\xi_1\xi_n + \sum_{k=2}^{n-2} b_{nk}\xi_k\xi_{n-k} \quad (5.38)$$

The first few of these can be written more explicitly using the relations already given for the coefficients b_{nk} in terms of the eigenvalues $\Lambda_n = n\bar{\lambda}_n$. It is found that

$$\dot{\xi}_2 + 2\bar{\lambda}_2\xi_2 = 2\xi_1\left(\xi_1 + \bar{\lambda}_2\xi_1\right)$$
$$\dot{\xi}_3 + 3\bar{\lambda}_3\xi_3 = 3\xi_2\left(\xi_1 + \bar{\lambda}_2\xi_1\right)$$
$$\dot{\xi}_4 + 4\bar{\lambda}_4\xi_4 = 4\xi_3\left(\xi_1 + \bar{\lambda}_2\xi_1\right) + 4\left(\bar{\lambda}_4 - \bar{\lambda}_2\right)\left(4\xi_1\xi_3 - 3\xi_2^2\right)$$
$$\dot{\xi}_5 + 5\bar{\lambda}_5\xi_5 = 5\xi_4\left(\xi_1 + \bar{\lambda}_2\xi_1\right) + 5\left(\bar{\lambda}_5 - \bar{\lambda}_2\right)\left(3\xi_1\xi_4 - 2\xi_2\xi_3\right)$$

$$(5.39)$$

The simple form of the first two equations reflects the degeneracy $\bar{\lambda}_2 = \bar{\lambda}_3$ we noted at Eq. (2.139), which holds for all Maxwell models irrespective of the form of the function $\mathcal{K}(s)$.

It is not difficult to spot the particular solution of the set (5.39) that takes the form

$$\xi_1(t) = - \eta e^{-\bar{\lambda}_2 t}, \qquad \xi_n(t) = 0; \qquad n \geqslant 2 \qquad (5.40)$$

This therefore only involves a first-degree polynomial and is in fact

$$F^S(x,t) = \frac{x^{p-1}}{\Gamma(p)} \left(\frac{\exp[-x/(1-\xi_1)]}{(1-\xi_1)^p} \right) \left[1 - \left(\frac{\xi_1}{1-\xi_1} \right) L_1^{(p-1)} \left(\frac{x}{1-\xi_1} \right) \right]$$

$$(5.41)$$

On identifying $L_1^{(p-1)}(z) = (p - z)$ and associating $\psi = (1 - \xi_1)^{-1}$, $\lambda_2 = \frac{1}{2}\Lambda_2$, this then takes the form of Eq. (3.76) adjusted to the new scaling $(\varepsilon/p \equiv 1)$.

It is thus in no way coincidental that the structure of $F^S(x,t)$ is identical for the several models treated in Section III, given that all can be reduced to the form (5.29) with the appropriate variations of the function $\mathcal{K}(s)$. To complete the picture, it might be noted that the transform of $F^S(x,t)$ is $\Phi^S(y,t) = (1 - y\xi)\exp[- y(1 - \xi)]$ and that if this is perceived to be a solution of (5.21) it may be substituted directly to give, via (5.38), the dependence (5.40) for $\xi(t)$.

The above derivation, adapted here from Hendriks and Ernst,[58] was also arrived at by Alexanian,[75] who earlier obtained the equations (5.38) using an essentially different integral transform method.

Many attempts have been made to obtain useful similarity solutions other than the "BKW mode" given above. Such solutions can be found and shown to depend systematically on the higher eigenvalues Λ_3, Λ_4,..., etc., but every attempt so far to prove positivity appears to have failed.[2,51,75] Extensive numerical computations have also invariably demonstrated the existence of negative-going regions in the solutions.[66,70,76] While this work has undoubted mathematical interest and may have applications in other areas, it seems that we must reluctantly accept Ernst's recent[11] conclusion that the similarity modes other than the BKW mode necessarily violate positivity and are thus of no value in relation to physical models. We shall not, therefore, explore this aspect further here.

3. Discrete and Continuous Variable Formulations

One of the most remarkable things to emerge from the study of transformed model Boltzmann equations is a general formulation of the relationship between discrete- and continuous-variable version of the same model. Using the ideas previously developed here, it is possible systematically to derive the discrete analogue of a continuous model (in standard complexity) even without prior knowledge of the appropriate orthogonal functions and relationships in the finite-difference calculus.

We know from the limiting property (2.19) applied to the moment equations (2.91) that the scaled moments $M_n = \langle x^n \rangle / (p)_n$ and $\overline{M}_n = \varepsilon_0^n \langle i^{(n)} \rangle / (p)_n$ must satisfy identical evolution equations, and from this it follows that the corresponding moment-generating functions must also be identical. It is therefore natural to write the two alternatives

$$\Phi(y,t) = \sum_{n=0}^{\infty} \frac{(-y)^n \langle x^n \rangle}{(p)_n n!} = \langle {}_0F_1(p, -yx) \rangle \tag{5.42}$$

$$= \sum_{n=0}^{\infty} \frac{(-\varepsilon_0 y)^n \langle i^{(n)} \rangle}{(p)_n n!} = \langle {}_1F_1(-i, p; \varepsilon_0 y) \rangle \tag{5.43}$$

in the second of which the factorial moment is referred to a normalized discrete probability distribution $\overline{F}(i,t)$. The analogue of the Hankel transformation (5.27) for the discrete variable is then to be found by writing

$$\Phi(y,t) = \sum_{n=0}^{\infty} \frac{(-\varepsilon_0 y)^n}{(p)_n n!} \sum_{i=n}^{\infty} i^{(n)} \overline{F}(i,t)$$

or, since $i^{(n)} = (-1)^n (-i)_n$:

$$\Phi(y,t) = \sum_{i=0}^{\infty} \overline{F}(i,t) \, {}_2F_1(-i, p; \varepsilon_0 y) \tag{5.44}$$

This transformation has no accepted name. Its inverse may immediately be found on observing that ${}_2F_1(-i, p; \varepsilon_0 y) = L_i^{(p-1)}(\varepsilon_0 y)$ and using the orthogonality property (A5.4). In this way,

$$\overline{F}(i,t) = \frac{(i+1)_{p-1}}{\Gamma(p)^2} \varepsilon_0^p \int_0^{\infty} dy \, y^{p-1} e^{-\varepsilon_0 y} {}_1F_1(-i, p; \varepsilon_0 y) \Phi(y,t) \tag{5.45}$$

To understand the relationship of this to the continuous Hankel transform we need only form the usual limit $\varepsilon_0 \to 0$, $i \equiv dx/\varepsilon_0$. Thus,

$$F(x,t) = \frac{1}{\Gamma(p)^2} \int_0^{\infty} dy \, y^{p-1} \Phi(y,t)$$

$$\times \lim_{\varepsilon_0 \to 0} \left\{ \varepsilon_0^{p-1} \left(\frac{x}{\varepsilon_0} + 1 \right)_{p-1} e^{-\varepsilon_0 y} {}_1F_1\left(-\frac{x}{\varepsilon_0}, p; \varepsilon_0 y \right) \right\}$$

The limit is straightforward and, since $_1F_1(-x/\varepsilon_0, p; \varepsilon_0 y) \to {}_0F_1(p, -xy)$, the integral on the right reduces, as expected, to give expression (5.28).

We can now operate a procedure similar to that in the continuous variable case to obtain a polynomial expansion of the solution $\bar{F}(i, t)$. The function Φ of Eq. (5.43) is again inverted term by term, but this time by means of (5.45) rather than (5.28). The result is the expansion formula

$$\bar{F}(i, t) = \bar{F}_0(i)\left\{1 + \sum_{n=2}^{\infty} \frac{c_n(t)(p)_n}{n!}\left(\frac{1}{\varepsilon_0 + 1}\right)^n M_n\left(i, n, (\varepsilon_0 + 1)^{-1}\right)\right\}$$

$$(5.46)$$

with $M_n(i, p; a)$ the Meixner polynomials (A5.11). These, it will be seen, have arisen naturally through the use of the generating function (5.43), without prior knowledge that they were the appropriate set.

Similar considerations apply to the BKW-type similarity solution. By repeating the argument leading to Eq. (5.41) step by step and inverting the series (5.36) by means of (5.45), we arrive at the discrete variable analogue of (5.41) in the form

$$\bar{F}^S(i, t) = \frac{(i+1)_{p-1}\varepsilon_0^p(1-\xi_1)^i}{\Gamma(p)(1+\varepsilon_0-\xi_1)^{i+p}}\left\{1 - \frac{p\xi_1}{1+\varepsilon_0-\xi_1}M_1\left(i, p, \frac{1-\xi_1}{1+\varepsilon_0-\xi_1}\right)\right\}$$

$$(5.47)$$

Taking account of the fact that $M_1(i, p; a) = 1 + (i/p)(1 - a^{-1})$ and associating $\psi = (\varepsilon_0/(1 + \varepsilon_0))(1 - \eta\xi_1)^{-1}$, we find that this is just the rescaled version of Eq. (3.91) for the BKW solution in p degrees of freedom. On taking the limit $\varepsilon_0 \to 0$, we recover Eq. (5.41). Further aspects of this method and its application to the p-q persistent scattering model will be found in Hendriks and Ernst.[58]

E.   Ziff's Transformations

Ziff,[74] working from a somewhat different standpoint from that taken here, has discovered an intricate set of relationships between the familiar Bobylev-Krook-Wu types of model and a number of "artificial" scattering models which, in a sense, generalize them. I use the term *artificial* to stress that such models are not intended to reproduce conservation laws or other physical properties directly, but rather to lead via various integral transformations to the kernels of physical interest. Here I shall translate Ziff's formulation into a notation closer to my own and also link it with some of the formulas quoted in Section V.C from the work of Hendriks and Ernst.

In order to avoid confusion in talking about different types of distribution functions and moments in the same context, I shall follow Ziff in labeling the functions concerned with the model to which they are attached. Thus $P^{KW}(x, t)$ and $P^{TW}(x, t)$ will stand for the Krook-Wu and Tjon-Wu distribution functions in three and two dimensions, respectively, while $m_n^{KW}(t)$ and $m_n^{TW}(t)$ will be the corresponding moments.

Consider a three-variable kernel defined by

$$K^*(x, y; u + y) = \begin{cases} (u + y)^{-1} & y < x < u \\ 0 & \text{otherwise} \end{cases} \tag{5.48}$$

Its moments are

$$\mathcal{M}^*(y, u) = \int_0^\infty dx\, x^n K^*(x, y; u + y) \tag{5.49}$$

and may immediately be calculated as

$$\mathcal{M}^*(y, u) = \frac{1}{|u + y|} \frac{|u^{n+1} - y^{n+1}|}{n + 1} = \frac{1}{n + 1} \sum_{\nu = 0}^{n} y^\nu u^{n - \nu} \tag{5.50}$$

It follows that if we designate by $P^*(x, t)$ the energy-space distribution that evolves under K^*, then the moments $M^*(t)$ defined

$$M^*(t) = \int_0^\infty dx\, x^n P^*(n, t) \tag{5.51}$$

satisfy precisely the Krook-Wu convolution equation (4.33) without rescaling. We know from previous discussions [cf. (4.32), (4.33)] that the BKW moments and the TW moments satisfy the same moment equations provided we rescale such that

$$M_n^{KW} = \frac{(2n + 1)!!}{2^{n-1}} M_n^* \tag{5.52}$$

It is also clear that

$$M_n^{TW} = n! M_n^* \tag{5.53}$$

Consider now the Tjon-Wu case and examine the Laplace transform with respect to the energy x. It will be seen that

$$\int_0^\infty dx\, e^{-sx} P^{TW}(x, t) = \sum_{n=0}^{\infty} \frac{(-s)^n}{n!} M_n^{TW} = \sum_{n=0}^{\infty} (-s)^n M_n^*$$

$$= \int_0^\infty \frac{dz\, P^*(z, t)}{1 + zs}$$

$$= \int_0^\infty dx\, e^{-xs} \int_0^\infty dz\, P^*(z, t) z^{-1} e^{-x/z}$$

In this way

$$P^{TW}(x,t) = \int_0^\infty \frac{dz}{z} e^{-x/z} P^*(z,t) \tag{5.54}$$

emerges as the integral transformation connecting P^{TW} and P^*. (Note that normalization is preserved on integrating with respect to x.) By precisely similar steps we can arrive at another transform connecting P^{KW} and P^*. Thus, proceeding via the moments M^{TW} we can show that

$$P^{KW}(x,t) = \frac{2x^{1/2}}{\pi^{1/2}} \int_0^\infty dz \, z^{-3/2} e^{-x/z} P^*(z,t) \tag{5.55}$$

A similar analysis, based on the M^{TW} and M^{KW} interrelationship (5.52) and (5.53), can be shown to lead to the transform pair

$$P^{TW}(x,t) = \frac{1}{2} \int_x^\infty \frac{dz \, P^{KW}(z,t)}{(z-x)^{1/2} z^{1/2}} \tag{5.56}$$

$$P^{KW}(x,t) = -\frac{2}{\pi} x^{1/2} \frac{d}{dx} \int_x^\infty \frac{dz \, P^{TW}(z,t)}{(z-x)^{1/2}} \tag{5.57}$$

These are an Abel transform pair and were first derived by Barnsley and Turchetti.[77] Equation (5.54) was first discovered by Alexanian,[75] who referred to it as a "temperature transform" because the z in $e^{-x/z}$ is suggestive of a scaled temperature and of the representation of a distribution function as a continuous superposition of Boltzmann distributions at different temperatures, a procedure that finds practical use in astrophysics.

Alexanian further examined the use of transformation (5.54) in relation to the BKW similarity solutions. By assuming the similarity moments for the distribution $F^*(x,t)$ in the form (5.51), he demonstrated that $P_S^*(x,t)$ must be the singular function

$$P_S^*(x,t) = \delta(x - K(t)) - [1 - K(t)] \delta'(x - K(t)) \tag{5.58}$$

where $K(t)$ is again given by (4.43) and δ, δ' are the delta function and its first derivative, respectively. The stationary value of P^* is then $P_0^*(x) = \delta(1-x)$. On its insertion into the transformations (5.54) and (5.55), the equilibrium gamma distributions in the appropriate number of degrees of freedom are obtained. It is obvious that P^* and P_S^* cannot be said to be probability distributions, since the second term in the latter always has a negative component. Likewise, the quantities M^* cannot be moments in any probabilistic sense.

The preceding results for the most part only systematize more fragmentary discoveries by Tjon and Wu,[47] Alexanian,[75] Barnsley and Turchetti,[73] and others. However, Ziff went further and considered a more general class of integral transforms, of which the above are special cases. Let P^* itself be transformed by a Mellin-type transformation in a degree-of-freedom parameter m:

$$P^{(m)}(x,t) = \frac{x^{m-1}}{\Gamma(m)} \int_0^\infty dz\, P^*(z,t) z^{-m} e^{-x/z} \qquad (5.59)$$

Then this clearly gives P^{TW} for $m = \frac{1}{2}$ and P^{BKW} for $m = \frac{3}{2}$ and leads to the general gamma distribution as a stationary distribution:

$$P_0^{(m)}(x) = \Gamma(m)^{-1} x^{m-1} e^{-x} \qquad (5.60)$$

By using result (5.59), a generalized BKW similarity solution can also be found in the form

$$P_S^{(m)}(x,t) = \frac{x^{m-1} e^{-x/K}}{\Gamma(m) K^{m+1}} \left[(K - m + mK) + x\left(\frac{1-K}{K}\right) \right] \qquad (5.61)$$

the moments of which are related to the M^* by

$$M_n^{(m)} = (m)_n M_n^* \qquad (5.62)$$

The interesting question now is: What is the kernel $K^*(x, y; u)$ that gives rise to $P^{(m)}(x, t)$ in the same way as $P^{KW}(x, t)\,[= P^{(3/2)}(x, t)]$ derives from (5.55)? A somewhat complicated analysis, which we shall not reproduce here, leads to the result that

$$K^{(m)}(x, y; u+y) = \frac{(u+y)^{2m-3}}{(uy)^{m-1}} \times \begin{cases} q^{(m)}\left(\dfrac{x}{u+y}\right), & 0 < x < y \\[2mm] q^{(m)}\left(\dfrac{y}{u+y}\right), & y < x < u \\[2mm] q^{(m)}\left(1 - \dfrac{x}{u+y}\right), & u < x < u+y \end{cases}$$

$$(5.63)$$

TABLE II

Special Cases Obtained from Ziff's "Generating Model" Equations (5.63)

m	$q^{(m)}(w)$
$\frac{1}{2}$	$\dfrac{1-2w}{[w(1-w)]^{1/2}}$ (Kac model)
1	1 (TW model)
$\frac{3}{2}$	$\sin^{-1}w^{1/2}$ (BKW model)
2	w
$\frac{5}{2}$	$\frac{3}{8}\{\sin^{-1}w^{1/2}-(1-2w)[w(1-w)]^{1/2}\}$
3	$w^2-\frac{2}{3}w^3$

with $q^{(m)}$ closely related to the incomplete beta function:

$$q^{(m)}(x) = (m-1)\int_0^x [v(1-v)]^{m-2}\,dv \tag{5.64}$$

when m is an integer, $q^{(m)}$ is a simple polynomial; some of the cases for m a half-integer have been worked out and are given in Table II with the models to which they correspond.

Other relationships, including Laguerre expansions and transform equations connecting $P^{(m)}(x,t)$ and $P^{(m-1)}(x,t)$, will be found in the original paper.[74] Ziff refers to K^* and $K^{(m)}$ as defining *generating models*, which provide a method of synthesizing new soluble transport equations from simple prototypes. He has since extended this kind of treatment to the case where the implied scattering function $h(\cos\chi)$ is an arbitrary even function of the scattering angle.[78] As we indicated earlier, a somewhat similar approach by Hendriks and Ernst[60] has recently led to a family of kernels based on Eq. (5.26), of which the above are special cases. A number of open questions remain at the time of writing that concern the relationship between generating models in energy space and actual scattering cross sections in the velocity variable.

VI. CONJECTURES AND REFUTATIONS

We shall collect in this section a number of miscellaneous results, several of which take the form of conjectures about universal properties of Boltzmann equations, or rather their solutions. These have undoubtedly provided part of the stimulus to obtain soluble models and a justification, in some cases, for detailed attention to what would otherwise have been rather minor points of interest. In the undoubted fascination of discovering and exploring soluble models it is easy to forget that one of their more important functions is to provide counterexamples to universal conjectures which, in sound methodology, is of more importance than providing weak corroboration of them.

A. The Krook-Wu Conjecture: Universality of the Similarity Solution

A great stimulus to the search for further soluble models was undoubtedly provided by the original conjecture of Krook and Wu: "An arbitrary initial state tends first to relax towards a state characterized by the [BKW] similarity solution. The subsequent stage is essentially represented by the similarity solution with appropriate phase." If this were true it would lend obvious importance to similarity-type solutions of all kinds, whether or not they were derivable from particular soluble models. Unfortunately, subsequent history has shown the Krook-Wu conjecture to be in all reasonable interpretations false. Though a subclass of initial distributions would appear to behave in this way, sufficient counterexamples have been forthcoming to suggest that the simplistic statement of the original conjecture can have little, if any, practical or mathematical significance. Evidence against the conjecture takes the form of numerical computations that show nonuniform approach to equilibrium in various forms as well as theoretical arguments predicting the same.[66] We defer discussion of the numerical calculations to Section VI.C, where they are taken up in a more general context, that of enhancement and depletion effects.

B. The McKean Conjecture: "Super-H Theorems"

Some years ago a number of extensions to the standard form of the H theorem were proposed, notably by McKean[10] and Harris.[79] They speculated on the possibility that *all* derivatives of the H function should approach zero monotonically under the Boltzmann equation, in fact that

$$(-1)^n \frac{d^n H}{dt^n} \geq 0 \qquad \text{for all } n \tag{6.1}$$

Technically, this is known as the property of *complete monotonicity*. The McKean conjecture is an attractive one in that, if true, it could be used to justify the choice of the usual H function $P \log P$ as a unique Liapunov function, from the infinity of alternative convex functions that may be so used.[15]

A great deal of work of this kind has appeared, both before and since the discovery of exact solutions to Boltzmann models, some of it in the context of more general thermodynamic systems. A detailed review of this literature is given by Ziff, Merajver, and Stell.[21] Although counterexamples were discovered in artificial approximations to the Boltzmann equation, such as the Bhatnagar-Gross-Krook equation for Maxwell molecules,[80] the idea of (6.1) as a credible hypothesis appears to have gained ground over a period of years.

After the BKW solution became available, Ziff, Merajver, and Stell[21] showed that the derivatives $d^n H/dt^n$ for this case could be calculated in terms of relatively simple integrals generated sequentially. Although an explicit calculation for each n eluded them, they were able to confirm (6.1) for up to about $n = 30$ in all dimensions $1 \le d \le 6$.

The denouement in these developments came during the completion of this review and proved bizarre in the extreme. Olaussen[81] began an analytic investigation of the necessary integrals, obtaining asymptotic bounds to their values in the form of power series. The structure of these immediately indicated that the conjecture should fail for some critical value $n_{cr}(d)$ of n, one surprisingly large and only weakly dependent on dimensionality. The value of n_{cr} was found to be close to 100 for $1 \le d \le 6$, and a numerical computation showed that, subject to certain unrigorous error estimates, $(d^{102} H/dt^{102}) \le 0$! It is rare that the exclamation mark is so justified in theoretical physics.

At almost the same time, Lieb[82] turned his attention to the problem. He was able to show in a few lines, using a corollary to the main theorem on complete monotonicity,[83] that the conjecture (6.1) is definitely false for the BKW solutions in all dimensions. This would appear to put an end to all possibility of a "super-H theorem" of any generality. It remains to be seen whether the undoubtedly perverse behavior of the H function is of any deep significance and whether a more limited reformulation of the problem is of sufficient interest to warrant further investigations.

C. Enhancement and Depletion Effects

The possibility of anomalous behavior in the relaxation of the high-energy tail of the distribution function $P(x, t)$ was noted in the earliest report by Krook and Wu of their exact solution.[3] Their evidence came from an alternative model, with $I(g, \chi) \propto g^{-3}$, for which, although no exact solution was available, a high-velocity, exact asymptotic solution could be found. Examining this, they noticed that a transient enhancement of the high-energy tail during relaxation was to be expected, and immediately recognized the possible implications for the behavior of thermal chemical and nuclear reactions. This is indeed one aspect of Boltzmann equation theory that could prove to be of the greatest importance, for example in controlled fusion experiments. A certain sense of proportion must be kept, however, in remembering that in the energy ranges often under discussion the values taken by $P(x, t)$ may sometimes be as low as 10^{-50}.

The existence of high-velocity anomalies has two aspects: the occurrence of a nonuniform approach to equilibrium at different energies, and the more dramatic possibility of a substantial enhancement (or depletion) of population in certain energy ranges as a transient effect. Such overshoot effects may also call into question the validity of a linear approximation even under conditions relatively close to equilibrium.

Early numerical calculations by Tjon[48] confirmed the possibility of enhancement for certain initial conditions in the Tjon-Wu model, but later, doubt was cast on these because of the simple discretization method used, with a cutoff at high energies. More thorough work by Cornille and Gervois followed,[65] using Laguerre series expansions up to some 20 terms for true Maxwell molecules in several dimensionalities. They proved the absence of enhancement for the "fundamental solutions" (consisting of a single L_k term) but did find a weak effect with more complicated initial conditions where high-order polynomial terms could offset the effect of the exponential in the Maxwellian.

The most extensive calculations so far appear to be those of Barnsley and Turchetti.[84] They use Padé approximant and other techniques to improve accuracy and investigate all sorts of initial conditions, including those of both polynomial and rational function types. However, although enhancement is detected in some cases, it is very weak, and their conclusion is that "the observed phenomenon is not of a magnitude nor of a location for us to endow it with a possible physical importance."

D. Singular and Anomalous Solutions

A very new aspect of the subject of model Boltzmann equations that has opened up in the past two years has already generated more detailed attention than I can do justice to here and may well dominate the next phase of development of the subject. This concerns the discovery and analysis of solutions to the Boltzmann equation that do not conform to the simple polynomial expansions I have postulated so far and which, until recently, were widely assumed to exhaust the solutions of physical interest. In fact, wider classes of solutions than these certainly exist and cannot be ruled out on physical grounds; they inhabit a Hilbert space larger than that spanned by the polynomial sets, their series expansions may involve nonintegral powers of x, and some (or even all) of their moments may be infinite.

The ultimate importance of these solutions in physics remains, to an extent, conjectural—justifying our treatment under this heading. However, their potential interest and mathematical content, as will be seen, is of a different order than that of the rather facile conjectures we have just treated in subsections VI.A and VI.B.

The development of these ideas begins, once again, with Bobylev,[71] who pointed to the well-known property that the *linearized* Boltzmann equation in kinetic theory in general possesses a continuous spectrum, which, depending on the scattering model, may or may not show a gap between the zero (equilibrium) eigenvalue and a continuum threshold. The existence of such a gap determines whether or not there exists a transient that is proportional to $e^{-\lambda_{\min} t}$, giving the slowest possible relaxation, and dominant at large times. Bobylev pointed out that for Maxwell molecules there was no such gap. His argument, recast to apply to general Maxwell models by Ernst,[11] proceeds as follows.

Using the Hankel transform $\Phi(y, t)$ of (5.27), we consider deviations from equilibrium of the form

$$\Phi(y, t) = e^{-y}\{1 + e^{-\Lambda t}\tilde{\varphi}(y)\} \tag{6.2}$$

with the transient term "small." The inversion by (5.28) thus corresponds to

$$F(x, t) = F_0(x)\{1 + e^{-\Lambda t}\varphi(x)\} \tag{6.3}$$

[cf. (2.67)]. If the $\Phi(y, t)$ of (6.2) is substituted back into the transport equation (5.29) for general Maxwell models, the following eigenvalue condition for the spectrum $\{\Lambda_q\}$ emerges:

$$-\Lambda\tilde{\varphi}(y) = \int_0^1 ds\, K(s, 0; 1)\{\tilde{\varphi}(sy) + \tilde{\varphi}((1 - s)y) - \tilde{\varphi}(y) - \tilde{\varphi}(0)\} \tag{6.4}$$

It is clear by inspection that, in addition to the special points $\Lambda_0 = 0$ [$\tilde{\varphi}_0(y) = 1$] and $\Lambda_1 = 0$ [$\tilde{\varphi}_1(y) = y$], which correspond to conservation of particles and energy, respectively, there exists a continuum of solution with eigenfunctions

$$\tilde{\varphi}(y) = y^q \tag{6.5}$$

and eigenvalues

$$\Lambda_q = \int_0^1 ds\, K(s, 0; 1)\{1 - s^q - (1 - s)^q\} \tag{6.6}$$

Thus expression (2.139) is interpreted for general q, while the eigenfunctions in x space are obtained by inversion as

$$\varphi_q(x) = (p)_q {}_1F_1(-q, p; x) \tag{6.7}$$

[in this case with $(p)_q = \Gamma(p+q)/\Gamma(p)$]. This confluent hypergeometric function for general q reduces to $\varphi_n(x) = n! L_n^{(p-1)}(x)$ only when q is an integer n. Otherwise the function $\varphi_q(x)$ remains a solution of the linearized equation, having properties very different from those of the polynomials. In particular, it behaves for large x as

$$\varphi_n(x) = \underset{x \to \infty}{\sim} \frac{\Gamma(p+q)\Gamma(p)}{\Gamma(q)\Gamma(-q)} e^x x^{-q-p} \tag{6.8}$$

In the treatment of the linearized Boltzmann equation, such solutions have traditionally been regarded as unphysical, for no better reason than that the norm in the conventional Hilbert space $\mathscr{H}_{\mathrm{I}}$,

$$\|\varphi_n\|_{\mathrm{I}}^2 = \int_0^\infty dx\, F_0(x)\varphi_n(x)^2 \tag{6.9}$$

would be infinite. But, as Uhlenbeck and Ford[85] stressed in a crucial observation, such a norm has no inherent physical necessity and is more a matter of convenience in allowing use of the standard theory of self-adjoint operators in Hilbert space. A previous attempt[16] to justify the L_2 norm as guarantee of a finite entropy $\int dx\, F\log F$ was also shown to be unfounded.

More important in physical terms is the need to show that such solutions would lead to finite moments M_0 and M_1, thus preserving particle and energy conservation, as well as a correct evolution to F_0 with $t \to \infty$. Such solutions, if available for the full Boltzmann equation, would inhabit the extended Hilbert space $\mathscr{H}_{\mathrm{II}}$ with norm

$$\|\varphi_n\|_{\mathrm{II}}^2 = \int_0^\infty dx\, \varphi_n(x)^2 \tag{6.10}$$

and corresponding inner product. They would not possess the congenial properties of orthogonality and completeness taken for granted in $\mathscr{H}_{\mathrm{I}}$, but this would not preclude the construction of solutions from initial conditions in the same space by other methods. The well-known polynomial solutions would appear in the restricted subspace $\mathscr{H}_{\mathrm{I}}$ with the norm (6.9).

The demonstration by Cornille and Gervois[86,87] that such a prescription could be realized for true Maxwell molecules and that the physical character of the linearized solutions was inherited from the full Boltzmann equation proved to be the starting point for a whole series of new investigations. Later the same authors extended their demonstration to Maxwell molecules in arbitrary dimensionality, using as a starting point the d-dimensional form of the generating function equation (4.35).[88] Hendriks and Ernst[58] easily adapted the same proof to general Maxwell models.

These proofs are highly technical, and I can only indicate here in broadest outline what is involved. As usual, the Hankel-transformed solution in y space is the natural framework for the general Maxwell case. We assert that solutions $\Phi(y, t)$ of (5.29) are expressible as

$$\Phi(y, t) = e^{-y}\left\{1 + \sum_{\{a\}} c_a(t) y^a\right\} \tag{6.11}$$

the summation being over a set of values of a parameter a, indexed in a manner to be disclosed. The restriction to $a > 1$ guarantees the required conservation.

Cornille and Gervois succeeded in constructing series corresponding to the inversion of (6.11) in the form

$$F(x, t) = F_0(x)\left\{1 + \sum_{n=0}^{\infty} B_n(t) \, {}_1F_1(-a(n), p; x)\right\} \tag{6.12}$$

giving prescriptions for (1) the determination of the numbers $a(n)$ for integers n and (2) an algorithm for determining the coefficients $B_n(t)$ recursively and relating them to the eigenvalues $\Lambda_{a(n)}$ and quantities $b_{a(n), a(m)}$, these being, in effect, the reinterpretation for continuous index of the coefficients b_{nk} defined in (2.134). In the course of a very instructive working-out of the above scheme for the particular case of the Tjon-Wu model, Cornille and Gervois[86,88] give two prescriptions that satisfy the required conditions and show that the scheme is not empty. These are (A) $a = (1 + n)\eta$, where η is any real number $\eta > 1$ and n an integer $n \geq 0$, and (B) $a = q_0 + N_0^{-1}(n + 1)$ with q_0 and N_0 fixed integers, $q_0, N_0 \geq 1$. The resulting solutions in either case decay in the manner described as series of inverse powers in x and have the correct property $\|\varphi(x, t)\|_{\mathrm{II}} \to 0$ as $t \to \infty$, showing that $F(x, t) \to F_0(x)$ as required. For a detailed account we must refer to Refs. 86 and 88.

Since the demonstration that inverse-power solutions are meaningful for true Maxwell molecules in d dimensions, recent efforts have been directed toward establishing the physical character of other $\mathscr{H}_{\mathrm{II}}$ solutions for more general classes of nonisotropic scattering models. Cornille and Gervois have now considered cross sections of the form $I(g, \chi) = g^{-4/s} h(\chi)$ for the linearized Boltzmann equation in general dimensions. The case $s = \infty$, corresponding to hard spheres, has been shown to lead to nonphysical $\mathscr{H}_{\mathrm{II}}$ solutions because of violation of energy conservation,[87] and the same is true of the generalized VHP model with $s = -4$. In the intermediate case $4 < s < \infty$ the situation is unclear, and, in the absence of definite proofs, various conjectures have been made.[90] A useful review of known results at the time of writing is given in Ref. 89.

Perhaps the most striking recent discovery is that the Ernst-Hendriks general solution to the VHP model given in Section III.C requires a supplementary condition in order that uniqueness be guaranteed.[90] This is that the distribution $F(x, t)$ shall at all times decay faster than x^{-3}. Violation of uniqueness when this is not the case is connected with loss of energy conservation and has a simple physical interpretation: there exists a flux from infinite energy that cannot be accommodated in the finite region. The nonuniqueness is readily demonstrated on attempting a proof of energy conservation like that in Section II.B with $K = $ const. It is interesting that recognition of this effect came only with the discovery of the analogous phenomenon in polymerization kinetics, where the VHP has a precise counterpart. (See Section VII.C.)

A quite different aspect of anomalous solutions concerns the behavior of delta-function peaks and related discontinuities in evolving systems. Although proper distribution theory methods have yet to be applied to model Boltzmann equations (and would no doubt give interesting results), a general picture of the evolution of systems with initial delta-function and step-function singularities can be given by elementary methods. (See, e.g., Ernst,[11] Section 15.3.) It is easy to see that collisions of particles in a delta function at, say $\delta(x - x_0)$ will, by conservation of energy, cause a step discontinuity at $x = 2x_0$ and a derivative discontinuity at $3x_0$ as the system evolves. Numerical results illustrating this have been obtained using the exact solution (3.31) for the VHP model. A more difficult question concerns the nature of the decay of a pure delta function for different scattering laws. This might either decay without broadening, as must happen in strictly hard-sphere models (see, e.g., the Rayleigh model for test-particle relaxation[25]), or could be imagined to spread instantaneously to a narrow but regular peak. An investigation by Ernst (loc. cit.) using asymptotic properties of the Kac model shows that the second possibility is what occurs in this case. Work on general criteria for delta-function behavior in Maxwell and other models appears to be lacking.

VII.   OTHER QUADRATIC PROCESSES

In stressing the broad application of quadratic transport processes, we promised to return to cases outside the scope of statistical mechanics as usually understood. Of the alternative fields that share transport equations at least superficially similar to those classified in Section I, the subject of aggregation-fragmentation kinetics is the most practically important and rich in published literature (see, e.g., Ref. 91). Other fields alluded to already include collective games,[20] genetics and evolutionary theory,[13] and other aspects of population dynamics.[92,93] I cannot attempt a detailed survey of these subjects here, but in concluding this review I will attempt at least to indicate important points of connection with the Boltzmann equation studies covered so far, as well as important entry points into the relevant literature.

A. Aggregation-Fragmentation Processes

A vast area of both pure and applied science is concerned, in one way or another, with the kinetics of processes in which a "bimolecular" growth interaction competes with a "unimolecular" decay of the formed species. Such include nucleation processes in colloid and aerosol science, polymer growth, and the more complex types of growth and transformation to be found in meteorology, astrophysics, and even cosmology, over a size range from atomic to astronomical dimensions. While these processes fall short of a real isomorphism with the gas-kinetic Boltzmann equation, they have nevertheless enough in common with it, particularly in its scalar forms, to invite the transfer of various solution techniques (for example, similarity solutions) and unifying ideas (for example, analogues of the H theorem).

Consider the prototype situation in which aggregates of i basic units of whatever sort grow by accretion in binary encounters and decay by fission into smaller fragments. If a statistically large ensemble of such units is considered, the balance equation for these processes can be written

$$\frac{\partial \bar{c}(i,t)}{\partial t} = \frac{1}{2} \sum_{j=1}^{i} \overline{K}(i-j,j)\bar{c}(i-j)\bar{c}(j) - \bar{c}(i) \sum_{j=1}^{\infty} \overline{K}(i,j)\bar{c}(j)$$

$$+ \sum_{j=1}^{\infty} \overline{F}(i,j-i)\bar{c}(j) - \tfrac{1}{2}\bar{c}(i) \sum_{j=1}^{i} \overline{F}(j,i-j) \qquad (7.1)$$

Here the function $\bar{c}(i,t)$ represents the concentration of i units per standard volume, while $\overline{K}(i,j)$ and $\overline{F}(i,j)$ represent, respectively, the transition rate for the fusion of an i unit with a j unit to form an aggregate $(i+j)$ unit and the fragmentation rate for decay of an $(i+j)$ unit into separate i and j units. This process may be written symbolically as $(i)+(j) \rightleftarrows (i+j)$.

The continuous analogue is easily written as

$$\frac{\partial c(x,t)}{\partial t} = \frac{1}{2} \int_0^x dy\, K(x-y, y) c(x-y) c(y) - c(x) \int_0^\infty dy\, K(x, y) c(y)$$
$$+ \int_x^\infty dy\, F(x, y-x) c(y) - \tfrac{1}{2} c(x) \int_0^x dy\, F(y, x-y) \qquad (7.2)$$

in the variable $x \in (0, \infty)$. This version differs fundamentally from the discrete case in lacking any interpretation of the monomer concentration $\bar{c}(1, t)$ and the all-important *monodisperse* initial condition $\bar{c}(i, 0) = 0$, $i \neq 1$.

It will be clear that, while Eqs. (7.1) and (7.2) bear a superficial resemblance to the Type B and B′ model Boltzmann equations in our earlier context, they must also differ radically in the nature of the implied boundary conditions and conservation laws. In particular, it is the total *mass* of material

$$M(t) = \int_0^\infty dx\, x c(x, t) = \sum_{i=0}^\infty i\bar{c}(i, t) \qquad (7.3)$$

which will, in general, be conserved rather than the total number or particles

$$N(t) = \int_0^\infty dx\, c(x, t) = \sum_{i=0}^\infty \bar{c}(i, t) \qquad (7.4)$$

The higher moments M_n,

$$\overline{M}_n(t) = \sum_{k=1}^\infty k^n \bar{c}(k, t) \qquad (7.5)$$

are also of obvious interest. The kinetic equation for $N(t)$, and in certain cases $M(t)$, referred to as the "macroscopic equations", are obtained by operating directly on the transport equations (7.1) and (7.2) and can be written

$$\frac{dN}{dt} = M - N - N^2 \qquad (7.6)$$

$$\frac{dM}{dt} = 0 \qquad (7.7)$$

The second condition, that of mass conservation, is fulfilled only in cases where M_2, the second moment, exists. The time-dependent quantity $N(t)$ is clearly important in its own right and enters various explicit solutions for the mass distribution $\bar{c}(i, t)$, which we shall derive later. The simplification of Eqs. (7.1) and (7.2) for the cases of pure aggregation ($F = 0$) and pure fragmentation ($K = 0$) is evident. The case of pure fragmentation involves only a linear time evolution and is thus outside the scope of this article. It is by no means a trivial problem, however, and some interesting new model solutions have recently been obtained in this area.[94]

In the field of quadratic transport associated with aggregation-fragmentation kinetics, much less scope seems to exist for the discovery of elaborate soluble models than is the case with scalar Boltzmann equations. In fact, significant progress has been achieved only with a handful of kernels, virtually limited to the following:

CASE 1

$$\bar{K}(i, j) = 1 \qquad (7.8)$$

CASE 2

$$\bar{K}(i, j) = i + j \qquad (7.9)$$

CASE 3

$$\bar{K}(i, j) = (ij)^{\omega} \qquad (7.10)$$

CASE 4

$$\bar{K}(i, j) = \left(\frac{i}{j}\right)^{\omega} + \left(\frac{j}{i}\right)^{\omega} \qquad (7.11)$$

(the latter cases usually with $\omega = 1$), and extensions beyond this simple pattern are very difficult. However, limited progress can be made with the model combining the first three in the form

$$\bar{K}(i, j) = \bar{\alpha} + \bar{\beta}(i + j) + \bar{\gamma} ij \qquad (7.12)$$

The corresponding continuous-variable versions are simple transcriptions. The symmetry $K(i, j) = K(j, i)$ is obviously an essential feature of the problem.

In order to make any progress at all, considerable simplifications are still necessary. The first case, (7.8), was first investigated by Smoluchoswki[95,96]; the second, (7.9), is particularly associated with Melzak,[97] Scott,[98] and Safronov.[99] The third, (7.10), is considered by Trubnikov,[100] who also considers (7.12). If (7.8) is combined with the equally simple fragmentation model $\bar{F}(i) = 1$, the result is the early model of Blatz and Tobolski[101] for reversible polymerization. These references contain a large part of what might be called the classical theory of aggregation-fragmentation, the early development of which is comprehensively documented in Ref. 91. After a brief sketch of what is involved, I shall be content to mark the points at which theories of quadratic transport in general have been particularly influenced by discoveries arising from, or at any rate closely linked with, methods well known in Boltzmann equation studies.

B. Pure Aggregation Models

If fragmentation is neglected ($\bar{F} = 0$), the pure aggregation equations still present formidable problems. The resulting equation

$$\frac{\partial \bar{c}(i,t)}{\partial t} = \frac{1}{2} \sum_{j=1}^{i-1} \bar{K}(i-j, j)\bar{c}(i-j)\bar{c}(j)$$

$$- \bar{c}(i) \sum_{j=1}^{\infty} \bar{K}(i, j)\bar{c}(j) \qquad (7.13)$$

is known as the Smoluchoswki equation, although Smoluchoswki[95] originally studied it only for the case $K(i, j) = 1$. Much interest still attaches to this special case. Credit for various solutions to the aggregation kernels (7.8) to (7.11) is divided among workers in the aerosol and polymer science fields as well as mathematicians who have considered the difficult problems of existence and uniqueness. Rather than attempt a historical review, it is convenient here to use Trubnikov's account of the composite kernel (7.12) as an introduction.[100]

Considering first the monodisperse initial condition $\bar{c}(i,0) = 0$ for $i \geq 2$, it is easy to show that the total number time dependence $N(t)$ can be obtained explicitly from the macroscopic equation

$$- \frac{d\hat{N}}{d\tau} = \alpha \hat{N}^2 + 2\beta \hat{N} + \gamma \qquad (7.14)$$

in which we have used the scalings $\tau = \frac{1}{2}N(0)t$ and $\hat{N}(t) = N(t)/N(0)$. This may be solved to give two cases

$$\hat{N}(t) = \frac{\Delta^{1/2} - (\beta + \gamma)\tanh(\tau \Delta^{1/2})}{\Delta^{1/2} + (\beta + \gamma)\tanh(\tau \Delta^{1/2})}, \qquad \Delta = \beta^2 - \alpha\gamma > 0 \quad (7.15)$$

or

$$\hat{N}(t) = \frac{|\Delta|^{1/2} - (\beta + \gamma)\tan(\tau |\Delta|^{1/2})}{|\Delta|^{1/2} + (\beta + \gamma)\tan(\tau |\Delta|^{1/2})}, \qquad \Delta = \beta^2 - \alpha\gamma < 0 \quad (7.16)$$

according to the relative values of the coefficients. If the following form is assumed for the concentrations $\bar{c}(i, t)$:

$$\bar{c}(i,t) = \Gamma_i \varphi(t) e^{-i\psi(t)} \tag{7.17}$$

it is possible to show that necessarily

$$\varphi = \frac{\alpha \hat{N}^2 + 2\beta \hat{N} + \gamma}{1 - \hat{N}} \tag{7.18}$$

and

$$\psi = \ln(1 - \hat{N})^{-1} - 2\left(\frac{\Delta}{\alpha}\right)t + \frac{\beta}{\alpha}\ln\left(\frac{\alpha + 2\beta + \gamma}{\alpha \hat{N}^2 + 2\beta \hat{N} + \gamma}\right) \tag{7.19}$$

The coefficients Γ_i must be obtained from the recursion relation

$$\Gamma_i = \frac{1}{1-i}\sum_{k=1}^{i-1}\Gamma_k\Gamma_{i-k}\left[\frac{K(k, i-k)}{K(1,1)}\right], \qquad \Gamma_1 = 1 \tag{7.20}$$

It appears difficult to provide a more transparent result for the general kernel (7.12). However, a number of special cases due to other workers can be seen to drop out. In particular:

CASE 1 ($\alpha \neq 0$, $\beta = \gamma = 0$) (Smoluchowski)

$$\Gamma_k = 1, \qquad \hat{N}(t) = \left(1 + \tfrac{1}{2}\alpha N(0)t\right)^{-1}$$
$$\bar{c}(i,t) = \hat{N}(t)^2[1 - \hat{N}(t)]^{i-1} \tag{7.21}$$

CASE 2 ($\beta \neq 0$, $\alpha = \gamma = 0$) (Safronov[99])

$$\Gamma_k = \frac{k^{k-1}}{k!}, \qquad \hat{N}(t) = \exp(-\beta\hat{N}(0)t)$$
$$\bar{c}(i,t) = \frac{i^{i-1}}{i!}\frac{\hat{N}(t)}{1 - \hat{N}(t)}\exp\{-i[1 - \hat{N} - \log(1 - \hat{N})]\} \tag{7.22}$$

CASE 3 ($\gamma \neq 0$, $\alpha = \beta = 0$) (McLeod[102])

$$\Gamma_k = \frac{(2k)^{k-1}}{k!k}, \qquad \hat{N}(t) = 1 - \tfrac{1}{2}\gamma N(0)t \qquad (t < t_c)$$
$$\bar{c}(i,t) = \frac{(2i)^{i-1}}{i!i(1-\hat{N})}\{(1-\hat{N})\exp[-2(1-\hat{N})]\}^i \tag{7.23}$$

or, on scaling by $\bar{t} = \gamma N(0)t$,

$$\bar{c}(i, \bar{t}) = \frac{i^{i-2}\bar{t}^{i-1}}{i!}e^{-i\bar{t}} \tag{7.24}$$

In Case 3, unlike the previous two, the solution is valid only for times less than a certain critical value t_c. In the present scaling, $t_c = 1$ and corresponds to the so-called gel point in polymer growth. We shall return to the question of gelation and the existence of solutions for $t \geq t_c$ in Section 7E.

When initial distributions are not monodisperse, transform methods are called for, and it is as a rule possible to obtain solutions for $\bar{c}(i, t)$ or $c(x, t)$ only to within a generating-function inversion. The three cases distinguished above are again of particular interest and show markedly different behavior.

Melzak[97] first showed that the Case 1 discrete variable model could be solved by a generating function in the form

$$G(i, t) = \sum_{k=1}^{\infty} z^k \bar{c}(k, t) \tag{7.25}$$

with the end conditions

$$G(1, t) = N(t), \qquad G_z(1, t) = M = 1 \tag{7.26}$$

The function G is in this case the factorial moment-generating function by virtue of the Taylor expansion about $z = 1$:

$$G(z, t) = \sum_{n=0}^{\infty} \frac{(z-1)^n}{n!}\bar{m}_n(t) \tag{7.27}$$

with

$$\bar{m}_n(t) = \sum_{k=n}^{\infty} k^{(n)}\bar{c}(k, t) \tag{7.28}$$

Substitution into (7.13) leads to the differential equation

$$\frac{\partial G}{\partial t} = -G^2 - 2GN \tag{7.29}$$

which can be shown to have the general solution

$$G(z, t) = N(t) - \frac{1}{\phi(t) + t} \tag{7.30}$$

where $\phi(z)$ is an arbitrary function to be fitted from the initial condition $G(z,0)$. Using $N(t)$ from (7.21) with time scaled such that $\alpha = 2$, we find the explicit form of $G(z,t)$ to be

$$G(z,t) = \frac{\{1/[1+N(0)t]\}^2}{1/G(z,0)-t/[1+N(0)t]} \tag{7.31}$$

From this, all necessary information about $\bar{c}(i,t)$ can be recovered.

In the continuous-variable version of the same problem, the analogous equation

$$\frac{\partial c(x,t)}{\partial t} = \frac{1}{2}\int_0^n dy\, K(x-y,y)c(x-y)c(y) - c(x)\int_0^\infty dy\, K(x,y)c(y) \tag{7.32}$$

first considered by Schumann,[103] may be solved by a Laplace transform method in similar simplified cases. Defining

$$L(s,t) = \int_0^\infty dx\, e^{-xs}c(x,t) \tag{7.33}$$

Trubnikov[100] was able to show that, for the kernel analogous to (7.12),

$$K(x,y) = \alpha + \beta(x+y) + \gamma xy \tag{7.34}$$

the Laplace transform satisfies the partial differential equation

$$L_t = \alpha\left(\tfrac{1}{2}L^2 - NL\right) + \beta\left(NL_s - LL_s - L\right) + \gamma\left(\tfrac{1}{2}L_s^2 + L_s\right) \tag{7.35}$$

with $N(t)$ to be determined by solution of the macroscopic equation (7.14) in a form similar to (7.15).

Little can be done with this in general, however, and it is necessary to examine the three cases separately as before. Since Laplace inversions can nowadays be carried out numerically to great accuracy (see, e.g., Refs. 104 and 105), possession of the transform of the solution may as well be regarded as equivalent to an effective algorithm for the recovery of the mass spectrum $c(x,t)$ and all derived information. The only reason to examine the transforms analytically is that in some cases useful information about asymptotic behavior of the solution can be obtained by study of the poles of the integrand in the inversion integral. With this in mind we shall simply quote Trubnikov's results in the form of inverted transforms. Thus, if

$L_0(s) = \mathscr{L}\{c(X,0)\}$, we have in the three standard cases:

CASE 1 ($\alpha \neq 0$, $\beta = \gamma = 0$)

$$\hat{N}(t) = \left[1 + \tfrac{1}{2}\alpha N_0 t\right]^{-1}$$

$$c(x,t) = \hat{N}(t)^2 \mathscr{L}^{-1}\left\{\frac{L_0(s)}{1 - N_0^{-1}(1 - \hat{N})L_0(s)}\right\} \tag{7.36}$$

CASE 2 ($\beta \neq 0$, $\alpha = \gamma = 0$)

$$\hat{N}(t) = e^{-\beta t}$$

$$c(x,t) = \frac{1}{x}\left(\frac{\hat{N}(t)}{1 - \hat{N}(t)}\right)\mathscr{L}^{-1}\{\exp[x(1 - \hat{N})(L_0(s) - N_0)]\} \tag{7.37}$$

CASE 3 ($\gamma \neq 0$, $\alpha = \beta = 0$)

$$\hat{N}(t) = 1 - \frac{\gamma t}{2N_0} \qquad \left(t < \frac{2N_0}{\gamma}\right)$$

$$c(x,t) = \frac{1}{x^2 \gamma t}\mathscr{L}^{-1}\left\{\exp\left[-x\gamma t\left(1 + \frac{\partial L_0(s)}{\partial s}\right)\right]\right\} \tag{7.38}$$

The inverses in each case are to be understood as

$$\mathscr{L}^{-1}\{F(s)\} = \frac{1}{2\pi i}\int_{\sigma - i\infty}^{\sigma + i\infty} e^{sx}F(s)\, ds \tag{7.39}$$

in the usual way. Although not self-evidently so, the inverses in each case are in fact regular. The solution in Case 1 was originally found by Schumann[103] and Melzak.[97]

Various asymptotic results may be obtained by series expansion of the inversion integrands. It is found that

CASE 1

$$c(x,t) \underset{t \to \infty}{\sim} N(t)^2 e^{-\hat{N}(t)x} \tag{7.40}$$

CASE 2

$$c(x,t) \underset{t \to \infty}{\sim} \frac{N(t)}{2\pi\langle x^2\rangle_0}\left(\frac{\langle x\rangle_0}{x}\right)^{3/2}\exp\left(-x\frac{\hat{N}^2\langle x\rangle_0}{2\langle x^2\rangle_0}\right) \tag{7.41}$$

The behavior $c(x, t) \sim x^{-3/2}$ in Case 2 for small x is in agreement with the mass distribution for small objects in the solar system and is thought to have general astrophysical significance.

Lushnikoff[106] has investigated alternative power-series expansions for solving Eq. (7.32) without restrictions on the form of the function K. His most original contribution is to suggest scaling laws—in effect, similarity solutions—that may have considerable generality. His arguments show that provided that K is a homogeneous function in the sense that $K(\lambda i, \lambda j) = \lambda^p K(i, j)$, then the long-time behavior of $c(x, t)$ should be of the form

$$c(x, t) = t^{-2/(1-\lambda)}\psi\left(xt^{-1/(1-\lambda)}\right) \tag{7.42}$$

in which the function ψ is to be determind by a somewhat complicated integrodifferential equation, unfortunately comparable in complexity to the original transport equation. Lushnikoff also investigates the somewhat peculiar case 4 [Eq. (7.11)], obtaining recurrence relations for expansion coefficients, although not explicit transform solutions.

We have attempted to give an account of the essentially classical theory of aggregation kinetics, much of which predates by many years the growth of interest in scalar Boltzmann equations. It is already clear that several themes are common to both fields, the occurrence of convolution equations for expansion coefficients, similarity-type solutions, and the usefulness of generating function and transform techniques being obvious examples. Quite recently, however, certain even more direct mathematical connections with soluble Boltzmann models have been discovered, the implications of which are still being explored at the time of writing.

In the final sections we shall treat in broad outline two such areas in which very striking progress has recently been made, significantly enough by workers well versed in the theory of Boltzmann models. These are the derivation of new general solutions to aggregation-fragmentation equations of the Blatz-Tobolsky type and the study of the gel-point transition from the point of view of critical points in the solution of kinetic equations of type (7.1).

C. Aggregation-Fragmentation Models

Perhaps the most notable step toward the understanding of reversible polymerization is the demonstration of exact general solutions to the Blatz-Tobolski models in the recent work of Aizenman and Bak[107] and Ernst.[12] Although, as seems inevitable, the solutions found are in the form of Laplace transforms and generating functions, the knowledge of these represents a considerable advance on the various special solutions hitherto available and makes possible a systematic analysis of many subtle properties of the models.

We consider first the continuous variable case corresponding to the choice of kernels $K(x, y) = 2$, $F(x, y) = 2$. The latter assumption refers most naturally to simple polymer chains, for which the probability of fission should be proportional to the number of linked monomers. On simplifying (7.2), the transport equation is seen to read

$$\frac{\partial c(x, t)}{\partial t} = \int_0^x dy\, c(y) c(u - y) - 2c(x) \int_0^\infty dy\, c(y)$$
$$+ 2 \int_x^\infty dy\, c(y) - x c(x) \tag{7.43}$$

As with the pure aggregation equations, it is a relatively simple matter to solve the macroscopic rate equations for $M(t)$. Operating on (7.43) we obtain

$$\dot{N}(t) = M - N^2 \tag{7.44}$$
$$\dot{M}(t) = 0 \tag{7.45}$$

with the solutions

$$N(t) = \frac{1 - ce^{-2t}}{1 + ce^{-2t}} \tag{7.46}$$

with

$$c = \frac{1 - N_0}{1 + N_0} \leqslant 1$$

$$M(t) = \text{const.}, \qquad N(\infty) = 1$$

The stationary distribution conforming to $N(\infty) = 1$ can be seen to be

$$c(x, \infty) = e^{-x} \tag{7.47}$$

$N(\infty)$ may be approached from either above or below according to the relative initial value N_0, which may lead to either net polymerization or net depolymerization.

Aizenman and Bak proceed to the general solution by defining the Laplace transform

$$\gamma(y, t) = \int_0^\infty dx\, e^{-xy} c(x, t) \tag{7.48}$$

and [noting that $\gamma(0, t) = N(t)$] the auxiliary function

$$\chi(y, t) = \gamma(y, t) - N(t) \tag{7.49}$$

Transformation of the original equation then leads to the nonlinear partial differential equation

$$\frac{\partial \chi}{\partial t} - \frac{\partial \chi}{\partial y} = \frac{2\chi}{y} - \chi^2 - 1 \tag{7.50}$$

This in turn converts to a Ricatti equation on setting $\chi = y/(y+1) + g(y,t)^{-1}$. The solution follows in the form

$$g(y,t) = \frac{y^2 - 1}{2y^2} - e^{-2y}\left(\frac{y+1}{y}\right)^2 f(y+t) \tag{7.51}$$

with $f(y+t)$ an arbitrary function to be fitted by initial conditions. Given the transform of the initial data $\gamma(y,0)$, it emerges that the function f is fitted by

$$f(y) = \frac{e^{2y}[y/(y+1)]^2}{\gamma(y,0) - \gamma(0,0) + y/(y+1)} + \frac{1}{2}e^{2y}\frac{1-y}{1+y} \tag{7.52}$$

Equations (7.49), (7.51), and (7.52) then constitute a general solution to within a Laplace transform. There is no analytic method of inversion.

A later derivation by Ernst[12] leads to the essentially equivalent form

$$\chi(y,t) = \frac{y[e^{-2t} - \phi(y+t)]}{(y+1)\phi(y+t) + (z-1)e^{-2t}} \tag{7.53}$$

which we quote here because of its remarkable relationship to a Boltzmann equation model. According to Hellesøe (Ref. 12, Section 4.2), the function $\chi(y,t)$ can be put into simple relationship with the generating function of the continuous-variable VHP model of Section III.C. The required transformation is simply

$$\chi(y,t) = \frac{G(2y,2t) - 1}{G(2y,2t) + 1} \tag{7.54}$$

There is thus a genuine connection, rather than just a useful analogy, between the VHP model in kinetic theory and the Blatz-Tobolsky model in polymer kinetics. This is particularly revealing in terms of the relationship between uniqueness of solutions and conservation laws in each case.

The discrete-variable Blatz-Tobolsky transport equation takes the form

$$\frac{\partial \bar{c}(i,t)}{\partial t} = \sum_{j=0}^{i} \bar{c}(j)\bar{c}(i-j) - 2\bar{c}(i) \sum_{j=1}^{\infty} \bar{c}(j)$$

$$+ 2\sum_{j=i}^{\infty} \bar{c}(j) - i\bar{c}(i) \tag{7.55}$$

by obvious analogy with (7.43). Note that the macroscopic transport equations are different from those for the continuous case, being

$$\dot{N}(t) = M - N - N^2 \quad \text{and} \quad \dot{M}(t) = 0 \tag{7.56}$$

The stationary distribution is evidently

$$\bar{c}(i,\infty) = \beta^i \tag{7.57}$$

with β to be fixed by mass conservation. Thus,

$$M = \frac{\beta}{(1-\beta)^2}, \quad \beta < 1 \tag{7.58}$$

The number $N(t)$ then follows as

$$N(t) = \frac{\beta}{1-\beta}\left(\frac{D+e^{-\omega t}}{D-e^{-\omega t}}\right) \tag{7.59}$$

with

$$\omega = \frac{1+\beta}{1-\beta}, \quad N(\infty) = \frac{\beta}{1-\beta}$$

and

$$D = \frac{\beta N(0) + N(\infty)}{N(0) - N(\infty)}$$

The solution given by Ernst[12] to Eq. (7.55) follows a generating function method closely parallel to the continuous-variable solution. Defining

$$\bar{\chi}(z,t) = \sum_{k=1}^{\infty} z^k \bar{c}(k) - N(t) \tag{7.60}$$

and substituting back, we again obtain and solve a Riccati equation, with the final result

$$\bar{\chi}(z,t) = \frac{\beta(1-z)\left[\phi(ze^{-t})+e^{-\omega t}\right]}{(1-\beta)\left[(1-\beta z)\phi(ze^{-t})+(z-\beta)e^{-\omega t}\right]} \tag{7.61}$$

with $\phi(z)$ this time determined through the initial conditions

$$\frac{\phi(z)-1}{\phi(z)+1} = \frac{1-\beta}{1+\beta}\left(\frac{2M(1-z)+(1+z)\chi(z,0)}{(z-1)\chi(z,0)}\right) \tag{7.62}$$

A very thorough study of the properties of these soluions for various special initial conditions will be found in Ref. 12.

Though Aizenman and Bak[107] derive the general solution (7.61), their paper is mainly concerned with proofs of existence and uniqueness and the relationship of both to the mass-conservation law $\dot{M} = 0$. The latter proves to be intimately connected with the requirement that solutions decay no more slowly than $c(x,t) \sim x^{-3}$ or $\bar{c}(i,t) \sim i^{-3}$, without which the second moment of the distribution diverges and both mass conservation and uniqueness are lost. A precisely analogous effect occurs in the Boltzmann equation for the Ernst-Hendriks VHP model, as mentioned in Section VI.D; in fact, as already indicated, it was the discovery of this anomaly in the polymer kinetics case that first drew attention to and clarified the situation for the VHP equation.

Another original feature of Aizenman and Bak's paper is their proof of an "F theorem" analogous to the H theorem for the Boltzmann equation but governing instead the monotonic decay of the free-energy function $F(t)$ defined

$$F(t) = \int_0^\infty dx\, c(x,t)\left[\ln c(x,t)-1\right] = \sum_{i=1}^\infty \bar{c}(i,t)\left[\log c(i,t)-1\right] \tag{7.63}$$

The F theorem clarifies the interesting concept of the "most probable distribution," introduced many years ago by Flory.[108] This is obtained by minimizing F under the constraint that the macroscopic rate equations (7.56) must be satisfied. The most probable distribution has an interesting interpretation. In reversible polymerization reactions from general initial conditions in the discrete variable it represents that distribution which at a particular time t would have evolved from a hypothetical monodisperse initial condition at a negative time $t_0 < 0$, subject to the same transport equation. In general there is no guarantee that a given distribution $\bar{c}(i, t)$ at some time t could have evolved from any monodisperse condition $\bar{c}(i, 0) = \delta_{i0}$.

Very little progress has been made with reversible polymerization models other than the Blatz-Tobolsky case. In a recent paper, Barrow[109] combined aggregation kernels of the type (7.34) with a constant fragmentation kernel $F(i, k) = \eta$. This leads to a total number evolution of the form

$$\hat{N}(t) = \frac{\theta^{1/2} - (\beta + \gamma - p^2)\tanh(\theta^{1/2}\tau)}{\theta^{1/2} + (\beta + \alpha)\tanh(\theta^{1/2}\tau)} \qquad (7.64)$$

with

$$\theta = \beta^2 - \alpha\gamma + \alpha p^2, \qquad p^2 = \frac{\eta M(t)}{N(0)}$$

and α, β, γ having the same significance as in (7.34). Although in this way an extra term involving the parameter p can be added to Trubnikov's equation (7.35); very little else can be derived except some asymptotic mass distributions for special cases.

D. Aggregation with Deposition

A final topic of relevance in this section concerns the possibility of studying simultaneous aggregation and removal by some external agency. Williams has recently considered this by adding an additional term $- R(x)c(x, t)$ to the right-hand side of the continuous-variable aggregation equation (7.32).[110] In this kind of model, the origin of $R(x)$ may be either Stokes sedimentation or Brownian diffusion, or possibly some form of removal at a surface. Again similarity solutions are sought and obtained in the form of their Laplace transforms. Some simplification is possible by a series expansion that has a physical interpretation in terms of an accounting of successive coalescences by particles. This is particularly appropriate when the initial mass distribution for the problem is a gamma distribution. Much work on aerosol coagulation with deposition has been stimulated by nuclear technology.

E. Gelation Kinetics

The phenomenon of gelation, touched on earlier in connection with the solution (7.23), concerns the formation of an infinite cluster (the gel) at some finite time t_0 in the evolution of the system, after which the presence of finite-sized clusters (the sol) decreases asymptotically to zero. Beyond t_0 the conservation of mass $\dot{M} = 0$ breaks down as the sol increasingly loses material to the gel and there are corresponding singularities in the higher moments of $\bar{c}(i, t)$.

The theoretical importance of this phenomenon must be seen against the long and sometimes confusing history of the Flory-Stockmeyer theory of polymerization.[108,111] It will be recalled that in this approach, monomers with, in general, f functional groups link together at random, the probability of a linkage taking place being governed by a Poisson process with intensity p through which the element of time enters. Time remains, however, an implicit parameter, the natural description of the evolution being in terms of the degree of advancement of the reaction. For gelation to occur it is necessary that $f > 2$. This contrasts with the kinetic approach, introduced (although not originally for $f > 2$) by Dorstal and Raff[112] and based on kinetic equations of the type (7.1). Gelation from this point of view is to be seen through singularities in the solutions $\bar{c}(i, t)$ or their moments at the "gel point," with time an explicit parameter throughout.

More recently, lattice theories of percolation and computer simulations based on them have added a new element to the theory of gelation and provided a useful connection with general theories of phase transitions and critical phenomena.[113] In either the Flory-Stockmeyer or the kinetic approach, the onset of gel formation—in fact the possibility of a gel point at all—is extremely sensitive to the detailed assumptions made concerning the multiplicity of functional groups, the degree of cross-linking or ring formation allowed, and the possibility of sol-gel (as opposed to monomer-gel) interactions. Ziff and coworkers[114-115] in articles which, arguably, are the first to present a coherent account of the relation between random bonding and kinetic approaches, claim convincingly that only in the latter are the assumptions and controlling factors clearly exposed and the mathematical properties open to rigorous investigation. We shall not be able to discuss the physical justification of the various models here, but will briefly outline some of the very recent developments that have flowed from the use of kinetic equations to describe the gelation process.

As solution (7.21) indicates, occurrence of a gel point is likely to be associated with the presence of a term (ij) in the kernel K, and recent attention has concentrated on the model $K(i, j) = (ij)^\omega$ with $\omega > \frac{1}{2}$. There is some scope for modeling the steric characteristics of the growing aggregates by choice of ω; Hendriks, Ernst, and Ziff,[118] for example, postulate that $\omega = 1 - d^{-1}$ effectively models the gelation transition for real polymers in d dimensions, taking some account of the effects of cross-linking and steric hindrance neglected in the Flory-Stockmeyer theory. In general, one would hope to be able to deal with kernels of the form $K(i, j) = s_i s_j$, where s_i represents in some sense the active surface area available for polymerization, taking into account whatever geometrical and steric hindrance factors may apply. A few other cases such as $s_i = e^{\alpha(i-1)}$ have also been examined.[116,117]

The key papers in the recent development of kinetic models for gelation are those of Ziff,[114] Ziff and Stell,[115] Leyvraz and Tschudi,[119,120] and

Hendriks, Ernst, and Ziff.[118] Leyvraz and Tschudi in particular were able to demonstrate actual post-gelation solutions for $t > t_c$ and revived Lushnikov's earlier preoccupation with scaling laws in the context of the gelation transition. It is now possible to search for necessary and sufficient conditions for a kernel $K(i, j)$ to cause gelation at a finite t_c and to test whether there is any degree of universality in the critical exponents characterizing the approach to the gel-point singularity for different models.

The earliest and one of the few rigorous investigations of a gel-forming system is that of McLeod[102] on the kernel $K(i, j) = ij$. He showed that for kernels bounded above by ij there is a unique solution to (7.55) satisfying positivity and mass conservation in a finite interval $t \not> 1$, but could not extend this to the gel point $t_c = 1$ or the post-gelation regime. Quite recently, White[116,117] has proved the finiteness of all moments of kernels $K(i, j) < i + j$ [which includes the case $K(i, j) < (ij)^{1/2}$ also proven by Leyvraz and Tschudi[119]]. These authors, together with Hendriks, Ernst, and Ziff,[118] have published the most ambitious attempts so far to prove comprehensive results for a range of kernels and determine whether there are elements of universality in known results. Though their papers remain largely speculative, they are the clearest exposition so far of what seem to be the main points at issue.

As an example of an actual post-gelation solution, we may quote the solution found by Leyvraz and Tschudi[119] for the kernel $K(i, j) = ij$. They show that [cf. (7.24)]

$$\bar{c}(i, t) = \begin{cases} \dfrac{i^{i-2}}{i!} t^{i-1} e^{-it} & t \leqslant 1 \\[2mm] \dfrac{i^{i-2} e^{-i}}{i! t} & t \geqslant 1 \end{cases} \qquad (7.65)$$

The more complicated case $K(i, j) = (Ai + B)(Aj + B)$ related to (7.12) can also be solved and leads to a gelation transition at $t_c = 1/A(A + B)$. A partial solution to the post-gelation equation for the general kernel $K(i, j) = s_i s_j$ has also been given by Hendriks, Ernst, and Ziff, who show that

$$\bar{c}(i, t) = \bar{c}(i, t_c)[1 + b(t - t_c)]^{-1} \qquad (t > t_c) \qquad (7.66)$$

The gelation point t_c is, however, not determined, and the constant b must be obtained by the solution of recursion relations.

Progress in determining scaling laws and testing universality for classes of kernels is still largely at the stage of conjecture, though certain relationships between critical exponents have been identified. We shall quote the four conjectures of Leyvraz and Tschudi,[120] rewritten in our own notation. They are

I.   If gelation occurs, then for a fixed time t after gelation we have, for large i,

$$\bar{c}(i,t) \underset{i \to \infty}{\sim} i^{-\tau} \tag{7.67}$$

for some τ.

II. If $\omega > \frac{1}{2}$ and gelation takes place at a finite time t_c, then, for t near t_c but below it, we have

$$\bar{c}(i,t) \sim i^{-\omega-3/2}\exp\{G(iF(t))\} \tag{7.68}$$

where

$$F(t) \sim (t_c - t)^{\nu}, \qquad \nu = \frac{2}{2\omega - 1}$$

and thus

$$M_n(t) = \sum_{i=1}^{\infty} i^n\bar{c}(i,t) \sim (t_c - t)^{-\delta_n}$$

with

$$\delta_n = \frac{2(n - \omega) - 1}{2\omega - 1}$$

It is further conjectured that $G(x) \sim x^{\omega}$ for $x \ll 1$.

III. For $\omega > \frac{1}{2}$, Eq. (7.1) has solutions of the type

$$\bar{c}(i,t) \sim \frac{b_i}{t} \tag{7.69}$$

where

$$b_i \geqslant 0 \qquad \text{and} \qquad \sum_{i=1}^{\infty} ib_i < \infty$$

IV. Gelation occurs for $\omega > \frac{1}{2}$.

Somewhat similar conjectures have been made by Hendriks, Ernst, and Ziff for the same models. Although more rigorous proofs are certainly needed, there are sufficient results in their paper and the others cited to show that what universality exists in the gelation problem is somewhat restricted within each type of model in use. Thus it is definitely known that the "universality class" of the model $K(i, j) = (ij)^\omega$ is not the same as that of the Flory-Stockmeyer theory, which in turn is not that of the lattice percolation models of Stauffer. Justification for this will be found along with a much more detailed discussion of critical exponents in Ref. 118.

VIII. CONCLUSION

In the introduction I disclaimed any intention of surveying the literature on quadratic transport systematically and must again caution the reader that, while we have accumulated a sizable bibliography, it is by no means exhaustive. Emphasis throughout has deliberately been placed on those topics that unify the field, are productive of cross references, and, above all, place the subject in its rightful position as an interesting, and virtually new, chapter in stochastic process theory. If, in doing this, I have played down some of the physicist's traditional concern with Boltzmann's equation proper, this tendency is certainly compensated in other treatments, such as that in Ref. 11.

Perhaps the most appealing aspect of the subject treated here is that, while the unifying idea and its basic equations translate freely into widely differing contexts, each translation seems to bring new kinds of mathematical problems into play and is rarely just a simple shift of terminology.

Several aspects we have ignored altogether deserve at least a belated mention. The problem of spatially inhomogeneous solutions to model Boltzmann equations has not yet yielded results of comparable interest to those for the homogeneous case; nevertheless, some notable beginnings have been made by Nikol'skii[121] and Ruijgrok and Wu.[122] The same may be said for solutions with non-isotropic initial conditions, a subject recently examined by Hendriks and Nieuwenhuizen[123] and Vedenyapin.[124] We should also have given more explicit attention to the relevant but difficult work of Truesdell and Muncaster[5] in these connections. Further afield, the genetic models of Moran[13] and the population-dynamics models of Kesten[92,93] deserve more equal treatment, but we have been unable to digest these topics into a form suitable for brief presentation.

In preparing this review, published literature up to about June 1982 has been considered, and certain preprint material available at that time has also been taken into account.

NOTE ADDED IN PROOF (DECEMBER 1983)

Since this review was completed several further works from outside the main literature of statistical physics have come to light which change the historical record outlined here in our introductory sections.

As first revealed to the physics community by C. Cercignani,[126] the Bobylev-Krook-Wu solution in velocity-space was first derived not as hitherto accepted in 1976 but in the unpublished M.S. Thesis of R. S. Krupp presented in 1967 in the M.I.T. Department of Aeronautics and Astronautics.[127] Krupp used the Fourier transformed equation in inverse velocity space to arrive at the equivalent of our Equation (4.47). His method, using Hadamard's representation theorem for the transform and making a single-pole approximation, is less heuristic than Bobylev's, more elegant than Krook and Wu's.

A second notable literature discovery is that of the existence of numerous papers on the topic of "Dispersed phase mixing" in chemical engineering applications. These deal with theory and experiment of the processes of coalescence and redisipersion of droplets of immiscible liquid phases in well-stirred reactors. A good introduction to this lesser-known variety of quadratic transport is the paper by R. L. Curl.[128] A subsequent paper by Bajpai, Ramkrishna, and Prokop[129] includes various models for the coalescence/redispersion process one of which effectively anticipates the Ernst-Hendriks VHP model treated here in Section III.C. Interest in dispersed phase mixing appears, however, to be confined to steady-state distributions and few results have been obtained on transient behaviour.

Acknowledgments

I am indebted to Professor H. Haken and the Institut für Theoretische Physik, Stuttgart, for their hospitality during the period in which my interest in the subject was kindled. I am particularly grateful to M. H. Ernst and E. M. Hendriks of the Institut voor Theoretische Fysica, Utrecht, for collaboration and stimulus and for generously making available notes and preprints.

APPENDIX

We shall collect here the main formulas and definitions used in the text that involve finite difference calculus and the simpler special functions. More comprehensive background may be obtained in Refs. 34 and 124.

1. Finite-Difference Functions

$$i! = i(i-1)\cdots 3\cdot 2\cdot 1 \qquad (A1.1)$$

$$i!! = i(i-2)\cdots 5\cdot 3\cdot 1 \qquad (A1.2)$$

$$i^{(p)} = i(i-1)\cdots(i-p+1) \qquad (A1.3)$$

$$(i)_p = i(i+1)\cdots(i+p-1) \qquad (A1.4)$$

$$\binom{i}{k} = \frac{i!}{(i-k)!k!} \tag{A1.5}$$

$$(-i)_k = (-1)^k \binom{i}{k} k! \tag{A1.6}$$

2. Continuous Functions

$$\Gamma(p) = \int_0^\infty dx\, x^{p-1} e^{-x} \tag{A2.1}$$

$$B(p,q) = \frac{\Gamma(p)\Gamma(q)}{\Gamma(p+q)} = \int_0^1 dx\, x^{p-1}(1-x)^{q-1} \tag{A2.2}$$

$$(x)_k = \frac{\Gamma(x+k)}{\Gamma(x)} \tag{A2.3}$$

$$x^{(k)} = \frac{\Gamma(x)}{\Gamma(x-k+1)} \tag{A2.4}$$

3. Series Formulas

$${}_0F_1(a;x) = \sum_{\nu=0}^\infty \frac{x^\nu}{(a)_\nu \nu!} \tag{A3.1}$$

$${}_1F_1(a,c;x) = \sum_{\nu=0}^\infty \frac{(a)_\nu x^\nu}{(c)_\nu \nu!}$$

(Confluent hypergeometric function)   (A3.2)

$${}_2F_1(a,b,c;x) = \sum_{\nu=0}^\infty \frac{(a)_\nu (b)_\nu x^\nu}{(c)_\nu \nu!}$$

(Gauss hypergeometric function)   (A3.3)

$$\sum_{i=0}^N \binom{N}{i} \alpha^i \beta^{N-i} = (\alpha+\beta)^N \qquad \text{(Positive binomial theorem)} \tag{A3.4}$$

$$\sum_{i=0}^\infty (i+1)_{p-1} a^i = \Gamma(p)(1-a)^{-p}$$

(Negative binomial theorem)   (A3.5)

$$\sum_{i=0}^N (i+1)_{p-1}(N-i+1)_{q-1} = B(p,q)(N+1)_{p+q-1}$$

(Vandermonde theorem)   (A3.6)

$$\sum_{i=0}^{N} \binom{N}{i} \alpha^{(i)} \beta^{(N-i)} = (\alpha + \beta)^{(N)}$$

(Vandermonde theorem)   (A3.7)

$$\sum_{i=0}^{\infty} \frac{(a)_i (b)_i}{(c)_i i!} = \frac{\Gamma(c)\Gamma(c-a-b)}{\Gamma(c-a)\Gamma(c-b)} \qquad \text{(Gauss' theorem)}$$

(A3.8)

4. Probability Distributions

(a) The Gamma Distribution

$$W_p(x) = \Gamma(p)^{-1} x^{p-1} e^{-x} \tag{A4.1}$$

Normalization: By (A2.1).
Moments:

$$\langle x^n \rangle = (p)_n \tag{A4.2}$$

(b) The Beta Distribution

$$W_{p,q}(x,u) = B(p,q)^{-1} \left[\frac{x^{p-1}(u-x)^{q-1}}{u^{p+q-1}} \right] \tag{A4.3}$$

Normalization: By (A2.2).
Moments:

$$\langle x^n \rangle = \frac{(p)_n u^n}{(p+q)_n} \tag{A4.4}$$

(c) The Negative Binomial Distribution

$$\overline{W}_p(i) = \Gamma(p)^{-1}(i+1)_{p-1}(1-\alpha)^p a^i \tag{A4.5}$$

Normalization: By (A3.5).
Moments:

$$\langle i^{(n)} \rangle = (p)_n \left(\frac{a}{1-a} \right)^n \tag{A4.6}$$

(d) The Negative Hypergeometric Distribution

$$\overline{W}_{p,q}(i,k) = B(p,q)^{-1}\left[\frac{(i+1)_{p-1}(k-i+1)_{q-1}}{(k+1)_{p+q-1}}\right] \qquad \text{(A4.7)}$$

Normalization: By (A3.6).

Moments:

$$\langle i^{(n)}\rangle = \frac{(p)_n k^{(n)}}{(p+q)_n} \qquad \text{(A4.8)}$$

5. Orthogonal Polynomials

(a) The Laguerre Polynomials

Definition:

$$L_n^{(p-1)}(x) = \frac{(p)_n}{n!}\,{}_1F_1(-n,p;x) \qquad \text{(A5.1)}$$

$$= \sum_{\nu=0}^{n}\frac{(-x)^\nu}{\nu!}\binom{n+p-1}{n-\nu} \qquad \text{(A5.2)}$$

$$= n!^{-1}x^{-(p-1)}e^x\left(\frac{d}{dx}\right)^n[x^{n+p-1}e^{-x}] \qquad \text{(A5.3)}$$

Orthogonality and normalization:

$$\int_0^\infty dx\, x^{p-1}e^{-x}L_n^{(p-1)}(x)L_m^{(p-1)}(x) = (n+1)_{p-1}\delta_{nm} \qquad \text{(A5.4)}$$

Generating function:

$$\sum_{k=0}^{\infty}L_k^{(p-1)}(x)t^k = (1-x)^{-1}\exp\left[\frac{xt}{x-1}\right] \qquad \text{(A5.5)}$$

Addition formula:

$$L_n^{(p+q-1)}(x+y) = \sum_{\nu=0}^{n}L_{n-\nu}^{(p-1)}(x)L_\nu^{(q-1)}(y) \qquad \text{(A5.6)}$$

Ladder-operator formula (Kogbetliantz formula):

$$\frac{(q)_n}{(p+q)_n}L_n^{(p+q-1)}(x) = \int_0^x dy\, W_{q,p}(y,x)L_n^{(q-1)}(y) \qquad \text{(A5.7)}$$

Laplace transform:

$$\mathscr{L}\left\{x^{p-1}L_n^{(p-1)}(x)\right\} = \int_0^\infty dx\, e^{-sx}x^{p-1}L_n(x) = \frac{(n+1)_{p-1}(s-1)^n}{s^{p+n}}$$

(A5.8)

Relation to Hermite polynomials:

$$L_n^{(-1/2)}(x) = \frac{(-1)^n}{n!2^{2n}}H_{2n}(x^{1/2})$$

(A5.9)

$$H_{2n}(x) = \frac{(-1)^n(2n)!}{n!}\,{}_1F_1\left(-n,\tfrac{1}{2};x^2\right)$$

(A5.10)

(b)   The Meixner Polynomials

Definition:

$$M_k(i,p;a) = {}_2F_1\left(-i,-k,p;1-a^{-1}\right)$$

(A5.11)

$$= \sum_{\nu=0}\binom{i}{\nu}\binom{k}{\nu}\frac{\nu!(1-a^{-1})^\nu}{(p)_\nu}$$

(A5.12)

$$= \frac{a^{-i-k}}{(p)_k(i+1)_{p-1}}\Delta_i^k\left[a^i(i-k+1)_{p+k-1}\right]$$

(A5.13)

Symmetry:

$$M_k(i,p;a) = M_i(k,p;a)$$

(A5.14)

Orthogonality and normalization:

$$\frac{(1-a)^p}{\Gamma(p)}\sum_{i=0}^\infty (i+1)_{p-1}a^i M_n(i,p;a)M_m(i,p,a) = \frac{n!\delta_{nm}}{a^n(p)_n}$$

(A5.15)

Generating function:

$$\frac{1}{\Gamma(p)}\sum_{i=0}^\infty (i+1)_{p-1}t^i M_k(i,p;a) = \frac{(1-t/a)^k}{(1-t)^{k+p}}$$

(A5.16)

Addition formula:

$$M_k(i,p+q;a) = \sum_{j=0}^i \overline{W}_{p,p}(j,i)M_{k-l}(i-j,p;a)M_l(j,q;a)$$

(A5.17)

Ladder-operator formula:

$$M_k(i, p+q; a) = \sum_{j=0}^{k} \overline{W}_{p,q}(j,k) M_i(j, p; a) \qquad (A5.18)$$

Continuous variable limit:

$$\lim_{\varepsilon_0 \to 0} M_i\left(\frac{x}{\varepsilon_0}, p; (\varepsilon_0 + 1)^{-1}\right) = {}_1F_1(-k; p; x) = \frac{n!}{(p)_n} L_n^{(p-1)}(x)$$
$$(A5.19)$$

Gottlieb polynomials:

$$l_k(i, a) = a^k M_k(i, 1; a) = a^k \sum_{\nu=0}^{k} \binom{k}{\nu}\binom{i}{\nu}(1 - a^{-1})^{\nu}$$
$$= a^{-i} k!^{-1} \Delta_i^k [a^k i^{(k)}] \qquad (A5.20)$$

REFERENCES

1. A. V. Bobylev, *Sov. Phys. Dokl.* **20**, 822 (1976).
2. A. V. Bobylev, *Sov. Phys. Dokl.* **21**, 632 (1976).
3. M. Krook and T. T. Wu, *Phys. Rev. Lett.* **36**, 1107 (1976).
4. M. Krook and T. T. Wu, *Phys. Fluids* **20**, 1589 (1977).
5. C. Truesdell and R. G. Muncaster, *Fundamentals of Maxwell's Kinetic Theory of a Simple Monatomic Gas*, Academic Press, New York, 1980.
6. M. Kac, *3rd Berkeley Symposium on Mathematics, Statistics and Probability*, University of California Press, 1955, p. 71. (Reprinted in *Probability Number Theory and Statistical Physics*, K. Baclawski and M. D. Donsker, Eds., M.I.T. Press, Cambridge, MA, 1979, p. 388.)
7. P. A. P. Moran, *Proc. Camb. Phil. Soc.* **57**, 833 (1961).
8. S. Nishimura, *J. Appl. Probab.* **11**, 266 (1974).
9. S. Nishimura, *J. Appl. Probab.* **11**, 703 (1974).
10. H. P. McKean, *Arch. Ration. Mech. Anal.* **21**, 343 (1966).
11. M. H. Ernst, *Phys. Rep.* **78**, 1 (1981).
12. M. H. Ernst, "Exact Solutions of the Nonlinear Boltzmann Equation and Related Kinetic Equations," in *Studies in Statistical Mechanics*, E. W. Montroll and J. L. Lebowitz, Eds., North Holland, Amsterdam, 1983.
13. P. A. P. Moran, *The Statistical Processes of Evolutionary Theory*, Clarendon Press, Oxford, 1962.
14. L. Takacs, *Stochastic Processes*, translated by P. Zadar, Methuens Monographs on Applied Probability and Statistics, Methuen, London, 1960, Ch. 2.
15. N. G. Van Kampen, *Stochastic Processes in Physics and Chemistry*, North-Holland, Amsterdam, 1981.

16. L. Waldman, "Transporterscheinungen in Gasen von mittelerem Druck" Sect. 30, p. 348 in *Handbüch der Physik*, S. Flügge, Ed., Springer, Berlin, 1958, Vol 12.

17. K. E. Shuler, *J. Chem. Phys.* **32**, 1692 (1960).

18. C. C. Rankin and J. C. Light, *J. Chem. Phys.* **46**, 1305 (1967).

19. K. Koura, in *Eighth International Symposium on Rarefied Gas Dynamics, Stanford, 1972*, K. Karamcheti, Ed., Academic Press, New York, 1974, p. 487.

20. M. R. Hoare, to be published.

21. R. M. Ziff, S. D. Merajver, and G. Stell, *Phys. Rev. Lett.* **47**, 1493 (1981).

22. F. Schlögl, *Z. Physik* **198**, 559 (1967).

23. P. Résibois and M. de Leener, *Classical Kinetic Theory of Fluids*, Wiley, New York, 1977.

24. M. R. Hoare, *Adv. Chem. Phys.* **20**, 135 (1971).

25. M. R. Hoare, M. Rahman, and S. Raval, *Phil. Trans. Roy. Soc.*, **A305**, 383 (1982).

26. I. Kuščer and M. M. R. Williams, *Phys. Fluids* **10**, 1922 (1967).

27. R. D. Cooper and M. R. Hoare, *J. Stat. Phys.* **20**, 597 (1979).

28. R. D. Cooper, M. R. Hoare, and M. Rahman, *J. Math. Anal. Appl.* **61**, 262 (1977).

29. M. R. Hoare and M. Rahman, *Physica*, **97A**, 1 (1979).

30. J. C. Maxwell, *Phil. Trans. Roy. Soc.*, **157**, 49 (1866); **170**, 231 (1879).

31. G. W. Bluman and J. D. Cole, *Similarity Methods for Differential Equations* (Applied Mathematical Sciences, No. 13), Springer, Berlin, 1974.

32. G. Tenti and W. H. Hui, *J. Math. Phys.* **19**, 774 (1978).

33. J. Meixner, *J. Lond. Math. Soc.* **9**, 6 (1934).

34. T. S. Chihara, *An Introduction to Orthogonal Polynomials*, Gordon and Breach, London, 1978, p. 161.

35. M. H. Ernst and E. M. Hendriks, *Phys. Lett.*, **81A**, 315 (1981).

36. H. S. Wall, *Analytic Theory of Continued Fractions*, Van Nostrand, New York, 1967, p. 267.

37. T. Carleman, *Problemes mathématiques dans la théorie des gaz*, Almqvist and Wiksells, Uppsala, 1957.

38. H. P. McKean, *Commun. Pure Appl. Math.* **28**, 435 (1975).

39. H. P. McKean, *J. Comb. Theory* **2**, 358 (1967).

40. F. Henin and I. Prigogine, *Proc. Natl. Acad. Sci. USA* **71**, 2618 (1974).

41. E. W. Montroll and K. E. Shuler, *J. Chem. Phys.* **26**, 434 (1957).

42. M. J. Gottlieb, *Am. J. Math.* **60**, 453 (1938).

43. L. Landau and E. Teller, *Physik. Z. der Sowjetunion* **10**, 34 (1936).

44. M. H. Ernst and E. M. Hendriks, *Phys. Lett.* **70A**, 183 (1979).

45. S. Rouse and S. Simons, *J. Phys. A: Math. Gen.* **11**, 423 (1978).

46. H. Freeman, *Discrete Time Systems*, Wiley, New York, 1974, Ch. 3.

47. J. Tjon and T. T. Wu, *Phys. Rev.* **A19**, 883 (1979).

48. J. A. Tjon, *Phys. Lett.* **70A**, 369 (1979).

49. M. R. Hoare, *Mol. Phys.* **4**, 465 (1961).

50. E. Futcher, M. R. Hoare, E. M. Hendriks, and M. H. Ernst, *Physica* **101A**, 185 (1980).

51. E. M. Hendriks, M. H. Ernst, E. Futcher, and M. R. Hoare, *Physica* **101A**, 375 (1980).

52. M. R. Hoare, *J. Chem. Phys.* **41**, 2356 (1964).

53. M. H. Ernst, *Phys. Lett.* **69A**, 390 (1979).

54. M. H. Ernst in *Mathematical Problems in the Kinetic Theory of Gases* (Oberwolfach Conference 1979), D. C. Pack and H. Neunzert, Eds., *Methoden und Verfahren der Mathematischen Physik*, Band 19, Lang, Frankfurt am Main, 1980, p. 83.

55. E. J. Futcher and M. R. Hoare, *Phys. Lett.* **75A**, 443 (1980).

56. E. J. Futcher, "Idealised Models for Closed System Relaxation," Ph.D. thesis, University of London, 1980.

57. E. J. Futcher and M. R. Hoare, *Physica* **122A**, 516 (1983).

58. E. M. Hendriks and M. H. Ernst, *Physica* **112A**, 101 (1982).

59. M. Rahman, *SIAM J. Math. Anal.* **7**, 92, 386 (1976).

60. E. M. Hendriks and M. H. Ernst, *Physica* **112A**, 119 (1982).

61. C. Cercignani, *Theory and Application of the Boltzmann Equation*, Scottish Academic Press, Edinburgh, 1979.

62. R. G. Muncaster, *Arch. Ration. Mech. Anal.* **70**, 79 (1979).

63. N. Corngold and J. Mathews, private communication, 1978.

64. R. M. Ziff, G. Stell, and P. T. Cummings, *Physica* **111A**, 288 (1982).

65. H. Cornille and A. Gervois, *J. Stat. Phys.* **23**, 167 (1980).

66. E. H. Hauge and E. Prestgaard, *J. Stat. Phys.* **24**, 121 (1981).

67. Z. Alterman, K. Frankowski, and C. L. Pekeris, *Astrophys. J. Suppl.* **7**, 291 (1962).

68. H. Cornille, *J. Stat. Phys.* **23**, 149 (1980).

69. M. Barnsley and H. Cornille, *Proc. Roy. Soc. London* **A374**, 371 (1981).

70. A. V. Bobylev, *Sov. Phys. Dokl.* **25**, 257 (1980).

71. A. V. Bobylev, *Sov. Phys. Dokl.* **20**, 820 (1976).

72. M. H. Ernst and E. M. Hendriks, *Phys. Lett.* **81A** 371 (1981).

73. M. Barnsley and G. Turchetti, *Lett. Nuovo Cimento* **30**, 359 (1981).

74. R. M. Ziff, *Phys. Rev.* **A23**, 916 (1981).

75. M. Alexanian, *Phys. Lett.* **74A**, 1 (1979).

76. M. Barnsley and H. Cornille, *J. Math. Phys.* **21**, 1176 (1980).

77. M. Barnsley and G. Turchetti, *Phys. Lett.* **72A**, 417 (1979).

78. R. M. Ziff, *Phys. Rev.* **A24**, 509 (1981).

79. S. Harris, *J. Math. Phys.* **8**, 2407 (1967).

80. P. L. Bhatnagar, E. P. Gross, and M. Krook, *Phys. Rev.* **94**, 511 (1954).

81. K. Olaussen, *Phys. Rev.* **A25**, 3393 (1982).

82. E. H. Lieb, *Phys. Rev. Lett.* **48**, 1057 (1982).

83. S. Bernstein, *Acta Math.* **52**, 1057 (1928).

84. M. F. Barnsley and G. Turchetti, *Nuovo Cimento* **65B**, 1 (1981).

85. G. Uhlenbeck and G. Ford, *Lectures in Statistical Mechanics* (Proceedings Seminar, Boulder, Colorado, 1960), Am. Math. Soc., Providence, R.I. 1963, p. 88.

86. H. Cornille and A. Gervois, *Phys. Lett.* **79A**, 291 (1980).

87. H. Cornille and A. Gervois, *Compt Rend. Acad. Sci. Paris* **291B**, 101 (1980).

88. H. Cornille and A. Gervois, *J. Stat. Phys.* **26**, 181 (1981).

89. H. Cornille and A. Gervois, *Physica* **113A**, 559 (1982).

90. M. H. Ernst, K. Hellesøe, and E. H. Hauge, *J. Stat. Phys.* **27**, 677 (1982).

91. R. L. Drake, in *Topics in Current Aerosol Research*, Vol III, Pt. 2, G. M. Hidy and J. R. Brock, Eds., Pergamon, Oxford, 1972.

92. H. Kesten, *Adv. Appl. Probab.* **2**, 1 (1970).

93. H. Kesten, *Adv. Appl. Probab.* **2**, 179 (1970).

94. M. R. Hoare, to be published.

95. M. v. Smoluchowski, *Physik Z.* **17**, 585 (1916).

96. M. v. Smoluchowski, *Z. Phys. Chem.* **92**, 129 (1917).

97. Z. A. Melzak, *Quart. J. Appl. Math.* (*Oxford*) **11**, 231 (1953).

98. W. T. Scott, *J. Atmos. Sci.* **25**, 54 (1968).

99. V. S. Safronov, *Sov. Phys. Dokl.* **7**, 967 (1963).

100. B. A. Trubnikov, *Sov. Phys. Dokl.* **16**, 124 (1971).

101. P. L. Blatz and A. V. Tobolsky, *J. Phys. Chem.* **49**, 77 (1945).

102. J. B. McLeod, *Quart. J. Appl. Math.* (*Oxford*) **13**, 119; 193 (1962).

103. T. E. W. Schumann, *Quart. J. Roy. Met. Soc.* **66**, 195 (1940).

104. H. Dubner and J. Abate, *J. Am. Soc. Comp. Manuf.* **15**, 115 (1968).

105. F. Durbin, *Comp. J.* **17**, 371 (1974).

106. A. A. Lushnikov, *J. Colloid Interfacial Sci.* **45**, 549 (1973).

107. M. Aizenman and T. A. Bak, *Commun. Math. Phys.* **65**, 203 (1979).

108. P. J. Flory, *J. Am. Chem. Soc.* **58**, 1877 (1936).

109. J. D. Barrow, *J. Phys. A: Math. Gen.* **14**, 729 (1981).

110. M. M. R. Williams, *J. Phys. A: Math. Gen.* **14**, 2037 (1981).

111. W. H. Stockmayer, *J. Chem. Phys.* **11**, 45 (1943): **12**, 125 (1944).

112. H. Dorstal and R. Raff, *Z. Physik. Chem.* **32B**, 117 (1936).

113. D. Stauffer, *Phys. Rep.* **54**, 1 (1979).

114. R. M. Ziff, *J. Stat. Phys.* **23**, 241 (1980).

115. R. M. Ziff and G. Stell, *J. Chem. Phys.* **73**, 3492 (1980).

116. W. White, *J. Colloid Interfacial Sci.* **87**, 204 (1982).

117. W. White, *Proc. Am. Math. Soc.* **80**, 273 (1980).

118. E. M. Hendriks, M. H. Ernst, and R. M. Ziff, *J. Phys. A: Math. Gen.* **15**, L743 (1982).

119. F. Leyvraz and H. R. Tschudi, *J. Phys. A: Math. Gen.* **14**, 3389 (1981).

120. F. Leyvraz and H. R. Tschudi, *J. Math. Phys. A: Math. Gen.* **15**, 1951 (1982).

121. A. A. Nikol'skii, *Sov. Phys. Dokl.* **8**, 633 (1964).

122. Th. W. Ruijgrok and T. T. Wu, *Physica* **113A**, 401 (1982).

123. E. M. Hendriks and T. M. Nieuwenhuizen, *J. Stat. Phys.* **29**, 591 (1982).

124. V. V. Vedenyapin, *Sov. Phys. Dokl.* **26**, 26 (1981).

125. A. Erdelyi, Ed. *Higher Transcendental Functions*, Vols I and II, McGraw-Hill, New York, 1953.

126. M. H. Ernst (Private communication).

127. R. S. Krupp "A Nonequilibrium Solution to the Fourier Transformed Boltzmann Equation", M. Sc. thesis, M.I.T. (1967).

128. R. L. Curl, *A.I.Ch.E. Journal* **9**, 175 (1963).

129. R. K. Bajpai, D. Ramkrishna and A. Prokop, *Chem. Eng. Sci.* **31**, 913 (1976).

THE STATISTICAL MECHANICS OF THE
ELECTRICAL DOUBLE LAYER

STEVEN L. CARNIE

Department of Chemistry
University of Toronto, Toronto, Ontario, Canada

GLENN M. TORRIE

Department of Mathematics and Computer Science
Royal Military College, Kingston, Ontario, Canada

CONTENTS

> *Here the boundaries meet and all*
> *Contradictions exist side by side*
>
> Dostoyevsky, *The Brothers Karamazov*

I. INTRODUCTION

The term "double layer" is used in its widest sense to refer to any surface region in which there is a separation of charge. At electrolyte/solid interfaces a major component of this charge separation can arise from the response of the free charges in the electrolyte to the electric field produced by a charge that is bound to the surface. This inhomogeneous region of electrolyte near a charged surface, which we shall henceforth call the *double layer*, is the subject of this review.

Double layers are of great importance in the fields of electrochemistry (the electrode/electrolyte interface[1]), colloid science and soil chemistry (the role of double layers in promoting colloidal stability[2]), and physiology and biophysics (electrolyte adjacent to biological membranes[3]). Such radically different systems will have quite different chemical processes occurring at the surface. However, they exhibit certain common features associated chiefly with the presence of a charged surface and the consequent presence of double layers. The simplest and most well-characterized experimental system, the liquid mercury/aqueous electrolyte interface, has yielded the most detailed and reproducible measurements of double layer properties[4] and has become the traditional exemplar of double layer behavior. Yet even this system shows behavior not found for aqueous electrolytes against other metal surfaces[5] or the mercury/nonaqueous electrolyte interface[6]. The details of ion-solvent, surface-solvent (e.g., solvent orientation), and ion-surface (e.g., specific ionic adsorption) interactions are clearly important for a quantitative description of double layer behavior covering the entire range of conditions of experimental interest.

The theoretical treatment of such complex effects in the immediate vicinity of the charged surface has taken two routes. In phenomenological theories, effects such as dielectric saturation, electrostriction, and ionic polarizability are treated by extending macroscopic constitutive relations to the microscopic regime.[7] The other approach is to describe the interactions through a discrete-state Hamiltonian (the molecules are constrained to have a small set of states, each of which has an associated energy[8,9].) Both approaches usually treat only the first monolayer of ions and solvents.

A more fundamental approach to the double layer problem was initiated by Buff and Stillinger,[10] who started from the exact equations of statistical mechanics applied to a molecular Hamiltonian model of the electrolyte and surface. The past ten years or so have seen the resurgence of this kind of theoretical treatment, with several new theories being developed for the double layer. This is the approach that we adopt in this review, and only the results of such theories are discussed here.

By a molecular Hamiltonian model we mean one in which the separation and orientation dependence of all intermolecular and external potentials are explicitly specified. For the aqueous electrolyte/mercury surface, even this is a daunting task. In order to proceed with such a program, we must simplify the Hamiltonian while retaining the essential features. Of course, comparison with experimental results for any particular system will suffer as a result.

Solvent effects are important determinants of double layer behavior for certain experimental conditions, but the explicit treatment of solvent molecules produces formidable difficulties in concocting accurate interaction potentials, not to mention purely computational difficulties. As a first step toward a molecular theory of the double layer, such explicit solvent effects must be neglected.

Progress can be made thereafter by using a McMillan-Mayer level Hamiltonian model for the electrolyte. Just as in bulk electrolyte theory,[11] solvent coordinates are averaged over to produce a Hamiltonian that contains only ionic coordinates. The solvent-averaged ion-ion and ion-surface potentials are, strictly speaking, temperature- and ion-concentration-dependent effective potentials. The solvent is parameterized solely by its dielectric constant ε. No attempt is made to incorporate nonlinear dielectric effects. A brief discussion of results for models containing solvent molecules explicitly is given in Section VIII, but the bulk of this review is devoted to McMillan-Mayer level double layers.

The most well-studied model of bulk electrolyte is the simple primitive model (PM) (hard spheres with embedded point charges) which shows solution thermodynamic behavior in qualitative agreement with experimental results on simple aqueous electrolytes.[12] This model has therefore become the usual representation of the electrolyte in double layer theory. By using such a model, we necessarily neglect ionic polarizability and hydration effects (except insofar as they appear through the choice of ion sizes). To be consistent with such a simple electrolyte, we model the surface as an impenetrable polarizable wall with uniform surface charge density. While losing the more realistic features of any actual surface, this choice of model allows us to test unambiguously several theories of the double layer against Monte Carlo simulations for the same model.

In this review we present the fundamental statistical mechanical theories of the double layer, starting with the work of Buff and Stillinger on the point-ion model and culminating in several theories of the primitive model double layer. The role of the classical double layer theory of Gouy[13] and Chapman[14] is elucidated, and its regions of quantitative validity in the primitive model defined. The recent rapid growth of theoretical activity on this model has quickly brought the state of knowledge to a level comparable to that of bulk electrolytes at the same level of modeling. The next important steps in improving the model will require substantial computational effort. Now, then, seems to be an appropriate time for the appearance of a summary of these recent achievements. Our discussion of how different approaches have fared for the relatively simple primitive model may serve as a guide for future efforts in theoretical treatment of more sophisticated models of the double layer.

II. FUNDAMENTAL EQUATIONS AND EXACT CONDITIONS

The statistical mechanical theory of the electrical double layer is one topic in the general theory of nonuniform fluids, its chief distinguishing features being the presence of long-range Coulomb interactions, noncentral image interactions, and the response to an external electric field. Although many of the treatments presented in this section are therefore applicable to situations such as the liquid-vapor interface, wetting of a solid surface, or gas adsorption, we shall focus on their application to the double layer problem. For discussion of these other topics, the reader is referred to recent review articles and books.[15-18]

A. The Hamiltonian

Given the restrictions described in Section I, the Hamiltonian can be written in terms of one- and two-body potentials referring to ionic coordinates:

$$\mathscr{H} = \sum_{\alpha=1}^{\nu} \sum_{i_\alpha} v_\alpha(\mathbf{r}_{i_\alpha}) + \frac{1}{2} \sum_{\alpha,\beta=1}^{\nu} \sum_{i_\alpha \neq j_\beta} u_{\alpha\beta}(\mathbf{r}_{i_\alpha}, \mathbf{r}_{j_\beta}) \tag{2.1}$$

$\alpha, \beta, \ldots, \nu$ denote ionic species, $v_\alpha(\mathbf{r}_{i_\alpha})$ is the one-body potential acting on an ion of species α at position $\mathbf{r}_{i_\alpha}$, and $u_{\alpha\beta}(\mathbf{r}_{i_\alpha}, \mathbf{r}_{j_\beta})$ is the two-body potential acting between ions of species α and β at positions $\mathbf{r}_{i_\alpha}$ and $\mathbf{r}_{j_\beta}$, respectively. We do not specialize to central two-body potentials, so as to include image potentials.

The simplest contributions are due to short-range interactions: dispersion forces, short-range repulsive forces, and the like. We denote by $v_\alpha^s(r)$ the short-range interaction between the ion and the surface, for example, a hard interaction that prevents ions from penetrating the surface, or a short-range attractive potential that may lead to specific adsorption. $u_{\alpha\beta}^s(r_{12})$ is the short-range interaction between ions of species α and β.

The electrostatic part of the Hamiltonian can be written

$$\mathcal{H}^E = \frac{1}{2} \sum_\alpha \sum_{i_\alpha} q_\alpha V(\mathbf{r}_{i_\alpha}) \tag{2.2}$$

where $V(r_{i_\alpha})$ is the potential at $\mathbf{r}_{i_\alpha}$ due to all charges. It obeys Poisson's equation and the usual electrostatic boundary conditions. This potential arises from three sources: free charges in the electrolyte, induced surface charge from polarization of the surface, and the uniform surface charge fixed at the surface.

The direct Coulomb interaction from all ions leads to a term

$$\mathcal{H}^C = \frac{1}{2} \sum_\alpha \sum_{i_\alpha} q_\alpha \sum_\beta \sum_{j_\beta \neq i_\alpha} \frac{q_\beta}{\varepsilon |\mathbf{r}_{i_\alpha} - \mathbf{r}_{j_\beta}|} \tag{2.3}$$

with ε the dielectric constant of the solvent. In Eq. (2.3) we have neglected the terms with $i_\alpha = j_\beta$, which lead to a constant self-energy term of no importance for our purpose.

In the presence of a dielectric discontinuity between the solvent and the surface at $z = 0$, fulfillment of the usual electrostatic joining conditions implies a polarization of the previously uniform surface charge. The potential produced at $\mathbf{r}_{i_\alpha}$ by this surface polarization has the relatively simple form $\sum_{\beta, j_\beta} V^I(\mathbf{r}_{i_\alpha}, \mathbf{r}_{j_\beta})$, with

$$V^I(\mathbf{r}_{i_\alpha}, \mathbf{r}_{j_\beta}) = \frac{q_\beta f}{\varepsilon |\mathbf{r}_{i_\alpha} - \mathbf{r}_{j_\beta}^*|} \tag{2.4}$$

V^I is the potential due to an image charge of magnitude $f q_\beta$,

$$f = \frac{\varepsilon - \varepsilon_w}{\varepsilon + \varepsilon_w} \tag{2.5}$$

at a position $\mathbf{r}_{j_\beta}^* = \mathbf{r}_{j_\beta} - 2z_{j_\beta}\hat{\mathbf{z}}$ (see Fig. 1). The image contribution to the Hamiltonian then becomes

$$\mathcal{H}^I = \frac{1}{2} \sum_\alpha \sum_{i_\alpha} q_\alpha \sum_\beta \sum_{j_\beta} \frac{q_\beta f}{\varepsilon |\mathbf{r}_{i_\alpha} - \mathbf{r}_{j_\beta}^*|}$$

$$= \frac{1}{2} \sum_\alpha \sum_{i_\alpha} \sum_\beta \sum_{j_\beta \neq i_\alpha} \frac{q_\alpha q_\beta f}{\varepsilon |\mathbf{r}_{i_\alpha} - \mathbf{r}_{j_\beta}^*|} + \sum_\alpha \sum_{i_\alpha} \frac{q_\alpha^2 f}{4\varepsilon z_{i_\alpha}} \qquad (2.6)$$

The first term in Eq. (2.6) is due to the interactions of the ions with images of all other ions; the second term can be formally written as a set of one-body potentials and is due to the interaction of each ion with its own image (self-image interactions). The happy fact that in plane geometry the surface polarization can be represented by one-body and noncentral pair interactions enables us to use a Hamiltonian of the form given in Eq. (2.1); for a spherical surface, for example, we are not so fortunate.[19] The one- and two-body interactions are closely intertwined; this fact becomes significant in Section II.C, because $\mathcal{H}^I$ is not of the form usually assumed in some applications of density functional theory (i.e., one cannot arbitrarily vary the one-body part of $\mathcal{H}^I$ while keeping the two-body part constant). The sign of the self-image interactions depends solely on f. For a surface of low dielectric constant relative to ε (e.g., air/water, hydrocarbon/water), $f \sim 1$, whereas for a metal surface ($\varepsilon_w \to \infty$) $f = -1$. These two limiting cases are referred to as repulsive and attractive images, respectively, because of the sign of both the self-image force and, as we shall see, the image component of the mean force.

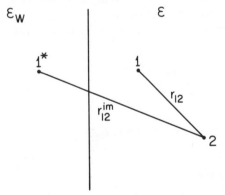

Fig. 1.   Geometry for the ion-image interaction.

Finally, the uniform surface charge density σ gives rise to a potential

$$V^{\sigma}(z_1) = -\frac{4\pi\sigma}{\varepsilon+\varepsilon_w}|z_1| = -2\pi\sigma|z_1|\frac{1+f}{\varepsilon} \tag{2.7}$$

This may be most easily derived by computing the field due to a sheet of charge of periodic density and taking the long-wavelength limit.[20] It is equivalent to the potential due to the sheet of surface charge σ plus the potential due to the image sheet of the surface charge.

In summary, it is convenient in what follows to separate the Coulombic, image, and short-range contributions to the pair interaction

$$u_{\alpha\beta}(\mathbf{r}_1,\mathbf{r}_2) = q_{\alpha}q_{\beta}u^C(r_{12}) + q_{\alpha}q_{\beta}u^I(\mathbf{r}_1,\mathbf{r}_2) + u_{\alpha\beta}^S(r_{12})$$

$$\equiv q_{\alpha}q_{\beta}u^E(\mathbf{r}_1,\mathbf{r}_2) + u_{\alpha\beta}^S(r_{12}) \tag{2.8}$$

and the surface charge, self-image, and short-range contributions to the one-body potential

$$v_{\alpha}(\mathbf{r}_1) = q_{\alpha}v^C(z_1) + q_{\alpha}^2 v^I(z_1) + v_{\alpha}^S(z_1) \tag{2.9}$$

For the simple model that we refer to for most of this review—the primitive model electrolyte against a hard, planar, uniformly charged surface—the Hamiltonian has the following form:

$$u_{\alpha\beta}^S(r_{12}) = \begin{cases} \infty & r_{12} < \tfrac{1}{2}(d_{\alpha}+d_{\beta}) \\ 0 & r_{12} > \tfrac{1}{2}(d_{\alpha}+d_{\beta}) \end{cases} \tag{2.10}$$

where d_{α}, d_{β} are the diameters of the hard-sphere ions of species α and β;

$$u^C(r_{12}) = \frac{1}{\varepsilon r_{12}} \tag{2.11}$$

and

$$u^I(\mathbf{r}_1,\mathbf{r}_2) = \frac{f}{\varepsilon r_{12}^{\text{im}}} \tag{2.12}$$

where

$$r_{12}^{im} = \left(r_{12}^2 + 4z_1 z_2 \right)^{1/2} \tag{2.13}$$

$$v_\alpha^S(z_1) = \begin{cases} \infty & z_1 < d_\alpha/2 \\ 0 & z_1 > d_\alpha/2 \end{cases} \tag{2.14}$$

$$v^I(z_1) = \frac{f}{4\varepsilon z_1} \tag{2.15}$$

$$v^C(z_1) = -2\pi\sigma z_1 \frac{1+f}{\varepsilon} \tag{2.16}$$

The term "restricted primitive model" (RPM) denotes the still simpler model obtained from Eqs. (2.10)–(2.16) by setting $d_\alpha = d$ for all α, that is, all ions have the same size.

B. Distribution Functions and Electrostatic Potentials

Once the system is specified through $\varepsilon, f, \sigma, \{ q_\alpha, \rho_\alpha, d_\alpha \}$, and the temperature T, the statistical mechanical problem is to determine the structure and thermodynamics of the interface. We shall be content with the amount of information that is provided by the one- and two-particle distribution functions, from which we can obtain all surface thermodynamic properties of the interface.

The one-particle density is given by

$$\rho_\alpha(\mathbf{r}_1) = \left\langle \sum_{i_\alpha} \delta(\mathbf{r}_1 - \mathbf{r}_{i_\alpha}) \right\rangle \tag{2.17}$$

and the two-particle density by

$$\rho_{\alpha\beta}(\mathbf{r}_1, \mathbf{r}_2) = \left\langle \sum_{i_\alpha} \delta(\mathbf{r}_1 - \mathbf{r}_{i_\alpha}) \sum_{j_\beta} \delta(\mathbf{r}_2 - \mathbf{r}_{j_\beta}) \right\rangle - \rho_\alpha(\mathbf{r}_1) \delta_{\alpha\beta} \delta(\mathbf{r}_1 - \mathbf{r}_2) \tag{2.18}$$

where $\langle \; \rangle$ denotes the ensemble average in the grand canonical (μ, V, T) ensemble.[21]

At large distances from the surface, the ionic densities $\rho_\alpha(\mathbf{r}_1)$ tend to their bulk values ρ_α. The bulk solution is neutral, so that

$$\sum_\alpha q_\alpha \rho_\alpha = 0 \tag{2.19}$$

From $\rho_\alpha(\mathbf{r}_1)$ and $\rho_{\alpha\beta}(\mathbf{r}_1, \mathbf{r}_2)$, we define the following quantities.

(i)

$$\frac{\rho_\alpha(\mathbf{r}_1)}{\rho_\alpha} \equiv g_\alpha(\mathbf{r}_1) \equiv 1 + h_\alpha(\mathbf{r}_1) \tag{2.20}$$

We shall refer to $\rho_\alpha(\mathbf{r}_1)$ or $g_\alpha(\mathbf{r}_1)$ as ionic density profiles and to $h_\alpha(\mathbf{r}_1)$ as the ion-wall total correlation function. $h_\alpha(\mathbf{r}_1)$ reflects the relative deviation in the local density of species α from its bulk value. Clearly,

$$h_\alpha(\mathbf{r}_1) \to 0, \quad z_1 \to \infty. \tag{2.21}$$

(ii)

$$\rho_{\alpha\beta}^T(\mathbf{r}_1,\mathbf{r}_2) \equiv \rho_{\alpha\beta}(\mathbf{r}_1,\mathbf{r}_2) - \rho_\alpha(\mathbf{r}_1)\rho_\beta(\mathbf{r}_2) \tag{2.22}$$

$$\equiv \rho_\alpha(\mathbf{r}_1)h_{\alpha\beta}(\mathbf{r}_1,\mathbf{r}_2)\rho_\beta(\mathbf{r}_2) \tag{2.23}$$

$$\equiv \rho_\alpha(\mathbf{r}_1)\left[g_{\alpha\beta}(\mathbf{r}_1,\mathbf{r}_2)-1\right]\rho_\beta(\mathbf{r}_2) \tag{2.24}$$

$\rho_{\alpha\beta}^T(\mathbf{r}_1,\mathbf{r}_2)$ is called the truncated two-particle density, $g_{\alpha\beta}(\mathbf{r}_1,\mathbf{r}_2)$ the pair correlation function, and $h_{\alpha\beta}(\mathbf{r}_1,\mathbf{r}_2)$ the ion-ion total correlation function. We must have

$$h_{\alpha\beta}(\mathbf{r}_1,\mathbf{r}_2) \to h_{\alpha\beta}^{\text{bulk}}(r_{12}), \quad z_1, z_2 \to \infty \tag{2.25}$$

where $h_{\alpha\beta}^{\text{bulk}}(r_{12})$ is the total correlation function of the bulk electrolyte.

For the present model (planar geometry and uniform surface charge density), there is cylindrical symmetry about the normal to the surface. We have

$$y_\alpha(\mathbf{r}_1) = y_\alpha(z_1), \quad y = \rho, g, \text{ or } h \tag{2.26}$$

$$y_{\alpha\beta}(\mathbf{r}_1,\mathbf{r}_2) = y_{\alpha\beta}(z_1, z_2, R), \quad y = \rho, \rho^T, g, \text{ or } h \tag{2.27}$$

where R is the transverse separation between $\mathbf{r}_1$ and $\mathbf{r}_2$,

$$R = \left[(x_1 - x_2)^2 + (y_1 - y_2)^2\right]^{1/2} \tag{2.28}$$

Two quantities that frequently arise in double layer theory are the mean electrostatic potential and the fluctuation potential. They are related to the distribution functions in the following way. The mean electrostatic potential $\psi(z_1)$ at a point z_1 is simply given by

$$\psi(z_1) = v^C(z_1) + \int d\mathbf{r}_2 \, u^E(\mathbf{r}_1,\mathbf{r}_2) \sum_\alpha q_\alpha \rho_\alpha(z_2) \tag{2.29}$$

The first term is the external potential due to the surface charge (and its image), the second term is the potential due to the distribution of charges in the double layer. From Eq. (2.29) we calculate the Maxwell field in the double layer:

$$
\begin{aligned}
E(z_1) &= -\frac{d}{dz_1}\psi(z_1) \\
&= \frac{2\pi\sigma}{\varepsilon}(1+f)+\frac{2\pi}{\varepsilon}\int_0^\infty dz_2\left[\mathrm{sgn}(z_1-z_2)+f\right]\sum_\alpha q_\alpha\rho_\alpha(z_2) \\
&= -\frac{4\pi}{\varepsilon}\int_{z_1}^\infty dz_2\sum_\alpha q_\alpha\rho_\alpha(z_2)
\end{aligned} \tag{2.30}
$$

where we have used the condition of overall electroneutrality of the interfacial region

$$
\int_0^\infty dz_2\sum_\alpha q_\alpha\rho_\alpha(z_2)=-\sigma \tag{2.31}
$$

Equation (2.30) is simply Poisson's equation, since $\sum_\alpha q_\alpha\rho_\alpha(z_1)$ is just the charge density in the double layer. It can be integrated to give

$$
\psi(z_1)=\frac{4\pi}{\varepsilon}\int_{z_1}^\infty dz_2(z_1-z_2)\sum_\alpha q_\alpha\rho_\alpha(z_2) \tag{2.32}
$$

The mean potential is thus a measure of the charge separation in the double layer. In particular, the surface potential

$$
\psi_0\equiv\psi(0)=-\frac{4\pi}{\varepsilon}\int_0^\infty dz_2\,z_2\sum_\alpha q_\alpha\rho_\alpha(z_2) \tag{2.33}
$$

is proportional to the dipole moment in the normal direction of the double layer and so provides a measure for the thickness of the double layer region where $\sum_\alpha q_\alpha\rho_\alpha(z_1)\neq 0$.

Suppose that instead of ensemble averaging over the positions of all ions to determine the mean potential at z_1, we average over the positions of all ions but one. This procedure defines the mean electrostatic potential at $\mathbf{r}_1$, given an ion of species α fixed at $\mathbf{r}_2$,

$$
\psi_\alpha(\mathbf{r}_1,\mathbf{r}_2)=v^C(\mathbf{r}_1)+q_\alpha u^E(\mathbf{r}_1,\mathbf{r}_2)+\int d\mathbf{r}_3\,u^E(\mathbf{r}_1,\mathbf{r}_3)\sum_\gamma q_\gamma\rho_\gamma(z_3)g_{\gamma\alpha}(\mathbf{r}_3,\mathbf{r}_2) \tag{2.34}
$$

The difference between $\psi_\alpha(\mathbf{r}_1,\mathbf{r}_2)$ and $\psi(z_1)$, that is, the change in potential at $\mathbf{r}_1$ on fixing an ion of species α at $\mathbf{r}_2$, is termed the fluctuation potential $\phi_\alpha(\mathbf{r}_1,\mathbf{r}_2)$

$$\phi_\alpha(\mathbf{r}_1,\mathbf{r}_2) \equiv \psi_\alpha(\mathbf{r}_1,\mathbf{r}_2) - \psi(z_1)$$

$$= q_\alpha u^E(\mathbf{r}_1,\mathbf{r}_2) + \int d\mathbf{r}_3 \, u^E(\mathbf{r}_1,\mathbf{r}_3) \sum_\gamma q_\gamma \rho_\gamma(z_3) h_{\gamma\alpha}(\mathbf{r}_3,\mathbf{r}_2) \quad (2.35)$$

Higher analogues of the fluctuation potential have been discussed by Outhwaite.[22]

The appropriate boundary conditions on $\psi(z_1)$ and $\phi_\alpha(\mathbf{r}_1,\mathbf{r}_2)$ follow from electrostatics. Both $\psi(z_1)$ and $\phi_\alpha(\mathbf{r}_1,\mathbf{r}_2)$ vanish as $z_1 \to \infty$. At the surface, the mean electrostatic potential satisfies (in electrostatic units)

$$\frac{d\psi(z_1)}{dz_1} = -\frac{4\pi\sigma}{\varepsilon}, \quad z_1 = 0 \quad (2.36)$$

which is equivalent to Eq. (2.31). The fluctuation potential satisfies the conditions of continuity at the surface

$$\phi_\alpha(\mathbf{r}_1,\mathbf{r}_2)|_{z_1=0^-} = \phi_\alpha(\mathbf{r}_1,\mathbf{r}_2)|_{z_1=0^+} \quad (2.37)$$

and the condition

$$\varepsilon_w \left.\frac{\partial\phi_\alpha(\mathbf{r}_1,\mathbf{r}_2)}{\partial z_1}\right|_{z_1=0^-} = \varepsilon \left.\frac{\partial\phi_\alpha(\mathbf{r}_1,\mathbf{r}_2)}{\partial z_1}\right|_{z_1=0^+} \quad (2.38)$$

Equations (2.36) and (2.38) combine to give the correct electrostatic boundary condition for $\psi_\alpha(\mathbf{r}_1,\mathbf{r}_2)$.

The knowledge of $g_\alpha(z_1)$, $\rho_{\alpha\beta}(\mathbf{r}_1,\mathbf{r}_2)$, and hence $\psi(z_1)$ then enables us to determine the thermodynamic properties of the interface. For example, the surface tension can be found from $g_\alpha(z_1)$ and $g_{\alpha\beta}(\mathbf{r}_1,\mathbf{r}_2)$,[10,23,24] and the surface adsorption excesses

$$\Gamma_\alpha \equiv \int_0^\infty dz_1 \, h_\alpha(z_1) \quad (2.39)$$

from the ionic profiles. From the $\psi_0 - \sigma$ relationship, one can find the change in surface tension on charging the interface:

$$\Delta\gamma = -\int_0^{\psi_0} \sigma(\psi') \, d\psi' \quad (2.40)$$

(where we have assumed $\psi_0 = 0$ at $\sigma = 0$) and the differential capacitance

$$C \equiv \frac{\partial \sigma}{\partial \psi_0} \tag{2.41}$$

For complete thermodynamic relationships between γ, Γ_α, σ, and ψ_0, we refer to other reviews of double layer thermodynamics.[7,25,26]

C. Exact Equations

Although there has been a recent flurry of interest in nonuniform fluids, most of the fundamental equations were known some twenty years ago. The BBGY hierarchy for nonuniform systems was applied to the double layer by Soviet workers.[27] The method employed by Kirkwood in the theory of bulk electrolytes[28] was generalized to nonuniform fluids and applied to the double layer by Buff and Stillinger.[10] The cluster expansion and functional derivative techniques, which led to the integral equation theories of simple fluids, were equally applicable to the nonuniform case.[29-32] The integral equations arising from functional differentiation have since come under the umbrella of density functional theory, the formalism stemming from the work of Kohn, Sham, and Hohenberg on the inhomogeneous electron gas.[33-36]

1. The BBGY Hierarchy

For a system at temperature T with a Hamiltonian of the form given by Eq. (2.1), the familiar BBGY hierarchy gives the following equations for $\rho_\alpha(\mathbf{r}_1)$ and $\rho_{\alpha\beta}(\mathbf{r}_1,\mathbf{r}_2)$[37]:

$$kT\nabla_1\rho_\alpha(\mathbf{r}_1) = -\rho_\alpha(\mathbf{r}_1)\nabla_1 v_\alpha(\mathbf{r}_1) - \int d\mathbf{r}_2 \sum_\beta \nabla_1 u_{\alpha\beta}(\mathbf{r}_1,\mathbf{r}_2)\rho_{\alpha\beta}(\mathbf{r}_1,\mathbf{r}_2)$$

$$\tag{2.42}$$

$$kT\nabla_1\rho_{\alpha\beta}(\mathbf{r}_1,\mathbf{r}_2) = -\rho_{\alpha\beta}(\mathbf{r}_1,\mathbf{r}_2)\left[\nabla_1 v_\alpha(\mathbf{r}_1) + \nabla_1 u_{\alpha\beta}(\mathbf{r}_1,\mathbf{r}_2)\right]$$

$$- \int d\mathbf{r}_3 \sum_\gamma \nabla_1 u_{\alpha\gamma}(\mathbf{r}_1,\mathbf{r}_3)\rho_{\alpha\beta\gamma}(\mathbf{r}_1,\mathbf{r}_2,\mathbf{r}_3) \tag{2.43}$$

Although exact, these equations are to some extent formal; an approximate closure (usually of unknown validity) is required to yield closed equations for the distribution functions.

Equation (2.42) provides an expression for the potential of mean force on an ion of species α, $W_\alpha(z_1)$.

$$kT\nabla_1\ln g_\alpha(z_1) = F_\alpha(z_1) = -\nabla_1 W_\alpha(z_1) \tag{2.44}$$

Using the definitions of the previous sections, we can rewrite Eq. (2.42) in the form $[\beta = (kT)^{-1}]$

$$
\begin{aligned}
\nabla_1 \ln g_\alpha(z_1) &= -\beta \nabla_1 W_\alpha(z_1) \\
&= -\beta q_\alpha \nabla_1 \psi(z) - \beta q_\alpha^2 \nabla_1 v^I(z_1) - \beta \nabla_1 v_\alpha^S(z_1) \\
&\quad - \beta \int d\mathbf{r}_2 \sum_\beta \nabla_1 u_{\alpha\beta}^S(\mathbf{r}_1, \mathbf{r}_2) \rho_\beta(z_2) g_{\alpha\beta}(\mathbf{r}_1, \mathbf{r}_2) \\
&\quad - q_\alpha \beta \int d\mathbf{r}_2 \nabla_1 u^E(\mathbf{r}_1, \mathbf{r}_2) \sum_\beta q_\beta \rho_\beta(z_2) h_{\alpha\beta}(\mathbf{r}_1, \mathbf{r}_2)
\end{aligned}
\tag{2.45}
$$

Using the definition of the fluctuation potential in Eq. (2.35), we can rewrite the first BBGY equation in the form

$$
\begin{aligned}
\nabla_1 \ln g_\alpha(z_1) &= -\beta q_\alpha \nabla_1 \psi(z_1) - \beta q_\alpha^2 \nabla_1 v^I(z_1) - \beta \nabla_1 v_\alpha^S(z_1) \\
&\quad - \beta \int d\mathbf{r}_2 \sum_\beta \nabla_1 u_{\alpha\beta}^S(\mathbf{r}_1, \mathbf{r}_2) \rho_\beta(z_2) g_{\alpha\beta}(\mathbf{r}_1, \mathbf{r}_2) \\
&\quad - \beta q_\alpha \lim_{\mathbf{r}_2 \to \mathbf{r}_1} \nabla_1 \left[\phi_\alpha(\mathbf{r}_1, \mathbf{r}_2) - q_\alpha u^E(\mathbf{r}_1, \mathbf{r}_2) \right]
\end{aligned}
$$

or, since

$$
\nabla_1 v^I(z_1) = \lim_{\mathbf{r}_2 \to \mathbf{r}_1} \nabla_1 u^I(\mathbf{r}_1, \mathbf{r}_2)
\tag{2.46}
$$

we have finally[38]

$$
\begin{aligned}
\nabla_1 \ln g_\alpha(z_1) &= -\beta q_\alpha \nabla_1 \psi(z_1) - \beta \nabla_1 v_\alpha^S(z_1) \\
&\quad - \beta \int d\mathbf{r}_2 \sum_\beta \nabla_1 u_{\alpha\beta}^S(\mathbf{r}_1, \mathbf{r}_2) \rho_\beta(z_2) g_{\alpha\beta}(\mathbf{r}_1, \mathbf{r}_2) \\
&\quad - \beta q_\alpha \lim_{\mathbf{r}_2 \to \mathbf{r}_1} \nabla_1 \left[\phi_\alpha(\mathbf{r}_1, \mathbf{r}_2) - q_\alpha u^C(r_{12}) \right]
\end{aligned}
\tag{2.47}
$$

2.   The Kirkwood Hierarchy

This hierarchy of equations is derived by introducing a coupling constant λ that couples one ion of species α with the rest of the system. The Hamiltonian thus takes the form

$$\mathcal{H}_\lambda = \mathcal{H}_0 + \lambda\left[v_\alpha(\mathbf{r}_{1_\alpha}) + \sum_{\beta,\,j_\beta} u_{\alpha\beta}(\mathbf{r}_{1_\alpha},\mathbf{r}_{j_\beta})\right] \tag{2.48}$$

and $\Omega(\lambda)$, $\rho_\alpha(\mathbf{r}_1|\lambda)$, $\rho_{\alpha\beta}(\mathbf{r}_1,\mathbf{r}_2|\lambda)$, etc., are the grand potential and distribution functions corresponding to the Hamiltonian $\mathcal{H}_\lambda$.

The first two equations in the resulting hierarchy are[10]

$$\frac{\partial}{\partial\lambda}\rho_\alpha(z_1|\lambda) = \rho_\alpha(z_1|\lambda)\left[\beta\frac{\partial\Omega(\lambda)}{\partial\lambda} - \beta v_\alpha(z_1)\right]$$

$$-\beta\int d\mathbf{r}_2\sum_\beta u_{\alpha\beta}(\mathbf{r}_1,\mathbf{r}_2)\rho_{\alpha\beta}(\mathbf{r}_1,\mathbf{r}_2|\lambda) \tag{2.49}$$

$$\frac{\partial}{\partial\lambda}\rho_{\alpha\beta}(\mathbf{r}_1,\mathbf{r}_2|\lambda) = \rho_{\alpha\beta}(\mathbf{r}_1,\mathbf{r}_2|\lambda)\left[\beta\frac{\partial\Omega}{\partial\lambda} - \beta v_\alpha(z_1) - \beta u_{\alpha\beta}(\mathbf{r}_1,\mathbf{r}_2)\right]$$

$$-\beta\int d\mathbf{r}_3\sum_\gamma u_{\alpha\gamma}(\mathbf{r}_1,\mathbf{r}_3)\rho_{\alpha\beta\gamma}(\mathbf{r}_1,\mathbf{r}_2,\mathbf{r}_3|\lambda) \tag{2.50}$$

The spatial integration required to obtain distribution functions from the BBGY hierarchy has been replaced by a coupling constant integration.

A slight modification of this procedure forms the basis of the modified Poisson-Boltzmann (MPB) theory.[22,39,40] In this case, the coupling constant λ couples only the electrostatic parts of the Hamiltonian:

$$\mathcal{H}'_\lambda = \mathcal{H}'_0 + \lambda\left[q_\alpha v^C(z_1) + q_\alpha^2 v^I(z_1) + \sum_{\beta,\,j_\beta} q_\alpha q_\beta u^E(\mathbf{r}_{1_\alpha},\mathbf{r}_{j_\beta})\right] \tag{2.51}$$

This procedure corresponds to charging particle 1_α; in the previous procedure, the particle must also be inserted at $\mathbf{r}_1$.

Equations (2.49) and (2.50) are affected by the modification in that only the electrostatic components of v_α and $u_{\alpha\beta}$ appear explicitly. Again we can collect terms and use the definition

$$\rho_{\alpha\beta}(\mathbf{r}_1,\mathbf{r}_2|\lambda) = \rho_\alpha(\mathbf{r}_1|\lambda)g_{\alpha\beta}(\mathbf{r}_1,\mathbf{r}_2|\lambda)\rho_\beta(\mathbf{r}_2) \tag{2.52}$$

to write an expression for the ionic potential of mean force

$$
\frac{\partial}{\partial \lambda} \ln g_\alpha(z_1|\lambda) = -\beta \frac{\partial}{\partial \lambda} W_\alpha(z_1|\lambda)
$$

$$
= \beta \frac{\partial \Omega(\lambda)}{\partial \lambda} - \beta q_\alpha \psi(z_1) - \beta q_\alpha^2 v^I(z_1) - \beta v_\alpha^S(z_1)
$$

$$
- \beta \int d\mathbf{r}_2 \sum_\beta u_{\alpha\beta}^S(\mathbf{r}_1,\mathbf{r}_2)\rho_\beta(z_2)g_{\alpha\beta}(\mathbf{r}_1,\mathbf{r}_2|\lambda)
$$

$$
- \beta \int d\mathbf{r}_2 q_\alpha u^E(\mathbf{r}_1,\mathbf{r}_2)\sum_\beta q_\beta \rho_\beta(z_2)h_{\alpha\beta}(\mathbf{r}_1,\mathbf{r}_2|\lambda)
$$

$$(2.53)$$

using $\mathscr{H}_\lambda$ given by Eq. (2.48), or

$$
\frac{\partial}{\partial \lambda} \ln g_\alpha(z_1|\lambda) = \beta \frac{\partial \Omega}{\partial \lambda} - \beta q_\alpha \psi(z_1) - \beta q_\alpha^2 v^I(z_1)
$$

$$
- \beta \int d\mathbf{r}_2 q_\alpha u^E(\mathbf{r}_1,\mathbf{r}_2)\sum_\beta q_\beta \rho_\beta(z_2)h_{\alpha\beta}(\mathbf{r}_1,\mathbf{r}_2|\lambda)
$$

$$(2.54)$$

using $\mathscr{H}_\lambda'$ given by Eq. (2.51).

Using the obvious definition for $\phi_\alpha(\mathbf{r}_1,\mathbf{r}_2|\lambda)$, we can write the first equation of the Kirkwood hierarchy in the form

$$
\frac{\partial}{\partial \lambda} \ln g_\alpha(z_1|\lambda) = \beta \frac{\partial \Omega}{\partial \lambda} - \beta q_\alpha \psi(z_1) - \beta v_\alpha^S(z_1)
$$

$$
- \beta \int d\mathbf{r}_2 \sum_\beta u_{\alpha\beta}^S(\mathbf{r}_1,\mathbf{r}_2)\rho_\beta(z_2)g_{\alpha\beta}(\mathbf{r}_1,\mathbf{r}_2|\lambda)
$$

$$
- \beta q_\alpha \lim_{\mathbf{r}_2 \to \mathbf{r}_1} \left[\phi_\alpha(\mathbf{r}_1,\mathbf{r}_2) - \lambda q_\alpha u^C(r_{12}) \right] \qquad (2.55)
$$

using $\mathscr{H}_\lambda$, or

$$
\frac{\partial}{\partial \lambda} \ln g_\alpha(z_1|\lambda) = \beta \frac{\partial \Omega}{\partial \lambda} - \beta q_\alpha \psi(z_1) - \beta q_\alpha \lim_{\mathbf{r}_2 \to \mathbf{r}_1} \left[\phi_\alpha(\mathbf{r}_1,\mathbf{r}_2|\lambda) - \lambda q_\alpha u^C(r_{12}) \right]
$$

$$(2.56)$$

using $\mathscr{H}_\lambda'$.

3. Cluster Expansions and Density Functional Theory

For Hamiltonian systems of the kind described in Eq. (2.1), we can write succinct graphical characterizations of the distribution functions as functionals of the intermolecular potentials, external potential, and one-particle density.[29-31] The one-particle density is given by

$$\ln\left[\frac{\rho_\alpha(z_1)}{\mathfrak{z}_\alpha}\right] = -\beta v_\alpha(z_1) + c_\alpha(z_1) \tag{2.57}$$

where

$$\mathfrak{z}_\alpha = \lambda_\alpha^{-3}\exp(-\beta\mu_\alpha) \tag{2.58}$$

is the fugacity, with λ_α the de Broglie wavelength of particle α $[= (h^2/2\pi m_\alpha kT)^{1/2}]$ and μ_α the chemical potential. The quantity $-kTc_\alpha(z_1)$ is an effective potential that incorporates the effects of pair interactions.

The complete graphical prescription of $c_\alpha(z_1)$ is given in Appendix A, where cluster expansions of $h_{\alpha\beta}(\mathbf{r}_1,\mathbf{r}_2)$ may also be found. The description of $\ln(\rho_\alpha(z_1)/\mathfrak{z}_\alpha)$ and $h_{\alpha\beta}(\mathbf{r}_1,\mathbf{r}_2)$ in terms of chain-sum bonds is also given, as preparation for some results in Section III.

These cluster expansion results led to variational principles for the grand potential of a classical system.[29-32] The simplest involves the quantity

$$\beta\Omega[\rho'] = \beta\int d\mathbf{r}_1 \sum_\alpha \rho_\alpha'(\mathbf{r}_1) v_\alpha(\mathbf{r}_1) + \int d\mathbf{r}_1 \sum_\alpha \rho_\alpha'(\mathbf{r}_1)\left\{\ln\left[\frac{\rho_\alpha'(\mathbf{r}_1)}{\mathfrak{z}_\alpha}\right] - 1\right\} + \Phi[\rho'] \tag{2.59}$$

where $\Phi[\rho]$ is obtained by functional integration of $c_\alpha(z_1)$:

$$\frac{\delta\Phi[\rho]}{\delta\rho_\alpha(\mathbf{r}_1)} = c_\alpha(\mathbf{r}_1) \tag{2.60}$$

(a graphical prescription of Φ is given in Ref. 30) and ρ' is a trial distribution function, not necessarily the correct equilibrium one. The variational principle states that $\Omega[\rho']$, for fixed $v_\alpha(\mathbf{r})$ and $u_{\alpha\beta}(\mathbf{r}_1,\mathbf{r}_2)$, is minimized when $\rho_\alpha'(\mathbf{r}_1)$ takes the equilibrium value $\rho_\alpha(\mathbf{r}_1)$ and that Ω is then the equilibrium grand potential.

$$\frac{\delta\Omega[\rho']}{\delta\rho_\alpha'(\mathbf{r}_1)} = 0 \qquad \text{at } \rho'(\mathbf{r}_1) = \rho_\alpha(\mathbf{r}_1) \tag{2.61}$$

which is simply equivalent to Eq. (2.57).

The functional $\Phi[\rho]$ generates the n-body direct correlation functions via functional differentiation with respect to $\rho_\alpha(\mathbf{r})$.[15,41] In particular, we have

$$\frac{\delta^2 \Phi[\rho]}{\delta \rho_\alpha(\mathbf{r}_1)\, \delta \rho_\beta(\mathbf{r}_2)} = \frac{\delta c_\alpha(\mathbf{r}_1)}{\delta \rho_\beta(\mathbf{r}_2)} = c_{\alpha\beta}(\mathbf{r}_1, \mathbf{r}_2) \qquad (2.62)$$

where the two-particle direct correlation function $c_{\alpha\beta}(\mathbf{r}_1, \mathbf{r}_2)$ is related to $h_{\alpha\beta}(\mathbf{r}_1, \mathbf{r}_2)$ by the nonuniform Ornstein-Zernike (OZ) equation

$$h_{\alpha\beta}(\mathbf{r}_1, \mathbf{r}_2) = c_{\alpha\beta}(\mathbf{r}_1, \mathbf{r}_2) + \sum_\gamma \int d\mathbf{r}_3\, c_{\alpha\gamma}(\mathbf{r}_1, \mathbf{r}_3)\, \rho_\gamma(\mathbf{r}_3)\, h_{\gamma\beta}(\mathbf{r}_3, \mathbf{r}_2) \qquad (2.63)$$

From Eqs. (2.59) and (2.62), the free energy functional Ω can be written in terms of functional integrals of $c_{\alpha\beta}(\mathbf{r}_1, \mathbf{r}_2)$ in the space of one-particle density functions (see Ref. 15). Once an approximation for $c_{\alpha\beta}(\mathbf{r}_1, \mathbf{r}_2)$ is supplied, the functional Ω may be minimized either explicitly [by parameterizing $\rho_\alpha(\mathbf{r}_1)$] or implicitly [by deriving the Euler-Lagrange equations, which must then be solved for $\rho_\alpha(\mathbf{r}_1)$].[23] Care must be taken with this approach, however, in the case that the Hamiltonian contains image terms, for reasons already discussed in Section II.A.

Two other approaches that feature the direct correlation function $c_{\alpha\beta}(\mathbf{r}_1, \mathbf{r}_2)$ are actually special cases of density functional theory and are only applicable to the case without images.

An exact integrodifferential equation relating $\rho_\alpha(\mathbf{r})$ to $c_{\alpha\beta}(\mathbf{r}_1, \mathbf{r}_2)$ has been derived by graphical methods,[42] by means of density functional theory[43] and the BBGY hierarchy[44,238] in the absence of images. It reads:

$$\nabla_1 \rho_\alpha(\mathbf{r}_1) = - \rho_\alpha(\mathbf{r}_1) \beta \nabla_1 v_\alpha(\mathbf{r}_1) + \rho_\alpha(\mathbf{r}_1) \sum_\beta \int d\mathbf{r}_2\, c_{\alpha\beta}(\mathbf{r}_1, \mathbf{r}_2) \nabla_2 \rho_\beta(\mathbf{r}_2)$$

$$(2.64)$$

where the region of integration includes possibly discontinuous changes in $\rho_\beta(\mathbf{r}_2)$, or, for a system with a hard surface,

$$\nabla_1 \ln g_\alpha(\mathbf{r}_1) = - \beta \nabla_1 v_\alpha(\mathbf{r}_1) - \int d\mathbf{r}_2 \sum_\beta \rho_\beta(\mathbf{r}_2) \nabla_2 c_{\alpha\beta}(\mathbf{r}_1, \mathbf{r}_2) \qquad (2.65)$$

where the region of integration now includes only the fluid region. Defining the short-range part of the direct correlation function by

$$c_{\alpha\beta}(\mathbf{r}_1, \mathbf{r}_2) = - \beta q_\alpha q_\beta u^C(r_{12}) + c_{\alpha\beta}^S(\mathbf{r}_1, \mathbf{r}_2) \qquad (2.66)$$

Eq. (2.65) can be written as

$$\nabla_1 \ln g_\alpha(z_1) = -\beta \nabla_1 v_\alpha^S(z_1) - \beta q_\alpha \nabla_1 \psi(z_1) - \int d\mathbf{r}_2 \sum_\beta \rho_\beta(\mathbf{r}_2) \nabla_2 c_{\alpha\beta}^S(\mathbf{r}_1, \mathbf{r}_2)$$

(2.67)

a form more appropriate to the double layer problem. Equation (2.64), originally due to Wertheim[42] and Lovett et al.,[43] cannot be applied to the case with images, because the interaction part of the Hamiltonian (namely $\mathscr{H}^I$) is not translationally invariant. If images are included, the derivation via density functional theory immediately introduces three-body correlations, and so this approach loses much of its simplicity.

The direct correlation function appears in another approach whereby a coupling constant is introduced that couples the external potential to the rest of the system:

$$\mathscr{H}_\lambda = \mathscr{H}_0 + \lambda \sum_\alpha \sum_{i_\alpha} v_\alpha(\mathbf{r}_{i_\alpha})$$

(2.68)

From Eq. (2.68) it immediately follows that[45]

$$\frac{\partial}{\partial \lambda} \rho_\alpha(\mathbf{r}_1|\lambda) = -\beta v_\alpha(\mathbf{r}_1) \rho_\alpha(\mathbf{r}_1|\lambda) - \beta \sum_\beta \int d\mathbf{r}_2 \, v_\beta(\mathbf{r}_2) \rho_{\alpha\beta}^T(\mathbf{r}_1, \mathbf{r}_2)$$

(2.69)

where $\rho_\alpha(\mathbf{r}_1|\lambda)$ and $\rho_{\alpha\beta}^T(\mathbf{r}_1, \mathbf{r}_2|\lambda)$ are distribution functions calculated with $\mathscr{H}_\lambda$, that is, the external potential is at strength λ. Using the appropriate OZ equation,

$$h_{\alpha\beta}(\mathbf{r}_1, \mathbf{r}_2|\lambda) = c_{\alpha\beta}(\mathbf{r}_1, \mathbf{r}_2|\lambda) + \int d\mathbf{r}_3 \sum_\gamma c_{\alpha\gamma}(\mathbf{r}_1, \mathbf{r}_3|\lambda) \rho_\gamma(\mathbf{r}_3|\lambda) h_{\gamma\beta}(\mathbf{r}_3, \mathbf{r}_2|\lambda)$$

(2.70)

Eq. (2.69) becomes

$$\ln g_\alpha(z_1) = -\beta v_\alpha^S(z_1) - \beta q_\alpha \psi(z_1) + \int_0^1 d\lambda \int d\mathbf{r}_2 \sum_\beta c_{\alpha\beta}^S(\mathbf{r}_1, \mathbf{r}_2|\lambda) \frac{\partial}{\partial \lambda} \rho_\beta(\mathbf{r}_2|\lambda)$$

(2.71)

where we have assumed that the state $\lambda = 0$ corresponds to uniform fluid. Once again, this approach is useful for the case where λ represents the surface charge density in the absence of images. The form of Eq. (2.68) is not appropriate for treating image potentials, for which both one-body and pair interactions must be coupled to $\mathcal{H}_0$. As before, the resultant expressions contain three-body correlation functions, and so this does not appear to be a promising route to the theoretical treatment of image effects in the double layer.

D. Exact Conditions

From the various formally exact equations of Section II.C it is possible to derive conditions on the free energy and distribution functions that any exact solution must satisfy. These conditions may be used as a guide to the construction of, or as criteria for testing the validity of, approximate theories.

The simplest conditions are those derivable from purely thermodynamic considerations such as Maxwell relations. These can be found in any thermodynamic treatment of electrochemical systems.[26] Clearly, such thermodynamic results are not restricted to any particular model such as the primitive model we consider here. We will mention only those reflecting the variation of free energy as the surface is charged. The thermodynamic potential Ω is the appropriate free energy for conditions of constant $\{\mu_\alpha\}$, V, T, A (surface area), and σ (external electric field). It then follows, from either thermodynamic considerations[46] or statistical mechanical arguments,[46,47] that

$$\frac{\partial \Omega}{\partial \sigma}\bigg|_{V,T,A,\{\mu_\alpha\}} = A\psi_0 \tag{2.72}$$

Similarly, we have

$$\frac{\partial \Omega}{\partial A}\bigg|_{V,T,\sigma,\{\mu_\alpha\}} = \gamma + \sigma\psi_0 \tag{2.73}$$

where γ is the surface tension of the interface. From Eqs. (2.72) and (2.73) we find

$$\gamma = \gamma_{\sigma=0} - \sigma\psi_0 + \int_0^\sigma \psi_0(\sigma')\, d\sigma' \tag{2.74}$$

so that

$$\frac{\partial \gamma}{\partial \psi_0} = -\sigma \tag{2.75}$$

which is Lippmann's equation, a fundamental equation of the theory of electrocapillarity.[25,26,46] The thermodynamic inequality following from the definition of Ω is[46,48]

$$\frac{\partial \psi_0}{\partial \sigma} \geq 0 \qquad (2.76)$$

that is, the total capacitance of the double layer is non-negative. The surface tension of the interface may be expressed in terms of distribution functions in at least three different ways[10,24]; a further requirement of exact one- and two-particle densities is that all routes lead to consistent thermodynamics, just as in the bulk case.

A special feature of fluids with Coulomb interactions is that they must obey electroneutrality conditions. The most obvious is that the charge on the surface must be balanced by the total charge in the double layer.

$$\sigma = - \int_0^\infty dz_1 \sum_\alpha q_\alpha \rho_\alpha(z_1) \qquad (2.77)$$

A further (and stronger) condition is the local electroneutrality (or perfect screening) condition: the total double layer charge around any given ion must balance the surface charge plus q_α.

$$q_\alpha + \sigma A = - \int_0^\infty d\mathbf{r}_2 \sum_\beta q_\beta \rho_\beta(z_2) g_{\alpha\beta}(\mathbf{r}_1, \mathbf{r}_2)$$

That is,

$$q_\alpha \rho_\alpha(z_1) + \int d\mathbf{r}_2 \sum_\beta q_\beta \rho_{\alpha\beta}^T(\mathbf{r}_1, \mathbf{r}_2) = 0 \qquad (2.78)$$

Higher analogues of Eq. (2.78) may be found in Ref. 49. A sufficient condition for Eq. (2.78) to hold is that the truncated two-particle density $\rho_{\alpha\beta}^T(\mathbf{r}_1, \mathbf{r}_2)$ decay no slower than r_{12}^{-d} (in spatial dimension d) in all directions.

New sum rules concerning the higher multipole moments of the charge distribution around an ion in the double layer have recently been derived.[49,50] Of particular interest are sum rules concerning the dipole moment of the excess charge distribution around an ion α:

$$\mu_\alpha(z_1) = \int d\mathbf{r}_2 (z_2 - z_1) \sum_\beta q_\beta \rho_{\alpha\beta}^T(\mathbf{r}_1, \mathbf{r}_2) \qquad (2.79)$$

If $\rho_{\alpha\beta}^T(\mathbf{r}_1, \mathbf{r}_2)$ decays faster than r_{12}^{-d} in all directions, $\mu_\alpha(z_1)$ vanishes.[50] From Eq. (2.69) however, one can derive[51]

$$\frac{\partial \rho_\alpha(z_1)}{\partial \sigma} = \frac{4\pi\beta}{\varepsilon} \mu_\alpha(z_1) \tag{2.80}$$

Since in general we expect the one-particle density to depend on the surface charge, Eq. (2.80) is consistent with the result of Ref. 50 only if the truncated two-particle density decays as r_{12}^{-d} (or more slowly) in at least one direction.

A sum rule predicting such behavior has been derived by linear response theory by Jancovici.[20] The sum rule can also be derived from the OZ equation [Eq. (2.63)] on the reasonable assumption that $c_{\alpha\beta}^S(\mathbf{r}_1, \mathbf{r}_2)$ is a short-range function.[52] The sum rule states, for one-plate geometry in two or three dimensions ($d = 2, 3$),

$$\int_0^\infty dz_1 \int_0^\infty dz_2 \sum_{\alpha\beta} q_\alpha q_\beta \rho_{\alpha\beta}^T(\mathbf{r}_1, \mathbf{r}_2) \sim \frac{-\varepsilon_w kT}{8\pi^2 R^d} \quad \text{as } R \to \infty \tag{2.81}$$

except for the case of a metal surface ($\varepsilon_w = \infty$).

Equation (2.80) leads to several new expressions and exact conditions. We mention a sum rule connecting $\rho_{\alpha\beta}^T(\mathbf{r}_1, \mathbf{r}_2)$ and the surface charge[44,51]:

$$\sigma = \sum_{\alpha\gamma} q_\gamma \int d\mathbf{r}_2 \left(z_2 - \frac{d_\alpha}{2} \right) \rho_{\alpha\gamma}^T \left(\frac{d_\alpha}{2}, z_2, R_2 \right) \tag{2.82}$$

and a nonuniform analogue of the Stillinger-Lovett condition,[53,54]

$$\beta \int d\mathbf{r}_2 \sum_\alpha q_\alpha \rho_\alpha(\mathbf{r}_2) \phi_\alpha(\mathbf{r}_1, \mathbf{r}_2) = 1 \tag{2.83}$$

For further examples, see Refs. 44 and 51.

Finally, from the condition of force balance in the double layer (or equivalently, from the BBGY equations), one finds a condition on the total ionic density at a hard surface. For the simplest case of a primitive model electrolyte in the absence of images, the contact condition takes the form[55,239]:

$$kT \sum_\alpha \rho_\alpha \left(\frac{d_\alpha}{2} \right) = p + \frac{2\pi\sigma^2}{\varepsilon} \tag{2.84}$$

where p is the pressure of the bulk electrolyte (i.e., in the McMillan-Mayer picture, p is the osmotic pressure of the solution) and z is measured from the hard surface. In the presence of some short-range one-body potential and image forces, the contact condition becomes $(d_\alpha = d)^{56}$

$$kT\sum_\alpha \rho_\alpha\left(\frac{d}{2}\right) = p + \frac{2\pi\sigma^2}{\varepsilon} + \int_0^\infty dz_1 \sum_\alpha \rho_\alpha(z_1)\frac{dv_\alpha^S(z_1)}{dz_1}$$

$$+ \int_0^\infty dz_1 \sum_\alpha \rho_\alpha(z_1)q_\alpha^2\frac{dv^I(z_1)}{dz_1}$$

$$+ \int_0^\infty dz_1 \int d\mathbf{r}_2 \sum_{\alpha\beta} q_\alpha q_\beta \rho_{\alpha\beta}^T(\mathbf{r}_1,\mathbf{r}_2)\frac{d}{dz_1}u^I(\mathbf{r}_1,\mathbf{r}_2) \quad (2.85)$$

III. THE POINT-ION MODEL AND GOUY-CHAPMAN THEORY

We first apply the exact equations of Section II to the simplest conceivable model, the point-ion model, which is obtained from the primitive model by setting $d_\alpha = 0$ in Eqs. (2.10) and (2.14). Of course, a real electrolyte must have finite ion size in order to prevent collapse. We will need to make approximations in the treatment of ion-ion correlations so as to avoid such unphysical behavior, much as is done in the Debye-Hückel theory of bulk electrolytes.

We first consider the case where images are absent and derive Gouy-Chapman theory, the simplest and oldest double layer theory. When considering image effects, it is necessary to go beyond Gouy-Chapman theory; we discuss such effects for the point-ion model in Section III.B.

A. Gouy-Chapman Theory

The classical theory of the double layer dates from around 1910,[13,14] some fifteen years before the corresponding work of Debye and Hückel on the theory of bulk electrolytes.[57] Both the Debye-Hückel (DH) and Gouy-Chapman (GC) theories are based on the Poisson-Boltzmann equation.

Starting from Poisson's equation,

$$\frac{d^2\psi(z)}{dz^2} = -\frac{4\pi}{\varepsilon}\sum_\alpha q_\alpha \rho_\alpha(z) \quad (3.1)$$

subject to the boundary conditions

$$\psi(z) \to 0, \quad z \to \infty$$
$$\frac{d\psi(z)}{dz} = -\frac{4\pi\sigma}{\varepsilon} \quad z = 0 \quad (3.2)$$

Gouy-Chapman theory proceeds by substituting an approximate expression for $\rho_\alpha(z)$, obtained from statistical mechanics, into Eq. (3.1). The density profile is taken to be given by a Boltzmann distribution with respect to the mean electrostatic potential energy in the double layer

$$\rho_\alpha(z) = \rho_\alpha \exp\left[-\beta q_\alpha \psi(z)\right] \tag{3.3}$$

or, equivalently, the potential of mean force is approximated by the mean electrostatic potential energy:

$$\beta W_\alpha(z) = -\ln\left[g_\alpha(z)\right] = \beta q_\alpha \psi(z) \tag{3.4}$$

The closure equation (3.4) can be derived as the lowest order approximation from all the exact equations of the previous section. Recalling that for the present model, $u^S_{\alpha\beta}$, v^I, u^I vanish and the effect of $v^S_\alpha(z)$ can be taken into account simply by setting

$$\rho_\alpha(z) = 0 \qquad (z < 0) \tag{3.5}$$

we recover Eq. (3.4) from the BBGY hierarchy by setting $h_{\alpha\beta}(\mathbf{r}_1, \mathbf{r}_2)$ to zero in Eq. (2.45). In the sense that ion-ion correlations are neglected except through the effect of the mean potential, Gouy-Chapman theory has the status of a mean field theory.

In a similar manner, Eq. (3.4) can be derived from the Kirkwood hierarchy, Eq. (2.56), or from the cluster expansions, (A.9), by neglecting all but the simplest terms. By setting $c^S_{\alpha\beta}(\mathbf{r}_1, \mathbf{r}_2)$ equal to zero, we recover the same result from Eqs. (2.67) and (2.71).

Substitution of Eq. (3.3) into Eq. (3.1) yields the Poisson-Boltzmann equation

$$\frac{d^2\psi(z)}{dz^2} = -\frac{4\pi}{\varepsilon} \sum_\alpha q_\alpha \rho_\alpha \exp\left[-\beta q_\alpha \psi(z)\right] \tag{3.6}$$

which must be solved numerically in general. A first integral of Eq. (3.6) gives, using Eq. (3.2),

$$\frac{2\pi\sigma^2}{\varepsilon} = kT\sum_\alpha \rho_\alpha\left[\exp(-\beta q_\alpha \psi_0) - 1\right] \tag{3.7}$$

for electrolytes of arbitrary composition. Thus the surface potential–surface charge relationship is determined by Eq. (3.7), which is known as Grahame's equation.[4]

Further progress for general electrolytes may be made by specializing to small potentials, so that we may linearize Eq. (3.6) to obtain

$$\frac{d^2\psi(z)}{dz^2} = \kappa^2\psi(z) \tag{3.8}$$

where κ^{-1} is the usual Debye-Hückel screening length

$$\kappa^2 = 4\pi\beta \sum_\alpha \frac{\rho_\alpha q_\alpha^2}{\varepsilon} \tag{3.9}$$

The solution of Eq. (3.8)

$$\psi(z) = \frac{4\pi\sigma}{\varepsilon\kappa}\exp(-\kappa z), \qquad \rho_\alpha(z) = \rho_\alpha[1 - \beta\psi_0 q_\alpha\exp(-\kappa z)] \tag{3.10}$$

is valid only for $\beta q_\alpha\psi_0 \le 1$; otherwise $\rho_\alpha(z)$ can become negative, which is unphysical. Equation (3.10) summarizes the results of linearized Gouy-Chapman (LGC) theory.

Clearly, ions of the same sign as the surface charge (co-ions) are depleted near the surface; ions of the opposite sign to the surface charge (counterions) are adsorbed. The exponential decay signals the screening of the surface charge by the intervening electrolyte, with the screening more complete as the ionic concentration increases. The linear relationship between ψ_0 and σ and the unrealistic antisymmetry between $h_+(z)$ and $h_-(z)$ are due to the linearization of Eq. (3.6).

Debye-Hückel (DH) theory may be most properly compared with LGC theory, since both rely on linearization of the Poisson-Boltzmann equation. A crucial difference is that DH theory is a self-consistent theory; the source of external potential is one particular ion among all the ions of the electrolyte. LGC theory does not have this feature; the source of external potential is quite different from the ions. It is perhaps for this reason that DH theory enjoys the status of an exact law in the limit of infinite dilution,[58,59] while LGC theory appears to have no such role, at least at fixed surface charge.

The various unphysical features of LGC theory may be eliminated by solving the full Poisson-Boltzmann equation, Eq. (3.6). (Notice that this may be done for the double-layer problem in the point-ion model, whereas for bulk electrolytes ion size must be introduced at this point.) In this sense, Gouy-Chapman theory is the analogue of the Poisson-Boltzmann theory for bulk electrolytes. Retention of the nonlinear closure equation (3.4) necessarily means that the thermodynamic integrability condition that is necessary for various charging processes to give consistent thermodynamics will not be obeyed.[60,61] While this does indicate a fundamental weakness in the theory, similar thermodynamic inconsistencies are routinely exhibited by successful integral equation theories of bulk electrolytes. For example, even though the Poisson-Boltzmann theory is inconsistent in this regard, it nevertheless represents a significant advance over DH theory in its description of thermodynamics and structure of electrolyte solutions.[62] For reasons that we shall see, the retention of nonlinear behavior is so crucial to double layer theory that the problem of thermodynamic inconsistency has been largely set aside; Gouy-Chapman theory has been almost universally adopted as the appropriate theory of the diffuse double layer by colloid scientists[2], electrochemists,[1] and physiologists.[3]

Analytic solutions of Eq. (3.6) have been found for the special cases of $2:1$,[63] $2:1:1$,[64] and, most importantly, binary symmetric $(q:q)$ electrolytes.[13,14] In this case, the Poisson-Boltzmann equation becomes

$$\frac{d^2}{dz^2}\beta q\psi(z) = \kappa^2 \sinh[\beta q\psi(z)] \tag{3.11}$$

with the solution

$$\beta q\psi(z) = 4\tanh^{-1}[u\exp(-\kappa z)]$$

where

$$u = \tanh\left(\frac{\beta q\psi_0}{4}\right) \quad \text{and} \quad \beta q\psi_0 = 2\sinh^{-1}\left(\frac{2\pi\beta q\sigma}{\epsilon\kappa}\right) \tag{3.12}$$

For small surface charge or high concentrations, the potential is low and Eq. (3.12) reduces to Eq. (3.10). The potential and ion density profiles are still monotonic, and asymptotically decay exponentially with decay length κ^{-1}.

In Fig. 2 we show the variation of surface potential with surface charge for both the GC and LGC theories for $1:1$ electrolytes. The sublinear dependence of surface potential on surface charge at fixed κ is a characteristic feature of GC theory. At fixed σ the surface potential increases strongly as the ion concentration is decreased. This can be easily understood by recognizing κ^{-1} as a qualitative measure of the thickness of the double layer.

The variation of the ionic densities through the double layer is shown in Fig. 3, again for both GC and LGC theories. In the GC theory, counterion and co-ion excess densities are no longer antisymmetric; the counterions are adsorbed more than co-ions are depleted, and the co-ion density remains physical at all surface charge densities.

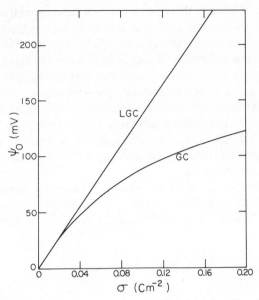

Fig. 2. Surface potential ψ_0, in millivolts, as a function of surface charge density σ, in C m^{-2}, for a 0.1 M 1:1 electrolyte, as predicted by Gouy-Chapman (GC) and linearized Gouy-Chapman (LGC) theory.

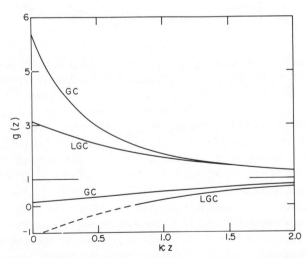

Fig. 3. Ion density profiles for a 0.1 M 1:1 electrolyte from GC and LGC theories at $\sigma = 0.04$ C m^{-2}. The dashed line shows the region where the LGC co-ion profile is unphysical.

In order to make a better comparison with theories based on the primitive model electrolyte, GC theory has been modified slightly to exclude ions from the region $0 < z < d/2$, that is, d_α is retained in Eq. (2.14) but set to zero in Eq. (2.10). For equal-sized ions, this modified GC (MGC) theory is obtained from GC theory by the transformation $\psi^{GC}(z) \to \psi^{MGC}(z + d/2)$, $z \geq 0$. The surface potential ψ_0 and the potential at the plane of closest approach $\psi_{d/2}$ are related by

$$\psi_0 = \psi_{d/2} + \frac{4\pi\sigma}{\varepsilon}\frac{d}{2} \tag{3.13}$$

since the field in the region $0 < z < d/2$ is uniform (no ions can penetrate this region) and is given by Eq. (3.2).

Thus we have

$$\beta q \psi_{d/2} = 2 \sinh^{-1}\left(\frac{2\pi\beta q\sigma}{\varepsilon\kappa}\right) \tag{3.14}$$

as in Eq. (3.12). We shall concentrate chiefly on comparisons of $\psi_{d/2}$ given by various theories; the extra potential in Eq. (3.13) is common to all theories based on the primitive model. However, ψ_0 remains the thermodynamic variable; $\psi_{d/2}$ has chiefly a microscopic significance.

As a first test of Gouy-Chapman theory, we note that Eq. (3.7) together with the closure Eq. (3.4) imply

$$kT\sum_\alpha \rho_\alpha(0) = kT\sum_\alpha \rho_\alpha + \frac{2\pi\sigma^2}{\varepsilon} \tag{3.15}$$

or, for MGC theory,

$$kT\sum_\alpha \rho_\alpha\left(\frac{d}{2}\right) = kT\sum_\alpha \rho_\alpha + \frac{2\pi\sigma^2}{\varepsilon} \tag{3.16}$$

for electrolytes of arbitrary composition. The contact density given by MGC theory has the correct dependence on surface charge [see Eq. (2.84)], but at zero charge it has the value $kT\sum_\alpha\rho_\alpha$ rather than the exact result p. This shows that the σ dependence of the contact condition is rather easy to satisfy. Since we are considering McMillan-Mayer level electrolytes where the deviation of p from $kT\sum_\alpha\rho_\alpha$ is not very strong, even the zero-charge contact density given by MGC theory is fairly accurate. Thus we see that the MGC contact density is close to the exact result and that, for dilute systems at least, the contact condition is quite a weak condition.

Although we shall only use the simple MGC theory to compare with more sophisticated theories, the reader should be warned that many further refinements (usually of the underlying model) are used in practical applications. The two most important extensions are ionic adsorption (by this we mean an ion-specific short-range adsorption potential, rather than the purely electrostatic adsorption discussed so far) and the incorporation of solvent effects through an "inner layer capacitance." Specific ionic adsorption is ubiquitous in electrochemical and physiological systems and is usually incorporated through an assumed ionic adsorption isotherm due to Stern.[65] The inner layer capacitance is essential for interpretation of differential capacitance measurements on metal electrodes.[4,7,8] In the MGC theory, we have

$$C^{-1} = \frac{d\psi_0}{d\sigma} = \frac{d\psi_{d/2}}{d\sigma} + \frac{2\pi d}{\varepsilon} = C_d^{-1} + C_i^{-1} \qquad (3.17)$$

where we have defined a diffuse layer capacitance C_d and an inner layer capacitance C_i. In real systems C_i exhibits complex variation with σ and differs widely from the simple form in Eq. (3.17), the major component of the difference being due to the breakdown of the primitive model and the need to account for solvent effects near the surface.[7,8]

Finally, we should ask what experimental tests GC theory (in its simplest form) has received and how it has fared in those tests. Even though we are focusing on the isolated double layer, the direct force measurements between mica surfaces immersed in electrolyte[66,67] provide persuasive evidence supporting simple GC theory predictions for the interaction between two charged surfaces. The requirement for indifferent (i.e., nonspecifically adsorbing) electrolyte would seem to rule out most physiological systems.[68,69] The surface potential of many colloidal particles can only be rather indirectly inferred (usually by mobility measurements[70]). By far the most direct measurements of single double-layer properties have come from differential capacitance measurements of the dropping mercury electrode.[4]

For this system the validity of GC theory has been tested by Grahame.[4] Differential capacitance measurements of aqueous NaF solutions (NaF is generally regarded as an indifferent electrolyte) show reasonable agreement with GC theory for low surface charge densities ($|\sigma| < 0.05$ C m^{-2}) and concentrations (< 0.1 M). The LGC theory is inadequate, however, except at the point of zero charge where $\sigma = 0$. As the surface charge or electrolyte concentration is increased, the diffuse layer capacitance given by GC theory increases and the total differential capacitance becomes determined by the inner layer capacitance and thus solvent effects.

In summary, the regime where Gouy-Chapman theory has been tested and found to be adequate is where the diffuse double layer is relatively thick (κd

$\ll 1$) and dominates the behavior of the total double layer. For concentrations where the diffuse layer is of molecular dimensions, no experimental test of GC theory has been made.

B. Weak-Coupling Theory for the Point-Ion Model

In this section we examine the first corrections to the mean field theory of the point-ion model discussed in Section III.A by employing a linearized, weak-coupling approximation for ion-ion correlations in the spirit of Debye-Hückel theory. This extension also allows us to incorporate image effects into the point-ion model. The same approximation can be derived from the BBGY equations, the Kirkwood hierarchy, and cluster methods.

Starting from the second BBGY equation, we close the hierarchy by using the Kirkwood superposition approximation

$$\rho_{\alpha\beta\gamma}(\mathbf{r}_1,\mathbf{r}_2,\mathbf{r}_3) = \rho_\alpha(\mathbf{r}_1)\rho_\beta(\mathbf{r}_2)\rho_\gamma(\mathbf{r}_3)g_{\alpha\beta}(\mathbf{r}_1,\mathbf{r}_2)g_{\beta\gamma}(\mathbf{r}_2,\mathbf{r}_3)g_{\gamma\alpha}(\mathbf{r}_3,\mathbf{r}_1)$$

$$(3.18)$$

whence Eq. (2.43) becomes

$$kT \nabla_1 \ln g_{\alpha\beta}(\mathbf{r}_1,\mathbf{r}_2) = -q_\alpha\nabla_1\phi_\beta(\mathbf{r}_1,\mathbf{r}_2) - q_\alpha\int d\mathbf{r}_3 \left[\nabla_1 u^E(\mathbf{r}_1,\mathbf{r}_3) \right]$$

$$\times \sum_\gamma q_\gamma\rho_\gamma(z_3)h_{\beta\gamma}(\mathbf{r}_2,\mathbf{r}_3)h_{\gamma\alpha}(\mathbf{r}_3,\mathbf{r}_1) \qquad (3.19)$$

For finite-sized ions, the analogue of Eq. (3.19), together with Eq. (2.47), could in principle be solved for $g_\alpha(z_1)$ and $g_{\alpha\beta}(\mathbf{r}_1,\mathbf{r}_2)$; such a calculation has never been attempted. For point ions, however, the equation must be linearized. The linearization can be done in two steps.

If we neglect the last term on the right-hand side of Eq. (3.19), we recover

$$\ln g_{\alpha\beta}(\mathbf{r}_1,\mathbf{r}_2) = -\beta q_\alpha\phi_\beta(\mathbf{r}_1,\mathbf{r}_2) \qquad (3.20)$$

an approximate equation that has come to be known as Loeb's closure.[22,71] In the bulk electrolyte, Loeb's closure leads directly to the Poisson-Boltzmann equation for $g_{\alpha\beta}(r_{12})$. The superposition closure, Eq. (3.18), states that the total potential of mean force between three ions may be replaced by the sum of one-body [ln $g_\alpha(z)$] and pair [ln $g_{\alpha\beta}(\mathbf{r}_1,\mathbf{r}_2)$] potentials of mean force. Loeb's closure is the much stronger statement that the potential of mean force between two ions is the sum of the one-body potentials of mean force and the change in electrostatic potential at $\mathbf{r}_1$ when an ion is fixed at $\mathbf{r}_2$. The analogues of Loeb's closure for higher correlation functions and their relationship to superposition approximations have been discussed by Outhwaite.[22]

Just as in Debye-Hückel theory, we must now linearize Eq. (3.20) to avoid infinite values of $g_{\alpha\beta}(\mathbf{r}_1,\mathbf{r}_2)$ corresponding to unlike point ions in contact and to restore the correct symmetry properties of $g_{\alpha\beta}(\mathbf{r}_1,\mathbf{r}_2)$. These approximations can be expected to be valid for weak ion-ion coupling, much as in Debye-Hückel theory. However, no assumption has been made about the strength of coupling between an ion and the external field. The term "weak coupling" refers only to the strength of ion-ion interactions.

We then obtain

$$h_{\alpha\beta}(\mathbf{r}_1,\mathbf{r}_2) = -\beta q_\alpha \phi_\beta(\mathbf{r}_1,\mathbf{r}_2) = -\beta q_\alpha q_\beta \phi(\mathbf{r}_1,\mathbf{r}_2) \qquad (3.21)$$

where $\phi(\mathbf{r}_1,\mathbf{r}_2)$ satisfies the equation [see Eq. (2.35)]

$$\phi(\mathbf{r}_1,\mathbf{r}_2) = u^E(\mathbf{r}_1,\mathbf{r}_2) - \beta \int d\mathbf{r}_3\, u^E(\mathbf{r}_1,\mathbf{r}_3) \sum_\gamma q_\gamma^2 \rho_\gamma(z_3) \phi(\mathbf{r}_3,\mathbf{r}_2) \quad (3.22)$$

Inserting Eq. (3.21) into Eq. (2.47), using the symmetry of $\phi(\mathbf{r}_1,\mathbf{r}_2)$ with respect to $\mathbf{r}_1$ and $\mathbf{r}_2$, we find

$$\ln g_\alpha(z_1) = -\beta q_\alpha \psi(z_1) - \tfrac{1}{2}\beta q_\alpha^2 [\eta(z_1) - \eta(\infty)]$$
$$\equiv -\beta q_\alpha \psi(z_1) - \tfrac{1}{2}\beta q_\alpha^2 \Delta\eta(z_1) \qquad (3.23)$$

where

$$\eta(z_1) = \lim_{\mathbf{r}_2 \to \mathbf{r}_1} \left[\phi(\mathbf{r}_1,\mathbf{r}_2) - u^C(r_{12})\right] \qquad (3.24)$$

Following a very similar procedure, we can derive the same equations from the Kirkwood hierarchy.[10,39,40] Again using the superposition approximation,

$$\rho_{\alpha\beta\gamma}(\mathbf{r}_1,\mathbf{r}_2,\mathbf{r}_3|\lambda) = \rho_\alpha(\mathbf{r}_1|\lambda)\rho_\beta(\mathbf{r}_2)\rho_\gamma(\mathbf{r}_3)g_{\alpha\beta}(\mathbf{r}_1,\mathbf{r}_2|\lambda)g_{\beta\gamma}(\mathbf{r}_2,\mathbf{r}_3)g_{\gamma\alpha}(\mathbf{r}_3,\mathbf{r}_1|\lambda)$$
$$(3.25)$$

the second member of the Kirkwood hierarchy becomes

$$\frac{\partial}{\partial\lambda}\ln g_{\alpha\beta}(\mathbf{r}_1,\mathbf{r}_2|\lambda) = -\beta q_\alpha \phi_\beta(\mathbf{r}_1,\mathbf{r}_2) - \beta q_\alpha \int dr_3\, u^E(\mathbf{r}_1,\mathbf{r}_3)$$
$$\times \sum_\gamma q_\gamma \rho_\gamma(z_3) h_{\beta\gamma}(\mathbf{r}_2,\mathbf{r}_3) h_{\gamma\alpha}(\mathbf{r}_3,\mathbf{r}_1|\lambda) \quad (3.26)$$

Making precisely the same arguments as above, we recover Loeb's closure

$$\ln g_{\alpha\beta}(\mathbf{r}_1,\mathbf{r}_2|\lambda) = -\beta\lambda q_\alpha \Phi_\beta(\mathbf{r}_1,\mathbf{r}_2) \tag{3.27}$$

and, in the weak-coupling approximation,

$$h_{\alpha\beta}(\mathbf{r}_1,\mathbf{r}_2|\lambda) = -\beta\lambda q_\alpha q_\beta \phi(\mathbf{r}_1,\mathbf{r}_2) \tag{3.28}$$

Inserting Eq. (3.28) into Eq. (2.56), performing the coupling constant integration, and removing the constant by evaluating the right-hand side at $z_1 = \infty$, we again recover Eq. (3.23).

Just as the Debye-Hückel theory follows from the summation of ring graphs in bulk electrolyte theory,[59] so the weak-coupling equations follow from the nonuniform version of ring graph summation. Consider the chain-sum bond function given in Appendix A.

$$\mathscr{C}_{\alpha\beta}(\mathbf{r}_1,\mathbf{r}_2) = \underset{1 \quad \ 2}{\circ\text{---}\circ} + \underset{1 \qquad 2}{\diagup\!\!\diagdown} + \underset{1 \qquad 2}{\square} + \cdots \tag{3.29}$$

where $\underset{1 \quad \ 2}{\circ\text{---}\circ} = \Phi_{\alpha\beta}(12) = -\beta q_\alpha q_\beta u^E(\mathbf{r}_1,\mathbf{r}_2)$. It follows that $\mathscr{C}_{\alpha\beta}(\mathbf{r}_1,\mathbf{r}_2)$ satisfies the integral equation

$$\mathscr{C}_{\alpha\beta}(\mathbf{r}_1,\mathbf{r}_2) = \Phi_{\alpha\beta}(\mathbf{r}_1,\mathbf{r}_2) + \int d\mathbf{r}_3 \sum_\gamma \Phi_{\alpha\gamma}(\mathbf{r}_1,\mathbf{r}_3)\rho_\gamma(z_3)\mathscr{C}_{\gamma\beta}(\mathbf{r}_3,\mathbf{r}_2)$$

$$= -\beta q_\alpha q_\beta u^E(\mathbf{r}_1,\mathbf{r}_2) - \beta q_\alpha \int d\mathbf{r}_3 \, u^E(\mathbf{r}_1,\mathbf{r}_3)\sum_\gamma q_\gamma \rho_\gamma(z_3)\mathscr{C}_{\gamma\beta}(\mathbf{r}_3,\mathbf{r}_2)$$

$$\tag{3.30}$$

so that

$$\mathscr{C}_{\alpha\beta}(\mathbf{r}_1,\mathbf{r}_2) = -\beta q_\alpha q_\beta \phi(\mathbf{r}_1,\mathbf{r}_2) \tag{3.31}$$

with $\phi(\mathbf{r}_1,\mathbf{r}_2)$ obeying Eq. (3.22). Thus the weak-coupling approximations discussed above are equivalent to the retention of the simplest graph (namely, the one with one $\mathscr{C}$ bond connecting two root points) in Eq. (A.8). In a similar manner, retention of the simplest terms in Eq. (A.9) leads once again to the weak-coupling equations.

The integral equation for $\phi(\mathbf{r}_1,\mathbf{r}_2)$, Eq. (3.22), can be written in differential form:

$$\nabla_1^2 \phi(\mathbf{r}_1,\mathbf{r}_2) - \kappa^2(z_1)\phi(\mathbf{r}_1,\mathbf{r}_2) = -\frac{4\pi}{\varepsilon}\delta(\mathbf{r}_1 - \mathbf{r}_2) \tag{3.32}$$

subject to the boundary conditions

$$\phi(\mathbf{r}_1, \mathbf{r}_2) \to 0 \qquad |z_1| \to \infty$$

$$\phi(\mathbf{r}_1, \mathbf{r}_2) \text{ continuous at } z_1 = 0 \tag{3.33}$$

$$\varepsilon \left. \frac{\partial \phi}{\partial z_1} \right|_{z_1 = 0^+} = \varepsilon_w \left. \frac{\partial \phi}{\partial z_1} \right|_{z_1 = 0^-}$$

where $\kappa^2(z_1)$ is a measure of the local ionic strength in the double layer,

$$\kappa^2(z_1) = \frac{4\pi\beta}{\varepsilon} \sum_\alpha q_\alpha^2 \rho_\alpha(z_1) \tag{3.34}$$

Equations (3.23), (3.24), and (3.32)–(3.34) constitute the most convenient form of the weak-coupling equations. They were first given by Loeb[71]; the first derivation from both the Kirkwood hierarchy[10] and cluster expansions[72] is due to Buff and Stillinger; their derivation from the BBGY hierarchy is given by Martynov.[27]

The weak-coupling equations have never been solved so as to obtain a completely self-consistent set of one- and two-particle densities. The most determined attempts have been those of Williams[73] and Gorelkin and Smilga.[74] The first step in such a program is the calculation of $\phi^{(0)}(\mathbf{r}_1, \mathbf{r}_2)$, given an assumed density profile $\rho_\alpha^{(0)}(z_1)$. From $\phi^{(0)}$ one obtains $\Delta\eta^{(0)}(z_1)$, and the next approximation for $\rho_\alpha(z)$ is obtained from

$$g_\alpha^{(1)}(z) = \exp\left[-\beta q_\alpha \psi^{(1)}(z_1) - \tfrac{1}{2}\beta q_\alpha^2 \Delta\eta^{(0)}(z_1) \right] \tag{3.35}$$

where the corrected potential is found by solving the equation (for $q:q$ electrolytes)

$$\frac{d^2}{d\dot{z}_1^2} \beta q \psi^{(1)}(z_1) = \kappa^2 \sinh\left[\beta q \psi^{(1)}(z_1) \right] \exp\left[-\tfrac{1}{2}\beta q^2 \Delta\eta^{(0)}(z_1) \right] \tag{3.36}$$

subject to Eq. (3.2). From $g_\alpha^{(1)}(z_1)$ one determines $\phi^{(1)}(\mathbf{r}_1, \mathbf{r}_2)$, and so on. The procedure should be continued to self-consistency; in practice, calculations of $\psi^{(1)}(z_1)$ are rare.[73,75]

The calculation of $\phi^{(0)}(\mathbf{r}_1, \mathbf{r}_2)$ and $\eta^{(0)}(z_1)$ has been performed for simple choices of $g_\alpha^{(0)}(z_1)$. For the case of zero surface charge and the choice

$$g_\alpha^{(0)}(z) = \theta(z) \tag{3.37}$$

where $\theta(z)$ is the Heaviside step function, one finds[27,76]

$$\phi^{(0)}(\mathbf{r}_1,\mathbf{r}_2) = \frac{1}{\varepsilon}\frac{e^{-\kappa r_{12}}}{r_{12}} + \frac{1}{\varepsilon}\int_0^\infty dk\,\frac{k}{p}J_0(kR)\frac{\varepsilon p - \varepsilon_w k}{\varepsilon p + \varepsilon_w k}\exp[-p(z_1 + z_2)]$$

(3.38)

where $p^2 = k^2 + \kappa^2$, so that

$$\Delta\eta^{(0)}(z_1) = \frac{1}{\varepsilon}\int_0^\infty dk\,\frac{k}{p}\frac{\varepsilon p - \varepsilon_w k}{\varepsilon p + \varepsilon_w k}\exp(-2pz_1)$$

(3.39)

The first term in Eq. (3.38) is simply the Debye-Hückel bulk correlation function, which we would expect to recover given the present weak-coupling assumptions. For the special cases $f = 1, -1$ ($\varepsilon_w = 0, \infty$), Eq. (3.38) has the form

$$\phi^{(0)}(\mathbf{r}_1,\mathbf{r}_2) = \frac{1}{\varepsilon}\frac{e^{-\kappa r_{12}}}{r_{12}} + \frac{f}{\varepsilon}\frac{\exp(-\kappa r_{12}^{im})}{r_{12}^{im}}$$

(3.40)

and

$$\Delta\eta^{(0)}(z_1) = \frac{f\exp(-2\kappa z_1)}{2\varepsilon z_1}$$

(3.41)

a result obtained by Wagner[77] and Onsager and Samaras[78] in their study of the surface tension of electrolytes. Their result shows the expected screening of the image potential by the electrolyte and the expected result that a metal surface ($\varepsilon_w = \infty$) gives rise to an extra attractive potential of mean force and an air-water interface ($\varepsilon_w \ll \varepsilon$) leads to repulsion of ions from the interface (and thus an increase in surface tension with electrolyte concentration). Even in the absence of images, $f = 0$, there is a small extra repulsion due to the fact that electrolyte is excluded from the region $z < 0$.[27,79] The quantity $\Delta\eta^{(0)}(z)$ as given by Eq. (3.39) is shown in Fig. 4 for various values of f. It is significant only for distances less than κ^{-1} from the surface.

Even though the quantity $\phi^{(0)}(\mathbf{r}_1,\mathbf{r}_2)$ gives a very crude representation of ion-ion correlations, it obeys some of the exact conditions of Section II.D. Taking $g_a(z)$ to be given by Eq. (3.37), $\phi^{(0)}(\mathbf{r}_1,\mathbf{r}_2)$ obeys the local electroneutrality condition [Eq. (2.78)] and shows long range transverse correlations $\sim R^{-3}$ that obey the sum rule Eq. (2.81).[79,80]

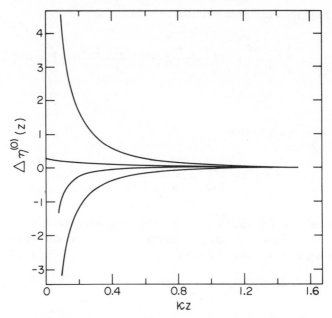

Fig. 4. The fluctuation effect assuming constant ion profiles. $\Delta\eta^0(z)$, given by Eq. (3.39), in units of κ/ε, as a function of κz. From top to bottom, the lines correspond to $f = 1.0, 0, -0.25$, and -0.75.

For symmetric electrolytes at zero surface charge, the extra potential of mean force due to images will act equally on both species of ions so that no separation of charge will occur, and the surface potential will remain zero. For asymmetric electrolytes, however, the image forces will affect, for example, divalent ions more than monovalent ions, so that a surface potential will be generated at zero surface charge.[81]

Calculations of $\phi^{(0)}(\mathbf{r}_1, \mathbf{r}_2)$ have also been carried out for the cases where $g_\alpha^{(0)}(z)$ is given by LGC theory, Eq. (3.10),[82] and where $g_\alpha(z)$ is given by GC theory,

$$g_\alpha^{(0)}(z_1) = \exp[-\beta q_\alpha \psi(z_1)] \tag{3.42}$$

where $\psi(z)$ is given by Eq. (3.12).[73,82] Again, $\phi^{(0)}(\mathbf{r}_1, \mathbf{r}_2)$ obeys the local electroneutrality and transverse correlation sum rules.[82] Finally, Williams numerically solved the resultant equation for $\psi^{(1)}$ and estimated the change in ψ_0 due to image effects to be ~ 5% for potentials less than 100 mV ($\beta q \psi_0 < 4$, $q = 1$ protonic charge).[73] As might be expected, there are computational problems with the point-ion model for attractive images. In this case, point ions can contact images of unlike sign; this is signaled by the divergence of $g_\alpha^{(1)}(z)$ at $z = 0$ from Eqs. (3.35) and (3.41). Such divergences can be removed by reintroducing ion size for ion-wall interactions, just as is done in MGC theory, at the expense of complicating the resultant expressions.

The extra potential of mean force represented by $\Delta\eta(z_1)$ in the weak-coupling equations [Eq. (3.23)] is called the fluctuation or self-atmosphere effect. It represents deviations from the mean field approximation of GC theory caused by purely electrostatic correlations between an ion and its own ionic atmosphere, correlations that are described by the fluctuation potential.

When considering the effect of $\Delta\eta(z_1)$ in producing deviations from GC theory, we must be careful to distinguish between deviations caused solely by a better statistical mechanical treatment for the same model (point ions with no images) and those caused by changes in the model (the presence of image potentials). Because of the complicated form for the fluctuation term in general, interpretation is facilitated by considering the WKB approximation[72] to $\Delta\eta(z_1)$:

$$\Delta\eta(z_1) \cong \frac{1}{\varepsilon}\left[\kappa - \kappa(z_1) + f\frac{\exp[-2\kappa(z_1)z_1]}{2z_1}\right] \qquad (3.43)$$

The terms independent of f give the extra potential of mean force for the same model as that used in GC theory. It is exactly the contribution that would be obtained from the method of local thermodynamic balance, assuming that the Debye-Hückel expression for the ionic activity coefficient holds locally. This effect is usually attractive, because in the region of locally high ionic strength near the surface, the local ionic activity coefficient is lower than in bulk solution.

The second term gives the extra potential of mean force to be expected upon changing the model by adding image interactions. Clearly, it has the same form as the Onsager-Samaras solution but with allowance for local variation in screening length. This term can be repulsive or attractive, depending on the type of images. Equation (3.43) has been tested against $\Delta\eta^{(0)}(z)$ (calculated assuming GC ionic profiles) by Williams,[73] who reported agreement to within 10% (40%) for $f = -1$ ($+1$) and $\psi_0 < 100$ mV.

In summary, even in the point-ion model we expect deviations from Gouy-Chapman theory. In the case of attractive images ($f < 0$), both effects serve to attract ions to the surface and so reduce the thickness of the double layer and decrease the surface potential. For repulsive images, the effects work in opposite directions, so without detailed calculations it is not possible to say in what direction the net deviation will be. The calculations of Williams, however, point to a decreased surface potential, even for repulsive images.[73]

IV. DOUBLE-LAYER THEORIES FOR PRIMITIVE
MODEL ELECTROLYTES

Starting with the work of Gouy,[13] Chapman,[14] Onsager and Samaras,[78] Loeb,[71] and Williams,[73] and culminating in the statistical mechanical investigations of Buff and Stillinger,[10,72] the fundamental equations of double-layer theory and their consequences for the point-ion model in the weak-coupling limit had been well studied by 1963. This is not to say that all the consequences had been realized; see, for example, the recent interest in the long-range transverse correlations exhibited by the weak-coupling equations.[79,80,82]

For some time other workers had felt the need to incorporate more detailed effects into the theory, among them:

1. Finite ion size.
2. Ion-concentration dependence of the solvent dielectric constant.
3. Ion polarizability.
4. Electrostriction and dielectric saturation of the solvent.

For a list of early work in the field, see Section 1 of Ref. 39. Clearly, some of these effects are beyond the compass of simple McMillan-Mayer Hamiltonian formulations.

A fundamental description of dielectric saturation and electrostriction in the external field requires a molecular treatment of the solvent (early approximate results for such models are due to MacDonald and Barlow[7]). Ionic polarizability introduces a many-body Hamiltonian at the McMillan-Mayer level.

Only effects 1 and 2 can be treated by the kind of Hamiltonian discussed in Section II.A. While the concentration dependence of the dielectric constant (as defined by Friedman[83]) should strictly be included in our model, the existence of dynamic contributions to experimental results[84] means that we have no reliable estimate of the concentration dependence of the equilibrium dielectric constant. We thus henceforth ignore this effect. The incorporation of finite-ion-size effects is the focus of this section.

The desire to incorporate the above-mentioned multitude of effects postponed for many years the development and numerical implementation of double layer theories for well-defined Hamiltonian models such as the primitive model electrolyte. The earliest work appears to be that of Stillinger and Kirkwood[85] and Bell and Levine,[86] followed by Soviet workers.[27,87] Cluster methods were first employed by Buff and Stillinger.[72] While the Kirkwood hierarchy was used by Bell and Levine,[39,40] who approximately incorporated all the effects listed above, it was left to Outhwaite[88] and Bell and Rangecroft[89] to use the Kirkwood hierarchy for the primitive model double layer problem. Numerical results were not published until 1980,[90] and further refinements to this modified Poisson-Boltzmann (MPB) theory have been made recently.[91,92]

Meanwhile the success of other integral equation theories of bulk primitive model electrolytes, especially the BGY and hypernetted chain (HNC) equations,[62,93,94] led to renewed interest in the double layer problem. Another class of theories was then developed, starting with the work of Blum.[95]

We first discuss the simplest additional terms in the ionic potential of mean force and the qualitatively new features of the ionic profiles that arise once finite ion size is introduced. In later sections we survey the various theories of the primitive model double layer.

A.   The Simplest Effects of Finite Ion Size

The introduction of finite ion size produces (at least) two new kinds of terms in the ionic potential of mean force. These terms produce qualitatively new behavior: oscillatory decay of the charge density in the double layer.

The finite-size effect that is simplest to understand is the excluded volume effect. The number density of ions near the surface is higher than in bulk (except possibly at zero surface charge). The geometric exclusion of an ion from the surface region because of the hard-core interactions with other ions leads to a repulsive contribution to the potential of mean force.

To lowest order in density, the additional potential of mean force for an RPM electrolyte is

$$\beta \Delta W^{\text{ex}}(z_1) = - \pi \int_{z_1 - d}^{z_1 + d} dz_2 \left[(z_1 - z_2)^2 - d^2 \right] \sum_\alpha \rho_\alpha h_\alpha(z_2) \qquad (4.1)$$

which can be written in the form[110]

$$\Delta W^{\text{ex}}(z_1) = \int_{z_1}^{\infty} dz_2 \int_{S_2} d\mathbf{S}_3 \cdot \hat{\mathbf{z}} \, kT \sum_\alpha \rho_\alpha g_\alpha(z_3) \qquad (4.2)$$

where S_2 is the sphere centered at z_2 of radius d. At low density, the stress tensor in a fluid is approximated by $kT \sum_\alpha \rho_\alpha g_\alpha(z_1) \mathbf{I}$. The surface integral over S_2 is the z component of the force exerted on a hard-sphere ion at z_2, and $\Delta W^{\text{ex}}(z_1)$ is the work done against this force. Clearly, such a simple interpretation only holds for the low density limit.

Equation (4.1) was given by Bell and Levine[86] and derived by cluster methods by Buff and Stillinger[72] [it corresponds to the graph

1

with one f^0 bond in Eq. (A.10)]. This additional term generally acts to repel all ions from the surface region. It will thus tend to thicken the double layer and so increase the surface potential relative to that given by MGC theory.

A second, perhaps less obvious, effect of finite ion size is a kind of cross-correlation between hard-core and electrostatic interactions. The mean electrostatic potential energy $q_\alpha \psi(z_1)$ incorporates the electrostatic interactions between an ion q_α at $\mathbf{r}_1$ and ions with centers anywhere in the double layer. The simplest correction due to finite ion size is equivalent to the subtraction of electrostatic interactions between ion α at $\mathbf{r}_1$ and ions within d of $\mathbf{r}_1$:

$$\Delta W_\alpha^{\mathrm{cav}}(z_1) = -q_\alpha \int_{V_1} d\mathbf{r}_2 \, u^E(\mathbf{r}_1,\mathbf{r}_2) \sum_\beta q_\beta \rho_\beta(z_2) \tag{4.3}$$

where V_1 is the sphere centered at $\mathbf{r}_1$ of radius d. This term was first given by Buff and Stillinger[72] and corresponds to the low density limit of the graph

with one f^0 bond and one $\mathscr{C}$ bond in Eq. (A.10). When $\Delta W_\alpha^{\mathrm{cav}}(z_1)$ is added to $q_\alpha \psi(z_1)$, it defines the cavity potential according to

$$q_\alpha \psi^{\mathrm{cav}}(z_1) = q_\alpha \psi(z_1) + \Delta W_\alpha^{\mathrm{cav}}(z_1) \tag{4.4}$$

The cavity term $\Delta W_\alpha^{\mathrm{cav}}(z_1)$ represents an extra attractive potential for counterions and a repulsive potential for co-ions. Since counterions predominate in the diffuse double layer, the net result is an attractive correction to the potential of mean force that would lead to a thinner double layer and lower surface potential.

For planar geometry and in the absence of images, the cavity potential takes the simple form

$$\psi^{\mathrm{cav}}(z_1) = \tfrac{1}{2}\left[\psi(z_1+d) + \psi(z_1-d)\right] \tag{4.5}$$

The potential of mean force is thus no longer a local function of the mean electrostatic potential once ion size effects are included. An immediate consequence of Eq. (4.5) is that the potential profile (and hence the charge density profile) can have oscillatory decaying solutions. To see this, ignore $\Delta W^{\mathrm{ex}}(z_1)$ for the moment, insert Eq. (4.5) into Poisson's equation, and specialize to large distances so that we may linearize

$$\frac{d^2\psi(z_1)}{dz_1^2} = \frac{\kappa^2}{2}\left[\psi(z_1+d) + \psi(z_1-d)\right] \tag{4.6}$$

Assuming exponential decay of the form $e^{-\lambda z_1}$, we see that Eq. (4.6) can be satisfied only provided $y = \lambda/\kappa$ satisfies the transcendental equation

$$y^2 - y \cosh \kappa d = 0 \qquad (4.7)$$

which has complex roots (signaling exponential oscillatory decay) for $\kappa d >$ 1.03. This qualitatively new behavior was first noted by Stillinger and Kirkwood.[85] Similar behavior occurs in the BBGY equations for $\kappa d > 1.72$,[38] in the MPB theory for $\kappa d > 1.24$,[88] and in the MSA/MSA and HNC/MSA theories to be discussed in Section IV.D for $\kappa d > 1.23$.[97] The most dramatic manifestation of such behavior is the existence of regions of charge inversion, where the co-ion density exceeds the counterion density; in such circumstances $\psi(z)$ may be nonmonotonic.

From the simple considerations of this and the previous section it is already clear that the ionic potential of mean force, to some low level of approximation, must take the form

$$W_\alpha(z_1) = q_\alpha \psi(z_1) + \Delta W^{HH}(z_1) + q_\alpha \Delta W^{HE}(z_1) + q_\alpha^2 \Delta W^{EE}(z_1). \qquad (4.8)$$

$\Delta W^{HH}(z_1)$ incorporates in some sense ionic correlations purely due to hard-core interactions. $\Delta W^{ex}(z_1)$ is the simplest such term; these effects would be present even for a hard-sphere fluid with no charges present. $\Delta W^{HE}(z_1)$ includes cross-correlations due to the effect of hard-core interactions on electrostatic correlations; $\Delta W^{cav}(z_1)$ is the prime example. Finally, $\Delta W^{EE}(z_1)$ represents purely electrostatic correlations. This term persists to zero ion size, where it becomes the self-atmosphere effects of Section III.B. In general, more complex dependence on q_α is to be expected, but Eq. (4.8) would seem to represent the minimum complexity required of a primitive model double layer theory.

Judging from the simple examples of each term in Eq. (4.8) that we have encountered so far, we would guess that $\Delta W^{HH}(z_1)$ would tend to increase the surface potential relative to MGC theory, $\Delta W^{HE}(z_1)$ would decrease it, and one part of $\Delta W^{EE}(z_1)$ would decrease it while another part could act in either direction, depending on the sign of image interactions. The net effect of all these terms and additional terms of higher order in density can only be gauged by the numerical solution of more complete theories of the primitive model double layer, which we now survey.

B. Modified Poisson-Boltzmann (MPB) Theory

The MPB theory of the double layer is based on the application of the Kirkwood hierarchy and the weak-coupling approximations of Section III.B to the case of primitive model (PM) electrolytes. The theory was initiated with the work of Bell and Levine,[39,40,86] who included many effects extraneous to the primitive model, but whose calculations and comments on the relative importance of the various effects discussed in the previous section remain largely valid for the PM double layer.[40] The chief contributors to the application of MPB theory to the standard PM double layer have been Bell and Rangecroft[89] and Outhwaite, Levine, and Bhuiyan.[88,90-92,96,98] These latter authors greatly refined the original theory and are responsible for a whole family of related theories: MPB1,...,MPB5.[90-92] Our comments will be focused largely on the MPB4[90,98] and MPB5[92] theories. Comparison with Monte Carlo simulations in Sections V and VI is exclusively with the MPB5 theory.

Starting from Eq. (2.56) we have

$$\ln g_\alpha(z_1) = \ln g_\alpha(z_1|\lambda = 0) - \beta q_\alpha \psi(z_1) - \beta q_\alpha \int_0^1 d\lambda \left\{ \eta_\alpha(z_1|\lambda) - \eta_\alpha(\infty|\lambda) \right\}$$

(4.9)

where

$$\eta_\alpha(z_1|\lambda) = \lim_{\mathbf{r}_2 \to \mathbf{r}_1} \left[\phi_\alpha(\mathbf{r}_1, \mathbf{r}_2|\lambda) - \lambda q_\alpha u^C(r_{12}) \right]$$

(4.10)

Specification of MPB theory requires a prescription for $\eta_\alpha(z_1|\lambda)$, just as in the point-ion model, as well as the term $\ln g_\alpha(z_1|\lambda = 0)$. In the point-ion model, $- kT \ln g_\alpha(z_1|\lambda = 0)$, the work required to insert an uncharged point ion at $\mathbf{r}_1$, obviously vanishes. The work required to insert an uncharged hard sphere at $\mathbf{r}_1$ is finite and represents the term ΔW^{HH} of Eq. (4.8).

The determination of $\phi_\alpha(\mathbf{r}_1, \mathbf{r}_2|\lambda)$, and hence $\eta_\alpha(z_1|\lambda)$, follows along lines similar to those outlined in Section III.B. The fluctuation potential satisfies the exact equations for the RPM:

$$\nabla_1^2 \phi_\alpha(\mathbf{r}_1, \mathbf{r}_2|\lambda) = 0 \qquad (z_1 < d/2)$$

(4.11)

$$\nabla_1^2 \phi_\alpha(\mathbf{r}_1, \mathbf{r}_2|\lambda) = -\nabla_1^2 \psi(z_1) - \frac{4\pi}{\varepsilon} \lambda q_\alpha \delta(\mathbf{r}_1 - \mathbf{r}_2) \qquad (z_1 > d/2, r_{12} < d)$$

(4.12)

$$\nabla_1^2 \phi_\alpha(\mathbf{r}_1, \mathbf{r}_2|\lambda) = -\frac{4\pi}{\varepsilon} \sum_\gamma q_\gamma \rho_\gamma(z_1) h_{\gamma\alpha}(\mathbf{r}_1, \mathbf{r}_2|\lambda) \qquad (z_1 > d/2, r_{12} > d)$$

(4.13)

subject to the boundary conditions ϕ_α, $\partial \phi_\alpha / \partial n$ (n is the normal) continuous at $r_{12} = d$ and $z_1 = d/2$, ϕ_α continuous at $z_1 = 0$, and

$$\varepsilon \left. \frac{\partial \phi_\alpha}{\partial z_1} \right|_{z_1 = 0^+} = \varepsilon_w \left. \frac{\partial \phi_\alpha}{\partial z_1} \right|_{z_1 = 0^-} \tag{4.14}$$

Equation (4.11) follows from Eq. (2.35) and the exact result

$$\rho_\alpha(z_1) = 0 \qquad (z_1 < d/2) \tag{4.15}$$

Equation (4.12) follows from Eq. (2.35) and the hard-core condition

$$h_{\alpha\beta}(\mathbf{r}_1, \mathbf{r}_2 | \lambda) = -1 \qquad (r_{12} < d) \tag{4.16}$$

In order to obtain a closed set of equations, we proceed as in Section III.B, use Loeb's closure between $\phi_\alpha(\mathbf{r}_1, \mathbf{r}_2)$ and $h_{\alpha\beta}(\mathbf{r}_1, \mathbf{r}_2)$, and linearize so that Eq. (4.13) becomes

$$\nabla_1^2 \phi_\alpha(\mathbf{r}_1, \mathbf{r}_2 | \lambda) = \kappa^2(z_1) \phi_\alpha(\mathbf{r}_1, \mathbf{r}_2 | \lambda) \qquad (z_1 > d/2, r_{12} > d) \tag{4.17}$$

Using Green's theorem, $\eta_\alpha(z_1 | \lambda)$ may be recast in terms of surface integrals of ϕ_α and $\partial \phi_\alpha / \partial n$ over the surface of the ionic exclusion sphere ($r_{12} = d$) or truncated sphere if $z_1 < d$.[90,92,96,98] In principle, Eqs. (4.12)–(4.14) and (4.17) could be solved to determine ϕ_α and $\partial \phi_\alpha / \partial n$ on the exclusion sphere. In practice, this constitutes an intractable set of boundary value problems, and only approximate solutions, along the line of the WKB solution to Eqs. (3.32)–(3.34), have been attempted. By using electrostatic boundary conditions at $z_1 = 0$, the MPB theory incorporates image effects in a much more natural way than any other current theory of the double layer. The calculation of ion-ion correlations through electrostatic boundary value problems is suitable for the dilute McMillan-Mayer level electrolytes for which the theory was designed. This very feature of the theory, together with the necessarily messy form of the resulting equations, has prevented a more widespread appreciation of the MPB theory by those more familiar with the theory of dense liquids.

Irrespective of the detailed approximations made in the various versions of MPB theory, certain general features appear. Because of the right-hand side of Eq. (4.12), $\eta_\alpha(z_1 | \lambda)$ has a term independent of λ and a term linear in λ. As a result, the equation for $g_\alpha(z_1)$ takes the form[90,92]

$$\ln g_\alpha(z_1) = \ln g_\alpha(z_1 | \lambda = 0) - \beta q_\alpha L[\psi(z)] - \frac{\beta q_\alpha^2}{2\varepsilon d}(F - F_0) \tag{4.18}$$

where $F_0 = \lim_{z_1 \to \infty} F(z_1)$. The term $F - F_0$ comes from the part of $\eta_\alpha(z_1|\lambda)$ linear in λ and plays the role of $\Delta W^{EE}(z_1)$. In the point-ion limit,

$$\frac{\beta q_\alpha^2}{2\varepsilon d}(F - F_0) \to \frac{\beta q_\alpha^2}{2\varepsilon}\left[\kappa - \kappa(z_1) + f\frac{\exp[-2\kappa(z_1)z_1]}{2z_1}\right] \quad (4.19)$$

that is, the WKB approximation to the weak-coupling equations [see Eq. (3.43)].

The term $L[\psi(z_1)]$ arises from the part of $\eta_\alpha(z_1|\lambda)$ independent of λ and corresponds to $q_\alpha[\psi(z_1) + \Delta W^{HE}(z_1)]$. It takes the form

$$L[\psi(z_1)] = \tfrac{1}{2}F[\psi(z_1 + d) + \psi(z_1 - d)] - \frac{F-1}{2d}\int_{z_1-d}^{z_1+d}\psi(z_2)\,dz_2$$

$$(4.20)$$

This term reduces to $\psi(z_1)$ as $d \to 0$; the resemblance to the cavity potential of Section IV.A is clear. The actual form for $F(z_1)$ depends on the detailed approximations, which distinguish the MPB4 theory, say, from MPB5.[90,92]

Although in a sense we see that $F - F_0$ describes self-atmosphere effects and $L[\psi(z)]$ cavity potential-like terms, the actual situation is more complex. Since F appears in $L[\psi(z)]$ and F is a function of κd, the interdependence of ion size and image/electrostatic effects is almost complete.

The specification of the term $\ln g_\alpha(z_1|\lambda = 0)$ completes the MPB theory. The simplest choice, which is the one used in MPB4 theory, is the term of lowest order in density and corresponds to the simple excluded volume term discussed in Section IV.A. Substituting $-\beta\Delta W^{\text{ex}}(z_1)$ for $\ln g_\alpha(z_1|\lambda = 0)$ and linearizing, we obtain the MPB4 form

$$g_\alpha(z_1) = [1 - \beta\Delta W^{\text{ex}}(z_1)]\exp\left\{-\beta q_\alpha L[\psi(z)] - \frac{\beta q_\alpha^2}{2\varepsilon d}(F - F_0)\right\}$$

$$(4.21)$$

A more refined estimate can be found from the BBGY hierarchy.[91] By noting that the hard-sphere potentials give rise to impulsive forces of the form

$$-\beta\frac{d}{dz_1}u_\alpha^S(z_1) = \delta\left(z_1 - \frac{d_\alpha}{2}\right) \quad (4.22)$$

$$-\beta\frac{d}{dr}u_{\alpha\beta}^S(r) = \delta(r - d_{\alpha\beta}) \quad (4.23)$$

the first BBGY equation gives, for $q_\alpha = 0$,

$$g_\alpha(z_1 | \lambda = 0) = \theta\left(z_1 - \frac{d}{2}\right) \exp\left[\int_\infty^{z_1} dz_2 \int d\mathbf{r}_3 (z_2 - z_3) \delta(r_{23} - d)\right.$$

$$\left. \times \sum_\gamma \rho_\gamma(z_3) g_{\gamma\alpha}(z_3, z_2, r_{23} = d; q_\alpha = 0)\right] \qquad (4.24)$$

Finally, using Loeb's closure once again,

$$g_{\alpha\gamma}(z_1, z_2, r_{12} = d; q_\alpha = 0) = \exp\left[-\beta q_\gamma \phi_\alpha(z_1, z_2, r_{12} = d; q_\alpha = 0)\right] \qquad (4.25)$$

Equation (4.25) completes the specification of MPB5 theory, since $\phi_\alpha(r_{12} = d)$ has already been determined in the process of finding $\eta_\alpha(z_1 | \lambda)$.

The numerical solution of the MPB equations proceeds by inserting Eq. (4.18) [using Eq. (4.24)] into Poisson's equation and solving for the potential $\psi(z_1)$. Since the boundary condition Eq. (2.36) is imposed, the condition of electroneutrality about the wall is automatically satisfied. The nonlinear integrodifferential equation for $\psi(z_1)$ is solved by a quasi-linearization technique that reduces the problem to an iteration process, each step of which requires the solution of a linear integrodifferential equation. By expressing integrals in terms of simple quadrature rules and derivatives in terms of finite differences, this linear equation reduces to a system of linear algebraic equations that can be solved by standard numerical methods. At each step, $g_\alpha(z)$ can be found by using Eq. (4.18). Since $g_\alpha(z)$ appears explicitly in Eq. (4.24), the solution consists of two iteration processes, one for $\psi(z)$ and one for $g_\alpha(z)$. Although the surface charge σ may be specified, it has been found that specifying the surface potential ψ_0 produces faster convergence.[90]

Since the term $L[\psi(z)]$ includes advanced arguments, special care must be taken to represent accurately the asymptotic behavior of $\psi(z)$, which is obtained from an analysis of the linearized equation similar to that leading to Eq. (4.7).

In the determination of $\eta_\alpha(z)$, MPB theory has implicitly assumed a form for the ionic pair correlation functions. In principle, many of the exact conditions of Section II.D could be tested, especially Eqs. (2.78) and (2.80)–(2.83). In the point-ion limit, Eq. (2.83) reduces to Eq. (2.78), which is satisfied.[91] The pair correlation function does show algebraic decay parallel to the surface,[99] but numerical checks in the primitive model case have yet to be carried out. The only tabulation of contact densities is for a hybrid theory between MPB4 and MPB5[91]; it shows heartening, although not perfect, agreement with the exact contact condition.

The MPB equations have been solved for $1:1$, $1:2$, $2:1$, and $2:2$ electrolytes[90-92,98] (the reader should check carefully which version of MPB theory is solved in each paper) for attractive and repulsive images and in the absence of images. Solutions have been obtained for most conditions of physical interest, the numerical algorithm proving very efficient. A detailed comparison of structural and thermodynamic ($\psi_{d/2}$ vs. σ) predictions of the MPB5 equations with Monte Carlo simulations is given in Sections V and VI.

C.   Theories Based on the BBGY Equations

By inserting the forms for the hard-sphere forces given in Eq. (4.22) and (4.23), the first BBGY equation becomes

$$
\nabla_1 \ln g_\alpha(z_1) = \delta\left(z_1 - \frac{d_\alpha}{2}\right) - \beta q_\alpha \nabla_1 \psi(z_1) - \beta q_\alpha^2 \nabla_1 v^I(z_1)
$$
$$
+ \int d\mathbf{r}_2 (z_1 - z_2) \sum_\beta \rho_\beta(z_2) g_{\alpha\beta}(z_1, z_2, r_{12} = d_{\alpha\beta}) \delta(r_{12} - d_{\alpha\beta})
$$
$$
- q_\alpha \beta \int d\mathbf{r}_2 \nabla_1 u^E(\mathbf{r}_1, \mathbf{r}_2) \sum_\beta q_\beta \rho_\beta(z_2) h_{\alpha\beta}(\mathbf{r}_1, \mathbf{r}_2) \qquad (4.26)
$$

where the hard-core condition requires

$$
h_{\alpha\beta}(\mathbf{r}_1, \mathbf{r}_2) = -1 \qquad r_{12} < d_{\alpha\beta} \qquad (4.27)
$$

Rather than closing the BBGY hierarchy at the second equation to obtain an equation for $g_{\alpha\beta}(\mathbf{r}_1, \mathbf{r}_2)$, Croxton and McQuarrie[100] have applied Eq. (4.26) directly to the double layer by making approximations on $g_{\alpha\beta}(\mathbf{r}_1, \mathbf{r}_2)$. The simplest approximation is

$$
g_{\alpha\beta}(\mathbf{r}_1, \mathbf{r}_2) \simeq g_{\alpha\beta}^{\text{bulk}}(r_{12}) \qquad (4.28)
$$

which is a kind of superposition approximation in that the existence of an external field is assumed not to affect $\rho_{\alpha\beta}(\mathbf{r}_1, \mathbf{r}_2)$ except through the factors $\rho_\alpha(\mathbf{r}_1)$ and $\rho_\beta(\mathbf{r}_2)$. Equations (4.26) and (4.28) constitute the BGY theory of the double layer.[100]

Close to the surface, Eq. (4.28) must fail; the pair distribution function must be expected to depend on the distance of the ions from the surface. A better approximation, therefore, consists of writing

$$
g_{\alpha\beta}(\mathbf{r}_1, \mathbf{r}_2) = 1 + f_\alpha(z_1) h_{\alpha\beta}^{\text{bulk}}(r_{12}) f_\beta(z_2) \qquad (4.29)
$$

where the function $f_\alpha(z_1)$ is determined by requiring the local electroneutrality condition

$$q_\alpha = -\int d\mathbf{r}_2 \sum_\beta f_\alpha(z_1) h_{\alpha\beta}^{\text{bulk}}(r_{12}) \rho_\beta(z_2) f_\beta(z_2) q_\beta \tag{4.30}$$

to be satisfied. Equations (4.26), (4.29), and (4.30) constitute the BGY + EN theory of the double layer.[100] The method of ensuring local electroneutrality was introduced by Pastor and Goodisman[101] in their study of molten salts.

Although Eq. (4.29) seems a more realistic approximation for $g_{\alpha\beta}(\mathbf{r}_1, \mathbf{r}_2)$, it implies that the pair distribution function is not explicitly dependent on image interactions [there may be an indirect dependence through $f_\alpha(z_1)$]. In order to correct this shortcoming, Croxton and McQuarrie have suggested the approximation[102]

$$h_{\alpha\beta}(\mathbf{r}_1, \mathbf{r}_2) = f_\alpha(z_1) f_\beta(z_2) \left\{ g_{\alpha\beta}^{\text{bulk}}(r_{12}) \exp\left[-\beta W_{\alpha\beta}^{\text{im}}(r_{12}^{\text{im}})\right] - 1 \right\} \tag{4.31}$$

where the image contribution to the potential of mean force is taken to be

$$W_{\alpha\beta}^{\text{im}}(r_{12}^{\text{im}}) = \frac{q_\alpha q_\beta f}{\varepsilon} \exp(-\kappa r_{12}^{\text{im}}) / r_{12}^{\text{im}} \tag{4.32}$$

by extension from the Onsager-Samaras expression for $\phi^{(0)}(\mathbf{r}_1, \mathbf{r}_2)$ [see Eq. (3.40)].

Croxton and McQuarrie have solved the BGY and BGY + EN theories numerically for the cases of 1:1 RPM electrolyte in the absence of images[100] and 1:1 RPM electrolyte with repulsive images ($\varepsilon_w = 1$) at zero surface charge.[103] The bulk distribution functions are taken from the solution of the second BBGY equation under superposition.[94] The functions $g_\alpha(z)$ and $f_\alpha(z)$ are represented by uniformly spaced points out to a distance of about seven Debye lengths from the surface. The integrations are done numerically, and the equations are solved by iteration with mixing of input and output required for convergence, much as in the numerical solution of integral equation theories of bulk liquids.[104]

It immediately follows from the derivation of the contact condition, Eq. (2.84), via the BBGY hierarchy,[56] that, provided $f_\alpha(z) \to 1$ as $z \to \infty$ in the case of the BGY + EN theory, the contact condition will be exactly satisfied in the BGY and BGY + EN theories in the absence of images. In particular,

$$kT \sum_\alpha \rho_\alpha \left(\frac{d_\alpha}{2}\right) = p^{\text{BGY}} + \frac{2\pi\sigma^2}{\varepsilon} \tag{4.33}$$

where p^{BGY} is the pressure obtained from the bulk BGY theory via the virial equation. The numerical test of the contact condition in Table 1 of Ref. 100 therefore yields a test of the accuracy of the method of solution, rather than a test of the theory. Equation (4.33) is satisfied only to within 2% for the BGY theory, but to within 0.3% for the BGY + EN theory. The condition of electroneutrality about the wall is satisfied to within 0.1% for both theories. It must be noted that the BGY-based theories require rather heavy computational effort, because of multiple integrals that must be performed numerically. Because an explicit form for $h_{\alpha\beta}(\mathbf{r}_1, \mathbf{r}_2)$ is given in these theories, many other exact conditions could be (but have not been) tested numerically.

It is difficult to compare the form of BGY-based theories with our earlier discussion in Section IV.A. The cavity potential and excluded volume effect do appear as terms of low order in density that are obtained by replacing $g_{\alpha\beta}(\mathbf{r}_1, \mathbf{r}_2)$ by a step function in r_{12}. Since in general the dependence of $g_{\alpha\beta}(\mathbf{r}_1, \mathbf{r}_2)$ on q_α is not simple, we cannot explicitly separate the ionic potential of mean force as in Eq. (4.8). Furthermore, the point-ion limit is ill-defined for the BGY theories, since it is not clear that the bulk BGY equations have solutions in the limit of point ions, without some further approximations (such as linearization).

D. Hypernetted-Chain (HNC) Theories

Consider a mixture of particles of densities $\rho_0, \rho_1, \ldots, \rho_\nu$, and diameters $d_0, d_1, \ldots, d_\nu$. The bulk OZ equation for this system is (assuming central pair potentials)

$$h_{\alpha\beta}(r_{12}) = c_{\alpha\beta}(r_{12}) + \sum_\gamma \rho_\gamma \int d\mathbf{r}_3 \, c_{\alpha\gamma}(r_{13}) h_{\gamma\beta}(r_{32}) \qquad (\alpha, \beta, \gamma = 0, 1, \ldots, \nu)$$

(4.34)

If we let one species become dilute by taking the limit $\rho_0 \to 0$, Eq. (4.34) separates into two sets of equations, one

$$h_{\alpha\beta}(r_{12}) = c_{\alpha\beta}(r_{12}) + \sum_\gamma \rho_\gamma \int d\mathbf{r}_3 \, c_{\alpha\gamma}(r_{13}) h_{\gamma\beta}(r_{32}) \qquad (\alpha, \beta, \gamma = 1, \ldots, \nu)$$

(4.35)

being the bulk OZ equations describing the correlations between the particles of species $1,\ldots,\nu$, the second

$$h_{\alpha 0}(r_{12}) = c_{\alpha 0}(r_{12}) + \sum_\gamma \rho_\gamma \int d\mathbf{r}_3 \, c_{\alpha\gamma}(r_{13}) h_{\gamma 0}(r_{32}) \qquad (\alpha, \gamma = 1,\ldots,\nu)$$

$$(4.36)$$

being a set of equations describing the correlations between the bulk species $1,\ldots,\nu$ and the dilute species 0. Only the bulk direct correlation functions appear under the integral in Eq. (4.36). By taking the planar limit $d_0 \to \infty$ (strictly, $d_0 \to \infty$ as $\rho_0 d_0^3 \to 0$), the correlations with the dilute species become density profiles against a planar surface. In this planar limit, Eq. (4.36) takes the form

$$h_\alpha(z_1) = c_{\alpha 0}(z_1) + 2\pi \sum_\gamma \rho_\gamma \int_{-\infty}^\infty dz_2 \, h_\gamma(z_2) \int_{|z_1 - z_2|}^\infty dr \, r c_{\alpha\gamma}(r) \quad (4.37)$$

where $c_{\alpha\gamma}(r)$ is the bulk direct correlation function between species α and γ. $c_{\alpha 0}(z_1)$ should not be confused with $c_\alpha(z_1)$ of Eq. (2.57).

Equation (4.37) can be regarded as an integrated form of the usual Ornstein-Zernike equations and as such an appropriate starting point for integral equation approximations of the double layer, similar to those successfully employed in the theory of electrolyte solutions. The limiting procedure leading to Eq. (4.36) was first considered by Perram and White[105]; the planar limit was first given by Henderson, Abraham, and Barker.[106]

In order to extend this formalism to Coulomb systems, let species $1,\ldots,\nu$ have charges $q_1, q_2,\ldots,q_\nu$. The uniform charge density on the surface can be obtained by taking the limit $q_0 \to \infty$ such that $q_0/d_0^2 = \pi\sigma$. The appropriate extension was first given by Henderson and Blum.[107] It can be written in the form

$$h_\alpha(z_1) = \left[c_{\alpha 0}(z_1) + \beta q_\alpha v^C(z_1) \right] - \beta q_\alpha \psi(z_1)$$
$$+ 2\pi \sum_\gamma \rho_\gamma \int_{-\infty}^\infty dz_2 \, h_\gamma(z_2) \int_{|z_1 - z_2|}^\infty dr \, r c_{\alpha\gamma}^S(r) \qquad (4.38)$$

where $c_{\alpha\gamma}^S(r)$ is the short-range part of the bulk direct correlation function

$$c_{\alpha\gamma}^S(r) = c_{\alpha\gamma}(r) + \beta q_\alpha q_\gamma u^C(r) \qquad (4.39)$$

A closure between $h_\alpha(z)$ and $c_{\alpha 0}(z_1)$ is now required before Eq. (4.38) can be used to calculate $h_\alpha(z)$. Obvious choices are the approximate closures employed in bulk electrolyte theory, the mean spherical approximation

(MSA)

$$c_{\alpha 0}(z_1) = -\beta u_{\alpha 0}(z_1) = -\beta q_\alpha v^C(z_1) \qquad (z_1 > d_\alpha/2)$$
$$h_\alpha(z_1) = -1 \qquad\qquad\qquad\qquad (z_1 < d_\alpha/2)$$

(4.40)

and the hypernetted-chain closure (HNC)

$$c_{\alpha 0}(z_1) = -\beta q_\alpha v^C(z_1) + h_\alpha(z_1) - \ln[1 + h_\alpha(z_1)] \qquad (z_1 > d_\alpha/2)$$
$$h_\alpha(z_1) = -1 \qquad\qquad\qquad\qquad\qquad\qquad\qquad (z_1 < d_\alpha/2)$$

(4.41)

We then obtain the OZ equation with MSA wall-ion closure

$$h_\alpha(z_1) = -\beta q_\alpha \psi(z_1) + 2\pi \sum_\gamma \rho_\gamma \int_{-\infty}^{\infty} dz_2\, h_\gamma(z_2) \int_{|z_1 - z_2|}^{\infty} dr\, r c_{\alpha\gamma}^S(r)$$

(4.42)

and with the HNC wall-ion closure

$$\ln g_\alpha(z_1) = -\beta q_\alpha \psi(z_1) + 2\pi \sum_\gamma \rho_\gamma \int_{-\infty}^{\infty} dz_2\, h_\gamma(z_2) \int_{|z_1 - z_2|}^{\infty} dr\, r c_{\alpha\gamma}^S(r)$$

(4.43)

Equation (4.42) is recovered by linearizing Eq. (4.43).

The choice of HNC closure may seem somewhat arbitrary, but several alternative derivations of Eq. (4.43) show the natural appearance of the HNC wall-ion closure. Starting from Eq. (2.57) and using Eq. (2.62), we may functionally expand $c_\alpha(z_1)$ about the field-free state in powers of $\rho_\alpha(\mathbf{r}) - \rho_\alpha$:

$$\ln\left[\frac{\rho_\alpha(z_1)}{\delta_\alpha}\right] + \beta v_\alpha(z_1) = \ln\left[\frac{\rho_\alpha}{\delta_\alpha}\right] + \int d\mathbf{r}_2 \sum_\beta \left(\frac{\delta c_\alpha(\mathbf{r}_1)}{\delta \rho_\beta(\mathbf{r}_2)}[\rho_\beta(\mathbf{r}_2) - \rho_\beta]\right)$$

$$+ \frac{1}{2}\int d\mathbf{r}_2\, d\mathbf{r}_3 \sum_{\beta\gamma}\left(\frac{\delta^2 c_\alpha(\mathbf{r}_1)}{\delta\rho_\beta(\mathbf{r}_2)\,\delta\rho_\gamma(\mathbf{r}_3)}[\rho_\beta(r_2) - \rho_\beta]\right.$$

$$\left.\times[\rho_\gamma(\mathbf{r}_3) - \rho_\gamma]\right) + \cdots \quad (4.44)$$

Truncating at the second term in the spirit of Percus,[108] we obtain

$$\ln g_\alpha(z_1) = -\beta v_\alpha(z_1) + \sum_\beta \rho_\beta \int d\mathbf{r}_2\, h_\beta(z_2) c_{\alpha\beta}(\mathbf{r}_{12})$$

(4.45)

which is equivalent to Eq. (4.43). Applying the approximation

$$c_{\alpha\beta}^{S}(\mathbf{r}_1, \mathbf{r}_2) \simeq c_{\alpha\beta}^{S,\text{bulk}}(r_{12}) \tag{4.46}$$

to Eq. (2.67), or the approximation

$$c_{\alpha\beta}^{S}(\mathbf{r}_1, \mathbf{r}_2 | \lambda) \simeq c_{\alpha\beta}^{S}(\mathbf{r}_1, \mathbf{r}_2 | \lambda = 0) = c_{\alpha\beta}^{S,\text{bulk}}(r_{12}) \tag{4.47}$$

to Eq. (2.71) again produces Eq. (4.43). These last two derivations make the approximations inherent in the HNC wall-ion closure somewhat more transparent; the effect of the external potential and inhomogeneous environment on the direct correlation function has been entirely neglected. One might hope that this approximation is less severe than a similar approximation on $h_{\alpha\beta}(\mathbf{r}_1, \mathbf{r}_2)$ as found in the BGY theory.

So far the bulk direct correlation functions appearing in Eq. (4.42) and (4.43) have been left unspecified. Ideally, one would employ the exact functions (determined from simulations, for example). In practice, however, approximate values of $c_{\alpha\beta}^{S}(r)$ are used. The particular choice for $c_{\alpha\beta}^{S}(r)$ must be stated before the theory is completely specified. [In this sense, the theory of Ref. 100 might be termed the BGY/BGY theory, since values of $h_{\alpha\beta}^{\text{bulk}}(r_{12})$ have been calculated using the BGY theory of bulk electrolytes.]

Three approximate theories have arisen from this approach: the MSA wall-ion closure [Eq. (4.42)] together with the MSA $c_{\alpha\beta}^{S}(r)$ (MSA/MSA)[95,107]; the HNC wall-ion closure [Eq. (4.43)] together with the HNC $c_{\alpha\beta}^{S}(r)$ (HNC/HNC),[109] and a hybrid theory derived from Eq. (4.43) together with $c_{\alpha\beta}^{S}(r)$ given by the mean spherical approximation (HNC/MSA).[110]

The MSA/MSA theory for RPM double layer has two notable features: it is analytically soluble, and it is an inherently linear theory. The potential at the plane of closest approach is[95]

$$\psi_{d/2} = \frac{4\pi\sigma}{\varepsilon\kappa} \left(\frac{1 + (1 + 2\kappa d)^{1/2}}{2} \right) \tag{4.48}$$

which reduces to the LGC value as $d \to 0$. The contact condition is given by[107]

$$\sum_{\alpha} g_{\alpha}\left(\frac{d}{2}\right) = \left(kT \sum_{\alpha} \frac{\rho_{\alpha}}{\chi_T} \right)^{1/2} \tag{4.49}$$

where χ_T is the isothermal compressibility of the electrolyte determined via the compressibility equation[21]

$$\chi_T^{-1} = kT\sum_\alpha \rho_\alpha - kT\sum_{\alpha\beta} \rho_\alpha\rho_\beta \int d\mathbf{r}\, c_{\alpha\beta}^S(r) \qquad (4.50)$$

in the mean spherical approximation. Thus the MSA/MSA theory is completely lacking the required σ^2 dependence of the contact density. The ionic profiles have the same unrealistic antisymmetry as seen in LGC theory, and the co-ion density profiles become unphysical for moderate surface potentials. These features rule out the MSA/MSA theory as a viable theory of the double layer, except possibly at the point of zero charge.

Most of these undesirable consequences of linearization are avoided in the two theories based on Eq. (4.43). They contain MGC theory as the first term in the ionic potential of mean force and satisfy the equation[110]

$$kT\sum_\alpha \rho_\alpha\left(\frac{d}{2}\right) = \frac{1}{2}\left(kT\sum_\alpha \rho_\alpha + \chi_T^{-1}\right) + \frac{2\pi\sigma^2}{\varepsilon} \qquad (4.51)$$

with χ_T determined via the compressibility equation from the appropriate bulk direct correlation functions. Since for typical ionic densities in electrolytes, the first term in Eq. (4.51) is slightly different from p (or $kT\sum_\alpha\rho_\alpha$), the HNC/MSA and HNC/HNC theories give contact values that are essentially exact (and essentially equal to the MGC value) except at low σ.

Given the accuracy of Eq. (4.46) or (4.47), one would expect the HNC/HNC theory to be superior to the HNC/MSA because, for bulk electrolytes, the HNC results are superior to the MSA for both structural and thermodynamic quantities at all conditions of physical interest.[93] Although the theoretical direct correlation functions are never tested explicitly, it would be difficult to believe that the superiority of the HNC theory did not extend to this quantity as well. The same difficulties in extracting simple terms of the form of Eq. (4.8) encountered for the BGY theories arise once more for the HNC/HNC theory, since presumably $c_{\alpha\beta}^{S,\text{HNC}}(r)$ depends in a complicated way on q_α. Again the point-ion limit is ill-defined, there being no solution to the HNC equations for point-ion electrolytes. It is not even clear how the theory should behave at infinite dilution, since there is no general way to estimate the second term of Eq. (4.43) in that limit.

Since the $c_{\alpha\beta}^{S,\text{MSA}}(r)$ have an analytic form,[111] the HNC/MSA equations can be analyzed much more easily. In particular, the HNC/MSA equations reduce to MGC theory in the point-ion limit or at infinite dilution. To lowest order in density, the ionic potential of mean force has the form[110]

$$W_\alpha^{\text{HNC/MSA}}(z_1) = q_\alpha\psi(z_1) + \Delta W^{\text{ex}}(z_1) + \Delta W_\alpha^{\text{cav}}(z_1) \qquad (4.52)$$

and in general, it has the form

$$W_\alpha^{\text{HNC/MSA}}(z_1) = q_\alpha \psi(z_1) + \Delta W^{HH}(z_1) + q_\alpha \Delta W^{HE}(z_1) \qquad (4.53)$$

where the specific formulas for ΔW^{HH} and ΔW^{HE} can be readily deduced from the equations of Ref. 110. From both Eq. (4.53) and the point-ion limit, we see that the HNC/MSA equations entirely neglect the purely electro-static correlations denoted in Eq. (4.8) by ΔW^{EE}, which appear in MPB theory as the term $F - F_0$ and which give rise to the self-atmosphere effects discussed in Section III.B. Since the neglected term is proportional to q_α^2, one would expect this deficiency of the HNC/MSA theory to worsen as the va-lence of the electrolyte is increased.

A sobering feature of these HNC theories is the difficulty in obtaining numerical solutions. The HNC/MSA equations have been solved for $1:1,$[110,112] $2:1, 1:2,$[113] and $2:2$[112] electrolytes for various conditions, but the HNC/HNC equations have been solved only for 1 M, $1:1$ electrolytes[109] or, at lower concentrations, only for surface charges so low that deviations from MGC theory are very small.[114] Even the HNC/MSA equations require heroic efforts to ensure convergence.[112]

The troublesome feature of these integral equations is the need to solve Eq. (4.43) subject to the condition of electroneutrality, Eq. (2.77). In Ref. 110 this was achieved by rescaling the ionic profiles so that Eq. (2.77) was satisfied at each iteration step. In later work, the surface potential rather than the surface charge is specified, but very high mixing of input and output is re-quired for convergence.[112] It may be that other algorithms would handle the electroneutrality constraint in a better way. In particular, the use of a Lagrange multiplier as in the HNC/MSA theory for two-plate geometry[115] may be profitable (although a decaying Lagrange multiplier function is re-quired for the single-plate case). Alternatively, by solving Poisson's equation subject to the usual boundary conditions, the constraint is automatically satisfied. This suggests reformulating the integral equations so that a differ-ential equation must be solved at each iteration step, much as in MPB the-ory[90] or recent work on the one-component plasma.[116]

Since the ionic pair distribution function never enters explicitly into the HNC/MSA equations, we are not able to test the various exact condi-tions that connect the one- and two-particle densities [Eqs. (2.78) and (2.80)–(2.83)]. The contact condition can be tested; in Ref. 110 it was satisfied to within 1%. The claimed accuracy for $2:2$ electrolytes is 0.05% for $g_\alpha(z_1)$ and 0.2% for $\psi_{d/2}$.[112]

Finally we mention a striking feature of the HNC-based theories: their fundamental inability to handle image interactions. In the derivation from the bulk OZ equations, this restriction arises because of the need for pair potentials at the outset. In the derivations from Eq. (2.67) or (2.71), it arises because the presence of images fundamentally changes the nature of the interaction part of the Hamiltonian (such as the property of translation invariance) so that the usual tenets of density functional theory are not valid. No theoretically satisfactory way to overcome these obstacles has been proposed to date.

V. COMPARISON OF THEORY WITH SIMULATION: SYMMETRIC ELECTROLYTES

A. Monte Carlo Calculations

In double-layer theory, as in condensed matter physics generally, computer experiments fulfill their customary role of providing numerically exact results for the microscopic structure and thermodynamic properties of systems with precisely specified Hamiltonians. Such results are useful in several ways. First, by observing the effect on these results of systematic variations of the Hamiltonian, the relevance of various features of the model can be examined. It is often easier and usually more reliable to do this with computer simulations than by extensions of one theory or another. In addition, many intrinsically interesting microscopic properties not readily accessible to experimental probes in real systems are routinely obtained in computer experiments; the variation of ion densities as a function of distance from the charged surface is a good example in the present context. Finally, by carrying out computer experiments for exactly the models to which theoretical approximations are applied, direct tests of these competing theoretical approaches are possible, free of the ambiguities that would otherwise arise from a direct comparison of an approximate theory of a particular model with experimental data on a real system. This is especially important for the double layer problem, because the present level of modeling of these systems (particularly the use of a continuum solvent and the naive treatment of the surface) is far too crude to allow much meaningful comparison to be made with real experimental data.

There are two principal kinds of computer experiments that are carried out on model systems. In molecular dynamics,[117] an actual trajectory of a system of a few hundred particles is followed by numerical integration of the classical equations of motion for the given microscopic Hamiltonian. In addition to static equilibrium properties, transport phenomena can be studied in this way. In electrolyte solutions, however, the latter are determined largely by ion-solvent interactions and hence are not of interest for the continuum solvent models being considered here.

Furthermore, almost all theoretical work on such continuum solvent models of bulk electrolytes has focused on the charged hard-sphere primitive models of this review.[62,93,118,119] The combination of a hard-core repulsion with the continuous electrostatic term in the pair potentials of such models introduces formidable technical difficulties into the molecular dynamics algorithm. For these reasons, virtually all computer experiments on such models in the bulk[12,120-122] have been of the Monte Carlo type. The flexibility in the choice of thermodynamic constraints allowed in the Monte Carlo method makes it still more appropriate for the study of interfacial phenomena, so it is not surprising that this method has been used in all computer experiments so far carried out on model double layers.

The essence of a Monte Carlo experiment is a random walk through the phase space of the model system, in the course of which a sequence of microscopic states of the system is generated. The prescription governing this random walk can be designed to maintain fixed values for any of the usual sets of thermodynamic variables; the microscopic states may then be treated as members of the corresponding statistical mechanical ensemble for the system. Static equilibrium properties are then computed as averages over the sequence. The most common application of the technique has been to bulk fluids, using the "Metropolis" method in the canonical ensemble. This has been reviewed elsewhere.[123-125] The embellishments of the standard Monte Carlo technique that have been developed for the double layer problem are discussed in Appendix B. Here we give only a brief description of the simulation cell before specifying the model parameters that have been used. This central Monte Carlo cell, shown schematically in Fig. 5, is a rectangular prism $W \times W \times L$ closed by impenetrable planes at $z = 0$ and $z = L$. In order to minimize *unwanted* surface effects, this cell is considered to be replicated periodically in the x and y directions to create an infinite slab of thickness L. The central cell contains an assembly of N_+ cations of diameter d_+, each carrying a charge of q_+, and N_- anions of diameter d_- and charge q_-. As discussed in Appendix B, a *grand canonical* Monte Carlo method is used so that N_+ and N_- fluctuate as the calculation proceeds but in such a way that the excess charge $q_+ N_+ + q_- N_-$ remains fixed. Electroneutrality is maintained by a compensating surface charge which, in most cases, is divided equally between the two hard walls at $z = 0$ and $z = L$ (an exception is discussed in Appendix B). The length L must then be large enough to ensure complete separation of the two identical double layers that form at each end of the cell. Essential to the Monte Carlo computation are accurate values for those changes in the energy of the system attributable to central cell charges. This energy comprises, first, straightforward coulombic terms involving the ionic and surface charges within the central cell. In addition, because there is a net mean charge separation throughout the infinite slab in which the central cell is embedded, there will be a net force exerted on each ion by the charge distribution external to the central cell with a corresponding contri-

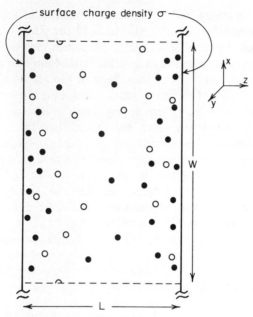

Fig. 5.   Schematic representation of a typical computer experiment "cell" containing an excess of anions (●) exactly neutralized by the uniform surface charges at $z = 0$ and $z = L$.

bution to the energy. The precise form of each of these energy terms depends, of course, on the details of the model such as ionic charge and presence of image forces. A discussion of these matters can be found in Appendix B.

All the results we discuss in this section concern the simplest case, a *symmetric restricted* primitive model electrolyte next to a uniformly charged plane surface bounding a semiinfinite region ($z < 0$) whose dielectric constant ε_w matches that of the continuum solvent, ε. Hence $f = 0$ in Eq. (2.5), there are no image forces, and the surface is characterized solely by its charge density σ which, for symmetric electrolytes, we take to be positive. We will use a dimensionless surface charge σ^* defined by

$$\sigma^* \equiv \sigma(4.25 \times 10^{-10} \text{ m})^2/e \qquad (5.1)$$

where e is the charge on a proton, an admittedly arbitrary reduction historically rooted in the choice of 4.25 Å as the ion diameter in much of the Monte Carlo work. Then $\sigma \doteq \sigma^* \times 0.887$ C m^{-2} and $1/\sigma \doteq 18.2/\sigma^*$ Å^2 per elementary charge. The electrolyte is characterized by the chemical potential driving the grand canonical Monte Carlo procedure (this implicitly fixes the bulk concentration with which the double layer is in equilibrium) and a dimensionless interaction strength β^* defined by

$$\beta^* \equiv \frac{q^2}{\varepsilon kTd} \tag{5.2}$$

where $q_+ = -q_- = q$, $d_+ = d_- = d$, T is the absolute temperature, and k is Boltzmann's constant. For a given bulk electrolyte and fixed surface charge, we are interested in the ionic density profiles $g_+(z)$ and $g_-(z)$ defined in Eq. (2.20) and the mean electrostatic potential $\psi(z)$ defined in Eq. (2.32). The dimensionless potential ψ^* is defined by

$$\psi^*(z) \equiv \psi(z)\frac{e}{kT} \tag{5.3}$$

At $T = 298°K$, $\psi \doteq 25.7 \text{ mV} \times \psi^*$. As discussed in Section III.A, the imposition of a finite distance of closest approach to the surface of $\frac{1}{2}d$ leads to a trivial linear variation of $\psi(z)$ between $z = 0$ and $z = \frac{1}{2}d$. We shall be interested in the behavior of ψ only for $z \geq \frac{1}{2}d$; of special importance is the diffuse layer potential $\psi_{d/2} = \psi(\frac{1}{2}d)$.

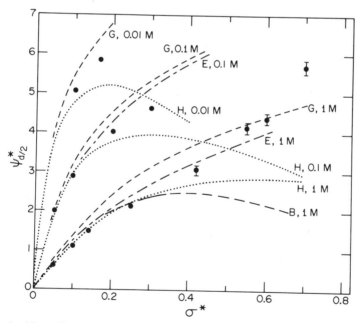

Fig. 6.   Monte Carlo and various theoretical results for the diffuse layer potential $\psi^*_{d/2}$ (Eq. (5.3)) as a function of surface charge density σ^* (Eq. (5.1)) for a model $1:1$ electrolyte. $G(\text{---})$, MGC theory; $B(\text{——})$, BGY theory; $E(\text{----})$, BGY + EN theory; $H(\cdots)$, HNC/HNC theory. The data points ($\bullet$) are Monte Carlo results.

B. 1:1 Aqueous Electrolytes

We consider first the results of both computer experiments[126,127] and theories[92,100,109,110,112,128] for a choice of parameters appropriate to a 1:1 aqueous electrolyte at room temperature: $q = e$, $\varepsilon = 78.5$, $T = 298°K$. Except where noted below, $d = 4.25$ Å and hence $\beta^* = 1.6809$. In Fig. 6 and 7 are shown Monte Carlo and various theoretical results for the diffuse layer potential $\psi_{d/2}^*$ as a function of surface charge for a series of concentrations, the double-layer "equation of state." We focus our attention first on the predictions of the modified Gouy-Chapman (MGC) theory of Section III.A, (labeled G). The differences between these MGC potentials and the Monte Carlo results increase rather gradually with surface charge, and, for concentrations up to 0.1M and surface charges up to $\sigma^* = 0.05$ ($\doteq 0.044$ C m^{-2}), the MGC potential is within 3 or 4 percent of the Monte Carlo value. Although our emphasis here is on the behavior of the primitive model in a broader context as a prototype electrified interface, we point out that, at least for 1:1 aqueous systems, this narrower range of parameters spans most of the conditions applicable to real double layers and covers all of the situations in which the diffuse layer makes an important contribution to, say, the total differential capacitance. Throughout this region the structure of the in-

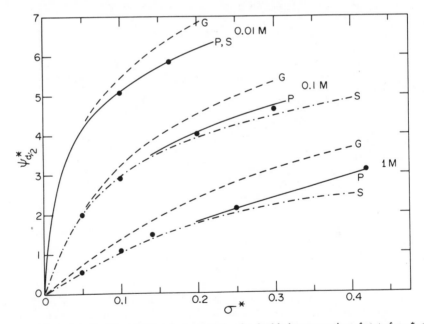

Fig. 7.   Monte Carlo (●) and theoretical results for the double-layer equation of state for $\sigma^* \leq$ 0.40. G(---), MGC theory; S(—·—), HNC/MSA theory; P(——), MPB5 theory.

terface found in the Monte Carlo experiments is essentially that already displayed in Fig. 3 for the MGC theory. This quantitative success of MGC theory extends well beyond potentials for which the linearization of the Poisson-Boltzmann equation (the interfacial analogue of Debye-Hückel theory) would be a sensible approximation. A similar situation exists in the application of the Poisson-Boltzmann equation to the bulk electrolyte problem,[62] as might be expected from the discussion in Section III.A of the relationship between the bulk and surface problems. In both cases the internal inconsistencies of the Poisson-Boltzmann equation seem to have small numerical consequences well beyond the weak-coupling regime.

As the surface charge increases beyond $\sigma^* = 0.05$, however, MGC theory begins to overestimate $\psi_{d/2}^*$ by an appreciable amount. Keeping in mind the identification of $\psi_{d/2}^*$ as a measure of the thickness of the interfacial regime made previously [cf. the discussion accompanying Eqs. (2.32) and (2.33)], we see that MGC theory is predicting too extended a double layer. Of course, this is a theory for point ions, that is, for a different *model* than is (or could be) used in the Monte Carlo work. Thus, in analyzing the cause of these deviations, we would like to be able to distinguish those that are due to finite ion size from those that arise primarily from the theoretical treatment of the point-ion model, the mean field approximation. From this point of view, both the excluded volume and cavity effects of Section IV.A result from correcting shortcomings of the model used in MGC theory in the sense that they vanish in the point-ion limit. On the other hand, the local depletion of charge near a fixed ion that results when electrostatic contributions to interionic correlations are no longer neglected is primarily a correction to the mean field

TABLE I

Monte Carlo Results for the Diffuse Layer Potential at $\sigma^* = 0.20$ in a 0.1 M 1 : 1 Electrolyte for Various Ion Diameters d

d (Å)	$\psi_{d/2}^*$ [a]
0 (MGC)	4.54
1.0	4.15 (.05)
2.76	3.94 (.04)
4.25	4.02 (.05)

[a] The quantities in parentheses are the statistical uncertainties in the Monte Carlo potentials.

approximation and is present in a correct treatment even of the point-ion model. Since both this fluctuation effect and the cavity effect operate to reduce the screening and so to constrict the double layer, the results shown in Fig. 6 imply that together they have overcome the simple excluded volume effect for $\sigma^* < 0.5$. This is hardly surprising considering that a typical ionic density in these systems is only a few percent of that of a normal liquid. This fact, together with the observation that the errors in the MGC potential do not depend very strongly on the concentration at fixed σ, suggests that the cavity effect is less important than the fluctuation effect, at least where the error in the MGC potential is more than a few percent. Although no systematic study of the ion size dependence of the behavior of double layers in the primitive model has been made, some support for the view that the net

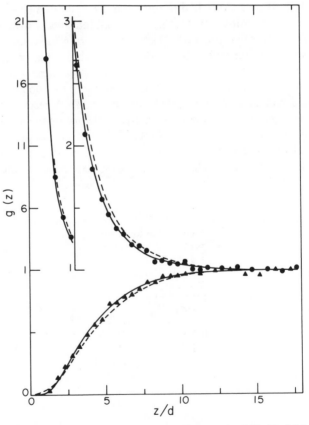

Fig. 8. Ion density profiles for a 0.1 M 1:1 electrolyte at $\sigma^* = 0.30$. ($\bullet$,$\blacktriangle$) Monte Carlo results. All other symbols as in Figure 7.

effect of ion size is small is provided by the data in Table I, which show the Monte Carlo results for $\psi^*_{d/2}$ at $0.10M$ and $\sigma^* = 0.20$ to be a very mild function of ion size.

Even when the error in the total diffuse layer potential predicted by MGC theory has become quite significant, however, the effect on the structure of the interfacial region is not very striking. For example, in Fig. 8 are shown the Monte Carlo and MGC ion density profiles at $0.10M$ for $\sigma^* = 0.30$. Superficially, the agreement appears to be quite good. Because the small differences in these profiles are systematic, however, the errors in the theory are easier to see once these profiles have been integrated to give the mean electrostatic potential shown in Fig. 9 for the same system. In double layers, then, we have the somewhat unusual situation that an integrated quantity is the more sensitive test of theory.

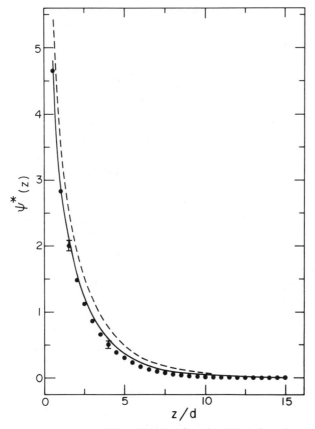

Fig. 9.   Mean electrostatic potential profile $\psi^*(z)$ for a 0.1 M 1 : 1 electrolyte at $\sigma^* = 0.30$. All symbols as in Fig. 7.

As the concentration of a bulk electrolyte is increased, the structure of the system gradually builds up until the pair correlation function between oppositely charged ions becomes oscillatory.[62] The discussion of Section IV.A leads us to expect analogous behavior of the ion density profiles of the double layer for suitable extremes of concentration and surface charge. For $1:1$ aqueous systems, however, such oscillations in the double layer profiles turn out to be quite difficult to obtain and are always convoluted with excluded volume effects that are largely absent in the bulk problem. In Fig. 10 are shown the ion density profiles for a $1.0\ M$ electrolyte at $\sigma^* = 0.25$. There are significant discrepancies between the MGC predictions and the Monte Carlo results, to be sure, particularly in the shape of the co-ion profile, but there is no evidence of any tendency toward oscillatory behavior in the Monte Carlo profiles. Again the small differences in the counterion density profiles are magnified in taking the first moment of the charge distribution leading to the substantial discrepancy between the MGC and MC values for $\psi^*(z)$ seen in Fig. 11. At still higher σ^*, the accumulation of counterions at the surface eventually leads to local densities large enough to produce large excluded volume effects that extend the double layer. At $\sigma^* = 0.70$ these effects are sufficient to cause the Monte Carlo diffuse layer potential to lie *above* the MGC prediction (cf. Fig. 6). The ion density profiles for this surface charge are shown in Fig. 12. The counterion profile is no longer monotone but shows instead a secondary maximum near $z = \frac{3}{2}d$. The excluded volume constraint has "interfered" with the purely electrostatic response of the system to the large field, causing a second layer of counterions to form against the initial near-monolayer. Rather than the charge inversion that might have been expected under these extreme conditions, there occurs a layering of counterions instead. If the bulk concentration is increased still further to $2.0\ M$, then a small but well-defined charge inversion does become visible in the ion profiles. An example of this behavior at $\sigma^* = 0.396$ is shown in Fig. 13, where a charge inversion is evident between $z = 2d$ and $z = 3d$. The position of the shoulder in the counterion profile near $z = \frac{3}{2}d$, reflecting once again the packing problems at the surface, suggests that the form and even the presence of the charge inversion would be strongly dependent on the size assigned to the ions in the Hamiltonian. If sufficiently pronounced, such an inversion can produce a nonmonotonic mean electrostatic potential; if the Monte Carlo $\psi(z)$ for this state, shown in Fig. 14, does in fact have a minimum, near $z = 1.75d$ say, it is too shallow to be resolved from the statistical noise in the data.

We turn now to a discussion of the predictions of the more sophisticated statistical mechanical theories described in Section IV. We begin by considering the BGY theory of the double layer, described in Section IV.C. The BGY equation of state at $1.0\ M$ is labeled B in Fig. 6. Unfortunately, these

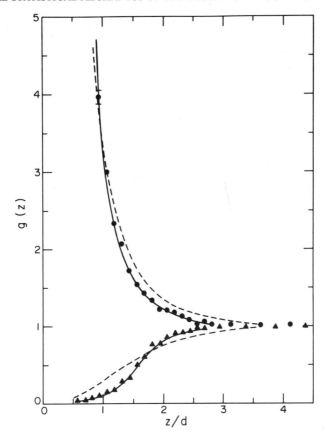

Fig. 10.   Ion density profiles for a 1.0 M 1:1 electrolyte at $\sigma^* = 0.25$. All symbols as in Fig. 7. (The HNC/MSA profiles are not shown separately because they are almost identical to the MPB5 profiles.)

results are for an ion diameter of 2.76 Å; however, it is clear from the discussion above regarding the role of ion size that where the deviations of the BGY potential from the Monte Carlo results are greatest, ion size is not an important factor. Below $\sigma^* = 0.20$ the BGY theory is fairly successful in correcting the shortcomings of the MGC equation of state. At higher surface charges, however, the BGY theory greatly exaggerates the curvature in the equation of state and predicts a maximum in $\psi_{d/2}^*(\sigma^*)$. Although such behavior of the *diffuse* layer potential does not, of itself, contradict the exact thermodynamic constraint Eq. (2.76) places on the *total* potential, such a violation will occur at extremely large (unphysical) σ^* if the curvature evident in Fig. 6 persists until the slope $d\psi_{d/2}^*/d\sigma^*$ becomes more negative than

$-2\pi\beta^*$ [cf. Eq. (3.13)]. We will return to the question of a maximum in $\psi^*_{d/2}(\sigma^*)$ several times in the subsequent discussion. For now, we note that there is no suggestion of such behavior in the Monte Carlo results for $1:1$ aqueous systems and that prediction of a maximum with pronounced curvature for such systems is a serious shortcoming of a theory. In the case of the BGY theory, Croxton and McQuarrie[100] have suggested that this excessive curvature is a consequence of exaggerating charge oscillations (indeed there are none at 1.0 M in the Monte Carlo results for $d = 4.25$ Å) and that this in turn is connected with the failure of the BGY theory to satisfy the *local* electroneutrality condition, Eq. (2.78). One way of remedying this defect results in the BGY + EN theory described in Section IV.C. The results of the BGY + EN theory for the equation of state at 0.1 and 1.0 M are also shown (labeled E) in Fig. 6. Where the BGY theory was accurate, $\sigma^* < 0.20$, the

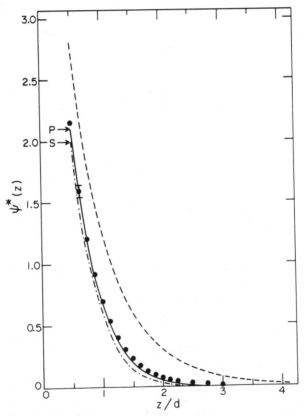

Fig. 11. Mean electrostatic potential profile for a 1.0 M $1:1$ electrolyte at $\sigma^* = 0.25$. All symbols as in Fig. 7. The arrows labeled P and S mark the MPB5 and HNC/MSA values of $\psi^*_{d/2}$.

BGY + EN theory agrees with it. Beyond this surface charge, however, the effect of imposing local electroneutrality is striking. The maximum previously predicted for $\psi^*_{d/2}$ disappears, and the potential now lies just below the MGC potential under all conditions. Indeed, the BGY + EN results for both the equation of state and the structure of the interface now follow the MGC predictions altogether too closely and so share most of the defects of the latter at high surface charge. Incidentally, the Monte Carlo potential at 0.1 M for $\sigma^* = 0.20$ is *lower* for $d = 2.76$ Å than for $d = 4.25$ Å (Table I), so the errors of the BGY + EN theory for 2.76 Å are probably slightly larger than suggested by their comparison with 4.25 Å Monte Carlo results in Fig. 6. It is somewhat ironic that two levels of improvement upon the complete neglect of interionic correlations in the MGC theory have brought about such modest improvements in the equation of state.

We now turn our attention to the HNC-based theories of Section IV.D, beginning with the HNC/HNC theory of the double layer. To those who are familiar with the success of the HNC theory of bulk electrolytes but have not followed closely the recent work on double layers, the performance of the HNC/HNC theory of the double layer may come as somewhat of a dis-

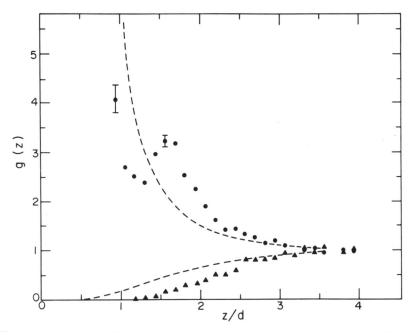

Fig. 12.   Ion density profiles for a 1.0 M 1 : 1 electrolyte at $\sigma^* = 0.70$. All symbols as in Fig. 7.

appointment. We have already mentioned in Section IV.D the severe numerical difficulties relative to the bulk electrolyte problem that have thus far prevented solutions to the double layer HNC/HNC equations below 1.0 *M*. It is not clear whether this is an inherent difficulty of the combination of one-dimensional geometry and the infinite-range Coulomb interactions or is simply a consequence of the particular formulations of the equations for which solutions have been sought thus far. (In this connection we wish to mention some interesting calculations of Patey,[129] who used the HNC approximation to compute the potential of mean force between two large charged spheres immersed in a primitive model electrolyte of much smaller spheres. As an intermediate step in this calculation, the large sphere–small ion correlation functions are found by numerical solution of

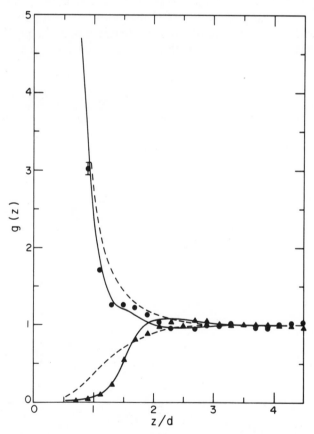

Fig. 13.   Ion density profiles for a 2.0 *M* 1 : 1 electrolyte at $\sigma^* = 0.396$. All symbols as in Fig. 7.

the HNC equations for what remains, of course, a *three*-dimensional mixture. These functions approach those for the one-dimensional problem considered here in the limit of infinite large-sphere radius, so it is encouraging to note Patey's success in obtaining solutions with relative ease at 0.10 M for $\sigma^* = 0.16$ with a large-sphere radius close to the planar limit.)

In the face of these difficulties, two different alternatives have been explored. Some of the numerical difficulties can be removed by using instead the mean spherical approximation (MSA) to compute the necessary bulk correlation functions[110,112]; this is the HNC/MSA theory of Section IV.D. Unlike the HNC closure, the MSA for the bulk can be solved analytically,[111] and some advantage can be taken of this fact to treat the particularly troublesome long-range parts of the correlation functions more efficiently in

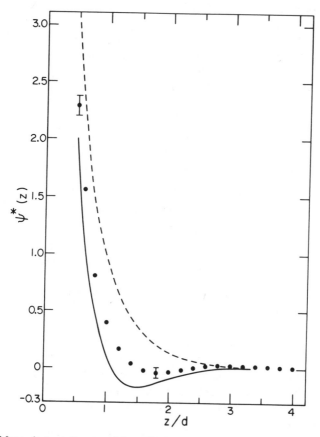

Fig. 14. Mean electrostatic potential profile for a 2.0 M 1:1 electrolyte at $\sigma^* = 0.396$. All symbols as in Fig. 7.

the numerical solution of the double layer problem. Happily, this turns out to be sufficient to allow solutions to be obtained for all conditions used in the Monte Carlo calculations. Of course, this HNC/MSA theory of the double layer will be inherently inferior to the HNC/HNC theory because of the well-known deficiencies of the MSA bulk correlation functions.[93]

A second response to the problem of solving the full HNC/HNC equations is a method proposed by Henderson and Blum[130] for *estimating* the HNC/HNC diffuse layer potential without actually solving (by iteration) the relevant integral equations. Of course, the results of this approximate procedure are of interest only insofar as they might suggest whether the performance of the HNC/HNC theory is likely to be good enough to warrant additional effort to overcome the numerical difficulties of obtaining a full solution.

This approximate HNC/HNC potential is labeled H in Fig. 6, while the predictions of the HNC/MSA theory (labeled S) are compared with other theories and the Monte Carlo results in Fig. 7. At high concentrations, both theories follow the Monte Carlo results more closely than either of the BGY theories. The approximate HNC/HNC potential, however, eventually falls below the Monte Carlo results until a maximum occurs in the equation of state, just as in BGY theory. This maximum occurs at a very large surface charge for 1.0 M, but shows an alarming trend toward appearing at quite modest surface charges at lower concentrations. (The HNC/MSA potential also shows such a maximum at 1.0 M but at still larger surface charge.) At 1.0 M, where the full HNC/HNC solution can be obtained, the approximate potential of Fig. 6 has been tested [for $d = 2.76$ Å (Ref. 130)] and found to *underestimate* the curvature in the true HNC/HNC equation of state, seriously so beyond the point at which the maximum occurs. Assuming this is true for the lower concentrations as well, it is hard to be enthusiastic about further attempts to solve the numerical difficulties of a theory whose physical content is at once obscure and apparently inappropriate for the interfacial problem.

The HNC/MSA predictions, on the other hand, are scarcely distinguishable from the Monte Carlo results at 0.01 and 0.10 M for $\sigma^* \leq 0.3$. This is as high a surface charge as is usually found in real double layers. This accuracy in the equation of state is naturally a reflection of a faithful description of the structure of the interfacial region under these conditions, so in Figs. 8 and 9, for example, no difference between the HNC/MSA and Monte Carlo profiles would likely be visible, although there are no HNC/MSA data to match the simulation conditions. At 1.0 M the HNC/MSA equation of state begins to bend below the Monte Carlo results beyond $\sigma^* = 0.20$. Although at $\sigma^* = 0.25$ the HNC/MSA ion profiles cannot be distinguished from the Monte Carlo results in Fig. 10, there is a small but statistically significant underestimation of the mean potential evident in the HNC/MSA theory for this state, as shown in Fig. 11. We have already commented on the growing importance of excluded volume effects at higher surface charges that can lead to a shoulder (Fig. 13) or even a second peak (Fig. 12) in the counterion profile. As might be expected on the basis of its performance when applied to dense fluids in the bulk, the HNC equation fails to account properly for such effects. Of course, even if the surface charges and concentration of Figs. 12 and 13 could be realized in an experimental situation, the continuum solvent/planar surface components of the present model are obviously inadequate once excluded volume effects become as important as those observed in these figures. At the same time, these very defects of the present model, if remedied, will clearly require theories capable of treating both high electrostatic coupling strengths and dense-liquid packing effects, so the shortcomings of the present theories in this regard are not without interest. In any case, at low concentration where there are no packing effects, the HNC/MSA predictions evidently lie much closer to the correct equation of state than do those of the HNC/HNC theory. This is a puzzling result in view of our earlier observations on the superiority of the HNC bulk correlation functions. In principle, one might hope to learn something of the nature of the failure of the HNC closure for the wall-ion correlation functions this implies, from the observation that it is largely offset by the errors in the MSA bulk correlation functions. Unfortunately, this type of insight is notoriously difficult to extract from Ornstein-Zernike-based theories. For the same reason, one cannot be certain of any connection between the excess curvature of the HNC/HNC equation of state and the same defect of the physically more transparent BGY theory. The dilemma is in no way lessened by the absence of any information about the HNC/HNC ion density profiles below 1.0 M, where only the approximate solution for $\psi^*_{d/2}$ is known.

For all its lack of sophistication, the modified Gouy-Chapman theory does as well as far more elaborate theories for a significant range of concentration and surface charge. Furthermore, where it works less well the defects can be understood in terms of corrections to the mean field approximation, and attempts can be made to estimate these. This is the philosophy of the modified Poisson-Boltzmann (MPB) theory of Section IV.B. In the most recent work on this theory[88-92,96,98] (roughly speaking, over the last decade), various versions have been spawned (MPB1, MPB2, and so forth) that are characterized by minor variations in their treatment of the fluctuation potential and excluded volume. To avoid the confusion this can cause, we display and discuss results of only that version of MPB theory described in

Ref. 92, denoted MPB5 by Outhwaite and Bhuiyan. MPB5 differs from its most recent precursor[91] in its treatment of the fluctuation potential in the charge free region between $z = 0$ and $z = d/2$.

The solid lines P in Fig. 7 show the MPB5 results for the equation of state. At 0.01 M the MPB5 potential cannot be distinguished from the HNC/MSA curve and hence from the Monte Carlo results. At 0.1 M the two theories are of comparable accuracy up to the largest surface charge for which there are Monte Carlo results, $\sigma^* = 0.30$. At this surface charge the MPB5 potential is 2 or 3 percent too high, while the HNC/MSA value is low by a similar amount, but these discrepancies are scarcely larger than the statistical uncertainty in the Monte Carlo data. At 1.0 M a similar agreement between the two theories persists to $\sigma^* = 0.20$, but the predilection of the HNC wall-ion closure for exaggerating the curvature of the equation of state is not shared by the MPB5 theory, which remains accurate to $\sigma^* = 0.42$. This fidelity of the MPB5 equation of state extends, of course, to the structure of the interface as well, as shown for $\sigma^* = 0.30$ at 0.1 M, where the MPB5 profiles agree very well with the Monte Carlo results (Figs. 8 and 9), and for $\sigma^* = 0.25$ at 1.0 M, where, although the MPB5, HNC/MSA, and Monte Carlo ion density profiles are not resolved in Fig. 10, the MPB5 potential profile is even closer to the Monte Carlo results than is the HNC/MSA $\psi(z)$ (Fig. 11). At 2.0 M and $\sigma^* = 0.396$, the MPB5 theory is qualitatively correct in predicting a charge inversion (Fig. 13) with a corresponding oscillatory mean potential (Fig. 14). There are significant excluded volume effects as well in this double layer, and the MPB5 counterion profile, unlike those of earlier versions of the theory,[90] does have a slight undulation near $1\frac{1}{2}d$, where the pronounced shoulder in the Monte Carlo profile is located. This may, however, be a fortuitous coincidence of a known artifact in the MPB5 solutions[92] with the actual physical effect observed in the simulations. In any case, precise quantitative agreement between the MPB5 theory and the computer results is now lacking; this is hardly surprising in a potential-based theory in which excluded volume plays a secondary role. Beyond this surface charge it does not seem possible to obtain solutions to the MPB equations,[128] so there is no question of applying the theory to the primitive model under conditions where excluded volume effects are even larger, such as in Fig. 12.

In summary, for 1 : 1 aqueous systems the MGC theory works well over a surprisingly large range of concentrations and surface charges. The theory seems easiest to improve at high concentrations and modest surface charges; all alternative theories seem to cluster more closely around the simulation results below $\sigma^* = 0.30$ at 1.0 M than is the case for higher σ or lower concentration. Relative to the other theories, BGY + EN differs the least from the MGC predictions despite a considerable increase in complexity, while the BGY theory fails altogether. The HNC/HNC theory also fails at high surface charge for 1.0 M and likely fails even for modest surface charges at lower concentrations, although it is extremely difficult to solve. Both the HNC/MSA and MPB theories are of comparably high accuracy over the entire range of concentrations and surface charges where excluded volume effects are small, although the agreement in the former case would be more encouraging were it not for the unexpected failure of the HNC/HNC theory. When excluded volume effects do become large, the HNC/MSA theory does not take them into account properly, while the MPB equations become insoluble.

C. 2:2 Aqueous and 1:1 Nonaqueous Electrolytes

In this section we discuss results for double layers in electrolytes whose ionic interactions are characterized by a coupling strength $\beta^* = 4 \times 1.6809$ [cf. Eq. (5.2)]. This has an obvious interpretation as a 2:2 aqueous electrolyte; that is, $q = 2e$ with the same values of d, T, and ε as specified at the beginning of Section V.B. Precisely because the ionic interactions depend only on the ratio q^2/ε, however, the primitive model with the value of β^* considered in this section can also be regarded as a 1:1 electrolyte in a non-aqueous solvent with a dielectric constant of about 20. (In the primitive model of a symmetric salt with q^2/ε fixed, $q\psi$ depends only on the ratio σ/q [cf. Eqs. (2.31) and (2.32)], so the ψ^* and σ^* we discuss below for $q = 2e$ must be doubled and halved, respectively, to obtain the 1:1 nonaqueous interpretation.) For bulk electrolytes the 2:2 case is well known to be a much more stringent test of theoretical approximations.[62,93] This turns out to be so for the double layer problem as well.

In modified Gouy-Chapman theory the product $q\psi$ is independent of q at fixed ε [cf. Eqs. (3.9) and(3.12)], so the MGC equation of state changes in a particularly simple way when the ionic charge is doubled: ψ^* is halved. A glance at the Monte Carlo results[131] for the equation of state in Fig. 15, however, shows that something quite different from this actually happens. There is a broad but distinct maximum in the Monte Carlo diffuse layer potential as a function of surface charge for both 0.50 and 0.05 M salts, and possibly for 0.005 M as well, although the simulations have not been carried out at high enough σ to verify this. The MGC theory is qualitatively wrong in this respect, of course, but even where $d\psi_{d/2}/d\sigma$ is positive the mean field theory seriously overestimates the potential—by more than a third, for example, at 0.005 M and $\sigma^* = 0.05$. For the similar but milder overestimation of $\psi_{d/2}$ in 1:1 systems, we identified two contributing factors: the cavity effect (i.e., the impossibility of there being any screening inside an ion's hard core) and the electrostatically driven depletion of screening charge about a fixed ion once interionic correlations are admitted. We argued on the basis of the weak ion-size dependence of the potential for the 1:1 case that the second effect was the more important factor whenever the total error in the MGC potential was large. At the higher electrostatic coupling parameter considered here, despite the absence of any results for different ion diameters, we would expect the ion-size dependence to be still milder, since the stronger electrostatic repulsions in the 2:2 system effectively preempt the hard-core exclusion. Moreover, the neglect of this fluctuation effect now becomes a serious defect of MGC theory at concentrations and surface charges so low that even the modest success of the theory for 1:1 electrolytes is lost altogether.

Only the two most successful theories discussed in Section V.B for 1:1 electrolytes—the HNC/MSA and MPB5 theories—have been solved for the 2:2 case. Roughly speaking, these theories are of comparably high accuracy for the 1:1 systems over the range of conditions for which both have been solved. In Fig. 15 we see that neither theory enjoys the same quantitative

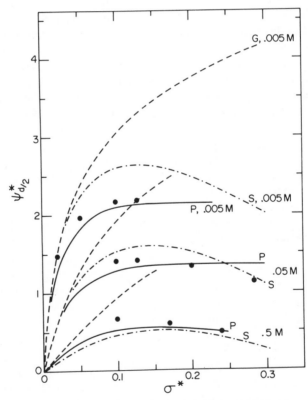

Fig. 15.   Results for the diffuse layer potential as a function of surface charge density for a model 2:2 electrolyte. All symbols as in Fig. 7.

success for the 2:2 case. The tendency of the HNC wall-ion closure to exaggerate the curvature in the equation of state at large σ can be seen in the HNC/MSA version for 1:1 systems only at 1.0 M at very high, unphysical surface charge. In the 2:2 case, where the Monte Carlo results show that there *is* a maximum in $\psi_{d/2}(\sigma)$, this tendency of the HNC/MSA theory is evident for physically reasonable σ at all concentrations and actually becomes more pronounced as the bulk electrolyte becomes more *dilute*. The MPB5 equation of state, on the other hand, shows every sign of becoming progressively more accurate as the concentration is lowered, a pleasing if somewhat predictable consequence of the physical basis of the theory. At 0.05 M the MPB5 theory does not predict the maximum found in the Monte Carlo equation of state (actually, the MPB5 potential at $\sigma^* = 0.284$ is 0.01

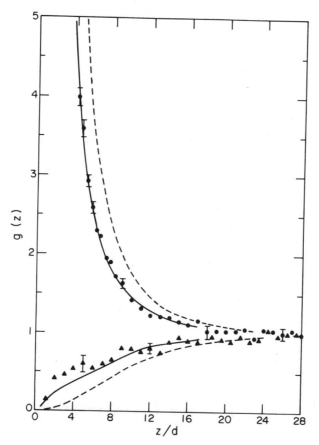

Fig. 16.   Ion density profiles for a 0.005 M 2:2 electrolyte at $\sigma^* = 0.0975$. All symbols as in Fig. 7.

lower than at $\sigma^* = 0.20$!^{128}) and so fares no better than HNC/MSA for this concentration, although the shortcomings are quite different in the two cases. At 0.50 M we observe the near coincidence of the MPB5 and HNC/MSA potentials, a quite remarkable result considering that in each case the lack of agreement with the Monte Carlo result at low σ implies a cancellation of errors has taken place to produce the good agreement at higher surface charge.

The potential curves of Fig. 15 conceal a wealth of fascinating structural features in these 2:2 systems. We begin with the fairly straightforward case of low concentration and moderate surface charge, $\sigma^* = 0.0975$ at 0.005 M, for which the ion density profiles are shown in Fig. 16. Our discussion of the equation of state has anticipated the principal feature of this figure, the overestimation of the thickness of the double layer by MGC theory. This is reflected in the mean electrostatic potential as well, of course, which is shown in Fig. 17. The solid lines in each case are the MPB5 predictions, which are

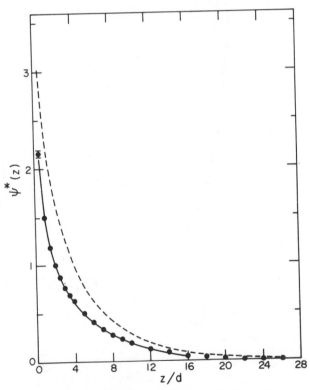

Fig. 17.   Mean electrostatic potential profile for a 0.005 M 2:2 electrolyte at $\sigma^* = 0.0975$. All symbols as in Fig. 7.

generally excellent. The co-ion profile may rise a little more sharply near the wall than is predicted by MPB5, but the error is not large and in any case is of little consequence in determining the potential. Much the same sort of constriction of the double layer, relative to MGC theory, occurs for similar surface charges at 0.05 M, but as σ increases, structure gradually builds up until at $\sigma^* = 0.284$ we find the profiles shown in Fig. 18. Nearly all the counterion charge has condensed into a very thin layer within a diameter of the wall, largely unhindered by the kind of excluded volume effects manifested in Fig. 12 for a 1:1 system. The result is essentially a dipole layer to which the co-ion profile responds by rising to almost twice the bulk density within two diameters of the wall. One curious side effect in this particular system appears to be a charge-neutral region in the double layer between 2.0 and 2.5d. The MPB5 theory does a reasonable job of reproducing the counterion condensation, but misses the co-ion response altogether, predicting a monotone profile instead. A less exaggerated version of this same behavior also occurs at $\sigma^* = 0.20$, but both the MPB5 and HNC/MSA[132] co-ion profiles are always monotone at this concentration. This confirms what might have been concluded anyway from the shapes of the potential-charge relationships at 0.05 M in Fig. 15: that the intersections of the theories with the

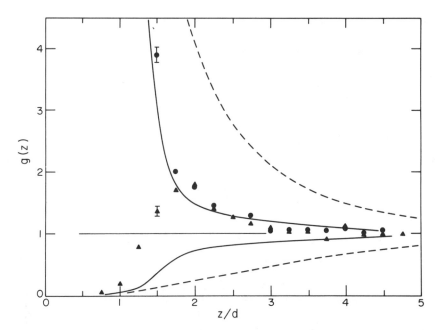

Fig. 18.   Ion density profiles for a 0.05 M 2:2 electrolyte at $\sigma^* = 0.284$. All symbols as in Fig. 7.

Monte Carlo equation of state at $\sigma^* = 0.284$ (HNC/MSA) and $\sigma^* = 0.20$ (MPB5) are to some extent accidental.

Despite the oscillation in the co-ion profile in Fig. 18, there is no charge inversion, and so $\psi(z)$, shown in Fig. 19, is necessarily monotone. It exhibits some interesting behavior nonetheless: the precipitous drop between $z = \frac{1}{2}d$ and $z = d$ (equivalent to nearly 50,000 V/cm) is another indication of the extremely compact nature of this double layer, for which the term "diffuse" is somewhat of a misnomer. The persistence of this behavior over a range of surface charge at 0.05 and 0.5 M results in the very flat equations of state in Fig. 15. Thermodynamically, it implies that the diffuse layer contribution to the total differential capacitance [cf. (Eq. (3.17)] is nearly zero, a behavior reminiscent of Helmholtz' original naive model of the double layer in which thermal fluctuations have no role.

Charge inversion does occur in 2:2 systems at 0.50 M. This is shown for $\sigma^* = 0.17$ in Fig. 20, where the co-ion density sensibly exceeds the counter-ion density between $z = 1\frac{1}{2}d$ and $z = 2\frac{1}{2}d$. Both the HNC/MSA and MPB5 theories do rather well at predicting this charge inversion. The Monte Carlo results show the counterions to be pushed out a little farther than MPB5 or, to a lesser extent, HNC/MSA predicts, so that the counterion minimum is

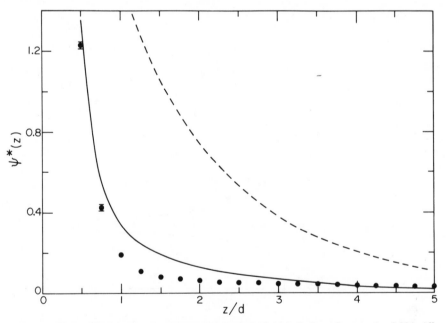

Fig. 19. Mean electrostatic potential profile for a 0.05 M 2:2 electrolyte at $\sigma^* = 0.284$. All symbols as in Fig. 7.

farther from the wall and not as deep as in the theoretical curves. This is reminiscent of similar behavior of the 2 M 1:1 system of Fig. 13 and is presumably once again primarily an excluded volume effect for which neither theory is particularly well suited. The underestimation by theory of the rise in the co-ion profile near the wall is by now a familiar feature of 2:2 salts. The mean potential for this system is shown in Fig. 21; as might be expected from the charge inversion evident in the profiles, $\psi(z)$ is oscillatory with a minimum of about -5 mV just beyond one diameter from the wall. The agreement of both the HNC/MSA and MPB5 theories with the Monte Carlo results is very impressive; the only difference between the two theories is a slight exaggeration of the dip in $\psi(z)$ in the HNC/MSA case. The somewhat faster decay of the theoretical potentials is presumably a consequence of each theory's underestimation of excluded volume effects in the counterion profile already mentioned. The monotone Gouy-Chapman potential profile for this system is also shown, largely for its entertainment value.

Possibly the most remarkable feature of Figs. 20 and 21 is the close agreement between the HNC/MSA and MPB5 theories. As suggested by the equation of state in Fig. 15, they are perhaps in even closer agreement with each other than either is with the Monte Carlo results for 0.5 M 2:2 sys-

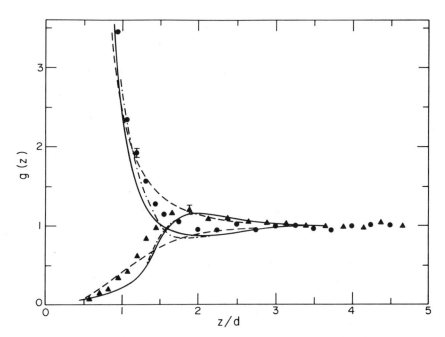

Fig. 20.   Ion density profiles for a 0.5 M 2:2 electrolyte at $\sigma^* = 0.17$. All symbols as in Fig. 7.

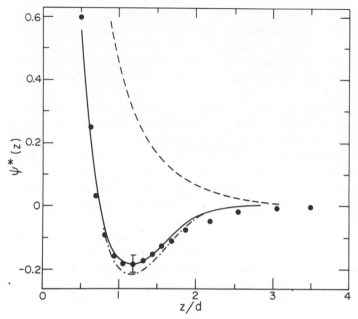

Fig. 21.   Mean electrostatic potential profile for a 0.5 M 2:2 electrolyte at $\sigma^* = 0.17$. All symbols as in Fig. 7.

tems. This is especially difficult to rationalize for the HNC/MSA theory, which here improves with *increasing* concentration, directly contrary to its behavior in the 1:1 case.

Certainly, it would be of interest to see the behavior of the HNC/HNC theory at this higher valence. Although the prospects of obtaining such solutions for the strictly planar geometry are not encouraging, it may be possible to approach the problem in the large-radius limit of the spherical charged surface as has been done for the 1:1 case.[129]

VI.   ELABORATIONS ON THE BASIC MODEL

A.   The Surface: Discrete Charges and Image Forces

In this section we discuss results for a number of variations on the basic double layer model of Section V. The first of these involves modifications to the modeling of the charged surface.

Monte Carlo calculations have been carried out by van Megen and Snook[133,134] for a primitive model electrolyte near a plane surface in which are embedded discrete point charges in a regular array instead of the uniform charge density we have considered thus far. This is a more realistic picture of the way surface charge is actually distributed in many real double layers, of course, and may be expected to provoke the largest differences in the diffuse layer at low overall surface charge. Most of these will be manifest only in the R dependence of the distribution functions (i.e., parallel to the surface), which is very difficult to resolve from thermal noise in the computer experiments. Van Megen and Snook originally reported that, in the direction *normal* to the surface, there was a marked enhancement of the counterion density close to the wall, relative to their continuous-σ results, but this large effect has since been found to be a consequence of an error in their continuous-σ program;[135] their corrected program shows there to be little, if any, difference between the discrete and continuous surface charge cases. In some of their discrete surface charge results the ionic densities outside the double layer exceed the correct bulk value; this effect results from a different programming error,[135] although in most cases it appears to be so small as to have no great effect on the ionic profiles near the wall. Although discrete surface-charge effects for periodic arrays of point surface charges have been investigated using the LGC theory,[136-138] there have been no attempts to incorporate a discrete surface charge into any of the more sophisticated theories we consider in this review, nor are there likely to be; the resulting loss of cylindrical symmetry about the surface normal just increases the complexity of these theories to an unmanageable level.

The other modification of the surface model that has been considered is the nonuniformity of the surface charge that results when ε_w, the dielectric constant of the wall material, is different from that of the continuum solvent. As explained in Section II.A, electrostatic boundary conditions at $z = 0$ in this case are satisfied only by polarization of the previously uniform surface charge, whose effect can be represented by a set of fictitious image charges in the region $z < 0$.

We do not regard the introduction of these image effects into the Hamiltonian as the first or the most important step in a systematic improvement of the model toward a more faithful representation of real double layers. There are many other aspects of charged surfaces—surface roughness, ion-specific interactions, and so forth—whose effect is doubtless greater than that of images. Image forces do, however, generalize the surface in a fundamentally interesting and important way: the pair interactions between ions near the surface are no longer central but depend on the positions of the ions relative to the surface [cf. Eqs. (2.12) and (2.13)]. Such noncentral forces are intrinsically interesting in their own right; moreover, noncentral pair forces will be present in any realistic model double layer, and images appear to be the simplest such effect on which to test simulation methods and theories. There are some points of technical interest in the generalization of the Monte

Carlo technique to handle image forces. These are touched on briefly in Appendix B and are discussed in detail in Ref. 139. On the theoretical side, we have already discussed in Section IV.D the difficulties of incorporating image forces into an Ornstein-Zernike theory such as HNC/MSA. Both the MPB5 and BGY + EN theories have been generalized to include images, although calculations in the latter case have been carried out only for repulsive images at zero surface charge.

Because the strength of the image charges is independent of σ, the effect on the structure of the double layer ought to be greatest at low surface charges. If the concentration is also low, then the counterions near the wall will on average be well separated from each other and experience a local surface charge that is dominated by the self-image component. Thus when

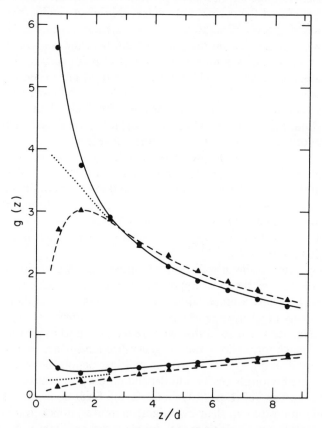

Fig. 22.   Ion density profiles for a 0.01 M 1:1 electrolyte at $\sigma^* = 0.01$. The points are Monte Carlo results for repulsive ($\blacktriangle$) and attractive ($\bullet$) images. The curves are the results of the MPB5 theory for no images ($\cdots$), repulsive images (---), and attractive images (——).

$f > 0$, for example, a counterion very close to the surface will find the nearby charge on the wall to be of opposite sign to the overall surface charge. The resulting repulsion turns out to be sufficient under certain conditions to bring about a local maximum in the counterion density profile. Similarly, for attractive images, the phenomenon can be sufficient to produce a minimum in the co-ion profile. Both these effects are evident for $\sigma^* = 0.01$ at $0.01M$, as shown in Fig. 22. In the attractive image case there is also a very pronounced enhancement of the counterion density near the wall. Very similar effects are also found for comparable values of σ at $0.1M$.[139] Because these dramatic changes in the ion-density profiles are confined to the immediate vicinity of the wall, the change in the first moment of the charge distribution, the electrostatic potential, is less marked. The potentials corresponding to the ion densities of Fig. 22 are shown in Fig. 23; there is a small upward shift in $\psi(z)$ for repulsive images, reflecting the additional work required to

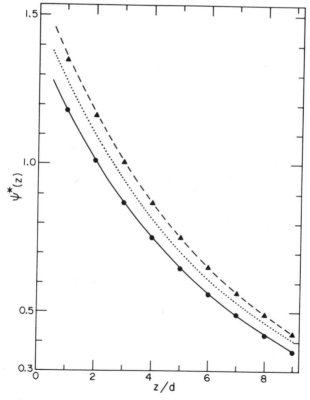

Fig. 23.   Mean electrostatic potential profiles of a $0.01\ M$ 1:1 electrolyte at $\sigma^* = 0.01$. All symbols as in Fig. 22.

bring an ion up to the surface in this case, and a corresponding reduction in the potential in the attractive image case. In both these figures the continuous curves are the predictions of the MPB5 theory. Clearly, within the statistical uncertainty in the simulation data, this theory is exact just as it is for these kinds of surface charges and concentrations in the absence of images.

At higher surface charges, the structure is less affected by images. For example, at $0.1M$ and $\sigma^* = 0.098$, both the attractive and repulsive image profiles are resolutely monotonic and differ very little from each other (Fig. 24). Nevertheless, both simulation and MPB5 predict a gradually increasing effect on the diffuse layer potential, the spread between the repulsive and attractive image values for $\psi_{d/2}^*$ increasing with σ as shown for $0.1M$ in Fig. 25. Not surprisingly, for a $2:2$ electrolyte the MPB5 theory predicts the relative effect of images on the equation of state to be much greater.[92] This is shown

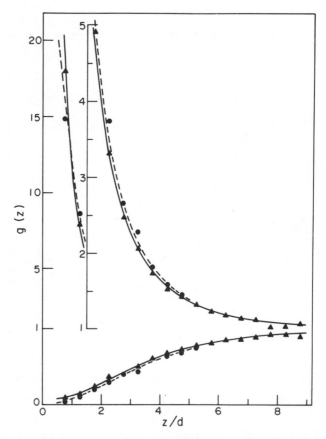

Fig. 24.   Ion density profiles for $0.1\ M$ $1:1$ electrolyte at $\sigma^* = 0.098$. All symbols as in Fig. 22.

in Fig. 26 for a 0.05 M 2:2 electrolyte. There are no computer experiments with image forces for these systems.

The effect of image forces at *zero* surface charge has also been studied by both simulation and theory.[92,103] As long as the electrolyte model is symmetrical in every respect, there can be no charge separation (i.e., no double layer) at an uncharged surface, since cations and anions must be affected

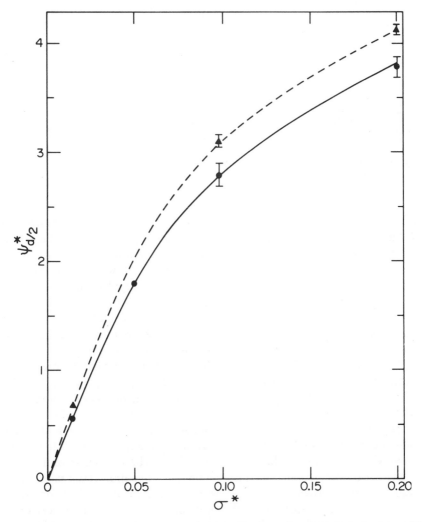

Fig. 25.   Diffuse layer potential as a function of surface charge for a 0.1 M 1:1 electrolyte. All symbols as in Fig. 22.

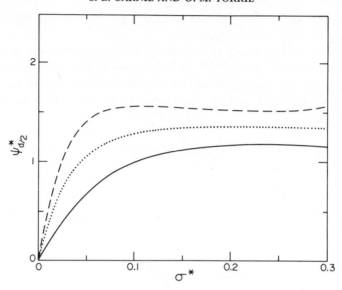

Fig. 26.   Diffuse layer potential as a function of surface charge for a 0.05 M 2:2 electrolyte. All symbols as in Fig. 22.

equally. For $\varepsilon_w < \varepsilon$, there is an equal depletion of both ion species near the surface caused by the repulsive images. Something like this must happen at the free surface of an electrolyte (where $\varepsilon_w \approx 1$) and is generally accounted responsible for the increase in the surface tension of water upon the addition of a strong electrolyte. While the addition of repulsive images at $\sigma = 0$ to a restricted primitive model confined by an impenetrable plane cannot be taken seriously as a model of the surface of a real salt solution, once again it does constitute the simplest possible model incorporating one important feature of such systems: noncentral pair forces at an uncharged surface. In Fig. 27 are shown Monte Carlo results and various theoretical predictions for the ion profile at $\sigma = 0$, $f \doteq 0.97484$ ($\varepsilon_w = 1$) for a 1:1 primitive model at $0.1M$. The BGY + EN theory is seen to underestimate the depletion somewhat. We mention as yet another example of the cancellation of errors that is the curse (or the salvation) of integral equation theories of the double layer that the BGY predictions (not shown) are virtually in quantitative agreement with the Monte Carlo results in Fig. 27.[103] The solid curve in the figure is the MPB5 prediction, which lies below the simulation result close to the wall. This is a particularly interesting application of the MPB theory: aside from very small ion-size effects, the entire effect here is a consequence of the fluctuation potential, since the mean potential is everywhere zero. The upper (dotted) curve in the figure is the original theory for the electrolyte solution surface of Wagner[77] and Onsager and Samaras[78] (strictly speaking for $f = 1$, not 0.97484). An important approximation made in solving this weak-coupling theory is the replacement of the local concentration-dependent screening length with the (shorter) bulk screening length. The resulting over-estimation of the screening of the repulsive image forces is evident in Fig. 27 as an underestimation of the ion depletion near the surface.

As discussed in Section III.B, the Onsager-Samaras theory predicts a potential of mean force on an ion near the wall that has the appearance of the coulombic interaction of the ion with its own image only, screened by a familiar exponential factor [cf. Eq. (3.41)]. Because this is a result of a weak-coupling theory, it will become exact in the limit of low concentration. This has led to the idea of introducing this potential of mean force directly into the Hamiltonian as a substitute for the complete set of interactions of each ion with all the images—that is, to the construction of a different, less sophisticated *model* of the surface polarization effect using only this effective potential.[103] This is obviously a retrograde step from the standpoint of studying noncentral pair forces! In removing such effects, however, this screened self-image (SSI) model has also restored the tractability of theories such as HNC/MSA and HNC/HNC. Comparisons with Monte Carlo results for the SSI model itself show the HNC/HNC theory (at low σ, where it can be solved) to be an accurate theory *for the model*. Unfortunately, the success of the SSI model as a good approximation to the full-image Hamil-

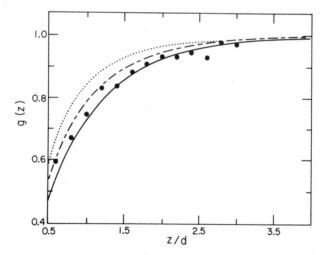

Fig. 27.   Ion density profile at an uncharged surface in the presence of repulsive images for a 0.1 M 1:1 electrolyte. (●) Monte Carlo results; (———) BGY + EN theory; (——) MPB5 theory; (· · ·) Onsager-Smaras theory.

tonian is uneven; it appears to be quite good at finite surface charge[139] but is noticeably less so at $\sigma = 0$.[103] The rather mixed success of integral equation theories of the double layer in the absence of images renders such an ad hoc approximation to image effects in such theories even less attractive, especially in view of the success of the MPB theory in treating images.

B. Asymmetric Electrolytes

Another aspect of double layer behavior that can be examined within the context of the primitive model is the effect of asymmetry in the bulk electrolyte obtained, for example, by assigning different sizes or different valences to the cations and anions. Some consequences of such modifications will be common to all types of asymmetries in the electrolyte model and are not difficult to anticipate. The first of these stems from the physically obvious notion that, at high surface charge, the behavior of the double layer will be dominated by the properties of the counterions alone, there being very few co-ions in the interfacial region under these conditions. For a given asymmetric electrolyte, then, the double layer equation of state for large $|\sigma|$ will approach that of the symmetric electrolyte corresponding to the anion for $\sigma > 0$ and the cation for $\sigma < 0$. A second more subtle effect is the inevitability in an asymmetric electrolyte of a charge separation (i.e., a double layer) at an *uncharged* surface simply because the presence of the wall will affect the different ionic species in different ways. Such a potential of zero charge is a very common feature of real double layers and cannot occur in the symmetric model considered in Sections V and VI.A. The size of this charge separation in the primitive model will depend on the nature of the asymmetry, of course, but will always be further magnified if image forces are present.

1. Charge Asymmetry

The only charge-asymmetric model that has been studied in any detail is the $2:1$ electrolyte ($q_+ = 2e$, $q_- = -e$) with both positive and negative surface charges, corresponding to singly and doubly charged counterions, respectively. (Because of the symmetry of the model in all other respects, it can be interpreted equally well as a $1:2$ electrolyte by interchanging these correspondences between the sign of the surface charge and the counterion valence. Thus the case of divalent counterions which we discuss here as $\sigma < 0$ for a $2:1$ salt is often referred to in the literature as a $1:2$ electrolyte with $\sigma > 0$.) For both cases, analytic solutions of the Poisson-Boltzmann equation have been obtained[63] and show just the behavior forecast in the preceding discussion. When the counterions are divalent ($\sigma < 0$), the MGC equation of state for the asymmetric electrolyte is almost identical to that of $2:2$ electrolyte at the same concentration. Because of the longer screening length of the $2:1$ electrolyte, the mean field theory predicts a slightly more extended diffuse layer (i.e., a larger $|\psi^*_{d/2}|$) than it does for the $2:2$ case, but these differences are very small and vanish in the limit of low concentration or high surface charge. Similarly, for singly charged counterions ($\sigma > 0$), the 2:1 MGC potential curves lie on or just below those for the corresponding 1:1 electrolyte (i.e., the 1:1 electrolyte at twice the molar concentration). The net effect is the highly asymmetric equation of state shown by the dashed line in Fig. 28. Also shown in this figure are the results of both the HNC/MSA[113] and MPB5[92] theories, generalized to this charge-asymmetric case. Like MGC, both predict an equation of state nearly identical to that of the corresponding theory when applied to the appropriate symmetric electrolyte.

The general observations we have made previously concerning the performance of these various theories for 1:1 and 2:2 electrolytes could be repeated here, then, for the right ($\sigma > 0$) and left ($\sigma < 0$) halves, respectively, of Fig. 28. Chief among these, of course, is the breakdown of MGC theory for divalent counterions even at low concentrations and surface charges. In fact, the first solid numerical evidence of this phenomenon in a primitive model electrolyte was the solution of the MPB4 equation for the charge-asymmetric case.[98] The improvements contained in the MPB5 version of the theory have led to much better agreement with the Monte Carlo results as shown in Fig. 28 than was the case for the MPB4 theory.[98] There are, how-

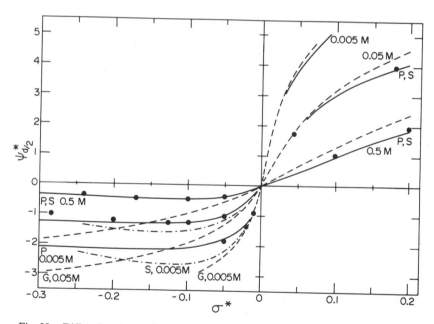

Fig. 28. Diffuse layer potential as a function of surface charge density for a model 2:1 electrolyte. For $\sigma > 0$ the counterions are singly charged; for $\sigma < 0$ the counterions are doubly charged. (●) Monte Carlo results; (— · —) HNC/MSA theory; (——) MPB5 theory. The predictions of these two theories are indistinguishable at 0.5 M and for $\sigma > 0$ at all concentrations.

ever, some small differences between the $2:2$ and $2:1$ divalent counterion cases seen in the Monte Carlo results that are not handled well by any theory. These differences are easiest to see as changes in the co-ion behavior, an example of which is shown in Fig. 29 for an $0.05M$ $2:1$ system at $\sigma^* = 0.284$. The dotted line in this figure is a smooth curve through the Monte Carlo co-ion profile for the $2:2$ system at the same concentration and $|\sigma^*|$ (cf. Fig. 18). In both the $2:2$ and $2:1$ systems there has occurred a condensation of the divalent counterions at the surface, creating nearly identical dipole layers. Not surprisingly, the response of the singly charged co-ions leads to a less compact profile than in the $2:2$ case, although the co-ion density in both cases exceeds its bulk value for $z \gtrsim \frac{3}{2}d$. One consequence of the more extended co-ion profile in the $2:1$ case is a charge inversion and a smaller overall potential drop. This is the opposite of the difference between the MGC potentials for $2:2$ and $2:1$ electrolytes, which we have already noted was in turn consistent with the difference in the screening lengths of these two systems. Of course, the bulk screening length is not the relevant distance scale for systems in which the double layer is so compact. Neither the HNC/MSA nor the MPB5 theory predicts co-ion profiles well in the presence of divalent counterions; the MPB5 co-ion profiles in both Fig. 18 and Fig. 29 are monotone. Both theories do agree with the Monte Carlo result

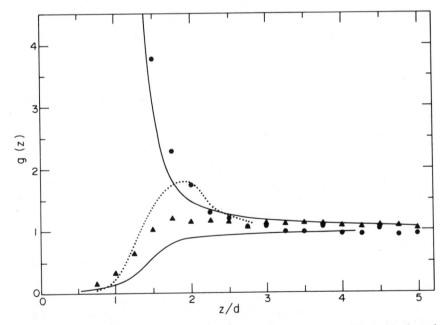

Fig. 29.   Ion density profiles for a 0.05 M $2:1$ electrolyte at $-\sigma^* = +0.284$ (doubly charged counterions). The dotted line is a smooth curve drawn through the Monte Carlo points for the co-ion profile of the corresponding $2:2$ system (cf. Fig. 18). All other symbols as in Fig. 28.

that $|\psi^*_{d/2}|$ for 2:1 electrolytes is smaller than for 2:2 systems, but both greatly underestimate this admittedly small effect.

The MPB5 equations have also been solved for 2:1 electrolytes at $0.05M$ in the presence of images.[92] For large $|\sigma|$ the results are, of course, similar to image effects in the relevant symmetric electrolytes (cf. Figs. 25 and 26). At $\sigma = 0$, however, there is a distinct nonzero potential of zero charge for both repulsive and attractive images of about $+0.05$ and -0.05, respectively, in ψ^*. These signal the existence of small charge separations induced in the absence of an applied field arising solely from the asymmetric response to the polarizable surface of the two different ion species. A still smaller charge separation must also develop even in the absence of any image forces. All of these effects are, of course, outside the scope of MGC theory.

2. Size Asymmetry

Marked asymmetry about the potential of zero charge is a characteristic feature of the properties of real double layers, even for 1:1 salts, signaling the presence of very different noncoulombic interactions of the cations and anions with the surface. A naive way of introducing this type of effect into the primitive model is to assign unequal distance of closest approach (UDCA) to the surface for the two ion species as suggested by Fig. 30. In

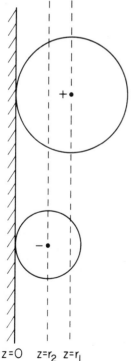

$z=0$ $z=r_2$ $z=r_1$

Fig. 30. A schematic view of a double layer in which anions can approach the surface more closely than cations. In the UDCA model the ionic radii shown apply only to the ion-wall interactions.

this way the effect of different wall-ion interactions can be studied independently of the additional complications that would be introduced by changes in the interionic potentials were different hard-sphere radii to be imputed to the ions as well. Such different distances of closest approach can obviously be introduced into the point-ion model too, and many of the consequences of such a modification are contained in the Gouy-Chapman theory without recourse to more sophisticated theories or computer simulation of finite-sized ion models. Surprisingly, although the solution of the Poisson-Boltzmann equation (PBE) for the UDCA model is straightforward, it does not seem to have been written down until 1982.[140] This was followed by a similar calcu-

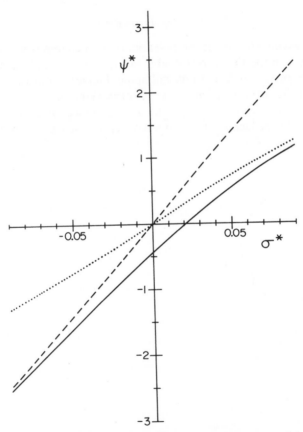

Fig. 31.   Diffuse layer potential at the small distance of closest approach (r_2 in Fig. 30) as a function of surface charge density for a 1.0 M 1:1 UDCA electrolyte. The dashed and dotted curves are the MGC potentials at r_2 for equal distances of closest approach corresponding to r_1 and r_2, respectively.

lation in which both size and charge asymmetries were allowed[141] and by a solution of the linearized PBE for the UDCA model.[142] In Fig. 31 is shown the resulting MGC equation of state for a $1:1$ UDCA electrolyte at $1.0M$ in which the anions may approach to within $r_2 = 1.0$ Å of the wall while the cations have a distance of closest approach of $r_1 = 3.6$ Å (cf. Fig. 30). The solid line is the UDCA potential at 1.0 Å (i.e., the diffuse layer potential), while the dotted and dashed curves show the MGC potentials, also at 1.0 Å, for the symmetric electrolytes corresponding to a distance of closest approach of both ions of 1.0 and 3.6 Å, respectively. The UDCA potential curve shows both the characteristics we anticipated for asymmetric electrolytes in general. First, it approaches the (MGC) equation of state for the appropriate symmetric electrolyte at large positive and negative σ. Second, it exhibits a substantial potential of zero charge of about -11 mV, representing the charge separation resulting from the exclusion of the cations from the region between 1.0 and 3.6 Å. This charge separation would obviously be greatly enhanced by attractive image forces in a more sophisticated theory.

The MGC theory of the point-ion UDCA model also exhibits an interesting variety of behavior in several quantities of experimental interest, although the effects are generally much smaller than in the real systems, owing to the oversimplification of the model. A full discussion of these effects can be found in Ref. 140.

VII. RELATED MODELS

Our attention has focused in Sections III–VI on the isolated planar primitive model double layer. The recent development of a variety of theories for this model enabled us to explore the subject in depth and to make detailed comparisons of the theoretical expectations with Monte Carlo calculations.

In this section we turn to a number of closely related models that have been relatively less well studied. While this situation will doubtless change in the near future, a detailed review of these topics would be somewhat premature given the present state of knowledge. Our discussion is necessarily at a simpler level and designed chiefly to alert the reader to the connections between the double layer theories described in earlier sections and related work in other fields. We apologize to those whose work we have overlooked.

An obvious extension to our standard model is to consider the effect of curvature of the surface on the structure of the double layer. If both principal radii of curvature are decreased at the same rate from infinity, the surface becomes a charged spherical particle of progressively smaller diameter. At some value of the diameter, this geometry represents an intermediate situation between the planar double layer (infinite radii of curvature) and the bulk electrolyte (radii of curvature equal to $d/2$). This case is also of practical interest in the study of electrolytes around charged micelles (diameter ~ 40 Å) and unilamellar vesicles (diameter ~ 250 Å).

The term Gouy-Chapman theory is normally (but not always[143]) restricted to the Poisson-Boltzmann (PB) equation in plane geometry. The PB equation, however, can be applied to any geometry, but only in special cases can analytic solutions be found. In the spherical case, it can be linearized to give for the mean electrostatic potential

$$\psi(r) = \frac{Q}{\varepsilon r} \frac{e^{-\kappa(r-a)}}{1 + \kappa a} \qquad (7.1)$$

where r is the radial coordinate and Q is the total charge on the sphere of radius a. The ion-density profiles consistent with the linearization leading to Eq. (7.1) are

$$h_\alpha(r) = -\beta q_\alpha \psi(r) \qquad (7.2)$$

Sometimes, however, the inconsistent equation

$$g_\alpha(r) = \exp[-\beta q_\alpha \psi(r)] \qquad (7.3)$$

is used. When applied to primitive model electrolytes (i.e., a is an ionic radius), Eqs. (7.1) and (7.2) constitute the Debye-Hückel (DH) theory. The combination of Eqs. (7.1) and (7.3) is called the DHX theory.[93] No analytic solution to the Poisson-Boltzmann equation for spherical geometry has been found, but extensive tabulations of numerical solutions have been prepared.[144] For approximate formulas that fit these numerical calculations, the reader is directed to the references listed in Ref. 145.

The limiting procedure that leads to Eq. (4.36) can be used to derive OZ equations for PM electrolyte surrounding an isolated hard, uniformly charged, nonpolarizable spherical particle. As in Section IV.D, approximate closures (for both bulk correlations and ionic profiles) can lead to a variety of integral equations. The only published results are the HNC/HNC results of Patey.[129] He calculates ionic profiles at $a = 50$, 100, and $200d$ for various values of surface charge and electrolyte concentration. Comparison is made with DHX theory [Eqs. (7.1) and (7.3)] and with MGC theory (to test the effect of curvature). It is clear from the discussion in Sections II and IV that image effects cannot be included in such theories.

In order to study the primitive model double layer at an isolated spherical surface by computer simulation, some difficult boundary condition problems will have to be addressed. Monte Carlo experiments have been performed, however, for a collection of ions confined between two concentric spheres, the smaller of which is charged.[146,147] The focus of this work is the cell model of micellar solutions[148]; this different emphasis has led to choices of model parameters and simulation conditions that are generally quitedifferent from those discussed in Sections V and VI. For example, the calculations are in the canonical ensemble, and in some cases no co-ions are present, so that no equilibrium of the interface with a well-defined bulk system is implied. Nevertheless, some parallels can be drawn with the work we have reviewed here. The solution of the Poisson-Boltzmann equation for the simulation geometry and conditions shows a more extended double layer than in the Monte Carlo results, and this effect increases dramatically in the divalent counterion case. The effect of varying the counterion size down to zero radius has been examined in the absence of co-ions and is found to be quite small.[147]

If we decrease only one radius of curvature, we obtain electrolyte surrounding an infinite charged cylinder. This is a widely used model in the theory of ion-polymer interactions for rigid polyelectrolytes.[148,149] Once again the linearized PB equation can be solved to give

$$\psi(\rho) = \frac{2\lambda}{\varepsilon} \frac{K_0(\kappa\rho)}{\kappa a K_1(\kappa a)} \tag{7.4}$$

where ρ is the coordinate normal to the cylinder axis, K_0 and K_1 are modified Bessel functions, and λ is the linear charge density of the cylinder of radius a. The PB equation can be solved analytically if only one type of ion is present (i.e., the counterion).[150,151] For the more useful situation, the PB equation must be solved numerically.[143,152,153] Even the PB equation for this model shows interesting behavior, especially insofar as it suggests the limitations of Manning's condensation theory.

Only a preliminary Monte Carlo study of the counterion distribution about a charged infinite cylinder has appeared thus far.[154] Again, this work has been carried out in the context of the cell model and uses the canonical ensemble without any co-ions. The counterion profile observed in the simulations again shows deviations from the solution of the appropriate Poisson-Boltzmann equation of a type and magnitude very reminiscent of the planar surface results.

Integral equation theories are only now being applied to the cylindrical case. (The MPB theory of Ref. 154 is not entirely consistent with cylindrical geometry, because the fluctuation potential used is appropriate only for the planar surface.) Fixman independently derived the HNC/MSA equations in order to estimate the error involved in the Poisson-Boltzmann equation.[155] Numerical results of the application of the HNC/HNC equations to the cylindrical geometry are expected shortly.[156]

Another obvious extension is to consider the interaction of two double layers. The simplest case is that of electrolyte separating two planar surfaces. The solution of the PB equation for such geometry provides the basis for the DLVO theory of colloidal stability—when the double layers of the two surfaces overlap, the interaction free energy of the system is raised, and so a repulsive force is generated.[2,157] For surfaces of equal charge, the PB equation can be solved analytically in terms of elliptic functions.[2,157] For more general cases, it must be solved numerically.[158-160]

The work of Bell and Levine,[39,40] based on the Kirkwood hierarchy, has been extended to the case of two-plate geometry.[40,161] Many of the approximate treatments of effects such as ion polarizability and dielectric saturation have also been included, so that the calculation does not correspond to a molecular Hamiltonian model such as we have been considering in this review. The MPB theory could, of course, be applied to a PM electrolyte between two charged planar surfaces without these complicating effects, and would likely be quite successful given its accuracy in the isolated double layer case.

The equations of Section IV.D are represented in the two-plate case by the work of Grimson and Rickayzen. They solve numerically the MSA/MSA equations[162] and HNC/MSA equations[115] for symmetric and asymmetric[163] RPM electrolytes. They calculate potential profiles and the double-layer force acting between the two surfaces and compare them with the predictions of the PB equation for the two-plate case. Unfortunately, they exhibit no distribution functions and have studied only 0.5, 1, and $2M$ electrolytes and surface charge densities < 0.05 C m^{-2}, so that it is not possible to make very detailed comparisons with the single-plate data of Sections V and VI. However, certain qualitative features conform with our expectations based on the isolated double layer. At fixed surface-charge density, the potential decays more rapidly with distance than PB theory, and the surface potential is lower. The double layer force is also a faster-decaying function of plate separation than the PB result. Finally, the deviations from PB theory increase with asymmetry of the electrolyte.

Some grand canonical Monte Carlo calculations for the primitive model between two plates are in progress and will be reported in due course.[164] The most interesting regime of such systems is that of small plate separation and high surface charge where deviations from the Poisson-Boltzmann equation solution will be greatest. Progress is slow, because use of the grand ensemble requires that there be a respectable number of co-ions in the system; in this regime this condition requires in turn the use of a simulation cell with a very large cross section and hence a very large number of counterions. In a recent Monte Carlo study[165] intended to model an ionic system between lamellar liquid crystals, this technical impediment was avoided by omitting the co-ions from the model altogether. We have already remarked that for such a (canonical ensemble) simulation one forfeits the existence of a thermodynamic equilibrium of the interface with any well-defined bulk state. In these circumstances no thermodynamic meaning can be assigned to the operation of decreasing the plate separation at constant surface charge. One interesting feature of these calculations, however, is the advantage that was taken of the absence of oppositely charged ions to simulate a true point-ion model. Comparison of the Monte Carlo results with the Poisson-Boltzmann equation for this constant-charge case shows the point-ion profiles to be more compact than the theory predicts, just as for finite-sized ions, but the increase in this effect when the valence is doubled appears to be much smaller in the point-ion case.

Although the case of two planar surfaces is the paradigm for the calculation of double layer forces and may be shown to be the leading term for slightly curved surfaces,[166,167] calculations of the force between bodies of finite extent are nevertheless of interest. In particular, the Poisson-Boltzmann equation has been solved numerically for the case of two spherical particles.[168-171] The only comparable calculation for primitive model electrolytes is found in Ref. 129. Again, the HNC/HNC equations are solved; an interesting result is the existence, for certain conditions, of an attractive potential of mean force as the large spheres approach each other at constant surface-charge density. This is completely at variance with the results of the PB equation and remains a tantalizing result.

Very recently, the behavior of a finite concentration of large charged spheres in a neutralizing bath of smaller counterions has been studied by a Monte Carlo method.[172] This is essentially a highly asymmetric bulk primitive model in which $d_- = 10d_+$ and $q_- = -12q_+$ (or $q_- = -6q_+$ for divalent counterions). A striking feature of these results is the decrease of the range of the repulsive potential of mean force between two large spheres when the counterion valence is doubled. Very likely this is closely related to the behavior of isolated planar double layers for divalent counterions discussed in Section V.B. Unfortunately, no distribution functions between unlike ions were reported, so nothing is known directly about the double layer structure for this system.

At liquid densities and with $\varepsilon = 1$, the primitive model has been extensively investigated as a simplified representation of a molten salt.[173] This is an altogether different physical regime from the one we have been considering here, of course, corresponding to a high-density, strongly coupled plasma. Evidently, various surface models could be introduced in this regime, corresponding to interfaces between two condensed phases of a salt or between a condensed phase and its vapor. In much of the work on bulk models of molten salts, however, the hard-sphere repulsions have been replaced by both repulsive and attractive continuous potentials,[174,175] and it is only for these models that interfacial phenomena have been simulated.[176-178] The closest of these in spirit to the type of work we discussed in Sections V and VI is a molecular dynamics simulation of a model of molten KCl next to a planar wall, both with and without an applied electric field.[178] In this dense system the application of the field has virtually no effect on the total particle density. The shifts in the individual ion-density profiles near the surface in the presence of the field give rise to charge oscillations that persist throughout samples of a size convenient for simulation. Similar behavior would likely be present in any reasonable choice for a molecular solvent in a more sophisticated model of an aqueous electrolyte double layer. The molten salt interface may provide an opportunity to test theories and simulation methods for condensed phases at charged surfaces in the absence of the additional problems likely to arise from the dielectric properties of a realistic solvent model.

Remarkable progress has recently been made in a model slightly different from the usual electrolyte models we have considered so far. The one-component plasma (OCP) is a highly idealized model for the plasma state, consisting of point charges of equal magnitude and sign embedded in a uniform neutralizing background, which ensures thermodynamic stability. Since there is no ion size, the OCP is a particularly simple Coulomb system and has been thoroughly studied in the bulk in three dimensions.[179] Less realistic, of course, is the two-dimensional OCP [by which we mean that the interaction potentials obey the two-dimensional Laplace equation, so that $u(r) \sim \ln r$].

Nevertheless, this model has gained significance in double layer theory because of recent exact solutions of the model at the temperature defined by the equation $\beta q^2 / \varepsilon = 2$ (ε is the dielectric constant of the background). Following the exact solution for the bulk thermodynamics and low-order distribution functions,[180,181] the methods were extended to the inhomogeneous case: the two-dimensional OCP adjacent to a surface (i.e., a line) bearing a charge density. Both the cases without images[79,182-184] and those with perfect repulsive images[20,184] ($f = 1$) have been solved for this model, always at the above-defined temperature. Further work has led to exact solutions in strip geometry (the two-dimensional analogue of interacting planar double layers).[185,186] These developments have created an unprecedented situation: exact solutions for a continuum, nontrivial, inhomogeneous Hamiltonian model against which to test both theory and simulation methods.

This opportunity has not been spurned. All the exact conditions of Section II.D have been verified for the two-dimensional OCP at this temperature (some, such as the contact condition,[187,188] must be modified to allow for the neutralizing background). Indeed, the observation of algebraic transverse correlations $\sim r^{-2}$ in this system was the inspiration for the sum rule given by Eq. (2.81). A highly sophisticated version of density functional theory has recently been devised and tested against the exact result, with reasonable agreement found for the density profile.[116]

The detailed expressions for the distribution functions may be found in the original papers.[20,79,182–184] The density profiles decay to their bulk value in essentially a Gaussian manner (this Gaussian decay appears to be a special feature of the particular temperature for which the model can be solved[181]). The decay of $\rho^T(\mathbf{r}_1, \mathbf{r}_2)$ parallel to the surface depends on the value of f. For $f = 0$, the decay is algebraic: $\rho^T(\mathbf{r}_1, \mathbf{r}_2) \sim r_{12}^{-2}$ in agreement with Eq. (2.81). For $f = 1$ ($\varepsilon_w = 0$), the transverse decay is exponentially damped. This lack of algebraic decay is again in accordance with Eq. (2.81). In any oblique direction, $\rho^T(\mathbf{r}_1, \mathbf{r}_2)$ for both values of f decays in a Gaussian manner or faster.[20,79] Interest generated by these findings has led to the investigation of similar behavior in the weak-coupling limit of the three-dimensional point-ion model.[79,80,82]

The excess surface free energy for the two-dimensional OCP is given explicitly in Ref. 184. Differentiation with respect to the charge density produces the ψ_0-σ relationship shown in Fig. 32. The striking features of this

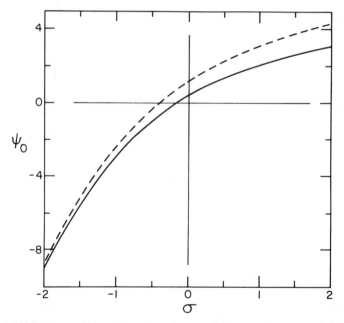

Fig. 32.   Surface potential ψ_0 (in units of $q^2/4\varepsilon$) as a function of surface charge [in units of $q(\rho/2\pi)^{1/2}$] for the two-dimensional one-component plasma, (——) without images, $f = 0$ and (---) with perfect repulsive images, $f = 1$.

graph are the existence of a potential of zero charge and the marked asymmetry between the regions of positive and negative surface charge density.

These features are reminiscent of the asymmetry observed in Section VI.B, especially for the case of size asymmetry. The two charged species differed in their response to the surface, thereby generating a charge separation and surface potential at zero surface charge. In the OCP, only one charged species can respond to the surface, the neutralizing background being totally inert. Thus the small repulsive potential of mean force acting on the charged species for $f = 0$ (Ref. 79) (similar to that shown in Fig. 4) is sufficient to cause a charge separation at zero σ. The addition of repulsive images to the model only serves to enhance this effect, as can be seen in Fig. 32.

VIII.   THE SOLVENT

An important step in any attempt to go beyond the McMillan-Mayer level of modeling is the explicit inclusion of solvent coordinates in the molecular Hamiltonian specifying the system. Solvent effects have long been regarded as of paramount importance in electrochemical systems, and a whole field of theory devoted to the modeling of the first solvent monolayer has been developed to describe the detailed behavior of the electrode/electrolyte interface.[6-8,189] In this section we briefly examine the achievements of more fundamental attempts to include solvent effects.

The simplest path from the primitive model of Sections II–VI is to aim for a more realistic model electrolyte by including the solvent while retaining a highly simplified surface. This in no way implies that the detailed structure of the surface does not contribute significantly to the electrical properties of the whole interface. Indeed, recent work on electrochemical systems suggests that the metal determines many features of double layer behavior.[190] Fundamental work on this aspect of double layer theory has begun from a viewpoint opposite to that adopted here: A detailed examination of the metal surface is made by applying density functional theory to the electron profile while the electrolyte is modeled in a highly simplified manner (as a dielectric slab).[191,192] The unfortunate truth is that a reliable picture of the region of electrolyte immediately adjacent to the surface will require realistic treatments of both solvent and surface, a distant prospect at the moment.

The choice of an appropriate Born-Oppenhiemer level model for the electrolyte (i.e., a Hamiltonian containing both ionic and solvent molecule coordinates) is no easy matter. In view of its technological and biological significance, water is the particular solvent of greatest interest. Models for water are by now legion, with none being outstandingly successful for all phases of water,[193] but some being quite successful in describing the properties of liquid water.[194-196] Although the field of modeling aqueous systems is an active one, very few studies have been performed on electrolytes containing such model solvents.

A different approach consists of employing a much simpler model electrolyte that may be easier to treat theoretically, may provide qualitative information for a variety of real electrolytes, and yet contains certain minimal physical requirements for an electrolyte, such as Coulomb interactions between the ions and dielectric response in the solvent. The simplest such model is the hard-sphere ion-dipole mixture (HSIDM), consisting of hard-sphere ions with embedded point charges and hard-sphere solvent molecules with embedded point dipoles. Analytic solutions in the mean spherical approximation[197,198] and numerical solutions of HNC-based theories[199] have been obtained for this model. The only Monte Carlo results for the HSIDM are restricted to a study of the ion-ion potential of mean force at infinite dilution.[200]

It has been suggested[201] that the HSIDM is too simplistic a model to exhibit realistic solution thermodynamic behavior. In particular, it is known that the dipolar hard-sphere solvent shows strong dipolar correlations that are greatly reduced once the model is refined through the introduction of short-range anisotropic interactions.[202-204] More complex solvent models, which include quadrupolar interactions and so presumably avoid the pathological problems of the dipolar hard-sphere fluid, have been studied so far only with ions at infinite dilution.[205,206]

In even the simplest model electrolyte, the solvent molecule has a permanent dipole moment, and so the one-particle density of the solvent $\rho_s(\mathbf{r}_1, \omega_1)$ depends on the orientation ω_1 of the molecule as well as its distance from the surface. The response of the solvent to the external field is reflected not only in the density profile, but also in the polarization density profile $P(z_1)$ and other angular components of $\rho_s(\mathbf{r}_1, \omega_1)$. The electrical properties of the interface are determined not by the charge-density profile and the solvent dielectric constant as in Eq. (2.32), but by the charge-density and polarization-density profiles. A similar fundamental description of the first monolayer of solvent was employed many years ago by MacDonald and Barlow,[7] but the consistent treatment of the entire double layer in this fashion is a recent development.

The more usual approach has been to employ a local dielectric constant, depending on the distance from the surface[207,208] or possibly on the local field strength[209] near the surface and proceed on the McMillan-Mayer level. The chief difficulty with this approach is that in the double layer the Maxwell field can vary rapidly on a molecular scale so that the very notion of a local dielectric constant becomes untenable.

The statistical mechanical calculation of ionic and solvent one-particle densities thus incorporates (in principle) the correct nonlocal nature of the dielectric response in the double layer. (Actually, the one-particle densities convey only the longitudinal dielectric response, the transverse dielectric response is described by the two-particle densities, which are much harder to calculate.) Nonlocal dielectric theory is commonly employed in other areas of interfacial condensed matter physics through the introduction of a wave-vector-dependent dielectric function $\varepsilon(\mathbf{k})$.[210] This formulation has been applied to double layer theory by the Soviet group of Kornyshev et al.,[211,212] and is equivalent in principle to the approach based on the calculation of distribution functions. In practice, the results differ. In one case, approximations must be made to enable the calculation of the charge and polarization density profiles; in the other, approximations are made directly on the form of $\varepsilon(\mathbf{k})$.

By dealing directly with $\rho_s(\mathbf{r}_1, \omega_1)$ we are also able to include nonlinear dielectric effects in a fundamental way. Electrostriction is manifested in a (μ, V, T) ensemble through the dependence of the solvent-density profile on the external electric field.[213] Similarly, the nonlinear dependence of $P(z_1)$ on the electric field corresponds to dielectric saturation in the double layer.

While image effects are, in general, just as difficult to handle in Born-Oppenheimer level models as in the primitive model electrolyte, in one instance they are in fact conceptually simpler than in the corresponding McMillan-Mayer level model. Since the electrostatic interactions among the ions and solvent molecules are not dielectrically screened (we are ignoring electronic polarizability here), the case $\varepsilon_w = 1$ is where all interactions are pair-additive and central. In reality, of course, the dipole moment of the solvent molecule gives rise to a dielectric constant ε of the solvent, so that the ions must respond to solvent-averaged potentials, which tend asymptotically to image potentials corresponding to repulsive images ($\varepsilon_w = 1$). An exact calculation for such a Born-Oppenheimer level Hamiltonian with central pair interactions would therefore automatically include the appropriate image interactions.

While the use of Born-Oppenheimer level models will clearly enable important new physical effects to be included in the fundamental statistical mechanical theory of the double layer, numerical implementation of theories at this level of modeling is difficult and at an early stage of development. Many of the exact conditions discussed in Section II.D can be generalized to the case where polar solvent molecules are present (see, for example, Refs. 44, 49, 53, 54, 56, and 214). It appears that these conditions may be much stronger tests of theory for the dense Born-Oppenheimer level electrolyte than the corresponding conditions in Section II.D are of the dilute McMillan-Mayer level electrolyte.

Given the impressive performance of the MPB theory of the primitive model double layer, it is perhaps not surprising that attempts are being made to extend the Kirkwood hierarchy to the HSIDM double layer.[215-217] A significant drawback to this program is the fact that little previous study has been made of the performance of such an approach for simple polar fluids or, indeed, dense fluids of any kind.

By contrast, integral equation theories have been applied extensively to simple polar fluids,[218] ion-dipole mixtures,[197-199] and polar fluids in external fields,[219-222] and their performance tested to some extent against simulation results. So far the only double layer results have come from analytically soluble theories of the HSIDM,[214,223-227] chiefly the MSA/MSA theory.[214,223-226] The strictly linear nature of the MSA/MSA theory precludes the existence of nonlinear dielectric effects and allows even qualitatively reasonable results to be obtained only near the point of zero charge. A further problem is that the MSA theory of the bulk HSIDM shows unphysical features, such as an ionic activity coefficient monotonically decreasing with increasing ionic concentration.[228]

Nevertheless, the results show certain features that must be expected from Born-Oppenheimer level models in general and are not seen in the primitive model results of Section V. The most striking of these is the pronounced oscillatory decay of the polarization density profile, and hence the mean electrostatic potential. This feature, a reflection of the nonlocal dielectric response of the solvent, is not merely a consequence of the oscillatory solvent density profile, since reducing the dipole moment of the solvent molecule produces a damping of the oscillations in $P(z_1)$ while not affecting the solvent density profile.[224,225] The thermodynamic implications of such solvent orientational ordering near the surface are discussed in Refs. 223–225. While the resultant capacitance is similar to that observed near $\sigma = 0$ in the mercury/ aqueous electrolyte system, such agreement is clearly fortuitous in view of the radical simplifications in the model, the neglect of contributions from the surface, and experimental dependence of C_i on the nature of the metal.

Many of the weaknesses of the MSA/MSA theory of the HSIDM double layer can be circumvented by employing a HNC-based closure.[229] The non-linear closure incorporates nonlinear dielectric effects in a mean field way and retains the correct nonlinear relationship between $g_\alpha(z_1)$ and $q_\alpha\psi(z_1)$. The contact density also has the correct dependence on surface charge density. Unfortunately, numerical solutions to the resulting equations are hard to obtain in regimes of interest. It is too early to say whether this failure is due to the artificial model (the HSIDM double layer), inadequate theory (the HNC closure), or simply a poor formulation of the numerical problem.

IX. CONCLUSIONS

The last five years of intense theoretical effort, then, have led to significant advances in our understanding of simplified models of the electrical double layer. Chief among these has been the derivation of a number of exact conditions for inhomogeneous Coulomb systems such as the double layer. It may be that the incorporation of such exact requirements will lead to accurate new theories.

Of the existing theories of the primitive model double layer, the modified Poisson-Boltzmann (MPB) and HNC/MSA theories clearly stand out in their ability to make quantitative predictions of the structural and thermodynamic properties of the double layer. The case of divalent electrolyte is a more stringent test of the theory. The MPB theory proves superior here and is furthermore distinguished by its ability to incorporate image effects accurately. In general, the MPB theory shows predictable trends as the extreme limits of high and low concentration or surface charge density are approached and as the valence of the electrolyte is increased. Less reassuring in this regard is the uneven performance of the HNC/MSA, particularly its concentration dependence for electrolytes of different valence. Whereas the HNC/HNC theory should by rights improve some of the defects of the HNC/MSA, what scanty evidence is available suggests just the opposite trend.

The use of Gouy-Chapman theory gives acceptable results for the surface potential ($\psi_{d/2}$) for $1:1$ electrolytes at concentrations less than 0.1 M and moderate surface charge densities. The range of validity of Gouy-Chapman theory, however, (as gauged by the primitive model results) is greatly reduced for divalent aqueous electrolytes or, equivalently, monovalent electrolytes in a solvent of low dielectric constant.

For $1:1$ electrolytes, the introduction of images produces striking structural effects for low concentrations and surface charge densities but has only a relatively small effect on $\psi_{d/2}$. Again, these effects are enhanced in the case of $2:2$ electrolytes.

While the behavior of the planar primitive model double layer has now been explored in great detail, the study of similar systems at the same level of modeling (such as those discussed in Section VII) will doubtless remain an area of active research. The incorporation of realistic solvent and surface interactions into models of the double layer represents a truly challenging task. It is difficult to see, however, how progress toward this goal can be made at a faster pace than the corresponding treatment of bulk electrolyte, for which only limited results are available with realistic models of the solvent.

APPENDIX A.　GRAPHICAL PRESCRIPTIONS
OF $\rho_\alpha(z_1)$ AND $h_{\alpha\beta}(r_1, r_2)$

The reader who is not conversant with the cluster notation of liquid state physics is urged to consult original papers,[29-31] textbooks,[37] or reviews.[230-232] In particular, the review of Andersen is extremely helpful and contains most of the results to be presented below. We follow his notation without comment.

The graphical prescriptions have the advantage of being able to incorporate naturally image interactions, unlike many of the integral equations discussed in Section IV. They have the disadvantage that convergence is almost never proved, especially for high-density systems (not of direct interest to us), and it is by no means obvious which sets of graphs contain the dominant behavior.

We have Eq. (2.57):

$$\ln\left(\frac{\rho_\alpha(z_1)}{\delta_\alpha}\right) = -\beta v_\alpha(z_1) + c_\alpha(z_1)$$

where $c_\alpha(z_1)$ is the sum of all topologically distinct irreducible simple graphs of black $\rho_\gamma(\mathbf{r}_1)$ vertices (field points) and $f_{\alpha\beta}(\mathbf{r}_i,\mathbf{r}_j)$ bonds, with one white root point labeled (α, z_1), one or more field points, and at least one bond:

$$c_\alpha(z_1) = \qquad\qquad\qquad\qquad\qquad\qquad\qquad\qquad + \cdots \qquad (A.1)$$

$f_{\alpha\beta}(\mathbf{r}_1,\mathbf{r}_2)$ is the usual Mayer function:

$$f_{\alpha\beta}(\mathbf{r}_1,\mathbf{r}_2) = \exp\left[-\beta u_{\alpha\beta}(\mathbf{r}_1,\mathbf{r}_2)\right] - 1 \qquad (A.2)$$

The two-particle total correlation function is given by $h_{\alpha\beta}(\mathbf{r}_1,\mathbf{r}_2)$, the sum of all topologically distinct irreducible simple graphs with $\rho_\gamma(\mathbf{r}_i)$ vertices and $f_{\alpha\beta}(\mathbf{r}_i,\mathbf{r}_j)$ bonds, with two white root points labeled $(\alpha,\mathbf{r}_1)$ and $(\beta,\mathbf{r}_2)$, any number of field points, and at least one bond:

$$h_{\alpha\beta}(\mathbf{r}_1,\mathbf{r}_2) = \qquad\qquad\qquad\qquad\qquad\qquad\qquad\qquad + \cdots \qquad (A.3)$$

It is worthwhile at this stage to point out a crucial difference between the practical status of Eq. (A.3) for uniform fluids and for nonuniform fluids. In the uniform case, the black vertices are weighted by constant factors ρ_γ, and $h_{\alpha\beta}(r_{12})$ is then a functional only of the pair potential. For the nonuniform case, the function $\rho_\gamma(\mathbf{r}_i)$ is associated with every black vertex. Equations (A.1) and (A.3) must be solved simultaneously. It follows from this that the simplest (nontrivial) approximation to Eqs. (A.1) and (A.3) must be a nonlinear integral equation in $\rho_\alpha(z_1)$.

The graphical resummations that are widely used in electrolyte theory[11,59,231] may be employed in Eqs. (A.1) and (A.3), although surprisingly little work has been done on such theories. The process of topological reduction by which these resummations are performed is described in detail by Andersen.[231]

A common procedure in electrolyte theory is to split the Mayer function into short-range and long-range parts:

$$f_{\alpha\beta}(\mathbf{r}_1,\mathbf{r}_2) = f^0_{\alpha\beta}(\mathbf{r}_1,\mathbf{r}_2) + \left[1 + f^0_{\alpha\beta}(\mathbf{r}_1,\mathbf{r}_2)\right] \sum_{n=1}^{\infty} \left[\Phi_{\alpha\beta}(\mathbf{r}_1,\mathbf{r}_2)\right]^n / n! \quad (A.4)$$

where

$$f^0_{\alpha\beta}(\mathbf{r}_1,\mathbf{r}_2) = \exp\left[-\beta u^S_{\alpha\beta}(\mathbf{r}_1,\mathbf{r}_2)\right] - 1 \quad (A.5)$$

and

$$\Phi_{\alpha\beta}(\mathbf{r}_1,\mathbf{r}_2) = -\beta q_\alpha q_\beta u^E(\mathbf{r}_1,\mathbf{r}_2) \quad (A.6)$$

We now replace each f bond in Eqs. (A.1) and (A.3) by at most one f^0 bond and any number of Φ bonds. We define the chain-sum bond function by analogy with the procedure in electrolyte theory[231]: $\mathscr{C}_{\alpha\beta}(\mathbf{r}_1,\mathbf{r}_2)$ is the sum of all chain graphs with two root points labeled $(\alpha,\mathbf{r}_1)$ and $(\beta,\mathbf{r}_2)$, no f^0 bonds, and one Φ bond ($\circ\text{----}\circ$) between each pair of black $\rho_\gamma(\mathbf{r}_i)$ vertices:

$$\mathscr{C}_{\alpha\beta}(\mathbf{r}_1,\mathbf{r}_2) = \quad (A.7)$$

Then by the process of topological reduction we can reexpress Eqs. (A.1) and (A.3) so that only f^0 bonds and $\mathscr{C}$ bonds appear. We find[231]: $h_{\alpha\beta}(\mathbf{r}_1,\mathbf{r}_2)$ is the sum of all topologically distinct irreducible graphs with two white points labeled $(\alpha,\mathbf{r}_1)$ and $(\beta,\mathbf{r}_2)$, any number of black $\rho_\gamma(\mathbf{r}_i)$ vertices, f^0 bonds ($\circ\text{---}\circ$) and $\mathscr{C}$ bonds ($\circ\text{ww}\circ$) at most one f^0 bond between each pair of points, any number of $\mathscr{C}$ bonds between each pair of points, at least one bond, and no pair of points whose residuals have chains of two or more $\mathscr{C}$ bonds:

$$h_{\alpha\beta}(\mathbf{r}_1,\mathbf{r}_2) = \quad (A.8)$$

After combining terms in $v_\alpha(z_1)$ and $c_\alpha(z_1)$, the prescription for $\rho_\alpha(z_1)$ in a form most suited to the double-layer problem becomes

$$\ln\left[\frac{\rho_\alpha(z_1)}{\mathfrak{z}_\alpha}\right] = -\beta v^S_\alpha(z_1) - \beta q_\alpha \psi(z_1)$$
$$+ \frac{1}{2} \lim_{\mathbf{r}_2 \to \mathbf{r}_1} \left[\mathscr{C}_{\alpha\alpha}(\mathbf{r}_1,\mathbf{r}_2) + \beta q_\alpha^2 u^C(r_{12})\right] + \bar{c}_\alpha(z_1) \quad (A.9)$$

where $\bar{c}_\alpha(z_1)$ is the sum of all topologically distinct irreducible graphs with one white root point labeled (α, z_1), one or more field points, at most one f^0 bond between each pair of points, any number of $\mathscr{C}$ bonds between each pair of points, no pairs of points whose residuals have chains of two or more $\mathscr{C}$ bonds, excluding the graphs with one field point, no f^0 bonds, and one or two $\mathscr{C}$ bonds.

$$\bar{c}_\alpha(z_1) = \underset{1}{\circ\!\!-\!\!\!-\!\!\bullet} + \underset{1}{\circ\!\!\sim\!\!\sim\!\!\bullet} + \underset{1}{\circ\!\!\sim\!\!\sim\!\!\bullet} + \cdots + \underset{1}{\triangle} + \cdots \quad (A.10)$$

The first two graphs in Eq. (A.10) are discussed in Section IV.A.

APPENDIX B.   SOME ASPECTS OF COMPUTER SIMULATION OF DOUBLE LAYERS

We summarize here the most important technical considerations in adapting standard Monte Carlo methods to the dilute inhomogeneous Coulomb system that is the primitive model double layer. For a more detailed, though necessarily somewhat disjointed, discussion of these matters, the original papers[126,131,139] should be consulted.

An attractive feature of the Monte Carlo method is the freedom it affords in the choice of thermodynamic constraints for the system. For an interfacial problem such as the double layer, the virtues of the grand ensemble are several. First, by fixing the chemical potential at the outset of the computer experiment, thermodynamic equilibrium of the interface with a precisely specified bulk state is automatically built into the calculation. By contrast, were the number of particles to be fixed instead (i.e., were the canonical ensemble to be used), the corresponding bulk concentration could be judged only from the apparent density far from the surface and could not be fixed in advance. Agreement of this observed "asymptotic" density with that corresponding to the applied chemical potential constitutes an important test of the correctness of the computation. Furthermore, it is sometimes desirable to carry out the computations in samples including *no* region with bulk properties; this would occur, for example, in studying the ionic distribution between two surfaces close together in connection with the colloid stability problem. In that case it is only by using the grand ensemble that one would have any knowledge of the bulk concentration to which the simulation corresponds. Of course, for dense liquids there will be problems in using the grand ensemble because of the difficulty of ensuring proper density fluctuations; this is not a problem with the primitive model below $2M$, where additions and subtractions of charge-neutral combinations of particles are easy to carry out. In this connection, we note that fluctuations in the ion-density profiles in the grand canonical simulation of a double layer are presumably

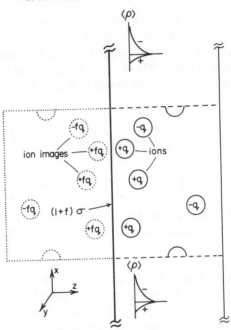

Fig. 33.   A schematic representation of a Monte Carlo simulation cell in which only one wall is charged and polarizable. f is defined in Eq. (2.5). Each central cell ion interacts with all charge—ionic, image, uniform surface—within its central cell and with the *mean* ionic charge density in the rest of the system.

more faithful to the behavior of a small portion of an infinite system than would be the case in a simulation with the number of particles fixed.

The regrettable presence of a second surface in any computer simulation of an interface seems inescapable. All simulations of the planar double layer have used a rectangular prism in which the ions are confined between two parallel impenetrable walls with periodic boundary conditions in the two directions parallel to the surfaces so created (cf. Fig. 33). In many of these experiments both hard walls are identically charged,[131,133,134] making a virtue of necessity. The two surfaces must then be separated by substantially more than the thickness of the resulting two double layers, as evidenced by an ample region in the middle of the box having the properties of the bulk system. Of course, useful data may then be obtained from both surfaces. In the special but most widely studied case of a symmetric restricted primitive model, however, there are advantages to treating the second surface as uncharged and inert.[126,139] The cation and anion densities will be slightly perturbed at this inert surface, but because of the symmetry of the model they will be affected *equally* and only in a very thin layer next to the wall. Consequently, it is easy to construct a cell long enough that the double layer at the charged wall is unaffected by the presence of the inert wall. Since most of the ions are usually in the dense region of the double layer, for a fixed maximum number of particles this arrangement allows one to maximize the dimensions of the sample parallel to the surface; these are precisely the dimensions to which some properties of the system are expected to be most sensitive. If image forces are present, this one-plate geometry has an additional advantage; if both confining walls are charged and have a dielectric constant different from that of the continuum solvent, the electrostatic boundary conditions can be satisfied only by an infinite line of image charges of varying strengths for *each* ion in the central cell. This is an unnecessary and expensive complication unless one is specifically interested in the problem of two overlapping layers.

Essential to Monte Carlo computations is the calculation of accurate energies for each configuration sampled. This might seem to offer insuperable difficulties where the range of the interparticle force is not small compared to the size of the basic simulation cell. Indeed, concerns of this nature have recently been the focus of intensive research and some controversy, particularly for dense polar fluids.[125,233-236] For the bulk primitive model at electrolyte concentrations, however, screening by the ionic atmosphere is sufficiently short-range that there is negligible correlation of ion-density fluctuations over distances comparable to typical "box" dimensions. This has meant that the nearest image convention, in which only the interaction of each ion with the nearest periodic image of any other is included, has proved a satisfactory method of computing the energy of a central cell imagined to be surrounded by a symmetric bulk of infinite extent. This procedure is entirely inappropriate, however, for the double layer problem or for the simulation of any other inhomogeneous system. Now the regions outside the central box will have a permanent mean charge separation near the walls (as opposed to mere fluctuations about zero), which provides a large unscreened field on the central cell particles under study. This "long-range correction" will typically contribute a substantial part of the total energy of the central cell, and its accurate estimation is *essential* to getting meaningful results. For simulations of the planar primitive model double layer, this external field has been computed by attributing to the surroundings a charge distribution equal to the *mean* charge distribution of the central cell to that point in the simulation. (The sketches of ion profiles in the neighboring cells in Fig. 33 are intended to represent this procedure.) This is consistent with the physical picture of the Monte Carlo cell as a small portion of a semiinfinite system. A concise expression for this external field is then

$$\phi_{ext}(z) = \psi(z) - \langle \psi_{cell}(z) \rangle \tag{B.1}$$

where $\psi(z)$ is the usual mean electrostatic potential obtained from the ion-density profiles via Eq. 2.32, and $\langle \psi_{cell}(z) \rangle$ is the electrostatic potential at z produced only by the mean charge distribution inside the central cell. In addition to the average Coulomb fields at z of all central cell ions, the quantity $\langle \psi_{cell} \rangle$, unlike ψ, also includes the uniform and mean image components of the surface charge as suggested in Fig. 33. An alternative way of approximating ϕ_{ext} is by an Ewald-like sum in which the mean potentials in Eq. (B.1) are replaced by their instantaneous values for the current configuration. The appropriate sums have been derived[176,237] and have been applied in simulations of molten salt surfaces.[176-178] Although some intuitive objections to their use for the dilute electrolyte interface have been raised on physical grounds,[126] no conclusive comparative studies have been made. The potential $\psi(z)$ in Eq. (B.1) has sometimes been replaced by the solution of the appropriate Poisson-Boltzmann equation.[154,165] This seems a needless approximation that will certainly cause errors when the behavior of the simulated system is in fact different from the Poisson-Boltzmann prediction.

Acknowledgments

S.L.C. thanks Derek Chan for introducing him to double layer theory and for the pleasure of a fruitful and happy collaboration ever since. We are grateful to Chris Outhwaite and Doug Henderson for providing us with numerical results of the MPB and HNC/MSA theories, respectively.

Finally, we owe a special debt to John Valleau, who participated as a coauthor in the vital planning stages of this undertaking and had drafted Section V.A and Appendix B when a more important commitment arose unexpectedly, requiring all his attention. Thereafter, he still found time to give invaluable administrative support, to read critically the various drafts, and to offer helpful suggestions for their improvement. Unfailingly generous with the credit for his own scientific endeavors, John was nevertheless adamant that it was somehow improper, despite all his efforts, that he continue to be listed as a coauthor. We have very reluctantly acceded to his insistence, if not his viewpoint, on the grounds that the article as it stands would have benefitted greatly from still further applications of John's keen analysis and clear writing. We, in our turn, insist on regarding him as a full partner in the enterprise in everything but name.

REFERENCES

1. J. O'M. Bockris and A. K. N. Reddy, *Modern Electrochemistry*, Vol. 2, Plenum, New York, 1970.
2. E. J. W. Verwey and J. Th. G. Overbeek, *Theory of the Stability of Lyophobic Colloids*, Elsevier, Amsterdam, 1948.
3. S. McLaughlin, in *Current Topics in Membranes and Transport*, Vol. 9, F. Bronner and A.

Kleinseller, Eds., Academic Press, New York, 1977.

4. D. C. Grahame, *Chem. Rev.* **41**, 441 (1947).

5. R. Parsons, in *Progress in Electrochemistry* (*Studies in Physical and Theoretical Chemistry*, Vol. 15), D. A. J. Rand, G. P. Power, and I. M. Ritchie, Eds., Elsevier, Amsterdam, 1981.

6. R. Parsons, *Electrochim. Acta* **21**, 681 (1976).

7. C. A. Barlow, Jr., in *Physical Chemistry: An Advanced Treatise*, Vol. 9A, H. Eyring and D. Henderson, Eds., Academic Press, New York, 1970.

8. W. R. Fawcett, *Isr. J. Chem.* **18**, 3 (1979).

9. S. K. Rangarajan, in *Electrochemistry*, Vol. 7, Specialist Periodical Reports, The Chemical Society, London, 1980.

10. F. P. Buff and F. H. Stillinger, *J. Chem. Phys.* **25**, 312 (1956).

11. H. L. Friedman and W. D. T. Dale, in *Statistical Mechanics, Part A: Equilibrium Techniques* (*Modern Theoretical Chemistry*, Vol. 5), B. Berne, Ed., Plenum, New York, 1977.

12. D. N. Card and J. P. Valleau, *J. Chem. Phys.* **52**, 6232 (1970).

13. G. Gouy, *J. Phys.* (*Paris*) **9**, 457 (1910).

14. D. L. Chapman, *Phil. Mag.* **25**, 475 (1913).

15. R. Evans, *Adv. Phys.* **28**, 143 (1979).

16. J. K. Percus, in *Studies in Statistical Mechanics*, Vol. 8, E. W. Montroll and J. L. Lebowitz, Eds., North-Holland, Amsterdam, 1982.

17. C. A. Croxton, *Statistical Mechanics of the Liquid Surface*, Wiley, New York, 1980.

18. J. S. Rowlinson and B. Widom, *Molecular Theory of Capillarity*, Clarendon, Oxford, 1982.

19. S. Levine, C. W. Outhwaite, and L. B. Bhuiyan, *J. Electroanal. Chem.* **123**, 105 (1981).

20. B. Jancovici, *J. Stat. Phys.* **29**, 263 (1982).

21. J. G. Kirkwood and F. P. Buff, *J. Chem. Phys.* **19**, 774 (1951).

22. C. W. Outhwaite, *Mol. Phys.* **27**, 561 (1974).

23. R. Evans and T. J. Sluckin, *Mol. Phys.* **40**, 413 (1980).

24. M. Baus and C. F. Tejero, *Mol. Phys.* **47**, 1211 (1982).

25. M. J. Spaarnay, *The Electrical Double Layer*, Pergamon, Oxford, 1972.

26. R. Parsons, in *Comprehensive Treatise of Electrochemistry*, Vol. 1, J. O'M. Bockris, B. E. Conway, and E. Yeager, Eds., Plenum, New York, 1980.

27. G. A. Martynov, in *Research in Surface Forces*, Vol. 2, B. V. Deryagin, Ed., New York Consultants Bureau, New York, 1966, pp. 75, 84, 94.

28. J. G. Kirkwood, *J. Chem. Phys.* **3**, 300 (1935).

29. C. de Dominicus, *J. Math. Phys.* **3**, 983 (1962).

30. T. Morita and K. Hiroike, *Prog. Theor. Phys.* **25**, 537 (1961).

31. F. H. Stillinger and F. P. Buff, *J. Chem. Phys.* **37**, 1 (1962).

32. J. L. Lebowitz and J. K. Percus, *J. Math. Phys.* **4**, 116 (1963).

33. P. Hohenberg and W. Kohn, *Phys. Rev. B* **136**, 864 (1964).

34. W. Kohn and L. J. Sham, *Phys. Rev. A* **140**, 1133 (1965).

35. N. D. Mermin, *Phys. Rev. A* **137**, 1441 (1965).

36. A. K. Rajagopal, *Adv. Chem. Phys.* **61**, 59 (1980).

37. J.-P. Hansen and I. R. McDonald, *Theory of Simple Liquids*, Academic Press, New York, 1976.

38. C. W. Outhwaite, *J. Chem. Soc. Faraday Trans. II* **74**, 1214 (1978).

39. G. M. Bell and S. Levine, in *Chemical Physics of Ionic Solutions*, B. E. Conway and R. G. Barradas, Eds., Wiley, New York, 1966.

40. S. Levine and G. M. Bell, *Disc. Faraday Soc.* **42**, 69 (1966).

41. G. Stell, in *Phase Transitions and Critical Phenomena*, Vol. 5B, C. Domb and M. S. Green, Eds., Academic Press, New York, 1976.

42. M. S. Wertheim, *J. Chem. Phys.* **65**, 2377 (1976).

43. R. A. Lovett, C. Y. Mou, and F. P. Buff, *J. Chem. Phys.* **65**, 570 (1976).

44. L. Blum, C. Gruber, D. Henderson, J. L. Lebowitz, and P. A. Martin, *J. Chem. Phys.* **78**, 3195 (1983).

45. S. L. Carnie, D. Y. C. Chan, D. J. Mitchell, and B. W. Ninham, *J. Chem. Phys.* **74**, 1472 (1981).

46. L. D. Landau and E. M. Lifshitz, *Electrodynamics of Continuous Media*, Pergamon, Oxford, 1960, pp. 53, 104.

47. D. J. Mitchell and P. Richmond, *J. Colloid Interface Sci.* **46**, 128 (1974).

48. L. Blum, J. L. Lebowitz, and D. Henderson, *J. Chem. Phys.* **72**, 4249 (1978).

49. L. Blum, C. Gruber, J. L. Lebowitz, and P. Martin, *Phys. Rev. Lett.* **48**, 1769 (1982).

50. C. Gruber, J. L. Lebowitz, and P. A. Martin, *J. Chem. Phys.* **75**, 944 (1981).

51. L. Blum, D. Henderson, J. L. Lebowitz, C. Gruber, and P. A. Martin, *J. Chem. Phys.* **75**, 5974 (1981).

52. M. Baus, *Mol. Phys.* **48**, 347 (1983).

53. S. L. Carnie and D. Y. C. Chan, *Chem. Phys. Lett.* **77**, 437 (1981).

54. S. L. Carnie, *J. Chem. Phys.* **78**, 2742 (1983).

55. D. Henderson, L. Blum, and J. L. Lebowitz, *J. Electroanal. Chem.* **102**, 315 (1979).

56. S. L. Carnie and D. Y. C. Chan, *J. Chem. Phys.* **74**, 1293 (1981); **78**, 3348 (1983).

57. P. Debye and E. Hückel, *Z. Physik* **24**, 185 (1923).

58. J. G. Kirkwood and J. C. Poirier, *J. Phys. Chem.* **58**, 591 (1954).

59. H. L. Friedman, *Ionic Solution Theory Based on Cluster Expansions*, Wiley-Interscience, New York, 1962.

60. R. H. Fowler and E. A. Guggenheim, *Statistical Thermodynamics*, Macmillan, New York, 1956, p. 387.

61. L. Onsager, *Chem. Rev.* **13**, 73 (1933).

62. C. W. Outhwaite, in *Statistical Mechanics*, Vol. 2, Specialist Periodical Reports, The Chemical Society, London, 1975.

63. D. C. Grahame, *J. Chem. Phys.* **21**, 1054 (1953).

64. B. Abraham-Schrauner, *J. Math. Biol.* **2**, 333 (1975); **4**, 201 (1977).

65. O. Stern, *Z. Elektrochem.* **30**, 508 (1924).

66. J. N. Israelachvili and G. E. Adams, *J. Chem. Soc. Faraday Trans. I* **74**, 975 (1978).

67. R. M. Pashley, *J. Colloid Interface Sci.* **83**, 531 (1981).

68. M. Eisenberg, T. Gresalfi, T. Riccio, and S. McLaughlin, *Biochemistry* **18**, 5213 (1979).

69. S. McLaughlin, N. Mulrine, T. Gresalfi, G. Vaio, and A. McLaughlin, *J. Gen. Physiol.* **77**, 445 (1981).

70. R. J. Hunter, *Zeta Potential in Colloid Science*, Academic Press, London, 1981.

71. A. L. Loeb, *J. Colloid Sci.* **6**, 75 (1951).

72. F. P. Buff and F. H. Stillinger, *J. Chem. Phys.* **39**, 1911 (1963).

73. W. E. Williams, *Proc. Phys. Soc. A* **66**, 372 (1953).

74. V. N. Gorelkin and V. P. Smilga, *Sov. Electrochem.* **2**, 454 (1966).

75. P. G. Ali-Zade, G. A. Martynov, and V. G. Melamed, *Dokl. Akad. Nauk. SSSR* **151**, 601 (1963).

76. A. L. Nicholls III and L. R. Pratt, *J. Chem. Phys.* **76**, 3782 (1982).

77. C. Wagner, *Physik. Z.* **25**, 474 (1924).

78. L. Onsager and N. N. T. Samaras, *J. Chem. Phys.* **2**, 528 (1934).

79. B. Jancovici, *J. Stat. Phys.* **28**, 43 (1982).

80. A. L. Nicholls III and L. R. Pratt, *J. Chem. Phys.* **77**, 1070 (1982).

81. G. M. Bell and P. D. Rangecroft, *Trans. Faraday Soc.* **67**, 649 (1971).

82. S. L. Carnie and D. Y. C. Chan, *Mol. Phys.* (in press).

83. H. L. Friedman, *J. Chem. Phys.* **76**, 1092 (1982).

84. P. G. Wolynes, *Ann. Rev. Phys. Chem.* **31**, 345 (1980).

85. F. H. Stillinger and J. G. Kirkwood, *J. Chem. Phys.* **33**, 1282 (1960).

86. S. Levine and G. Bell, *J. Phys. Chem.* **64**, 1188 (1960).

87. V. S. Krylov and V. G. Levich, *Russ. J. Phys. Chem.* **37**, 50, 1224 (1963).

88. C. W. Outhwaite, *Chem. Phys. Lett.* **7**, 636 (1970).

89. G. M. Bell and P. D. Rangecroft, *Mol. Phys.* **24**, 255 (1972).

90. C. W. Outhwaite, L. B. Bhuiyan, and S. Levine, *J. Chem. Soc. Faraday Trans. II* **76**, 1388 (1980).

91. C. W. Outhwaite and L. B. Bhuiyan, *J. Chem. Soc. Faraday Trans. II* **78**, 775 (1982).

92. C. W. Outhwaite and L. B. Bhuiyan, *J. Chem. Soc. Faraday Trans. II* **79**, 707 (1983).

93. J. P. Valleau, L. K. Cohen, and D. N. Card, *J. Chem. Phys.* **72**, 5942 (1980).

94. T. L. Croxton and D. A. McQuarrie, *J. Phys. Chem.* **83**, 1840 (1979).

95. L. Blum, *J. Phys. Chem.* **81**, 136 (1977).

96. S. Levine and C. W. Outhwaite, *J. Chem. Soc. Faraday Trans. II* **74**, 1670 (1978).

97. C. W. Outhwaite, L. B. Bhuiyan, and S. Levine, *Chem. Phys. Lett.* **64**, 150 (1979).

98. L. B. Bhuiyan, C. W. Outhwaite, and S. Levine, *Mol. Phys.* **42**, 1271 (1981).

99. C. W. Outhwaite, *J. Chem. Soc. Faraday Trans. II* **79**, 1315 (1983).

100. T. L. Croxton and D. A. McQuarrie, *Mol. Phys.* **42**, 141 (1981).

101. R. W. Pastor and J. Goodisman, *J. Chem. Phys.* **68**, 3654 (1978).

102. T. L. Croxton and D. A. McQuarrie, *Chem. Phys. Lett.* **68**, 489 (1979).

103. T. L. Croxton, D. A. McQuarrie, G. N. Patey, G. M. Torrie, and J. P. Valleau, *Can. J. Chem.* **59**, 1998 (1981).

104. R. O. Watts, in *Statistical Mechanics, Specialist Periodical Reports*, Vol. 1 The Chemical Society, London, 1973.

105. J. W. Perram and L. R. White, *Disc. Faraday Soc.* **59**, 29 (1975).

106. D. Henderson, F. F. Abraham, and J. A. Barker, *Mol. Phys.* **31**, 1291 (1976).

107. D. Henderson and L. Blum, *J. Chem. Phys.* **69**, 5441 (1978).

108. J. K. Percus, in *Equilibrium Theory of Classical Fluids*, H. L. Frisch and J. L. Lebowitz, Eds., Benjamin, New York, 1964.

109. D. Henderson, L. Blum, and W. R. Smith, *Chem. Phys. Lett.* **63**, 381 (1979).

110. S. L. Carnie, D. Y. C. Chan, D. J. Mitchell, and B. W. Ninham, *J. Chem. Phys.* **74**, 1472 (1981).

111. E. Waisman and J. L. Lebowitz, *J. Chem. Phys.* **56**, 3086, 3093 (1972).

112. M. Lozada-Cassou, R. Saavedra-Barrera, and D. Henderson, *J. Chem. Phys.* **77**, 5150 (1982).

113. M. Lozada-Cassou and D. Henderson, *J. Phys. Chem.* **87**, 2821 (1983).

114. G. N. Patey, private communication.

115. M. J. Grimson and G. Rickayzen, *Mol. Phys.* **45**, 221 (1982).

116. A. Alastuey and D. Levesque, *Mol. Phys.* **47**, 1349 (1982).

117. (a) J. J. Erpenbeck and W. W. Wood, in *Statistical Mechanics, Part B: Time Dependent Processes* (*Modern Theoretical Chemistry*, Vol. 6), B. Berne, Ed., Plenum, New York, 1977. (b) J. Kushick and B. J. Berne, *ibid*.

118. H. L. Friedman, *Ann. Rev. Phys. Chem.* **32**, 179 (1981).

119. H. C. Andersen, *Ann. Rev. Phys. Chem.* **26**, 145 (1975).

120. J. C. Rasaiah, D. N. Card, and J. P. Valleau, *J. Chem. Phys.* **56**, 248 (1972).

121. J. P. Valleau and L. K. Cohen, *J. Chem. Phys.* **72**, 5935 (1980).

122. W. van Megen and I. Snook, *Mol. Phys.* **39**, 1043 (1980).

123. W. W. Wood, in *Physics of Simple Liquids*, H. N. V. Temperley, G. S. Rushbrooke, and J. S. Rowlinson, Eds., North-Holland, Amsterdam, 1968.

124. F. H. Ree, in *Physical Chemistry. An Advanced Treatise*, Vol. 8A, H. Eyring, D. Henderson, and W. Jost, Eds., Academic Press, New York, 1971.

125. J. P. Valleau and S. G. Whittington, in *Statistical Mechanics, Part A: Equilibrium Techniques* (*Modern Theoretical Chemistry*, Vol. 5), B. Berne, Ed., Plenum, New York, 1977.

126. G. M. Torrie and J. P. Valleau, *J. Chem. Phys.* **73**, 5807 (1980).

127. G. M. Torrie, unpublished results.

128. C. W. Outhwaite, private communication.

129. G. N. Patey, *J. Chem. Phys.* **72**, 5763 (1980).

130. D. Henderson and L. Blum, *J. Electroanal. Chem.* **111**, 217 (1980).

131. G. M. Torrie and J. P. Valleau, *J. Phys. Chem.* **86**, 3251 (1982); G. M. Torrie, unpublished results.

132. D. Henderson, private communication.

133. W. van Megen and I. Snook, *J. Chem. Phys.* **73**, 4656 (1980).

134. I. Snook and W. van Megen, *J. Chem. Phys.* **75**, 4104 (1981).

135. I. Snook, private communication.

136. Yu. I. Yalamov, in *Research in Surface Forces*, Vol. 2, B. V. Deryagin, Ed., New York Consultants Bureau, New York, 1966.

137. P. Richmond, *J. Chem. Soc. Faraday Trans. II* **71**, 1154 (1975).

138. A. P. Nelson and D. A. McQuarrie, *J. Theor. Biol.* **55**, 13 (1975).

139. G. M. Torrie, J. P. Valleau, and G. N. Patey, *J. Chem. Phys.* **76**, 4615 (1982).

140. J. P. Valleau and G. M. Torrie, *J. Chem. Phys.* **76**, 4623 (1982).

141. L. B. Bhuiyan, L. Blum, and D. Henderson, *J. Chem. Phys.* **78**, 442 (1983).

142. J. J. Spitzer, *J. Colloid Interface Sci.* **92**, 198 (1983).

143. D. Stigter, *J. Colloid Interface Sci.* **53**, 296 (1975).

144. A. L. Loeb, J. T. G. Overbeek, and P. H. Wiersema, *The Electrical Double Layer Around a Spherical Colloidal Particle*, MIT Press, Cambridge, MA, 1961.

145. H. Ohshima, T. W. Healy, and L. R. White, *J. Colloid Interface Sci.* **90**, 17 (1982).

146. H. Wennerström, B. Jönsson, and P. Linse, *J. Chem. Phys.* **76**, 4665 (1982).

147. P. Linse, G. Gunnarsson, and B. Jönsson, *J. Phys. Chem.* **86**, 413 (1982).

148. A. Katchalsky, *Pure Appl. Chem.* **26**, 327 (1971).

149. G. S. Manning, *Ann. Rev. Phys. Chem.* **23**, 117 (1972).

150. T. Alfrey, P. W. Berg, and H. Morawetz, *J. Polymer Sci.* **7**, 543 (1951).

151. R. M. Fuoss, A. Katchalsky, and S. Lifson, *Proc. Nat. Acad. Sci. USA* **37**, 579 (1951).

152. J. A. Schellmann and D. Stigter, *Biopolymers* **16**, 1415 (1977).

153. M. Guéron and G. Weisbuch, *Biopolymers* **19**, 353 (1980).

154. D. Bratko and V. Vlachy, *Chem. Phys. Lett.* **90**, 434 (1982).

155. M. Fixman, *J. Chem. Phys.* **70**, 4995 (1979).

156. R. Bacquet and P. Rossky, *J. Phys. Chem.* (in press).

157. B. Deryagin and L. Landau, *Acta Phys.-Chim. USSR* **14**, 633 (1941).

158. O. F. Devereux and P. L. de Bruyn, *Interaction of Plane Parallel Double Layers*, MIT Press, Cambridge, MA, 1963.

159. E. P. Honig and P. M. Mul, *J. Colloid Interface Sci.* **36**, 258 (1971).

160. D. Y. C. Chan, R. M. Pashley, and L. R. White, *J. Colloid Interface Sci.* **77**, 283 (1980).

161. P. L. Levine, *J. Colloid Interface Sci.* **51**, 72 (1975).

162. M. J. Grimson and G. Rickayzen, *Mol. Phys.* **44**, 817 (1981).

163. M. J. Grimson, *Chem. Phys. Lett.* **86**, 38 (1982).

164. J. P. Valleau and G. M. Torrie, unpublished results.

165. B. Jönsson, H. Wennerström, and B. Halle, *J. Phys. Chem.* **84**, 2179 (1980).

166. B. V. Deryagin, *Kolloid Z.* **69**, 155 (1934).

167. G. Frens and J. T. G. Overbeek, *J. Colloid Interface Sci.* **38**, 376 (1972).

168. N. E. Hoskin and S. Levine, *Trans. Roy. Soc. Ser. A* **248**, 449 (1956).

169. L. N. Macartney and S. Levine, *J. Colloid Interface Sci.* **30**, 345 (1969).

170. J. E. Ledbetter, T. L. Croxton, and D. A. McQuarrie, *Can. J. Chem.* **59**, 1860 (1981).

171. T. A. Ring, *J. Chem. Soc. Faraday Trans. II* **78**, 1528 (1982).

172. P. Linse and B. Jönsson, *J. Chem. Phys.* **78**, 3167 (1983).

173. B. Hafskjold and G. Stell, in *Studies in Statistical Mechanics*, Vol. 8, E. W. Montroll and J. L. Lebowitz, Eds., North-Holland, Amsterdam, 1982.

174. L. V. Woodcock and K. Singer, *Trans. Faraday Soc.* **67**, 12 (1971).

175. L. V. Woodcock, in *Advances in Molten Salt Chemistry*, Vol. 3, G. Mamantov and G. P. Smith, Eds., Plenum, New York, 1975.

176. D. M. Heyes, M. Barber, and J. H. R. Clarke, *J. Chem. Soc. Faraday Trans. II* **73**, 1485 (1977).

177. D. M. Heyes and J. H. R. Clarke, *J. Chem. Soc. Faraday Trans. II* **75**, 1240 (1979).

178. D. M. Heyes and J. H. R. Clarke, *J. Chem. Soc. Faraday Trans. II* **77**, 1089 (1981).

179. M. Baus and J.-P. Hansen, *Phys. Rep.* **59**, 1 (1980).

180. A. Alastuey and B. Jancovici, *J. Physique* **42**, 1 (1981).

181. B. Jancovici, *Phys. Rev. Lett.* **46**, 386 (1981).

182. B. Jancovici, *J. Physique-Lett.* **42**, L223 (1981).

183. E. R. Smith, *Phys. Rev. A* **24**, 2851 (1981).

184. E. R. Smith, *J. Phys. A. Gen. Phys.* **15**, 1271 (1982).

185. P. J. Forrester and E. R. Smith, *J. Phys. A. Gen. Phys.* **15**, 3861 (1982).

186. P. J. Forrester, B. Jancovici, and E. R. Smith (preprint).

187. P. Choquard, P. Favre, and C. Gruber, *J. Stat. Phys.* **23**, 405 (1980).

188. H. Totsuji, *J. Chem. Phys.* **75**, 871 (1981).

189. S. Trasatti, in *Modern Aspects of Electrochemistry*, Vol. 13, J. O'M. Bockris and B. E. Conway, Eds., Plenum, New York, 1979.

190. S. Trasatti, in *Advances in Electrochemistry and Electrochemical Engineering* Vol. 10, H. Gerischer and C. W. Tobias, Eds., Wiley, New York, 1977.

191. J. P. Badiali, M. L. Rosinberg, and J. Goodisman, *J. Electroanal. Chem.* **130**, 31 (1981).

192. J. P. Badiali, M. L. Rosinberg, and J. Goodisman, *J. Electroanal.Chem.* **143**, 73 (1983).

193. J. R. Reimers, R. O. Watts, and M. L. Klein, *Chem. Phys.* **64**, 95 (1982).

194. F. H. Stillinger, *Adv. Chem. Phys.* **31**, 1 (1975).

195. G. C. Lie, E. Clementi, and M. Yoshime, *J. Chem. Phys.* **64**, 2314 (1976).

196. W. L. Jorgensen, *J. Chem. Phys.* **77**, 4156 (1982).

197. S. A. Adelman and J. M. Deutch, *J. Chem. Phys.* **60**, 3935 (1974).

198. L. Blum, *J. Chem. Phys.* **61**, 2129 (1974).

199. D. Levesque, J. J. Weis, and G. N. Patey, *J. Chem. Phys.* **72**, 1887 (1980).

200. G. N. Patey and J. P. Valleau, *J. Chem. Phys.* **63**, 2334 (1975).

201. H. L. Friedman, Pure Appl. Chem. **53**, 1277 (1981).

202. G. N. Patey, D. Levesque, and J. J. Weis, *Mol. Phys.* **38**, 219 (1979).

203. S. L. Carnie and G. N. Patey, *Mol. Phys.* **47**, 1129 (1982).

204. G. P. Morriss and P. T. Cummings, *Mol. Phys.* **45**, 1099 (1982).

205. F. Hirata, P. J. Rossky, and B. M. Pettit, *J. Chem. Phys.* **78**, 4133 (1983).

206. G. N. Patey and S. L. Carnie, *J. Chem. Phys.* **78**, 5183 (1983).

207. J. P. Clay, N. S. Goel, and F. P. Buff, *J. Chem. Phys.* **56**, 4245 (1972).

208. J. W. Perram and M. N. Barber, *Mol. Phys.* **28**, 131 (1974).

209. D. C. Grahame, *J. Chem. Phys.* **18**, 903 (1950).

210. K. L. Kliewer and R. Fuchs, *Adv. Chem. Phys.* **27**, 355 (1974).

211. A. A. Kornyshev, *Electrochim. Acta* **26**, 1 (1981).

212. A. A. Kornyshev and M. A. Vorotyntsev, *Can. J. Chem.* **59**, 2031 (1981).

213. J. C. Rasaiah. D. J. Isbister, and G. Stell, *J. Chem. Phys.* **75**, 4707 (1981).

214. D. Henderson and L. Blum, *J. Chem. Phys.* **74**, 1902 (1981).

215. C. W. Outhwaite, *Chem. Phys. Lett.* **76**, 619 (1980).

216. C. W. Outhwaite, *Can. J. Chem.* **59**, 1854 (1981).

217. C. W. Outhwaite, *Mol. Phys.* **48**, 599 (1983).

218. G. Stell, G. N. Patey, and J. S. Høye, *Adv. Chem. Phys.* **48**, 183 (1981).

219. D. J. Isbister and B. C. Freasier, *J. Stat. Phys.* **20**, 331 (1979).

220. J. M. Eggebrecht, D. J. Isbister, and J. C. Rasaiah, *J. Chem. Phys.* **73**, 3980 (1980).

221. J. C. Rasaiah, D. J. Isbister, and J. Eggebrecht, *J. Chem. Phys.* **75**, 5497 (1981).

222. S. L. Carnie and G. Stell, *J. Chem. Phys.* **77**, 1017 (1982).

223. S. L. Carnie and D. Y. C. Chan, *J. Chem. Phys.* **73**, 2949 (1980).

224. S. L. Carnie and D. Y. C. Chan, *Adv. Colloid Interface Sci.* **16**, 81 (1982).

225. S. L. Carnie and D. Y. C. Chan, *J. Chem. Soc. Faraday Trans. II* **78**, 695 (1982).

226. D. Henderson and L. Blum, *Faraday Symp.* **16**, 151 (1981).

227. F. Vericat, L. Blum, and D. Henderson, *J. Chem. Phys.* **77**, 5808 (1982)

228. D. Y. C. Chan, D. J. Mitchell, and B. W. Ninham, *J. Chem. Phys.* **70**, 2946 (1979).

229. S. L. Carnie and D. Y. C. Chan (to be published).

230. G. Stell, in *Equilibrium Theory of Classical Fluids*, H. L. Frisch and J. L. Lebowitz, Eds., Benjamin, New York, 1964.

231. H. C. Andersen, in *Statistical Mechanics, Part A: Equilibrium Techniques* (*Modern Theoretical Chemistry*, Vol. 5), B. Berne, Ed., Plenum, New York, 1977.

232. I. R. McDonald and S. P. O'Gorman, *Phys. Chem. Liquids* **8**, 57 (1978).

233. G. N. Patey, D. Levesque, and J.-J. Weis, *Mol. Phys.* **45**, 733 (1982).

234. S. W. de Leeuw, J. W. Perram, and E. R. Smith, *Proc. Roy. Soc. Lond.* **353A**, 27 (1980).

235. S. W. de Leeuw and J. W. Perram, *Physica* **107A**, 179 (1981).

236. C. S. Hoskins and E. R. Smith, *Mol. Phys.* **45**, 915 (1982).

237. S. W. de Leeuw and J. W. Perram, *Mol. Phys.* **37**, 1313 (1979).

238. A. R. Altenberger, *J. Chem. Phys.* **76**, 1473 (1982).

239. G. Bell and S. Levine, *Z. Physik. Chem.* (*Leipzig*) **231**, 289 (1966).

KERR EFFECT RELAXATION IN HIGH ELECTRIC FIELDS

HIROSHI WATANABE AND AKIO MORITA

Department of Chemistry, College of Arts and Sciences
University of Tokyo
Komaba, Meguro-ku, Tokyo 153, Japan

CONTENTS

I. INTRODUCTION

When an electric field is applied to a fluid or solution, molecules or (in the case of a colloidal solution) particles are oriented due to the torque caused by the action of the electric field on the anisotropy of the electric properties of the molecule or particle, mainly the permanent dipole moment and the induced dipole moment. The orientation of the molecules or particles in a preferential direction results in an anisotropy of the medium on a macroscopic scale, which can most easily be monitored by a light beam passing through the medium. This is called the electrooptical method.

Electric birefringence, the most familiar electro-optical effect, was discovered by John Kerr in 1875. The fluid or solution in which the molecules or particles are oriented in one direction by a homogeneous electric field behaves like a uniaxial crystal with the optical axis parallel to the direction of the field. If a linearly polarized light beam enters such a medium in a direction perpendicular to the plane containing the optical axis, the light beam suffers a phase difference between two components, one parallel to and one perpendicular to the electric field, after traveling the medium of length l.

The phase difference δ is related to the principal refractive indices of the medium parallel to and perpendicular to the axis, $n_\parallel$ and $n_\perp$, respectively, and to the wavelength λ of the monitoring light in vacuo by the equation

$$\delta = \frac{2\pi l}{\lambda}(n_\parallel - n_\perp) = \frac{2\pi l}{\lambda}\Delta n \qquad (1.1)$$

When the oriented molecules or particles absorb the monitoring light, the medium exhibits dichroism, which is defined as the difference in optical absorbance between the two components of the light:

$$\Delta A = A_\parallel - A_\perp \qquad (1.2)$$

In Appendix A it is shown how $\Delta n = n_\parallel - n_\perp$ is related to a quantity characterizing the motion of the particles.

Since the main subject of this article is the time-varying response of the electric birefringence, which is most interesting in the colloidal solution from the experimental point of view, the term "particle" will henceforth be used to represent either a molecule or a particle. It is assumed throughout this article, unless otherwise stated, that the particle has a common axis of symmetry for its electric, optical, and hydrodynamic properties, and therefore that the permanent dipole moment, if any, is along this axis. It is further assumed that the solution is so dilute that the interparticle interaction has no appreciable effect on the orientational movement of the particle.

Previous studies[1] were mainly concerned with the limiting low field for the dynamic birefringence, and the high field problem has been treated successfully only for the stationary state birefringence which is characterized by the Maxwell-Boltzmann distribution.

In this article we consider the Kerr-effect relaxation in high fields, whose results are obtained recently and successively published in the Journal of Chemical Physics. We also present some new unpublished results. We review earlier studies concerning Kerr-effect relaxation for the step-up and reversing electric fields in Section II and for the sinusoidal electric field in Section III. Section IV is devoted to an exact treatment of Kerr-effect relaxation in a high step-up field. Numerical treatments for the sudden application of time-independent fields are carried out in Section V. In Section VI, we consider Kerr-Effect relaxation in time-varying electric fields. We propose a new method using a rapidly rotating field by which a larger value for Δn may be obtained in comparison to other methods. We shall confine ourselves to this subject in Section VII. In Sections VIII–X, we discuss the theory of rotational Brownian motion, because this theory gives us not only a better understanding of the existing results in a consistent and transparent manner but also useful implications on how unsolved problems may be treated. In Section XI, we consider the behavior of the Kerr effect relaxation in extremely high electric fields based on considerations in Section X with a different approach from the previous one, obtaining some new results.

Peterlin and Stuart[2] obtained an expression of the birefringence as a product of an optical anisotropy factor, $g_1 - g_2$, and an orientation factor, Φ:

$$\Delta n = \frac{2\pi c_v}{n}(g_1 - g_2)\Phi \qquad (1.3)$$

where c_v is the volume fraction of the particle and n is the refractive index of the solution in the absence of the electric field.

The optical anisotropy factor $g_1 - g_2$ is the difference between the excess electric polarizabilities in the frequency of the monitoring light (optical polarizability) per unit volume of the particle parallel to and perpendicular to the symmetry axis. The orientation factor Φ is the ensemble average of the Legendre polynomials of degree 2, $P_2(\cos\theta) = (3\cos^2\theta - 1)/2$, and is given by (see Appendix A)

$$\Phi(t) = \frac{\int_0^\pi P_2(\cos\theta) f(\theta, t)\sin\theta\, d\theta}{\int_0^\pi f(\theta, t)\sin\theta\, d\theta} \tag{1.4}$$

where θ is the angle between the symmetry axis and the field direction, t is the time, and $f(\theta, t)$ is the angular distribution function for the orientation of the particle symmetry axis at any time, provided that the angular distribution of the particle axis is homogeneous with respect to the azimuth. A more general case of inhomogeneous distribution of the axis is considered in Section VII.

The dichroism caused by an external electric field is also an ensemble average of the particle axis and is given by

$$A_\| = A\left[1 + (3\cos^2\Psi - 1)\Phi\right] \tag{1.5a}$$

$$A_\perp = A\left[1 - \tfrac{1}{2}(3\cos^2\Psi - 1)\Phi\right] \tag{1.5b}$$

and

$$\Delta A = A_\| - A_\perp = \tfrac{3}{2}A(3\cos^2\Psi - 1)\Phi \tag{1.5c}$$

where A is the absorbance in the absence of the electric field and Ψ is the angle between the symmetry axis and the transition moment responsible for the absorption. Thus the electric dichroism is also governed by the same orientation function $\Phi = \langle P_2(\cos\theta)\rangle$ as the electric birefringence, its amplitude depending on the angle Ψ.

The main problem in the theoretical approach to the electrooptical effect is therefore how to obtain the ensemble average $\langle P_2(\cos(\theta, t))\rangle$, under the influence of an external electric field.

Upon the assumption that the particle is axially symmetric and non-interacting each other, the angular distribution function may be given by the rotational diffusion equation

$$\frac{\partial f(\theta, \tau)}{\partial \tau} = \frac{1}{\sin\theta}\frac{\partial}{\partial\theta}\left[\sin\theta\left(\frac{\partial f}{\partial\theta} + \frac{1}{k_B T}\frac{\partial V}{\partial\theta}f\right)\right] \tag{1.6}$$

where $\tau = Dt$ is the reduced time, D being the rotational diffusion constant about the transverse axis of the particle, $k_B T$ is the thermal energy, and V is the potential function. We may call this a Smoluchowski equation (see Section X). The potential function is given by

$$V = -\mu E(t)\cos\theta - \tfrac{1}{2}(\alpha_1 - \alpha_2)E^2(t)\cos^2\theta \qquad (1.7)$$

where μ is the permanent dipole moment; α_1 and α_2 are the electric polarizabilities parallel and perpendicular to the symmetry axis, respectively; and $E(t)$ is the external electric field. α_1 and α_2 may generally be functions of the electric field and can be expressed in a power series of E as follows[3]:

$$\alpha_i = \alpha_i + \beta_i E^2 + \gamma_i E^4 + \cdots \qquad (i=1,2) \qquad (1.8)$$

In the following, however, we regard the polarizability as a constant by neglecting effects due to the hyperpolarizability.

When the electric field is independent of time, the angular distribution reaches an equilibrium after a long enough time, and it is seen from Eq. (1.6) that in the limit $\partial f/\partial\tau = 0$, the distribution function is given by the Maxwell-Boltzmann distribution

$$f(\theta,\infty) = \frac{\exp(-V/k_B T)}{\int_0^\pi \exp(-V/k_B T)\sin\theta\, d\theta} \qquad (1.9)$$

For the stationary state birefringence, we have from Eq. (1.9) together with Eq. (1.7) that

$$\Phi = \frac{3\int_{-1}^1 u^2 \exp(eu + gu^2)\, du}{2\int_{-1}^1 \exp(eu + gu^2)\, du} - \frac{1}{2} \qquad (1.10)$$

where $u = \cos\theta$,

$$e = \frac{\mu E}{k_B T} \qquad (1.11a)$$

and

$$g = \frac{(\alpha_1 - \alpha_2)E^2}{2k_B T} \qquad (1.11b)$$

For sufficiently low fields, by expanding the exponentials in Eq. (1.10) in a power series and taking only the terms involving $1 + V/k_B T$, we have

$$\Phi = \tfrac{1}{15}(e^2 + 2g) = \frac{1}{15}\left(\frac{\mu^2}{k_B T} + \frac{\alpha_1 - \alpha_2}{k_B T}\right)E^2 \qquad (1.12)$$

Thus the birefringence is proportional to the square of the field strength for the limiting low field. This is Kerr's law. A general verification of Kerr's law is given in Appendix A.

The exact calculation of the integrals in Eq. (1.10) for arbitrary field strength was carried out by O'Konski, Yoshioka, and Oruttung[4] for the case of $(\alpha_1 - \alpha_2) > 0$, $g > 0$. The result is

$$\Phi = \frac{3}{4g}\left[\frac{\exp(e^2/4g + g)\left[2\sqrt{g}\cosh e - (e/\sqrt{g})\sinh e\right]}{\int_{e/2\sqrt{g} - g}^{e/2\sqrt{g} + g}\exp(x^2)\,dx} + \frac{e^2}{2g} - 1\right] - \frac{1}{2}$$

$$(1.13)$$

The case of $(\alpha_1 - \alpha_2) < 0$, $g < 0$, was solved by Shah,[5] with the result

$$\Phi = \frac{3}{4g}\left[\frac{\exp(e^2/4g + g)\left(e/\sqrt{-g}\sinh e + 2\sqrt{-g}\cosh e\right)}{\int_{-e/2\sqrt{-g} - \sqrt{-g}}^{-e/2\sqrt{-g} + \sqrt{-g}}\exp(-x^2)\,dx} + \frac{e^2}{2g} - 1\right] - \frac{1}{2}$$

$$(1.14)$$

Thus the analytical expressions of the stationary state birefringence include an evaluation of the error function.

The integral in Eq. (1.10) can be also given by

$$\Phi = \frac{3}{4g'}\left(\frac{e^2}{2g'} + 1\right) - \frac{3}{2g'}e^{-g'}\left(\cosh e + \frac{e^2}{2g'}\sinh e\right)\frac{1}{I} - \frac{1}{2} \qquad (1.15)$$

where $g' = -g$,

$$I = \int_{-1}^{1}\exp(eu - g'u^2)\,du$$

$$= \frac{1}{2}\sqrt{\frac{\pi}{g'}}\exp\left(\frac{e^2}{4g'}\right)\left[\operatorname{erf}\left(\sqrt{g'} - \frac{e}{2\sqrt{g'}}\right) + \operatorname{erf}\left(\sqrt{g'} + \frac{e}{2\sqrt{g'}}\right)\right] \qquad (1.16)$$

and the error function erf(z) is defined by

$$\text{erf}(z) = \frac{2}{\sqrt{\pi}} \int_0^z e^{-x^2} \, dx \tag{1.17}$$

For positive values of g, we note that

$$\text{erf}(iz) = \frac{2i}{\sqrt{\pi}} \int_0^z e^{x^2} \, dx \tag{1.18}$$

When the error function is expressed in terms of the continued fraction [cf. Eq. (6-1-38) of Jones and Thron][6],

$$\text{erf}(z) = \frac{2}{\sqrt{\pi}} e^{-z^2} \cfrac{z}{1 - \cfrac{2z^2}{3 + \cfrac{4z^2}{5 - \cfrac{6z^2}{7 + \cdot_{\cdot_{\cdot}}}}}} \tag{1.19}$$

Φ can be easily calculated numerically. The expression in Eq. (1.15) can be reduced to Eqs. (1.13) and (1.14) for positive and negative g, respectively. A more convenient numerical method for calculating the stationary state birefringence that is free from the evaluation of the error function will be presented in Section V.

For the special case of $g = 0$ (pure permanent dipole orientation), we have

$$\Phi = 1 - \frac{3L(e)}{e} = 1 - \frac{3(\coth e - 1/e)}{e} \tag{1.20}$$

where $L(x)$ is the Langevin function, whereas for the case of $e = 0$ (pure induced dipole orientation, $g > 0$), we have

$$\Phi = \frac{3}{4} \left[\frac{\exp(g)}{\sqrt{g} \int_0^{\sqrt{g}} \exp(x^2) \, dx} - \frac{1}{g} \right] - \frac{1}{2} \tag{1.21}$$

The orientation factor is plotted against $|e^2 + 2g|$ for various values of R in Figs. 1a and 1b, where

$$R = \frac{e^2}{2g} = \frac{\mu^2}{(\alpha_1 - \alpha_2) k_B T} \tag{1.22}$$

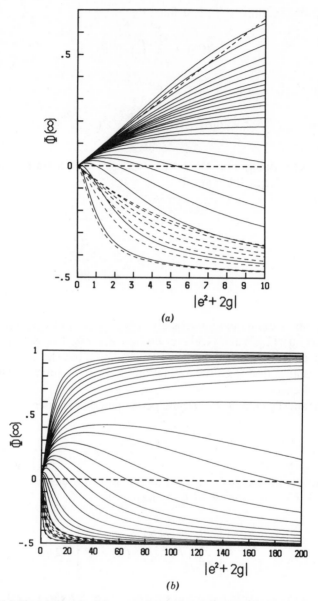

Fig. 1. (a) The initial part of the saturation curves of the electric birefringence. Full lines: from the top to the bottom, $R = 0, 0.5, 1, 2, 3, 5, 10, 20, \infty, -20, -10, -8, -7, -6, -5, -4, -3.5, -3, -2, -1.8, -1.6, -1.4, -1.2, -1.1$. Broken lines: from the top to the bottom, $R = 0, -0.3, -0.5, -0.6, -0.7, -0.8, -0.9$. Dashed line shows the Kerr law, $\Phi^0(\infty) = (e^2 + 2g)/15$. (b) Saturation curves of the electric birefringence. The values of R are the same as in Fig. 1a.

is the ratio of the contribution of the permanent dipole moment to the induced dipole moment and is independent of the field strength. Full lines are the orientation factors for $\infty \geq R \geq 0$ and $-1.1 \geq R \geq -20$, whereas broken lines are those for $-0 \geq R \geq -0.9$. (Note that $R = 0$ means that the induced dipole lies along the particle axis, whereas $R = -0$ means that it lies perpendicular to the axis.) These curves show how the orientation factor deviates from Kerr's law at high field, depending on the electric property of the particle. The deviation is the so-called saturation effect. It is also seen that the saturation curve is quite sensitive to the electric parameter, which provides a powerful experimental method for elucidating the electric properties of the particle.

It is seen that the values of Φ reach -0.5 eventually as $|e^2 + 2g| \rightarrow \infty$, as long as the value of R is negative. This is understandable, since the induced dipole moment, which is proportional to the square of the field strength and lies perpendicular to the particle axis, dominates the particle orientation for increasing values of the field strength. Since the particle axis, and therefore the axis of optical polarizability, is oriented toward the plane perpendicular to the external field without any restriction with respect to the azimuth, the saturation value is not -1 but -0.5. When $R \leq -1.2$, Φ is positive for the low values of $|e^2 + 2g|$ but changes sign at some value of $|e^2 + 2g|$, and the saturation curves cross over the zero line. This provides us a useful method for determining the value of R from the saturation measurements, since the zero point of the signals can be determined most easily in the measurements. The values of $|e^2 + 2g|$ at which the curves cross the zero line for values of R are presented in Table I.

TABLE I
Values of $e^2 + 2g$ At Which the Saturation
Curves Cross Over Zero for Several Values of R

R	$e^2 + 2g$	R	$e^2 + 2g$
-1.1	0	-3.0	18.8
-1.2	0.25	-4.0	39.23
-1.3	0.53	-5.0	66.1
-1.4	0.93	-6.0	99.3
-1.5	1.42	-7.0	138.7
-1.6	2.02	-8.0	184.3
-1.7	2.72	-9.0	235.7
-1.8	3.56	-10.0	293.3
-1.9	4.31	-11.0	356.3
-2.0	5.34	-12.0	426.1

We shall investigate the dynamic processes of Φ using Eq. (1.6) for various kinds of applied electric fields. Investigating the electric birefringence involves consideration of a dynamic nonlinear response problem with respect to the applied electric field. It is hoped that the results contained in this article will be useful for general treatments of nonlinear, nonequilibrium statistical mechanics, which are difficult and have not been developed significantly.

II. PREVIOUS TREATMENT OF THE RELAXATION PROCESSES OF ELECTRIC BIREFRINGENCE FOR THE STEP-UP AND REVERSING FIELDS

By solving the rotational diffusion equation, Eq. (1.6), we obtain the relaxation process of birefringence after the sudden change of the electric field and/or the steady state of birefringence for the time-varying electric field. Before stating our general treatment, we briefly review the development of the theoretical treatment of this problem.

Analytical equations for the transients of the electric birefringence caused by the sudden application and annihilation of a static homogeneous electric field on an ensemble of the symmetrical rigid particles were first derived by Benoit.[7] The distribution function $f(\theta, t)$ is expanded in a series of Legendre polynomials,

$$f(\theta, t) = \sum_{n=0}^{\infty} a_n(t) P_n(\cos \theta) \tag{2.1}$$

where $a_n(t)$ is a function of time and P_n is the Legendre polynomial of degree n, and then introduced into Eq. (1.6) with potential function Eq. (1.7). We then obtain a set of differential equations

$$\sum_n a_n(\tau) \left[-n(n+1) P_n - (1 - u^2) \frac{dP_n}{du} (e + 2gu) \right.$$

$$\left. + (2eu + (3u^2 - 1)2g) P_n \right] = \sum_n \frac{da_n(\tau)}{d\tau} P_n \tag{2.2}$$

where $\tau = Dt$ is the dimensionless reduced time, $u = \cos \theta$, $e = \mu E / k_B T$, and $g = (\alpha_1 - \alpha_2) E^2 / 2 k_B T$ [cf. Eqs. (1.11)]. By the use of the recursion relations

$$(2n + 1) u P_n = (n + 1) P_{n+1} + n P_{n-1} \tag{2.3a}$$

$$(1 - u^2) \frac{dP_n}{du} = n P_{n-1} - n u P_n \tag{2.3b}$$

Eq. (2.2) can be reduced to

$$\sum_n a_n \left[-n(n+1)P_n - e\frac{n(n-1)}{2n+1}P_{n-1} + e\frac{(n+1)(n+2)}{2n+1}P_{n+1} \right.$$

$$+2g\frac{(n+1)(n+2)(n+3)}{(2n+1)(2n+3)}P_{n+2} - 2g\frac{n(n-1)(n-2)}{4n^2-1}P_{n-2}$$

$$\left. -2g\left(1 + \frac{n^2(n-2)}{4n^2-1} - \frac{(n+1)^2(n+3)}{(2n+1)(2n+3)}\right)P_n \right] = \sum_n \frac{da_n}{d\tau}P_n \qquad (2.4)$$

Multiplying both sides of Eq. (2.4) by $P_m(u)$ and integrating from -1 to 1 with respect to u, Benoit obtained[7]

$$-a_m\left[m(m+1) + 2g\left(1 + \frac{m^2(m-2)}{4m^2-1} - \frac{(m+1)^2(m+3)}{(2m+1)(2m+3)}\right)\right]$$

$$+a_{m-1}e\frac{m(m+1)}{2m-1} - a_{m+1}e\frac{m(m+1)}{2m+3} + a_{m-2}2g\frac{m(m^2-1)}{(2m-3)(2m-1)}$$

$$-a_{m+2}2g\frac{m(m+1)(m+2)}{(2m+3)(2m+5)} = \frac{da_m}{d\tau} \qquad (2.5)$$

The orthogonal relation

$$\int_{-1}^{1} P_m(u)P_n(u)\,du = \begin{cases} 0 & m \neq n \\ \dfrac{2}{2m+1} & m = n \end{cases} \qquad (2.6)$$

is used here. [The coefficient of g in the first set of brackets on the left-hand side of Eq. (2.5) is not correct but happens to be insignificant for the cases of $m = 1$ and 2 at extremely low fields.] For $m = 0, 1, 2$, we find

$$\frac{da_0}{d\tau} = 0 \qquad (2.7a)$$

$$\frac{da_1}{d\tau} = 2ea_0 - 2a_1\left(1 - \frac{2g}{5}\right) - \frac{5}{2}ea_2 - \frac{12}{35}ga_3 \qquad (2.7b)$$

$$\frac{da_2}{d\tau} = 4ga_0 + 2ea_1 - 2a_2\left(3 - \frac{2g}{7}\right) - \frac{6}{7}ea_3 - \frac{16}{21}ga_4 \qquad (2.7c)$$

From the condition

$$\int_{-1}^{+1} 2\pi f \, du = 1 \tag{2.8}$$

we have

$$a_0 = \frac{1}{4\pi} \tag{2.9}$$

By neglecting the a_3 and a_4 terms, we obtain

$$a_1 = e\left[1 + \frac{4g}{15} - \frac{e^2}{15} - \left(1 + \frac{g}{5} - \frac{e^2}{20}\right)e^{-\gamma^{(1)}\tau} + \left(\frac{e^2}{60} - \frac{g}{15}\right)e^{-\gamma^{(2)}\tau}\right] \tag{2.10}$$

$$a_2 = \frac{e^2}{3} + \frac{2g}{3} - \frac{e^4}{45} + \frac{4g^2}{63} + \frac{38}{315}e^2 g - e^2\left(\frac{1}{2} + \frac{g}{14}\right)e^{-\gamma^{(1)}\tau}$$

$$+ \left(\frac{e^2}{6} - \frac{2g}{3} + \frac{e^4}{45} - \frac{4g^2}{63} - \frac{31}{630}e^2 g\right)e^{-\gamma^{(2)}\tau} \tag{2.11}$$

where

$$\gamma^{(1)} = 2 - \frac{6}{5}g + \frac{e^2}{5} \tag{2.12a}$$

$$\gamma^{(2)} = 6 - \frac{4}{7}g - \frac{e^2}{5} \tag{2.12b}$$

Neglecting terms higher than E^4, the normalized birefringence for electric fields of sufficiently low strength can be obtained. The results obtained by Benoit are as follows.

 a. *Buildup Process.* For the sudden application of a static electric field to the ensemble that has been free from any external field and has therefore been randomly oriented, the transient of the electric birefringence is

$$\frac{\Phi(\tau)}{\Phi(\infty)} = \frac{\Delta n}{\Delta n(\infty)}$$

$$= 1 - \frac{3R}{2(R+1)}\exp(-2\tau) + \frac{R-2}{2(R+1)}\exp(-6\tau) \tag{2.13}$$

where $\Delta n(\infty)$ is the birefringence at $\tau \to \infty$ and is equivalent to the stationary state value given by Eq. (1.13) or (1.14). It should be noted that the normalized birefringence diverges when $R = -1$. This is because of the null value of the stationary state birefringence when $R = -1$ and $E \to 0$.

  b. *Decay Process.* After a sudden removal of the electric field, the particles disorient by Brownian motion, and the birefringence decays. Since the particles are released from the orienting field, the diffusion equation (1.6) reduces to

$$\frac{\partial f}{\partial \tau} = \frac{1}{\sin \theta} \frac{\partial}{\partial \theta} \left(\sin \theta \frac{\partial}{\partial \theta} f \right)$$ (2.14)

In this case we can obtain the solution without solving the diffusion equation.[7] Multiplying both sides of Eq. (2.14) with $P_2(\cos \theta)$ and integrating 0 to π with respect to θ, we find after using Eq. (1.4) that

$$\begin{aligned}
\frac{d\Phi}{d\tau} &= \int_0^\pi P_2(\cos \theta) \frac{\partial f}{\partial \tau} 2\pi \sin \theta \, d\theta \\
&= \int_0^\pi P_2(\cos \theta) \frac{1}{\sin \theta} \frac{\partial}{\partial \theta} \left(\sin \theta \frac{\partial f}{\partial \theta} \right) 2\pi \sin \theta \, d\theta \\
&= -6 \int_0^\pi P_2(\cos \theta) f 2\pi \sin \theta \, d\theta = -6\Phi
\end{aligned}$$ (2.15)

The decay process of the normalized birefringence is given by

$$\frac{\Phi(\tau)}{\Phi(\infty)} = \frac{\Delta n}{\Delta n(\infty)} = \exp(-6\tau)$$ (2.16)

The relaxation time of the birefringence is therefore

$$\tau_{1/2} = \frac{1}{6} \qquad t_{1/2} = \frac{1}{6D}$$ (2.17)

  The equation for the transient birefringence in a rapidly reversing electric field was derived by Tinoco and Yamaoka.[8]

$$\frac{\Phi(\tau)}{\Phi(\infty)} = 1 - \frac{3R}{R+1} \exp(-2\tau) + \frac{3R}{R+1} \exp(-6\tau)$$ (2.18)

  Although the equations of both Benoit and Tinoco and Yamaoka are limited to the very low field, it was shown that the transients of birefringence are very sensitive to the electric parameters of the particle, and therefore an analysis of the transients of the birefringence for step-up and reversing fields is a very useful technique from both the theoretical and experimental points of view. Theoretical curves of the transient electric birefringence for the step-up and reversing electric fields of limiting low strength are presented for several values of R in Figs. 2 and 3, respectively.

O'Konski, Yoshioka, and Oruttung[4] have obtained equations of the transient electric birefringence resulting from a sudden application of infinitely high field in cases of pure permanent moment and pure induced dipole moment orientation, following the treatment of Schwarz[9] for pure induced dipole orientation, by neglecting the rotational diffusion. This problem will be discussed in detail in Section XI.

Nishinari and Yoshioka[10] proposed a theory for the rise of birefringence upon application of a step-up electric field of arbitrary strength which holds in the initial stage of buildup. To this end, by expressing Eq. (1.6) as

$$\frac{\partial f(u,\tau)}{\partial \tau} = \hat{F}f(u,\tau) \tag{2.19}$$

with

$$\hat{F} = (1-u^2)\frac{\partial^2}{\partial u^2} - 2u\frac{\partial}{\partial u} - (1-u^2)(e+2gu)\frac{\partial}{\partial u} + 2\left[eu + g(3u^2-1)\right] \tag{2.20}$$

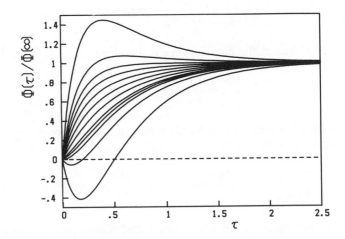

Fig. 2. Buildup of the normalized electric birefringence for the limiting low step-up field. From the top to the bottom, $R = -0.5, -0.2, 0, 0.5, 0.2, 1, 2, 5, 10, \infty, -5, -2$.

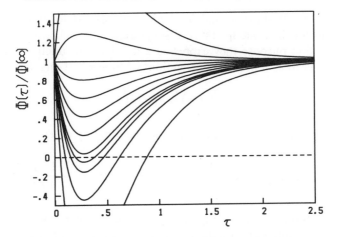

Fig. 3. Normalized electric birefringence for the limiting low reversing field. The values of R are the same as in Fig. 2.

they expressed the distribution function as

$$f(u,\tau) = \exp(\tau\hat{F})f(u,0) \qquad (2.21)$$

The orientation factor is then given by

$$\Phi = \frac{1}{2}\int_{-1}^{1} P_2(u)\exp(\tau\hat{F})P_0(u)\,du$$

$$= \frac{1}{2}\langle 2|\exp(\tau\hat{F})|0\rangle \qquad (2.22)$$

By expanding the operator $\exp(\hat{F}\tau)$ in the power series

$$\exp(\tau\hat{F}) = \sum_{n=0}^{\infty} \frac{\tau^n}{n!}\hat{F}^n \qquad (2.23)$$

Φ is expressed as

$$\Phi = \langle 2|\hat{F}|0\rangle\tau + \langle 2|\hat{F}^2|0\rangle(\tau^2/2!)$$
$$+ \langle 2|\hat{F}^3|0\rangle(\tau^3/3!) + \langle 2|\hat{F}^4|0\rangle(\tau^4/4!) + \cdots$$
$$= \tfrac{5}{4}g\tau + \left(\tfrac{2}{5}e^2 - \tfrac{12}{5}g + \tfrac{8}{35}g^2\right)\tau^2 - \left(\tfrac{16}{15}e^2 - \tfrac{24}{5}g + \tfrac{8}{21}e^2g + \tfrac{32}{35}g^2 + \tfrac{32}{105}g^3\right)\tau^3$$
$$+ \left(\tfrac{26}{15}e^2 - \tfrac{36}{5}g - \tfrac{2}{21}e^4 - \tfrac{72}{35}g^2 + \tfrac{12}{5}e^2g + \tfrac{32}{441}e^2g^2 + \tfrac{273}{105}g^3 - \tfrac{32}{231}g^4\right)\tau^4 + \cdots$$

$$(2.24)$$

Matsumoto, Watanabe, and Yoshioka[11] attempted to obtain analytical expressions for the buildup and reversing birefringences that hold for higher fields. For that purpose they expressed a distribution function that should satisfy Eq. (1.6) as a product of functions of time τ and angular space u:

$$f(u, \tau) = h(u)\rho(\tau) \tag{2.25}$$

Introducing Eq. (2.25) into Eq. (1.6) and dividing by $h(u)\rho(\tau)$, we obtain

$$\frac{1}{h}\frac{d}{du}\left[(1-u^2)\frac{dh}{du} - (1-u^2)(e+2gu)h\right] = \frac{1}{\rho}\frac{d\rho}{d\tau} \tag{2.26}$$

Since the right-hand side of Eq. (2.26) is a function of τ alone, and the left-hand side is a function of u alone, the value of the quantity to which each side is equal must be a constant, say $-\lambda$. Accordingly, Eq. (2.26) can then be written as two differential equations,

$$\frac{d}{du}\left[(1-u^2)\frac{dh}{du} - (1-u^2)(e+2gu)h\right] + \lambda h = 0 \tag{2.27}$$

and

$$\frac{d\rho}{d\tau} = -\lambda\rho \tag{2.28}$$

Equation (2.28) can be integrated at once to give

$$\rho(\tau) = C\exp(-\lambda\tau) \tag{2.29}$$

where C is a constant. The differential equation for $h(u)$ cannot be integrated in general. However, Eq. (2.27) is transformed into the self-adjoint form by multiplying it by $\exp(-eu - gu^2)$, namely,

$$\frac{d}{du}\left[(1-u^2)\exp(-eu - gu^2)\frac{dh}{du}\right] + (6gu^2 + 2eu - 2g)\exp(-eu - gu^2)h$$

$$+ \lambda\exp(-eu - gu^2)h = 0 \tag{2.30}$$

This is a kind of Liouville equation with two singular points at $u = 1$ and $u = -1$. Denoting the solutions by h_m for $\lambda = \lambda_m$ and h_n for $\lambda = \lambda_n$, we multiply the equation for h_n and h_m by h_m and h_n, respectively, and integrating from -1 to 1, after subtracting one from the other, we obtain

$$(\lambda_n - \lambda_m) \int_{-1}^{1} h_n h_m \exp(-eu - gu^2)\, du = 0 \qquad (2.31)$$

as long as h_m and h_n are finite at $u = 1$ and $u = -1$. Equation (2.31) indicates that h_m and h_n weighted by $\exp[-(eu + gu^2)/2]$ are mutually orthogonal and that λ_m and λ_n are eigenvalues. Making the substitution

$$y_n = h_n(u)\exp\left[-(eu + gu^2)/2\right] \qquad (2.32)$$

into Eq. (2.27), we have a set of differential equations with respect to y_n:

$$\frac{d}{du}\left[(1-u^2)\frac{dy_n}{du}\right] + \left\{ buE + \left[\left(\frac{b^2}{4} + 3c\right)u^2 - \left(\frac{b^2}{4} + c\right)\right]E^2 + bc(u^3 - u)E^3 \right.$$

$$\left. + c^2(u^4 - u^2)E^4 \right\} y_n + \lambda_n y_n = 0 \qquad (2.33)$$

where $b = e/E = \mu/k_B T$ and $c = g/E^2 = (\alpha_1 - \alpha_2)/2k_B T$. In the limit as $E \to 0$, Eq. (2.33) reduces to

$$\frac{d}{du}\left[(1-u^2)\frac{dy_n}{du}\right] + \lambda_n y_n = 0 \qquad (2.34)$$

From the condition that this equation should be finite in the range $-1 \leq u \leq 1$, $\lambda_n = n(n+1)$ with $n = 0, 1, 2, 3, \ldots$. The function y_n corresponding to λ_n is the Legendre polynomial of degree n, $P_n(u)$. The perturbation method was used to solve Eq. (2.33). Expanding the eigenvalue λ_n and the eigenfunction y_n in terms of E,

$$\lambda_n = n(n+1) + \alpha_n E + \beta_n E^2 + \gamma_n E^3 + \delta_n E^4 + \cdots \qquad (2.35)$$

$$y_n(u) = P_n(u) + \alpha_n'(u)E + \beta_n'(u)E^2 + \gamma_n'(u)E^3 + \delta_n'(u)E^4 + \cdots \qquad (2.36)$$

introducing Eqs. (2.35) and (2.36) into Eq. (2.33), and equating the terms of

E, E^2, E^3, and E^4 to zero, we obtain a set of differential equations:

$$\frac{d}{du}\left[(1-u^2)\frac{d\alpha'_n}{du}\right]+(bu+\alpha_n)P_n+n(n+1)\alpha'_n=0 \qquad (2.37)$$

$$\frac{d}{du}\left[(1-u^2)\frac{d\beta'_n}{du}\right]+\left[\left(\frac{b^2}{4}+3c\right)u^2-\left(\frac{b^2}{4}+c\right)+\beta_n\right]P_n$$

$$+(bu+\alpha_n)\alpha'_n+n(n+1)\beta'_n=0 \qquad (2.38)$$

$$\frac{d}{du}\left[(1-u^2)\frac{d\gamma'_n}{du}\right]+\left[bc(u^3-u)+\gamma_n\right]P_n$$

$$+\left[\left(\frac{b^2}{4}+3c\right)u^2-\left(\frac{b^2}{4}+c\right)+\beta_n\right]\alpha'_n+(bu+\alpha_n)\beta'_n+n(n+1)\gamma'_n=0$$

$$(2.39)$$

$$\frac{d}{du}\left[(1-u^2)\frac{d\delta'_n}{du}\right]+\left[c^2(u^4-u^2)+\delta_n\right]P_n+\left[bc(u^3-u)+\gamma_n\right]\alpha'_n$$

$$+\left[\left(\frac{b^2}{4}+3c\right)u^2-\left(\frac{b^2}{4}+c\right)+\beta_n\right]\beta'_n+(bu+\alpha_n)\gamma'_n+n(n+1)\delta'_n=0$$

$$(2.40)$$

These equations are solved by expanding $\alpha'_n(u)$, $\beta'_n(u)$, $\gamma'_n(u)$, and $\delta'_n(u)$ in terms of Legendre polynomials. Thus the general solution of Eq. (1.6) is

$$f(u,\tau)=\sum_{n=0}^{\infty}A_n h_n\exp(-\lambda_n\tau)$$

$$=\sum_{n=0}^{\infty}A_n y_n\exp\left(\frac{eu+gu^2}{2}\right)\exp(-\lambda_n\tau) \qquad (2.41)$$

where the A_n's are constants.

For the buildup process, the A_n's are determined so as to satisfy the initial condition

$$f(u,0)=\sum_{n=0}^{\infty}A_n h_n=\sum_{n=0}^{\infty}A_n y_n\exp\left(\frac{eu+gu^2}{2}\right)=\frac{1}{4\pi} \qquad (2.42)$$

resulting in

$$A_n=\frac{2n+1}{8\pi}\int_{-1}^{1}y_n\exp\left(-\frac{eu+gu^2}{2}\right)du \qquad (2.43)$$

The orthogonal condition of P_n, Eq. (2.6) is used. The solution of Eq. (2.27) for $\lambda = 0$ gives h_0, which is the Maxwell-Boltzmann distribution function. The normalized birefringence is then

$$\frac{\Phi(\tau)}{\Phi(\infty)} = \frac{\Delta n(\tau)}{\Delta n(\infty)} = 1 - \frac{X_1}{X_0}\exp(-\lambda_1\tau) + \frac{X_2}{X_0}\exp(-\lambda_2\tau) - \frac{X_3}{X_0}\exp(-\lambda_3\tau)$$

(2.44)

where λ_1, λ_2, λ_3, X_0, X_1, X_2, and X_3 are functions of e and g:

$$\lambda_1 = 2 + \frac{e^2}{5} - \frac{4g}{5}$$

(2.45a)

$$\lambda_2 = 6 + \frac{e^2}{7} - \frac{4g}{7}$$

(2.45b)

$$\lambda_3 = 12 + \frac{2e^2}{15} - \frac{8g}{15}$$

(2.45c)

$$X_0 = e^2 + 2g - \frac{2}{21}(e^4 - 2e^2g - 2g^2)$$

(2.46a)

$$X_1 = \frac{3e^2}{2} - \frac{3}{350}(9e^4 - e^2g)$$

(2.46b)

$$X_2 = \frac{1}{2}(e^4 - 4g) + \frac{1}{42}(e^4 - 11e^2g - 8g^2)$$

(2.46c)

$$X_3 = \frac{1}{175}(e^4 - 14e^2g)$$

(2.46d)

For the reverse process, the angular distribution function can be obtained by replacing E by $-E$ in the preceding treatment. The normalized birefringence is then given by

$$\frac{\Phi(\tau)}{\Phi(\infty)} = 1 - \frac{X_1'}{X_0'}\exp(-\lambda_1\tau) + \frac{X_2'}{X_0'}\exp(-\lambda_2\tau) - \frac{X_3'}{X_0'}\exp(-\lambda_3\tau)$$

(2.47)

where λ_1, λ_2, and λ_3 are the same as expressed in Eqs. (2.45), and

$$X_0' = e^2 + 2g + \frac{1}{14}(e^4 + 12e^2g + 12g^2)$$

(2.48a)

$$X_1' = 3e^2 + \frac{3}{175}(26e^4 + 71e^2g)$$

(2.48b)

$$X_2' = 3e^2 + \frac{1}{7}(4e^2 + 9e^2g)$$

(2.48c)

$$X_3' = \frac{2}{175}(11e^4 + 6e^2g)$$

(2.48d)

At limitingly low fields, Eqs. (2.44) and (2.47) reduce to Eqs. (2.13) and (2.18), respectively.

The above equations are no longer valid for large values of $|e^2 + 2g|$ beyond some value around 4, depending on the parameter. Koopmans, de Bore, and Greve[12] applied a numerical method for calculating the transient electric birefringence to the MWY theory. In doing so they expressed Eq. (2.41) as

$$|f\rangle = \frac{1}{\sqrt{2}} \sum Ex|y_i\rangle e^{\lambda_i \tau} \langle y_i|Ex^{-1}|0\rangle \qquad (2.49)$$

for the buildup and

$$|f\rangle = \frac{1}{\sqrt{2}} \sum_i Ex|y_i\rangle e^{\lambda_i(\tau - \tau_0)} \langle y_i|Ex^{-1} \sum_j |j\rangle(-1)^j \qquad (2.50)$$

$$\times \langle j| \sum_k E_x|y_k\rangle e^{\lambda k \tau_0} \langle y_k|E_x^1|0\rangle$$

for the processes after field reversal at τ_0, where $|y_i\rangle = Ex|h_i\rangle$ is the one defined by Eq. (2.32), for

$$Ex = \exp\left(-\frac{eu + gu^2}{2}\right) \qquad (2.51)$$

and $|h_i\rangle$ is the eigenvector corresponding to the eigenvalue λ_i. On the basis of completeness of $P_n(u)$, we have

$$\langle n|Ex^{-1}\hat{F}Ex|y_k\rangle = \sum_l \langle n|Ex^{-1}\hat{F}Ex|l\rangle\langle l|y_k\rangle \qquad (2.52)$$

and the calculation is reduced to a matrix multiplication after $\langle n|Ex^{-1}|\hat{F}|Ex|l\rangle$ has been diagonalized numerically.

III. PREVIOUS TREATMENT OF ELECTRIC BIREFRINGENCE FOR THE SINUSOIDAL ELECTRIC FIELD

The theoretical calculation of the electric birefringence caused by the time-varying electric field is a much harder task than the static field transient problem, and the theories have been successful only in the lowest field strength.

Peterlin and Stuart[2] obtained the solutions for a sinusoidal electric field $E(t) = E_0 \cos \omega t$ for the cases of pure induced dipole and pure permanent dipole orientation. The solutions are limited to the infinitely low field. The general case of both induced and permanent dipoles coexist on the particle was first solved by Ogawa and Oka[13] for a very low sinusoidal electric field $E(t) = E_0 \sin \omega t$. Later the same problem was treated by Thurston and Bowling[14] for a sinusoidal electric field $E(t) = E_0 \cos \omega t$.

On assuming that a very low sinusoidal electric field of angular frequency ω, $E(t) = E_0 \sin \omega t = E_0 \sin \omega' \tau$, is applied at $t = 0$, Ogawa and Oka expressed a_1 and a_2 in Eqs. (2.7b) and (2.7c) as

$$a_1 = a_{10} + a_{11} e_0 + a_{12} e_0^2 \tag{3.1a}$$

$$a_2 = a_{20} + a_{21} e_0 + a_{22} e_0^2 \tag{3.1b}$$

where

$$e_0 = \frac{\mu E_0}{k_B T} \tag{3.2}$$

[cf. Eq. (1.11a)] and $\omega' = \omega/D$. Neglecting the a_3 and a_4 terms, the coefficients for the e_0^0, e_0^1, and e_0^2 terms were equated to zero, respectively, to give

$$\frac{da_{10}}{d\tau} = -2a_{10} \tag{3.3a}$$

$$\frac{da_{11}}{d\tau} = \frac{1}{2\pi} \sin \omega' \tau - 2a_{11} - \frac{2}{5} a_{20} \sin \omega' \tau \tag{3.3b}$$

$$\frac{da_{12}}{d\tau} = -2\left(\frac{2}{15} \frac{a_{10}}{R} \sin^2 \omega' \tau + a_{12} \right) - \frac{2}{5} \cdot a_{21} \sin \omega' \tau \tag{3.3c}$$

$$\frac{da_{20}}{d\tau} = -6a_{20} \tag{3.3d}$$

$$\frac{da_{21}}{d\tau} = 2a_{10} \sin \omega' \tau - 6a_{21} \tag{3.3e}$$

$$\frac{da_{22}}{d\tau} = \frac{2a_0}{R} \sin^2 \omega' \tau + 2a_{11} \sin \omega' \tau - 2\left(3a_{22} - \frac{a_{20}}{7R} \sin^2 \omega' \tau \right) \tag{3.3f}$$

In these equations, $R = \mu^2/2(\alpha_1 - \alpha_2)k_B T = e_0^2/2g_0$ where $g_0 = 2(\alpha_1 - \alpha_2)/E_0^2 k_B T$. From Eq. (3.3a), we have

$$a_{10} = A \exp(-2\tau) \tag{3.4}$$

From the condition $f = 1/4\pi = a_0$ when $\tau = 0$, we have $a_1(0) = a_2(0) = 0$, and therefore $a_{10}(0) = a_{11}(0) = a_{12}(0) = 0$. And from Eq. (3.3d) we have

$$a_{20} = B\exp(-6\tau) \qquad (3.5)$$

Since $a_{20}(0) = a_{21}(0) = a_{22}(0) = 0$, B vanishes. Then Eq. (3.3b) reduces to

$$\frac{da_{11}}{d\tau} = -\frac{1}{2\pi}\sin\omega'\tau - 2a_{11} \qquad (3.6)$$

The solution is

$$a_{11} = Ce^{-2\tau} + \frac{1}{\pi}\frac{1}{4+\omega'^2}\left(\sin\omega'\tau - \frac{\omega'}{2}\cos\omega'\tau\right) \qquad (3.7)$$

Since $a_{11}(0) = 0$, C is given by

$$C = \frac{1}{2\pi}\frac{\omega'}{4+\omega'^2} \qquad (3.8)$$

Therefore,

$$a_{11} = \frac{1}{\pi}\frac{1}{4+\omega'^2}\left[\frac{\omega'}{2}\exp(-2\tau) + \sin\omega'\tau - \frac{\omega'}{2}\cos\omega'\tau\right] \qquad (3.9)$$

The integral constant for Eq. (3.3e) also vanishes, and $a_{21}(0) = 0$. Substituting Eq. (3.7) into Eq. (3.3f), we have

$$\frac{da_{22}}{d\tau} = -6a_{22} + \frac{1}{2\pi}\frac{1}{R}\sin^2\omega'\tau$$

$$+ \frac{1}{\pi}\frac{2}{4+\omega'^2}\left(\frac{\omega'}{2}e^{-2\tau}\sin\omega'\tau + \sin^2\omega'\tau - \frac{\omega'}{2}\cos\omega'\tau\sin\omega'\tau\right) \qquad (3.10)$$

The solution is given by

$$a_{22} = \frac{1}{24\pi}\frac{1}{R}\left[1 + \frac{36}{36+\omega'^2}\left(\cos 2\omega'\tau + \frac{\omega'}{3}\sin\omega'\tau\right)\right]$$

$$+ \frac{1}{\pi}\frac{4\omega'e^{-2\tau}}{(4+\omega'^2)(16+\omega'^2)}\left(\sin\omega'\tau - \frac{\omega'}{4}\cos\omega'\tau\right)$$

$$+ \frac{1}{24\pi}\frac{1}{4+\omega'^2}\left[1 - \frac{36}{36+\omega'^2}\left(\cos 2\omega'\tau + \frac{\omega'}{3}\sin\omega'\tau\right)\right]$$

$$+ \frac{1}{24\pi}\frac{2\times 36}{(4+\omega'^2)(36+\omega'^2)}\left(\sin 2\omega'\tau - \frac{\omega'}{3}\cos 2\omega'\tau\right) + Fe^{-6\tau} \qquad (3.11)$$

For a large value of τ, we have the solution for the steady state birefringence. From Eqs. (3.16) and (3.11) we then have

$$
\begin{aligned}
a_2 = \frac{2g_0}{24\pi} &\left[1 + \frac{1}{36 + \omega'^2} \left(\cos 2\omega'\tau + \frac{\omega'}{3} \sin 2\omega'\tau \right) \right] \\
&+ \frac{e_0^2}{24\pi} \left\{ \frac{4}{4 + \omega'^2} \left[1 + \frac{36}{36 + \omega'^2} \left(\cos 2\omega'\tau + \frac{\omega'}{3} \sin 2\omega'\tau \right) \right] \right. \\
&\left. + \frac{18\omega'}{(36 + \omega'^2)} \left(\sin 2\omega'\tau - \frac{\omega'}{3} \cos 2\omega'\tau \right) \right\} \quad (3.12)
\end{aligned}
$$

This equation agrees essentially with the equation derived by Morita and Watanabe[15] [cf. Eq. (6.52)]. Introducing φ and γ according to

$$
\tan \varphi = \frac{\omega'}{3} \quad (3.13a)
$$

$$
\tan \gamma = \frac{\omega'}{2} \quad (3.13b)
$$

into Eq. (3.12), we obtain

$$
\begin{aligned}
a_2 = \frac{1}{24\pi} &\left\{ \left(\frac{4e_0^2}{4 + \omega^2} + 2g_0 \right) \right. \\
&\left. - \frac{6}{(36 + \omega'^2)^{1/2}} \left[4g_0^2 + \frac{16e_0^2 g_0}{4 + \omega^2} + \frac{e_0^4}{4 + \omega^2} \right]^{1/2} \cos(2\omega'\tau - \varphi - \gamma) \right\}
\end{aligned}
$$
$$(3.14)$$

Thurston and Bowling[14] expressed the steady state birefringence Δn as

$$
\Delta n = \Delta n_{st} + \Delta n_{alt} \cos(2\omega'\tau - \delta) \quad (3.15)
$$

where Δn_{st} is the stationary state component and Δn_{alt} is the alternating component having a phase angle δ. Expressions for Δn_{st} and Δn_{alt} are with our notations,

$$
\Delta n_{st} = \Delta n_0 \left[1 + \frac{R}{1 + \dfrac{\omega'^2}{4}} \right] \frac{1}{R + 1} \quad (3.16)
$$

and

$$\Delta n_{\text{alt}} = \left(\Delta n_{\text{alt}}'^{2} + \Delta n_{\text{alt}}''^{2} \right)^{1/2} \tag{3.17}$$

with

$$\Delta n_{\text{alt}}' = \frac{\Delta n_0}{R+1} \left[\frac{R(1 - \omega'^2/6)}{(1 + \omega'^2/4)(1 + \omega'^2/9)} + \frac{1}{1 + \omega'^2/9} \right] \tag{3.18}$$

and

$$\Delta n_{\text{alt}}'' = \frac{\Delta n_0}{R+1} \left[\frac{R(\omega'/2 + \omega'/3)}{(1 + \omega'^2/4)(1 + \omega'^2/9)} + \frac{\omega'/3}{1 + \omega'^2/9} \right] \tag{3.19}$$

where

$$\Delta n_0 = \frac{2\pi c_v}{15n} (g_1 - g_2)(e_0^2 + 2g_0) \tag{3.20}$$

is the birefringence for $\omega' \to 0$, $E \to 0$. The phase drag is then given by

$$\tan \delta = \frac{\Delta n_{\text{alt}}'}{\Delta n_{\text{alt}}''} \tag{3.21}$$

Equations (3.15)–(3.20) essentially agree with the results arising from the expression of a_2 obtained by Ogawa and Oka,[13] Eq. (3.14).

Käs and Brückner[15] calculated the electric birefringence in the sinusoidal electric field $E(t) = E_0 \cos \omega t$ simply by taking the product of equations obtained by Peterlin and Stuart for pure induced and pure permanent dipole orientation, which are guaranteed up to E^2 terms. The result cannot be applied to the more general cases where both the induced and permanent dipoles are responsible for the particle orientation.

IV. AN EXACT TREATMENT OF KERR-EFFECT RELAXATION IN A STRONG UNIDIRECTIONAL ELECTRIC FIELD

The problem we treat in this section is the calculation of $a_n(\tau)$ in Eq. (2.1), which is equivalent to calculating the average $\langle P_2(\cos\theta(\tau)) \rangle$ [cf. Eq. (1.4)], following the sudden application of a static electric field E as exactly as possible by using the method proposed by Morita.[16,17] To this end, by putting the expansion of the distribution function, Eq. (2.1), into the diffusion equation, Eq. (1.6), with the potential function, Eq. (1.7), we find the recurrence relation for $a_n(\tau)$:

$$\frac{da_0(\tau)}{d\tau} = 0 \tag{4.1a}$$

$$\frac{1}{n(n+1)} \frac{da_n(\tau)}{d\tau} = -\left[1 - \frac{2g(\tau)}{(2n-1)(2n+3)}\right] a_n(\tau)$$
$$+ e(\tau)\left[\frac{1}{2n-1} a_{n-1}(\tau) - \frac{1}{2n+3} a_{n+1}(\tau)\right]$$
$$+ 2g(\tau)\left[\frac{n-1}{(2n-3)(2n-1)} a_{n-2}(\tau) - \frac{n+2}{(2n+3)(2n+5)} a_{n+2}(\tau)\right]$$
$$(n = 1,2,3,\ldots) \tag{4.1b}$$

where e and g are given by Eqs. (1.11). [We may recall that Eq. (2.5) should be identical to Eq. (4.1b), and it becomes clear that even the starting equation Eq. (2.5) has not been properly derived previously.]

Since e and g are independent of time after switching on the field E at a time $\tau = 0$, on taking the Laplace transform of both sides of Eq. (4.1b) we find for $n = 1,2,3,\ldots$ that

$$\left[\frac{s}{n(n+1)} + 1 - \frac{2g}{(2n-1)(2n+3)}\right] A_n(s)$$
$$= e\left[\frac{1}{2n-1} A_{n-1}(s) - \frac{1}{2n+3} A_{n+1}(s)\right]$$
$$+ 2g\left[\frac{n-1}{(2n-3)(2n-1)} A_{n-2}(s) - \frac{n+2}{(2n+3)(2n+5)} A_{n+2}(s)\right] \tag{4.2}$$

where

$$A_n(s) = \int_0^\infty a_n(\tau) e^{-s\tau} d\tau \tag{4.3}$$

Now the problem is to solve Eq. (4.2) for $A_n(s)$.

A. Exact Calculation of $A_n(s)$ in the Case $g = 0$

When the contribution from the induced dipole is very small in comparison with that from the permanent dipole [this is, in fact, the case for most polar molecules], putting $g = 0$ in Eq. (4.2), we have

$$[s + n(n+1)] A_n(s) = n(n+1)e\left[\frac{A_{n-1}(s)}{2n-1} - \frac{A_{n+1}(s)}{2n+3}\right] \tag{4.4}$$

Here it should be noted that a unidirectional field E is suddenly switched on at time $\tau = 0$, and then

$$\lim_{\tau \to 0} a_n(\tau) = 0 \qquad (n = 1, 2, 3, \ldots) \qquad (4.5)$$

Equation (4.4) may be written as

$$\frac{A_n(s)}{A_{n-1}(s)} = e\frac{n(n+1)}{2n-1}\left[s + n(n+1) + e\frac{n(n+1)}{2n+3}\frac{A_{n+1}(s)}{A_n(s)}\right]^{-1} \qquad (4.6)$$

Using the fact that a_0 is independent of time, along with Eq. (4.6), we deduce that

$$\pi(s) = \int_0^\infty \langle \cos\theta(t) \rangle e^{-s\tau} d\tau = \frac{A_1(s)}{3a_0} = \frac{2e}{3}\frac{\Lambda(s,e)}{s} \qquad (4.7)$$

where

$$\Lambda(s,e) = \cfrac{1}{s_1 + \cfrac{\gamma_1 e^2}{s_2 + \cfrac{\gamma_2 e^2}{s_3 + \cfrac{\gamma_3 e^2}{s_4 + \cdots}}}} \qquad (4.8)$$

in which

$$s_n = s + n(n+1) \qquad (4.9)$$

and

$$\gamma_n = \frac{n(n+1)^2(n+2)}{(2n+3)(2n+1)} \qquad (4.10)$$

It should be noted that Eq. (4.7) is the exact solution for $\pi(s)$. Furthermore, it is clear from Eq. (4.7) that $\langle \cos\theta(\tau) \rangle$ can be obtained by inverting the Laplace transform of $\pi(s)$.

Now, putting $g = 0$ and $n = 1$ in Eq. (4.4) and using the exact result for $A_1(s)/A_0(s)$, Eq. (4.7), we have

$$\int_0^\infty \langle P_2(\cos\theta) \rangle e^{-s\tau} d\tau = \frac{1}{s}[1 - (s+2)\Lambda(s,e)] \qquad (4.11)$$

Equation (4.11) is the exact result for the Laplace transform of $\langle P_2(\cos\theta) \rangle$.

By expanding Eq. (4.11) as a power series of e, we find that

$$\int_0^\infty \langle P_2(\cos\theta)\rangle e^{-s\tau}\,d\tau$$

$$= \frac{1}{s}\left[\frac{e^2\gamma_1}{s_1 s_2} - e^4\left(\frac{\gamma_1\gamma_2}{s_1 s_2^2 s_3} + \frac{\gamma_1^2}{s_1^2 s_2^2}\right)\right.$$

$$\left. + e^6\left(\frac{\gamma_1\gamma_2\gamma_3}{s_1 s_2^2 s_3^2 s_4} + \frac{\gamma_1\gamma_2^2}{s_1 s_2^3 s_3} + \frac{2\gamma_1\gamma_2}{s_1^2 s_2^3 s_3} + \frac{\gamma_1^3}{s_1^3 s_2^3 s_3}\right) - \cdots\right] \quad (4.12)$$

B. Exact Calculation of $A_n(s)$ in the Case $e = 0$

For nonpolar molecules where $\mu = 0$, putting $e = 0$ in Eq. (4.2) we have

$$\frac{A_n(s)}{A_{n-2}(s)} = \cfrac{2(n-1)g/(2n-3)(2n-1)}{\cfrac{s}{n(n+1)} + 1 - \cfrac{2g}{(2n-1)(2n+3)} + \cfrac{2(n+2)g}{(2n+3)(2n+5)}\cfrac{A_{n+2}(s)}{A_n(s)}}$$

$$(4.13)$$

Therefore, it follows by letting $n = 2, 3, 4, \ldots$ successively in Eq. (4.13) that

$$\int_0^\infty \langle P_2(\cos\theta)\rangle e^{-s\tau}\,d\tau = \frac{1}{s}\cfrac{2g/15}{s_2' + \cfrac{2\gamma_2'g}{s_4' + \cfrac{2\gamma_4'g}{s_6' + \cdots}}} \quad (4.14)$$

where

$$s_n' = \frac{s}{n(n+1)} + 1 - \frac{2g}{(2n-1)(2n+3)} \quad (4.15)$$

and

$$\gamma_n' = \frac{(n+1)(n+2)}{(2n+1)(2n+3)^2(2n+5)} \quad (4.16)$$

Equation (4.14) again is the exact result. As before, on expanding the continued fraction in Eq. (4.14), we find that

$$
\int_0^\infty \langle P_2(\cos\theta)\rangle e^{-s\tau}\,d\tau
$$

$$
= \frac{2g}{15s}\left(\frac{1}{s_2'} - \frac{2^2\gamma_2' g^2}{s_2'^2 s_4'} + \frac{2^4\gamma_2'\gamma_4' g^4}{s_2'^2 s_4'^2 s_6'} + \frac{2^4\gamma_2'^2 g^4}{s_2'^3 s_4'^2} \right.
$$

$$
\left. - \frac{2^6\gamma_2'\gamma_4'\gamma_6' g^6}{s_2'^2 s_4'^2 s_6'^2 s_8'} - \frac{2^6\gamma_2'\gamma_4'^2 g^6}{s_2'^2 s_4'^3 s_6'^2} - \frac{2^7\gamma_2'^2\gamma_4' g^6}{s_2'^3 s_4'^3 s_6'} - \frac{2^6\gamma_2'^3 g^6}{s_2'^4 s_4'^3} + \cdots \right)
$$

$$
(4.17)
$$

C. Calculation of $A_2(s)$ in the General Case $e \neq 0$ and $g \neq 0$

In the general case of both the e and g are contributing to the electric birefringence, it seems no longer possible to use the continued fraction technique for the calculation of $A_2(s)$ from Eq. (4.2). So we regard Eq. (4.2) as the matrix equation

$$
(\mathbf{L} + \lambda\mathbf{M} + \lambda^2\mathbf{N})\mathbf{A} = \mathbf{a} \tag{4.18}
$$

where

$$
\mathbf{A} = \begin{bmatrix} A_1(s) \\ A_2(s) \\ A_3(s) \\ \vdots \end{bmatrix} \tag{4.19a}
$$

$$
\mathbf{a} = \begin{bmatrix} \lambda e A_0(s) \\ \dfrac{\lambda^2 2g}{3} A_0(s) \\ 0 \\ 0 \\ \vdots \end{bmatrix} \tag{4.19b}
$$

$$
\mathbf{L} = \begin{bmatrix} p_1 & 0 & 0 & \cdots \\ 0 & p_2 & 0 & \cdots \\ 0 & 0 & p_3 & \cdots \\ \vdots & \vdots & \vdots & \vdots \end{bmatrix} \tag{4.19c}
$$

$$\mathbf{M} = \begin{bmatrix} 0 & r_1 & 0 & 0 & 0 & \cdots \\ q_2 & 0 & r_2 & 0 & 0 & \cdots \\ 0 & q_3 & 0 & r_3 & 0 & \cdots \\ \vdots & \vdots & \vdots & \vdots & \vdots & \vdots \end{bmatrix} \qquad (4.19d)$$

$$\mathbf{N} = \begin{bmatrix} z_1 & 0 & \beta_1 & 0 & 0 & 0 & \cdots \\ 0 & z_2 & 0 & \beta_2 & 0 & 0 & \cdots \\ \alpha_3 & 0 & z_3 & 0 & \beta_3 & 0 & \cdots \\ 0 & \alpha_4 & 0 & z_4 & 0 & \beta_4 & \cdots \\ \vdots & \vdots & \vdots & \vdots & \vdots & \vdots & \vdots \end{bmatrix} \qquad (4.19e)$$

In Eqs. (4.18) and (4.19b), λ indicates the order of E, and

$$p_n = 1 + \frac{s}{n(n+1)} \qquad (4.20a)$$

$$q_n = -\frac{e}{(2n-1)} \qquad (4.20b)$$

$$r_n = \frac{e}{2n+3} \qquad (4.20c)$$

$$z_n = -\frac{2g}{(2n-1)(2n+3)} \qquad (4.20d)$$

$$\alpha_n = -\frac{2(n-1)g}{(2n-3)(2n-1)} \qquad (4.20e)$$

$$\beta_n = \frac{2(n+2)g}{(2n+3)(2n+5)} \qquad (4.20f)$$

The inverse matrix of $(\mathbf{L} + \mathbf{M} + \lambda\mathbf{N})$ may be obtained noting the fact that $\mathbf{L}$ is a diagonal matrix. Then it follows that

$$(\mathbf{L} + \lambda\mathbf{M} + \lambda^2\mathbf{N})^{-1}$$
$$= \left[\mathbf{L}(\mathbf{I} + \lambda\mathbf{L}^{-1}\mathbf{M} + \lambda^2\mathbf{L}^{-1}\mathbf{N})\right]^{-1}$$
$$= (\mathbf{I} + \lambda\mathbf{L}^{-1}\mathbf{M} + \lambda^2\mathbf{L}^{-1}\mathbf{N})^{-1}\mathbf{L}^{-1}$$
$$= \left[\mathbf{I} - \lambda\mathbf{L}^{-1}\mathbf{M} - \lambda^2(\mathbf{L}^{-1}\mathbf{N} - \mathbf{L}^{-1}\mathbf{M}\mathbf{L}^{-1}\mathbf{M})\right.$$
$$+ \lambda^3(\mathbf{L}^{-1}\mathbf{M}\mathbf{L}^{-1}\mathbf{N} + \mathbf{L}^{-1}\mathbf{N}\mathbf{L}^{-1}\mathbf{M} - \mathbf{L}^{-1}\mathbf{M}\mathbf{L}^{-1}\mathbf{M}\mathbf{L}^{-1}\mathbf{M})$$
$$+ \lambda^4(\mathbf{L}^{-1}\mathbf{N}\mathbf{L}^{-1}\mathbf{N} - \mathbf{L}^{-1}\mathbf{M}\mathbf{L}^{-1}\mathbf{N}\mathbf{L}^{-1}\mathbf{M} - \mathbf{L}^{-1}\mathbf{N}\mathbf{L}^{-1}\mathbf{M}\mathbf{L}^{-1}\mathbf{M}$$
$$\left. - \mathbf{L}^{-1}\mathbf{M}\mathbf{L}^{-1}\mathbf{M}\mathbf{L}^{-1}\mathbf{N} + \mathbf{L}^{-1}\mathbf{M}\mathbf{L}^{-1}\mathbf{M}\mathbf{L}^{-1}\mathbf{M}\mathbf{L}^{-1}\mathbf{M}) + \cdots \right]\mathbf{L}^{-1}$$
$$(4.21)$$

where **I** stands for the unit matrix. This may be regarded as a perturbation expansion with respect to λ. Using Eq. (4.21) and calculating $A_2(s)$ up to the order of λ^4, we find that

$$
\int_0^\infty \langle P_2(\cos\theta)\rangle e^{-s\tau}\,d\tau
$$

$$
= -\frac{\lambda^2}{5s}\left[\frac{\alpha_2}{p_2} - \frac{q_1 q_2}{p_1 p_2} + \lambda^2\left(\frac{q_2 r_1 \alpha_2}{p_1 p_2^2} + \frac{q_3 r_2 \alpha_2}{p_2^2 p_3} + \frac{q_1 q_2 z_1}{p_1^2 p_2}\right.\right.
$$

$$
+ \frac{q_1 r_2 \alpha_3}{p_1 p_2 p_3} + \frac{q_1 q_2 z_2}{p_1 p_2^2} - \frac{\alpha_2 z_2}{p_2^2}
$$

$$
\left.\left. - \frac{q_1 q_2^2 r_1}{p_1^2 p_2^2} - \frac{q_1 q_2 q_3 r_2}{p_1 p_2^2 p_3} + \cdots\right)\right] \qquad (4.22)
$$

It is shown readily that as special cases of $g = 0$ and $e = 0$, Eq. (4.22) leads to Eqs. (4.12) and (4.17), respectively.

D.   Calculation of $\langle P_2(\cos\theta)\rangle$ in the Limit as $\tau \to \infty$

It is useful to calculate $\langle P_2(\cos\theta)\rangle$ in the limit of $\tau \to \infty$ independently from the methods in Sections IV.A–IV.C. In this manner we are able to check the results in the previous sections as $\tau \to \infty$. In this limit, $\langle P_2(\cos\theta)\rangle$ reaches the equilibrium value governed by the Maxwell-Boltzmann distribution function with the potential energy function, Eq. (1.7), thus leading to

$$
\langle P_2(\cos\theta)\rangle = \frac{\int_{-1}^1 \frac{3}{2}u^2\exp(eu + gu^2)\,du}{\int_{-1}^1 \exp(eu + gu^2)\,du} - \frac{1}{2} \qquad (4.23a)
$$

$$
= \tfrac{1}{15}\left[(e^2 + 2g) + \tfrac{2}{21}(2e^2 g + 2g^2 - e^4) + \cdots\right] \qquad (4.23b)
$$

It is readily seen by noting the relation

$$
\lim_{\tau\to\infty}\langle P_2(\cos\theta)\rangle = \lim_{s\to 0} s\int_0^\infty \langle P_2(\cos\theta)\rangle e^{-s\tau}\,d\tau \qquad (4.24)
$$

that $\langle P_2(\cos\theta)\rangle$ in Eq. (4.22) as $\tau \to \infty$ leads to Eq. (4.23b).

In the particular case where $g = 0$, Eq. (4.23a) directly gives

$$
\langle P_2(\cos\theta)\rangle = 1 - \frac{3}{e}L(e) \qquad (g = 0) \qquad (4.25)
$$

which have already been given in Eq. (1.20), where $L(e)$ is the Langevin function. In considering a new expression of $\langle P_2(\cos\theta)\rangle$ in the form of a continued fraction, it may be useful to examine the continued fraction, Eq. (4.11), in more detail. We note that the Langevin function $L(e)$ defined by

$$L(e) = \coth e - \frac{1}{e} = e\left(\frac{1}{3} - \frac{e^2}{45} + \frac{2}{945}e^4 - \cdots\right) \qquad (4.26)$$

may be expressed in terms of an infinite continued fraction [refer to Eq. (91.6) on page 349 of Wall[18]],

$$L(e) = \cfrac{e}{3 + \cfrac{e^2}{5 + \cfrac{e^2}{7 + \cfrac{e^2}{9 + \cdots}}}} \qquad (4.27)$$

Thus it follows that

$$\langle P_2(\cos\theta)\rangle = 1 - \cfrac{3}{3 + \cfrac{e^2}{5 + \cfrac{e^2}{7 + \cdots}}} \qquad (g = 0) \qquad (4.28)$$

In view of the relations

$$\lim_{\tau \to \infty} \langle P_2(\cos\theta)\rangle = 1 - 2\Lambda(0, e) \qquad (4.29)$$

which is readily seen by using Eq. (4.11) and Eq. (4.24), we have

$$\frac{2e}{3}\Lambda(0, e) = L(e) \qquad (4.30)$$

which is clear from Eqs. (4.8) and (4.27), we, in fact, find that $\langle P_2(\cos\theta)\rangle$ in Eq. (4.11) as $\tau \to \infty$ satisfies Eq. (4.25).

Now we shall show that $\langle P_2(\cos\theta)\rangle$ of Eq. (4.23a) in the particular case of $e = 0$ may be expressed as a continued fraction. It is seen by expanding

$\exp(gx^2)$ as a power series of (gx^2) that

$$\int_{-1}^{1} \exp(gx^2)\, dx = 2\,_1F_1(\tfrac{1}{2},\tfrac{3}{2};g)$$ (4.31)

$$\int_{-1}^{1} x^2\exp(gx^2)\, dx = \tfrac{2}{3}\,_1F_1(\tfrac{3}{2},\tfrac{5}{2};g)$$ (4.32)

where $_1F_1(b,c;z)$ is a hypergeometric function defined by

$$_1F_1(b,c;z) = 1 + \frac{b}{c}z + \frac{b(b+1)}{c(c+1)}\frac{z^2}{2!} + \frac{b(b+1)(b+2)}{c(c+1)(c+2)}\frac{z^3}{3!} + \cdots$$

(4.33)

In view of Eqs. (4.23a), (4.31), and (4.32), and the equation

$$\frac{_1F_1(b+1,c+1;z)}{_1F_1(b,c;z)} = \cfrac{1}{1 - \cfrac{\dfrac{c-b}{c(c+1)}z}{1 + \cfrac{\dfrac{b+1}{(c+1)(c+2)}z}{1 - \cfrac{\dfrac{c-b+1}{(c+2)(c+3)}z}{1 + \cfrac{\dfrac{b+2}{(c+3)(c+4)}z}{1 - \cdots}}}}}$$ (4.34)

[see Eq. (91.1) of Wall[18]], it follows after some rearrangement that

$$\langle P_2(\cos\theta)\rangle = \frac{1}{2}\left[\frac{_1F_1(\tfrac{3}{2},\tfrac{5}{2};g)}{_1F_1(\tfrac{1}{2},\tfrac{3}{2};g)} - 1\right]$$

$$= -\frac{1}{2} + \cfrac{\tfrac{3}{2}}{3 - \cfrac{4g}{5 + \cfrac{6g}{7 - \cfrac{8g}{9+\cdots}}}}$$ (4.35a)

$$= \tfrac{2}{15}\left(g + \tfrac{2}{21}g^2 - \tfrac{4}{315}g^3 \cdots\right).$$ (4.35b)

It is seen that the continued fraction in Eq. (4.35a) is also obtained after putting $e = 0$ in Eqs. (1.15)–(1.19). Equation (4.35b) agrees with $\langle P_2(\cos\theta)\rangle$ in Eq. (4.23b) for $e = 0$ up to the order of g^2.

$\langle P_2(\cos\theta)\rangle$ in Eq. (4.14) as $\tau \to \infty$ may be obtained from Eq. (4.24), resulting in

$$\lim_{\tau \to \infty} \langle P_2(\cos\theta)\rangle = \cfrac{2g/15}{1-\left(\dfrac{2g}{3\cdot7}\right)+\cfrac{4\gamma_2'g^2}{1-\left(\dfrac{2g}{7\cdot11}\right)+\cfrac{4\gamma_4'g^2}{1-\left(\dfrac{2g}{11\cdot15}\right)+\cdot\,.}}}$$

$$(4.36)$$

Although it is not directly evident that Eq. (4.35a) is equivalent to Eq. (4.36), it is shown that Eq. (4.36) gives Eq. (4.35b).

E. Analytical Expressions of the Relaxation Time of Kerr-Effect Buildup for the Step-up Electric Field

Taking the inverse Laplace transform of Eq. (4.22), we obtain

$$\langle P_2(\cos\theta)\rangle = \frac{1}{5}\left\{\frac{e^2}{3}\left(1-\frac{3}{2}\exp(-2\tau)+\frac{1}{2}\exp(-6\tau)\right)+\frac{2g}{3}(1-\exp(-6\tau))\right.$$

$$+\left[\frac{e^2}{350}(g-9e^2)+\frac{e^2}{5}\left(2g-\frac{e^2}{2}\right)\tau\right]\exp(-2\tau)$$

$$+\left[\frac{1}{63}\left(4g^2+\frac{11}{2}e^2g-\frac{e^4}{2}\right)\right.$$

$$+\frac{1}{21}\left(8g^2-4e^2g+\frac{e^4}{2}\right)\tau\left.\right]\exp(-6\tau)$$

$$\left.+\frac{1}{525}(e^4-14e^2g)\exp(-12\tau)+\frac{2}{63}(2g^2+2e^2g-e^4)\right\}$$

$$(4.37)$$

This agrees fully with the previous result by Matsumoto, Watanabe, and Yoshioka,[11] Eq. (2.46), who did not use Eq. (4.1b). By expanding the right-hand side of Eq. (4.12) as a power series of $1/s$ up to the order of $(1/s)^6$, and taking the inverse Laplace transform, we find that

$$\langle P_2(\cos\theta)\rangle = e\gamma_1\left[\frac{1}{2}\tau^2 - \frac{4}{3}\tau^3 + \frac{52}{4!}\tau^4 - \frac{320}{5!}\tau^5 + \frac{1936}{6!}\tau^6\right]$$
$$- e^4\gamma_1\left[(\gamma_1+\gamma_2)\frac{\tau^4}{4!} - 2(8\gamma_1+13\gamma_2)\frac{\tau^5}{5!} + 56(3\gamma_1+8\gamma_2)\frac{\tau^6}{6!}\right]$$
$$+ e^6\left(\gamma_1\gamma_2\gamma_3 + \gamma_1\gamma_2^2 + 2\gamma_1^2\gamma_2 + \gamma_1^3\right)\frac{\tau^6}{6!} - \cdots \qquad (4.38)$$

This equation agrees fully with Eq. (2.24) on letting $g = 0$ in Eq. (2.24).

We may be able to obtain the analytical expressions for the buildup processes of the electric birefringence for the pure permanent and pure induced dipole orientation that are valid for higher order of E, by obtaining the inverse Laplace transform of Eqs. (4.12) and (4.17), respectively.

Now we shall proceed to obtain analytical expressions for the buildup of the electric birefringence, which may be remarkably significant from the experimental point of view. For a large value of τ, it may be assumed in Eq. (4.14) that

$$s_n' = 1 - \frac{2g}{(2n-1)(2n+3)} \qquad (n = 4,6,8,\dots) \qquad (4.39)$$

which leads to

$$\frac{\langle P_2(\cos\theta)\rangle}{\langle P_2(\cos\theta)\rangle_\infty} = 1 - \exp\left(-\frac{2g}{15}\frac{6\tau}{\langle P_2(\cos\theta)\rangle_\infty}\right) \qquad (4.40)$$

where

$$\langle P_2(\cos\theta)\rangle_\infty = \lim_{\tau\to\infty}\langle P_2(\cos\theta)\rangle \qquad (4.41)$$

in Eq. (4.36) is the stationary state value. Equation (4.40) means that the buildup process of the electric birefringence for pure induced dipole orientation can be approximated by a single exponential term with the relaxation time

$$\tau_g^{(1)} = \frac{5}{4g}\langle P_2(\cos\theta)\rangle_\infty \qquad (4.42)$$

A plot of $6\tau_g^{(1)}$ versus $2g$ is shown in Fig. 4, together with the exact value of

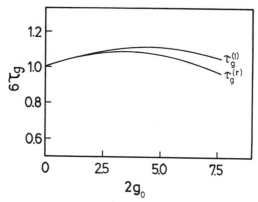

Fig. 4.   Plots of $\tau_g^{(1)}$ and τ_g^r, versus $2g_0$.

the relaxation time defined as

$$\frac{\langle P_2(\tau^r)\rangle}{\langle P_2\rangle_\infty} = 1 - \exp(-1) = 0.63212\ldots \qquad (4.43)$$

obtained by the numerical method whose details will be presented in the next section. It is seen that Eq. (4.42) shows explicitly how the relaxation time $\tau_g^{(1)}$ may depend on g, as long as the value of g is not too large. Here it is noted that Eq. (4.42) for $g \ll 1$ is given by

$$\langle P_2(\cos\theta)\rangle = \frac{2g}{15}\left[1 - \exp(-6\tau)\right] \qquad (4.44)$$

Therefore, $\langle P_2(\cos\theta)\rangle$ in Eq. (4.40) is sufficiently accurate as long as $g \ll 1$. In other words, for $g \ll 1$, $\langle P_2(\cos\theta)\rangle$ in Eq. (4.40) is not subject to the condition $\tau \gg \frac{1}{6}$, which is needed when Eq. (4.40) is derived for any value of g. Recalling Eq. (2.16), it is also noted that the buildup process of the electric birefringence for pure induced dipole moment is symmetrical with the decay process as long as $g \ll 1$, and Eq. (4.40) provides us a measure of the deviation of the buildup process from the symmetrical nature of the transient. Also we now understand from Eq. (4.40) the reason why the relaxation time for pure induced dipole orientation is larger than $\frac{1}{6}$ as long as the value of g is not too large, by reference to Fig. 1a, in which we see that the stationary state value of the birefringence $\langle P_2(\cos\theta)\rangle_\infty$ for $e = 0$ deviates upwards from Kerr's law.

Next, by assuming for $\Lambda(s, e)$ in Eq. (4.8) that

$$s + n(n+1) \simeq n(n+1) \qquad (n = 2,3,4,\ldots) \qquad (4.45)$$

we find that

$$\langle P_2(\cos\theta)\rangle = \left[1 - \frac{3L(e)}{e}\right][1 - \exp(-Z(e)\tau)] \qquad (4.46)$$

where

$$Z(e) = \frac{2e}{3L(e)} \qquad (4.47)$$

In Eq. (4.47) $Z(e)$ has been introduced by Morita for the analytical expression of the relaxation time of the dielectric polarization.[16] From Eq. (4.46) we define the relaxation time for pure permanent dipole orientation as

$$\tau_e^{(1)} = \frac{3L(e)}{2e} \qquad (4.48)$$

which is plotted against e in Fig. 5. In the figure τ_e^r obtained by the numerical method is also presented. Since the agreement is poor in comparison with the case of $e = 0$, $\tau_e^{(2)}$ will be defined shortly.

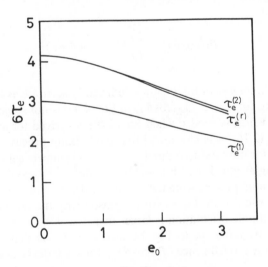

Fig. 5. Plots of $\tau_e^{(1)}$, $\tau_e^{(2)}$, and τ_e^r, versus e_0.

In order to obtain an improved analytical expression for the relaxation time $\tau_e^{(1)}$, the continued fraction Eq. (4.8) is approximated up to the second term of s:

$$\Lambda(s,e) = \frac{1}{s+2+\gamma_1 e^2/[s+Q(e)]} \tag{4.49}$$

Taking the inverse Laplace transform of Eq. (4.49), we obtain

$$\langle P_2(\cos\theta)\rangle = \frac{\gamma_1 e^2}{\alpha\beta}\left[1-\frac{\alpha}{\alpha-\beta}\exp(-\alpha\tau)+\frac{\beta}{\alpha-\beta}\exp(-\beta\tau)\right] \tag{4.50}$$

where

$$\frac{\alpha}{\beta} = \frac{1}{2}\left\{(Q+2)\mp\left[(Q+2)^2-4(2Q+\gamma_1 e^2)\right]^{1/2}\right\} \tag{4.51}$$

On letting $\tau \to \infty$, we have from Eqs. (4.50) and (4.51)

$$2Q = \gamma_1 e^2\left[\frac{1}{\langle P_2\rangle_\infty}-1\right] \tag{4.52}$$

which determines the value of Q. The values of $\tau_e^{(2)}$ obtained by Eq. (4.50) in the same manner as τ^r are compared with $\tau_e^{(1)}$ and τ_e^r in Fig. 5, and we see that the improvement is remarkable. For a limiting low field, $e \to 0$, we may let $Q = 6$, resulting in $\alpha = 2$ and $\beta = 6$. Thus Eq. (4.50) reduces to Benoit's equation, Eq. (2.13). This suggests that the expression of the Kerr effect transient for pure permanent dipole orientation by two exponential terms as in Benoit's equation is essential.

In exactly the same way, on writing Eq. (4.14) as

$$\int_0^\infty \langle P_2(\cos\theta)\rangle e^{-s\tau}d\tau = \frac{1}{s}\frac{2g/15}{s_2'+\dfrac{2\gamma_2' g}{s_4'+R_g}} \tag{4.53}$$

we can derive an improved expression for the pure induced dipole orientation:

$$\langle P_2(\cos\theta)\rangle = \frac{12g}{15}\left[\frac{q}{\alpha'\beta'}-\frac{q-\alpha'}{\alpha'(\beta'-\alpha')}\exp(-\alpha'\tau)\right.$$
$$\left. -\frac{q-\beta'}{\beta'(\alpha'-\beta')}\exp(-\beta'\tau)\right] \tag{4.54}$$

where

$$\begin{matrix} \alpha' \\ \beta' \end{matrix} = \frac{1}{2}\left\{(p+q) \mp \left[(p+q)^2 - 4(pq+240\gamma_2'g)\right]^{1/2}\right\} \qquad (4.55)$$

with

$$p = 6\left(1 - \frac{2g}{21}\right) \qquad (4.56)$$

and

$$q = 20\left[1 - \frac{2g}{7 \cdot 11} + R_g\right] \qquad (4.57)$$

On letting $\tau \to \infty$, we have from Eqs. (4.54) and (4.55)

$$\langle P_2(\cos\theta)\rangle_\infty = \frac{2g}{15} \frac{q/10}{(p/6)(q/10)+4\gamma_2'g} \qquad (4.58)$$

with $\gamma_2' = 3 \cdot 4/(5 \cdot 7^2 \cdot 9)$ [refer to Eq. (4.16)], which determines the value of R_g. Numerical values obtained by Eq. (4.54) are slightly smaller than that of τ_g' and the improvement is remarkable. For $2g > 2.5$, however, α' and β' become complex and we have an oscillating result, due to the truncation in Eq. (4.53).

For the pure permanent dipole orientation, when the field strength is very high so that the value of e is very large, we may assume in $\Lambda(s, e)$ of Eq. (4.8) that

$$s + n(n+1) \simeq s \qquad (4.59)$$

This assumption corresponds to the recurrence relation [cf. Eq. (4.2)]

$$sA_n(s) = n(n+1)e\left[\frac{1}{2n-1}A_{n-1}(s) - \frac{1}{2n+3}A_{n+1}(s)\right] \qquad (4.60)$$

According to the relation [see Eq. (94.5) on page 370 of Wall[18]]

$$\int_0^\infty {}_2F_1\left(a, b, \frac{a+b}{2}; -\sinh^2 u\right)e^{-zu}du = \cfrac{1}{z + \cfrac{d_1}{z + \cfrac{d_2}{z + \cdot^{\cdot^{\cdot}}}}} \qquad (4.61)$$

where $_2F_1(a, b, c; z)$ is the hypergeometric function defined by

$$_2F_1(a, b, c; z) = 1 + \frac{ab}{c} z + \frac{a(a+1)b(b+1)}{c(c+1)} z^2 + \cdots \qquad (4.62)$$

and

$$d_{n+1} = \frac{4(n+a)(n+b)(n+1)(n+a+b-1)}{(2n+a+b-1)(2n+a+b+1)} \qquad (4.63)$$

together with Eq. (4.9), we see that

$$4\gamma_{n+1} = d_{n+1} \qquad (4.64)$$

Therefore, $\Lambda(s, e)$ of Eq. (4.8) with assumption (4.59) leads, after some straightforward rearrangements, to

$$\Lambda(s, e) = \int_0^\infty {_2F_1}\left(2, 2, \tfrac{5}{2}; -\sinh^2\frac{e\tau}{2}\right)\exp(-s\tau)\, d\tau \qquad (4.65)$$

This immediately gives

$$\mathcal{L}^{-1}[\Lambda(s, e)] = {_2F_1}\left(2, 2, \tfrac{5}{2}; -\sinh^2\frac{e\tau}{2}\right) \qquad (4.66)$$

where $\mathcal{L}^{-1}[f(s)]$ represents the inverse Laplace transform. Therefore, it follows from Eqs. (4.60)–(4.66) that

$$q_2(\tau) = \langle P_2(\cos\theta)\rangle = 1 - {_2F_1}\left(2, 2, \tfrac{5}{2}; -\sinh^2\frac{e\tau}{2}\right) \qquad (4.67)$$

A plot of $\langle P_2(\cos\theta)\rangle$ in Eq. (4.67) versus $e\tau$ is shown in Fig. 6 by a broken line. Full lines are the normalized buildup for pure permanent dipole orientation with the values of e as indicated on the right-hand side of each curve. As was expected, the normalized buildup curve approaches asymptotically, with increasing value of e, the theoretical curve for $e \to \infty$.

It may be useful to show that $\langle P_2(\cos\theta)\rangle$ in Eq. (4.67) can be expressed in a closed form. Using the relations[19]

$$_2F_1(2, 2, \tfrac{5}{2}; z) = (1-z)^{-3/2} {_2F_1}(\tfrac{1}{2}, \tfrac{1}{2}, \tfrac{5}{2}; z) \qquad (4.68)$$

$$c(c-1)(z-1){_2F_1}(a, b, c-1; z)$$
$$+ c[c-1-(2c-a-b-1)z]{_2F_1}(a, b, c; z)$$
$$+ (c-a)(c-b)z {_2F_1}(a, b, c+1; z) = 0 \qquad (4.69)$$

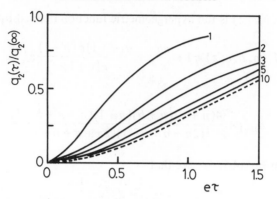

Fig. 6.   Plots of the normalized buildup of the electric birefringence versus $e\tau$ for pure permanent dipole orientation. Numerals on the right are the values of e. The broken line is the theoretical one from Eq. (4.67).

we find that

$$F\left(\frac{1}{2},\frac{1}{2},\frac{5}{2};z\right) = \frac{3}{4z}\left[(1-z)\,_2F_1\left(\frac{1}{2},\frac{1}{2},\frac{1}{2};z\right)\right.$$

$$\left. +(2z-1)\,_2F_1\left(\frac{1}{2},\frac{1}{2},\frac{3}{2};z\right)\right] \qquad (4.70)$$

It follows by using

$$_2F_1(a,b,b;z) = (1-z)^{-a} \qquad (4.71a)$$

$$_2F_1(\tfrac{1}{2},\tfrac{1}{2},\tfrac{3}{2};z^2) = z^{-1}\sin^{-1}z \qquad (4.71b)$$

that

$$\langle P_2(\cos\theta)\rangle = 1 + \frac{3}{\sinh^2 e\tau}(1 - e\tau\coth e\tau) \qquad (4.72)$$

which may be written in terms of Langevin's function

$$\langle P_2(\cos\theta)\rangle = 1 - \frac{3e\tau}{\sinh^2 e\tau}L(e\tau) \qquad (4.73)$$

This may be compared with $\langle P_2(\cos\theta)\rangle$ for the static field e:

$$\langle P_2(\cos\theta)\rangle = 1 - \frac{3}{e}L(e) \qquad (4.74)$$

Equation (4.73) does not lead to Eq. (4.74) in the limit of $\tau \to \infty$, since collisions between the particle and the surrounding molecules are neglected in making assumptions in Eq. (4.59). The physical significance of this extreme case will be discussed more fully in Section XI. It should be noted that Eq. (4.73) agrees with the result obtained from Eq. (11.19).

V. TIME-DEPENDENT BIREFRINGENCE FOLLOWING THE SUDDEN CHANGE OF A HOMOGENEOUS ELECTRIC FIELD

A. Expression of the Time Course of Electric Birefringence in the Expanded Form of a Matrix Equation

The analytical treatment of the transient electric birefringence in the previous section will be extended to the case where a homogeneous electric field is suddenly applied to a system in which a Maxwell-Boltzmann distribution of particle orientation has been established by another homogeneous electric field. For the most general case, where both the permanent dipole moment and the induced dipole moment contribute to the particle orientation under the electric field, we are no longer able to use the continued fraction technique. We therefore seek another compact expression of the set of the differential equations (4.1a) and (4.1b) that might be convenient for the numerical calculation.[20]

To that end, dividing both sides of Eq. (4.1b) by a_0 and using the relation

$$\frac{a_n(\tau)}{a_0} = (2n+1)\langle P_n(\cos\theta)\rangle = (2n+1)q_n(\tau) \tag{5.1}$$

we find that

$$\frac{dq_n(\tau)}{d\tau} = -d_n(\tau)q_n(\tau) + e(\tau)w_n'\left[q_{n-1}(\tau) - q_{n+1}(\tau)\right]$$
$$+ 2g(\tau)\left[v_n'q_{n-2}(\tau) - y_n'q_{n+2}(\tau)\right] \tag{5.2}$$

where

$$d_n(\tau) = n(n+1)\left[1 - \frac{2g(\tau)}{(2n-1)(2n+3)}\right] \tag{5.3a}$$

$$w_n' = \frac{n(n+1)}{2n+1} \tag{5.3b}$$

$$v_n' = \frac{(n-1)n(n+1)}{(2n-1)(2n+1)} \tag{5.3c}$$

$$y_n' = \frac{n(n+1)(n+2)}{(2n+1)(2n+3)} \tag{5.3d}$$

We now imagine that a unidirectional electric field $E(\tau) = E_0$ has been applied for a long time, so that the angular distribution function $f(u, \tau)$ is governed by the Maxwell-Boltzmann distribution function

$$f(u) = \frac{\exp(e_0 u + g_0 u^2)}{\int_{-1}^{+1} \exp(e_0 u + g_0 u^2)\, du} \qquad (5.4)$$

where $u = \cos\theta$, and $e_0 = \mu E_0 / k_B T$ and $g_0 = \Delta\alpha E_0^2 / 2k_B T$ are the parameters corresponding to E_0. Quite suddenly at $\tau = 0$, we apply another electric field $E(\tau) = E$ ($\tau > 0$) (see Fig. 7). So our problem now is to solve Eq. (5.2) under the initial condition

$$q_n^0 = \frac{\int_{-1}^{1} P_n(u)\exp(e_0 u + g_0 u^2)\, du}{\int_{-1}^{1} \exp(e_0 u + g_0 u^2)\, du} \qquad (5.5)$$

To this end we take the Laplace transform of both sides of Eq. (5.2) with respect to τ, and write

$$(s + d_n)\tilde{Q}_n(s) - ew_n'[\tilde{Q}_{n-1}(s) - \tilde{Q}_{n+1}(s)]$$
$$-2g[v_n'\tilde{Q}_{n-2}(s) - y_n'\tilde{Q}_{n+2}(s)] = q_n^0 \qquad (5.6)$$

where e and g are no longer functions of τ, and

$$\tilde{Q}_n(s) = \mathcal{L}[q_n(\tau)] = \int_0^\infty q_n(\tau)e^{-s\tau}\, d\tau \qquad (5.7)$$

Regarding the recursion relation of $\tilde{Q}_n(s)$ in Eq. (5.6) as a matrix equation, we may write

$$(\mathbf{B} + s\mathbf{I})\tilde{\mathbf{Q}} = \mathbf{Q}^0 + \frac{\mathbf{C}}{s} \qquad (5.8)$$

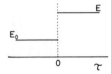

0   τ Fig. 7. A sudden change of the homogeneous electric field at $\tau = 0$.

where **I** is the unit matrix of infinite dimension, and

$$
\mathbf{B} = \begin{bmatrix}
d_1 & ew_1' & 2gy_1' & 0 & 0 & 0 & \cdots \\
-ew_2' & d_2 & ew_2' & 2gy_2' & 0 & 0 & \cdots \\
-2gv_3 & -ew_3' & d_3 & ew_3' & 2gy_3' & 0 & \cdots \\
0 & -2gv_4' & -ew_4' & d_4 & ew_4' & 2gy_4' & \cdots \\
\cdots & \cdots & \cdots & \cdots & \cdots & \cdots & \cdots
\end{bmatrix} \qquad (5.9)
$$

$$
\tilde{\mathbf{Q}} = \begin{bmatrix}
\tilde{Q}_1(s) \\
\tilde{Q}_2(s) \\
\tilde{Q}_3(s) \\
\vdots
\end{bmatrix} \qquad (5.10)
$$

$$
\mathbf{Q}^0 = \begin{bmatrix}
q_1^0 \\
q_2^0 \\
q_3^0 \\
\vdots
\end{bmatrix} \qquad (5.11)
$$

$$
\mathbf{C} = \begin{bmatrix}
ew_1' \\
2gv_2' \\
0 \\
0 \\
\vdots
\end{bmatrix} \qquad (5.12)
$$

By noting the relation

$$
(\mathbf{B} + s\mathbf{I})^{-1} = \sum_{l=0}^{\infty} \frac{(-1)^l \mathbf{B}^l}{s^{l+1}} \qquad (5.13)
$$

we find after the inverse Laplace transformation that

$$
\mathbf{Q}(\tau) = \left[\sum_{l=0}^{\infty} \frac{(-\tau)^l \mathbf{B}^l}{l!} \right] \mathbf{Q}^0 - \left[\sum_{l=0}^{\infty} \frac{(-\tau)^{l+1} \mathbf{B}^l}{(l+1)!} \right] \mathbf{C} \qquad (5.14)
$$

where

$$\mathbf{Q}(\tau) = \begin{bmatrix} q_1(\tau) \\ q_2(\tau) \\ q_3(\tau) \\ \vdots \end{bmatrix} \tag{5.15}$$

It is evident that $q_2(\tau)$ is equivalent to $\Phi(\tau)$ of Eq. (1.4). And it should be noted that Eq. (5.14) gives a kind of series expansion of the time-dependent course of the birefringence in terms of time τ and is a generalization of the Nishinari-Yoshioka method[10] introduced in Section II. Equation (5.14) has a form particularly convenient for computer programming, and we now are able to calculate the transients of the electric birefringence after a sudden change in the strength of the unidirectional electric field and/or its direction.

In order to estimate the accuracy of calculation, keeping the value of $E(\tau)$ same as that of E_0, deviations of $q_2(\tau)$ from the starting value $q_2(0)$ for increasing values of τ are estimated for different values of N (the dimension of the matrix $\mathbf{B}$) and l [the number of terms in the summation in Eq. (5.14)]. Generally, the deviations are on the order of 10^{-6}–10^{-2} as long as the value of τ does not exceed a certain limit. On increasing the value of τ beyond this limit, however, the deviation increases dramatically as shown in Fig. 8. The limit depends on the value of N and l, and also on the values of e and g. In Table II the values of τ_c at which the deviation of $q_2(\tau)/q_2(0)$ from unity exceeds 0.01 are shown for different values of N, l, e, and g. As can be seen in the table, $N = 4$ or 5 is sufficient for relatively small values of e and g. A further increase in N does not necessarily improve the result. For relatively

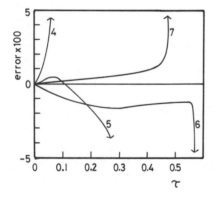

Fig. 8. Plots of the error of Eq. (5.14) with $l = 50$, $-e = e_0 = 4$, and $2g = 2g_0 = 4$ versus τ. The numeral on each curve is the dimension of matrix $\mathbf{B}$.

TABLE II
Values of τ_c

N \ l	10	30	50
$e = 0, g = 1$			
4	0.270	0.662	0.906
5	0.270	0.662	0.906
6	0.250	0.388	0.507
7	0.250	0.388	0.507
$e = 0, g = 5$			
4	0.049	0.049	0.050
5	0.049	0.049	0.050
6	0.148	0.443	0.430
7	0.148	0.443	0.430
$e = 1, g = 0$			
4	0.560	0.760	1.090
5	0.510	0.630	0.855
6	0.510	0.551	0.696
7	0.506	0.566	0.622
$e = 10, g = 0$			
4	0.045	0.042	0.041
5	0.082	0.081	0.081
6	0.115	0.165	0.400
7	0.123	0.165	0.300
$e = 1, g = 1$			
4	0.290	0.684	0.890
5	0.280	0.508	0.707
6	0.258	0.423	0.560
7	0.258	0.229	0.456
$e = 4, g = 4$			
4	0.024	0.024	0.024
5	0.150	0.149	0.150
6	0.120	0.114	0.120
7	0.153	0.324	0.430

large values of e and/or g, however, increasing N to 6 or 7 seems to extend the range of applicability of Eq. (5.14) to larger values of τ. Generally, the larger the number of terms l the better the accuracy, and also the larger the value of the limit τ_c.

The values of τ_c for some values of e and g and the corrseponding values of N are presented in Table III. The values of $r = q_2(\tau_c)/q_2(\infty)$ of rise

TABLE III

The Largest Values of τ_c and $r = q_2(\tau_c)/q_2(\infty)$
from Eq. (5.14) with $l = 50$ and Nishinari-Yoshioka (NY) Theory

			Eq. (5.14)		NY Theory	
e	g	N	τ_c	r	τ_c	r
0	1	4	0.906	0.992	0.11	0.45
0	10	6	0.430	0.964	0.07	0.35
1	0	4	1.090	0.859	0.36	0.35
10	0	7	0.300	0.931	0.11	0.36
1	1	4	0.890	0.885	0.14	0.38
4	4	7	0.430	0.827	0.08	0.32

processes are also presented. The quantity r is the degree of saturation of the buildup at which the inaccuracy of the numerical value reaches 0.01. The value of r, of course, depends on N, l, e, and g and is a measure of the range over which Eq. (5.14) is guaranteed. As a comparison, the values of τ_c and r obtained from the Nishinari-Yoshioka equation, Eq. (2.24), are also presented.

B. A Closed Form of the Matrix Representation of Birefringence Transients

We now proceed to obtain a closed form of Eq. (5.14) by obtaining the infinite sum of the power series of the matrix in the equation.[21] To this end we introduce the transformation

$$BT = T\Lambda \tag{5.16}$$

where

$$\Lambda = \begin{bmatrix} \lambda_1 & 0 & 0 & \cdots \\ 0 & \lambda_2 & 0 & \cdots \\ 0 & 0 & \lambda_3 & \cdots \\ \cdots & \cdots & \cdots & \cdots \end{bmatrix} \tag{5.17}$$

is the diagonal matrix whose elements are the eigenvalues of B, and T is the eigenmatrix of B. Then Eq. (5.14) can be reduced to

$$Q(\tau) = \sum_{l=0}^{\infty} T\left[\frac{(-\tau)^l \Lambda^l}{l!}\right] T^{-1} Q^0 - \sum_{l=0}^{\infty} T\left[\frac{(-\tau)^{l+1} \Lambda^l}{(l+1)!}\right] T^{-1} C$$

$$= T\exp(-\tau\Lambda)T^{-1}Q^0 + T(I - \exp(-\tau\Lambda))\Lambda^{-1}T^{-1}C \tag{5.18}$$

where

$$\exp(-\tau\Lambda) = \begin{bmatrix} \exp(-\lambda_1\tau) & 0 & 0 & \cdots \\ 0 & \exp(-\lambda_2\tau) & 0 & \cdots \\ 0 & 0 & \exp(-\lambda_3\tau) & \cdots \\ \cdots & \cdots & \cdots & \cdots \end{bmatrix}$$

(5.19)

Since $Q^0 = 0$ and $C = 0$ when $E_0 = 0$ and $E = 0$, respectively, it is seen immediately that the first term on the right-hand side of Eq. (5.18) is responsible for the buildup processes, whereas the second term is responsible for the decay processes. We shall now examine Eq. (5.18) for several special cases that are particularly interesting from the experimental point of view.

1. Buildup Processes

Letting $E_0 = 0$, and therefore $Q^0 = 0$, we have the rise processes. Then Eq. (5.18) reduces to

$$Q^b(\tau) = T[I - \exp(-\tau\Lambda)] \Lambda^{-1}T^{-1}C$$

(5.20)

For $\tau \to \infty$, we have

$$Q(\infty) = T\Lambda^{-1}T^{-1}C = B^{-1}C$$

(5.21)

From Eq. (5.21), Eq. (5.20) can be written as

$$Q^b(\tau) = T[I - \exp(-\tau\Lambda)]T^{-1}Q(\infty)$$

(5.22)

It should be noted that $Q(\infty)$ corresponds to the equilibrium value for E, whereas Q^0 corresponds to the stationary state value for E_0.

2. Transients for the Reversing Fields

Letting $E = -E_0$, that is, $e = -e_0$ and $g = g_0$ in Eq. (5.18), we have the transients for the reversing fields. With

$$Q^+ = B^{-1}(E)C(E) = Q(\infty)$$

(5.23)

and

$$Q^- = B^{-1}(-E)C(-E)$$

(5.24)

the transients for the reversing fields can be expressed as

$$Q'(\tau) = T\exp(-\tau\Lambda)T^{-1}(Q^+ - Q^-) + Q^+$$

(5.25)

3. Decay Processes

Letting $E = 0$, and therefore $C = 0$, we have the decay processes. Then Eq. (5.18) leads to

$$Q^d(\tau) = \exp(-\tau B^0)Q^0 \tag{5.26}$$

where

$$\exp(-\tau B^0) = \begin{bmatrix} \exp(-1 \cdot 2\tau) & 0 & 0 & \cdots \\ 0 & \exp(-2 \cdot 3\tau) & 0 & \cdots \\ 0 & 0 & \exp(-3 \cdot 4\tau) & \cdots \\ \cdots & \cdots & \cdots & \cdots \end{bmatrix} \tag{5.27}$$

is a diagonal matrix. Equation (5.27) is a general expression of the field-free relaxation of $q_n(\tau)$, and the relaxation time of $q_n(\tau)$ is

$$\tau_n^d = \frac{1}{n(n+1)} \tag{5.28}$$

Equation (5.18) is no longer a power series of the reduced time τ that is applicable only for the initial part of the transients but a closed form of the transients. A truncation of the matrix B in the numerical calculation at some dimension N is somewhat equivalent to truncating the expansion in E at some power of E. The numerical error caused by the truncation is therefore independent of time τ. In Table IV, the way in which the numerical errors at $\tau = 0.3$ in the buildup processes of the normalized birefringence $q_2(\tau)/q_2(\infty)$ depend on the dimension N is demonstrated for several values of e and $2g$. As can be noticed, calculation with $N = 4$ or 5 gives enough precision as long as the value of $e^2 + |2g|$ is not very large. On increasing the value of $e^2 + |2g|$, the matrix B becomes more and more degenerate, resulting in the smaller value of the determinant of the eigenmatrix T. The values of $\det T$ are listed in Table V for several values of parameters and N. In the table, the asterisk (*) indicates that the value of the determinant of T is smaller than 10^{-70}, making it impossible to obtain the inverse matrix of T. This difficult problem can be relaxed by increasing the value of N, however, because of less singularity at larger values of N.

We now are able to calculate the transients of the electric birefringence upon the sudden change in the unidirectional field. Figures 9 to 25 show how the normalized Kerr effect, $q_2(\tau)/q_2(\infty)$, for fixed values of the ratio $R = 0.1$, 0.2, 0.5, 1, 2, 5, 10, ∞, -0.1, -0.2, -0.5, -1.0, -2.0, -5.0, -10.0, $+0$

TABLE IV
Error at $\tau = 0.3$ in buildup[a]

e^2	$2g$	$N = 4$	$N = 5$	$N = 6$	$N = 7$	$N = 8$
1	0	3×10^{-6}	0	0	0	0
2	0	2×10^{-5}	2×10^{-7}	0	0	0
10	0	*	1×10^{-4}	6×10^{-6}	2×10^{-7}	0
15	0	*	*	2×10^{-5}	9×10^{-7}	1×10^{-7}
20	0	*	*	*	*	3×10^{-7}
2	1	6×10^{-4}	3×10^{-6}	1×10^{-6}	0	0
4	2	4×10^{-3}	1×10^{-4}	4×10^{-5}	2×10^{-6}	2×10^{-7}
6	3	7×10^{-3}	7×10^{-4}	2×10^{-4}	2×10^{-5}	2×10^{-6}
10	5	*	*	*	*	3×10^{-5}

[a] The 0 means the value is smaller than 10^{-7}.

(pure induced dipole with $g > 0$), and -0 (pure induced dipole with $g < 0$), respectively, depend on the field strength. It is noted that the normalized curves of the transient Kerr effect does not depend strongly on the field strength for positive values of R, whereas they depend remarkably on the field strength for negative values of R. What this means is that in the case of $R < 0$, the torques due to the permanent dipole moment and the induced dipole moment counteract each other, and the electric birefringence is a result of the difference of the counteracting contributions. This state of matters is particularly amplified when $R = -1$, where the contributions from the permanent and induced dipoles cancel out each other and the normalized birefringence diverges at the limiting low field. We have seen in Section I that the static birefringence $q_2(\infty)$ changes from positive to negative when $R < -1.2$. Therefore, the normalized birefringence $q_2(\tau)/q_2(\infty)$ reverses its sign in accordance with that of $q_2(\infty)$. For example, when $R = -2$, $q_2(\infty) = 0$ at $|e^2 + 2g| = 5.34$ and $q_2(\tau)/q_2(\infty)$ changes from negative to positive through

TABLE V
Value of det T[a]

e^2	$2g$	$N = 4$	$N = 8$	$N = 16$
4	0	0.431	-0.176	
10	0	2.2×10^{-15}	1.5×10^{-16}	-7×10^{-20}
8	4	0	0	3.3×10^{-5}
15	0	0	5×10^{-19}	2×10^{-21}
10	5	0	3×10^{-19}	-4×10^{-22}

[a] The 0 means the value is smaller than 10^{-70}.

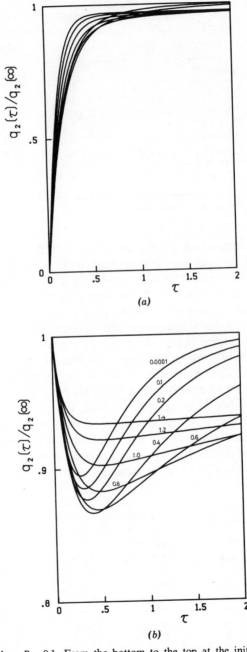

Fig. 9. (a) Buildup. $R = 0.1$. From the bottom to the top at the initial part of the rise, $e^2 = 0.0001, 0.2, 0.6, 0.8, 1, 1.2, 1.4$. (b) Reversing. The values of e^2 are shown on the curves.

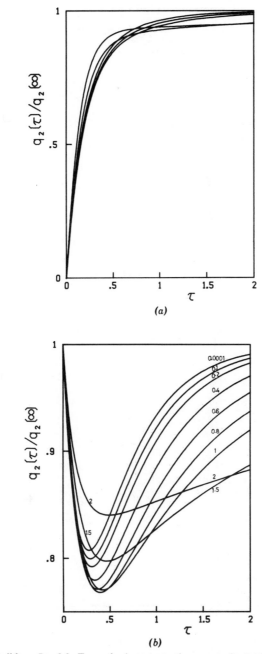

Fig. 10. (a) Buildup. $R = 0.2$. From the bottom to the top at the initial part of the rise, $e^2 = 0.0001, 0.1, 0.2, 1.5, 2$. (b) Reversing. The values of e^2 are shown on the curves.

305

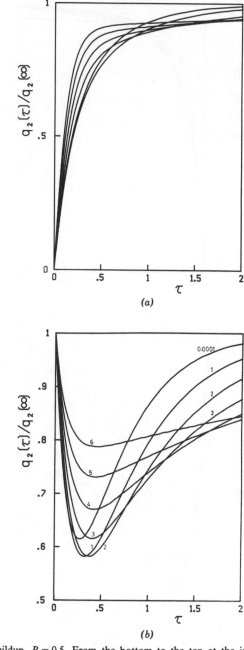

Fig. 11. (a) Buildup. $R = 0.5$. From the bottom to the top at the initial part of the rise, $e^2 = 0.0001, 1, 3, 4, 5, 6$. (b) Reversing. $R = 0.5$. The values of e^2 are shown on the curves.

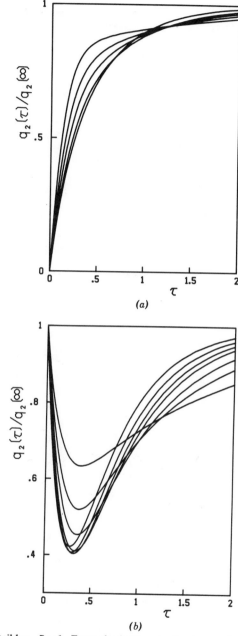

Fig. 12. (a) Buildup. $R = 1$. From the bottom to the top at the initial part of rise, $e^2 = 0.0001, 3, 5, 7, 10$. (b) Reversing. From the top to the bottom at the end of curves, $e^2 = 0.0001, 1, 2, 3, 5, 7, 10$.

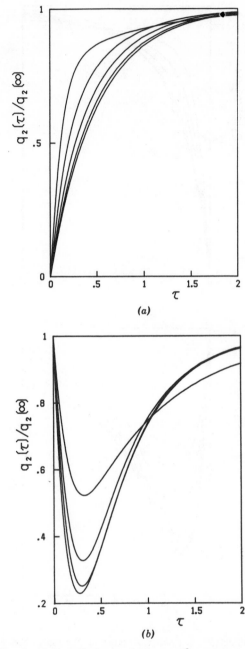

Fig. 13. (a) Buildup. $R = 2$. From the bottom to the top, $e^2 = 0.0001, 2, 5, 10, 20$. (b) Reversing. $R = 2$. From the bottom to the top in the middle, $e^2 = 0.0001, 2, 5, 10, 20$.

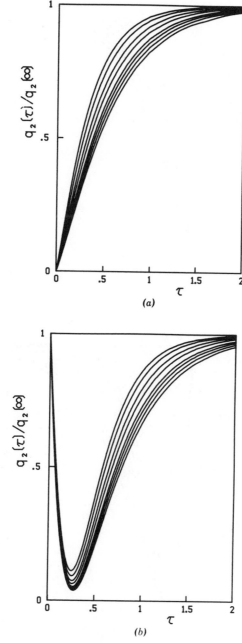

Fig. 14. (a) Buildup. $R = 5$. From the bottom to the top, $e^2 = 0.0001, 1, 2, 3, 5, 7, 10, 13$.
(b) Reversing. $R = 5$. From the bottom to the top, $e^2 = 0.0001, 1, 2, 3, 5, 7, 10, 13$.

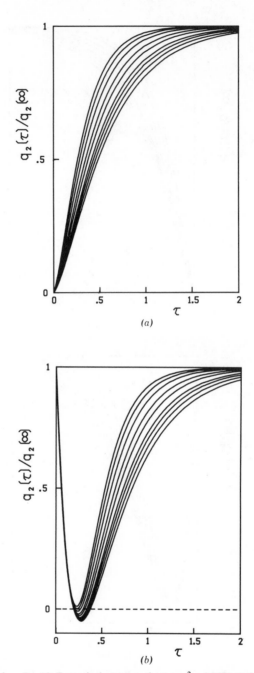

Fig. 15. (a) Buildup. $R = 10$. From the bottom to the top, $e^2 = 0.0001, 1, 2, 3, 5, 7, 10, 13, 16$.
(b) Reversing. $R = 10$. From the bottom to the top, $e^2 = 0.0001, 1, 2, 3, 5, 7, 10, 13, 16$.

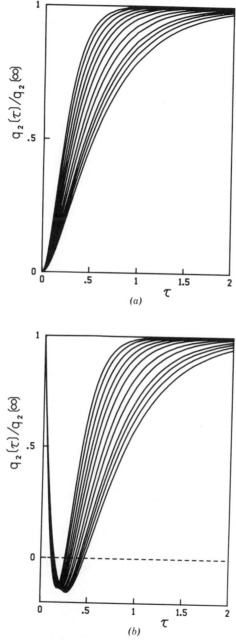

Fig. 16. (a) Buildup. $R = \infty$. From the bottom to the top, $e^2 = 0.0001, 1, 2, 3, 5, 7, 10, 13, 16,$ 20, 25, 30. (b) Reversing. $R = \infty$.

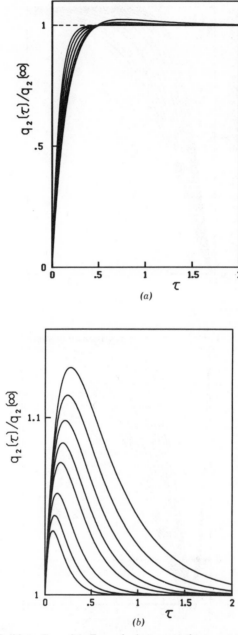

Fig. 17. (a) Buildup. $R = -0.1$. From the bottom to the top at the initial part of rise, $e^2 = 0.00001, 0.2, 0.4, 0.6, 0.8, 1$. (b) Reversing. From the top to the bottom, $e^2 = 0.00001, 0.1, 0.2, 0.3, 0.4, 0.6, 0.8, 1$.

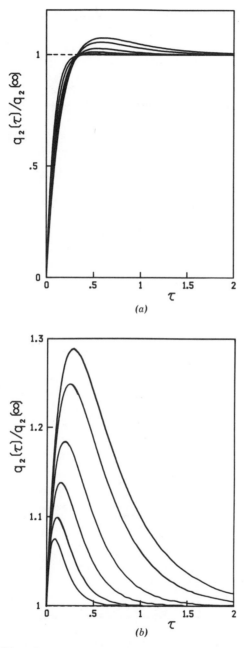

Fig. 18. (a) Buildup. $R = -0.2$. From the bottom to the top at the initial part of rise, $e^2 = 0.0001, 0.2, 0.6, 1, 1.5, 2$. (b) Reversing. From the top to the bottom, $e^2 = 0.0001, 0.2, 0.6, 1, 1.5, 2$.

313

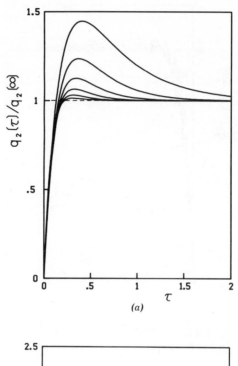

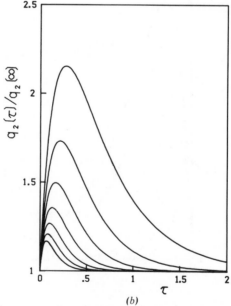

Fig. 19. (a) Buildup. $R = -0.5$. From the top to the bottom, $e^2 = 0.0001, 1, 2, 3, 4, 5$. (b) Reversing. $R = -0.5$. From the top to the bottom, $e^2 = 0.0001, 1, 2, 3, 4, 5, 6$.

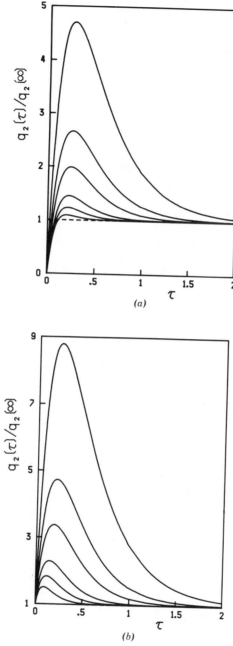

Fig. 20. (a) Buildup. $R = -1$. From the top to the bottom, $e^2 = 1, 2, 3, 5, 7, 10$. (b) Reversing.

315

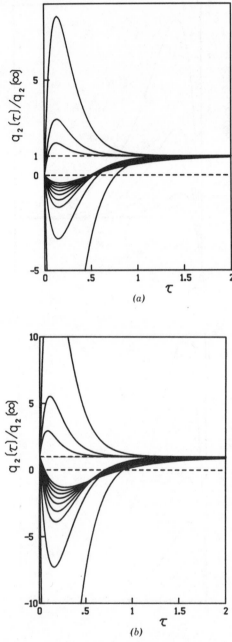

Fig. 21. (a) Buildup. $R = -2$. Lower half; from the top to the bottom, $e^2 = 0.0001, 1, 2, 3, 4,$ 5, 6, 8, 10. Upper half; from the top to the bottom, $e^2 = 12, 15, 20$. (b) Reversing.

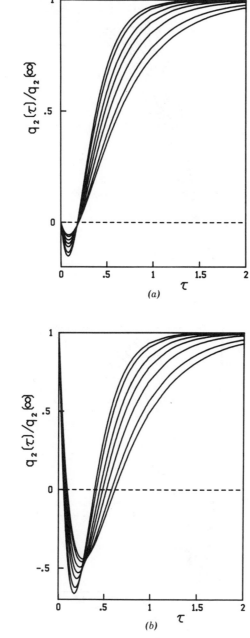

Fig. 22. (a) Buildup. $R = -5$. From the top to the bottom in the initial part, $e^2 = 0.0001, 1,$ 2, 3, 5, 7, 10, 12. (b) Reversing.

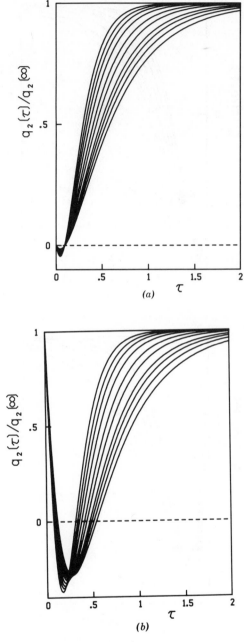

Fig. 23. (a) Buildup. $R = -10$. From the bottom to the top in the middle part, $e^2 = 0.0001$, 1, 2, 3, 5, 7, 10, 13, 16, 20. (b) Reversing.

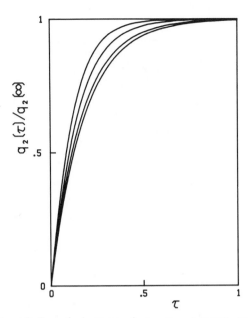

Fig. 24. Buildup. $R = +0$. From the bottom to the top, $2g = 3$, 0.0001, 10, 13. Note that the normalized buildup becomes slower in the initial increase of g. See also Fig. 4.

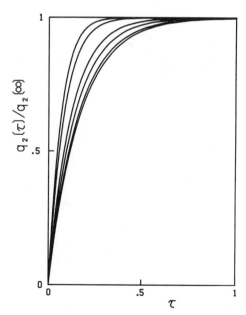

Fig. 25. Buildup. $R = -0$. From the bottom to the top, $2g = -0.0001$, -1, -3, -5, -10, -13. The slowdown of the rise does not show up.

319

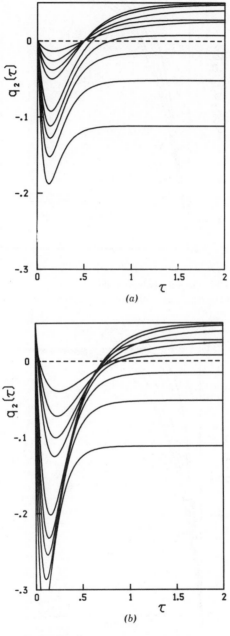

Fig. 26. (a) Unnormalized buildup. $R = -2$. From the top to the bottom in the initial part; $e^2 = 1, 2, 3, 4, 8, 10, 12, 15, 20$. The sign of the stationary state birefringence reverses in the interval $10 < e^2 < 12$. (b) Unnormalized reversing.

an infinitive. For the sake of clearer understanding, "unnormalized transients" $q_2(\tau)$ for $R = -2$ are plotted in Figs. 26a and b.

By setting the partial derivative of Eq. (2.18) with respect to τ to zero, we see that the time at which the birefringence for the reversing pulse takes the minimum value, τ_m, is

$$\tau_m = \frac{\ln 3}{4} \tag{5.29}$$

irrespective to the value of R.[8] In Fig. 27, dependencies of the minimum time τ_m on the field strength are shown for $R = \infty$. For the infinitely low field, $\Phi_m = \Phi(\tau_m)/\Phi(\infty)$ is given by

$$\Phi_m = 1 - \frac{2}{\sqrt{3}} \frac{R}{R+1} \tag{5.30}$$

However, Φ_m strongly depends on the field strength, except for the case where $|R|$ is very large.

C. The Stationary State Value and the Areas Surrounded by the Transient Curves

It may be noted that Eq. (5.21) supplies us with a convenient method of obtaining the stationary state value of $q_n(\infty)$. Since the matrix **B** is not singular for any value of the parameters, it is always possible to obtain the inverse matrix of **B** at any desired precision. Equation (5.21) is therefore much more convenient for obtaining the stationary state value numerically than the previous analytical results, Eqs. (1.13) to (1.15), which involve a calculation of the error function. For a very large value of $e^2 + |2g|$, however, we need to use a large number of dimensions, N, for the precise calculation, and the calculation becomes time consuming.

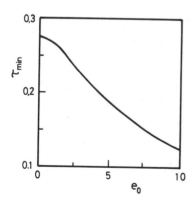

Fig. 27. A plot of $\tau_{\min}$ versus $-e = e_0$ for $R = \infty$.

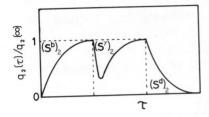

Fig. 28.   Areas $(S^b)_2$, $(S^r)_2$, and $(S^d)_2$.

The areas surrounded by the normalized transients of the birefringence denoted by $(S^b)_2$ and $(S^r)_2$ in Fig. 28 is given, for the limiting case of $E \to 0$, by[22,23]

$$\int_0^\infty \left[1 - \frac{\Phi^b(\tau)}{\Phi^b(\infty)} \right] d\tau = \int_0^\infty \left[\frac{3}{2} \frac{R}{R+1} \exp(-2\tau) - \frac{R-2}{2(R+1)} \exp(-6\tau) \right] d\tau$$

$$= \frac{4R+1}{6(R+1)} \tag{5.31}$$

and

$$\int_0^\infty \left[1 - \frac{\Phi^r(\tau)}{\Phi^r(\infty)} \right] d\tau = \frac{3R}{R+1} \int_0^\infty [\exp(-2\tau) - \exp(-6\tau)] \, d\tau$$

$$= \frac{R}{R+1} \tag{5.32}$$

These are simple functions of R and therefore provide us an additional method for determining the electrical parameters of the solute particles from transient birefringence. The area $(S^d)_2$ surrounded by the decay curve as defined in Fig. 28 is given by

$$\int_0^\infty \frac{\Phi^d(\tau)}{\Phi^d(\infty)} \, d\tau = \int_0^\infty \exp(-6\tau) \, d\tau = \tfrac{1}{6} \tag{5.33}$$

which provides us a useful method for determining the hydrodynamic parameter (the rotational diffusion constant D) of the solute particles.

For finite field strength, we have, from Eqs. (5.22) and (5.23),

$$(S^b)_m = \int_0^\infty \left\{ 1 - \left[\frac{[Q^b(\tau)]_m}{(Q^+)_m} \right] \right\} d\tau = \frac{[B^{-1}(E)Q^+]_m}{(Q^+)_m} \tag{5.34}$$

and from Eq. (5.25)

$$(\mathbf{S}^r)_m = \int_0^\infty \left\{ 1 - \left[\frac{[\mathbf{Q}^r(\tau)]_m}{(\mathbf{Q}^-)_m} \right] \right\} d\tau = \frac{[\mathbf{B}^{-1}(-E)(\mathbf{Q}^+ - \mathbf{Q}^-)]_m}{(Q^-)_m}$$

$$(5.35)$$

where [column matrix]$_m$ represents the mth element of the column matrix, and $\mathbf{S}^b$ and $\mathbf{S}^r$ are the column matrices whose second elements $(\mathbf{S}^b)_2$ and $(\mathbf{S}^r)_2$ correspond to the areas defined by the normalized birefringence, respectively. The areas surrounded by the decay curves $(\mathbf{S}^d)_m$ are, from Eq. (5.27),

$$(\mathbf{S}^d)_m = [m(m+1)]^{-1}$$

$$(5.36)$$

which does not depend on the field strength. Figures 29 and 30 show the dependencies of the areas $(\mathbf{S}^b)_2$ and $(\mathbf{S}^r)_2$ on the field strength for several values of R. It is interesting to note that the areas increase with increasing field strength when $0 < R < 1$. When $R < 0$, the effect of the field strength on the area is much more remarkable, as can be expected from the normalized transients shown in Figs. 9–25. Thus the field strength dependencies of the areas surrounded by the buildup and reversing curves also provide us useful information about the electric nature of the particle.

For the polydisperse system, assuming as before that the parameters except for the rotational diffusion constant D do not depend on the species of

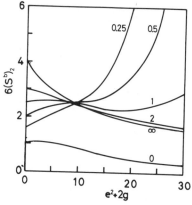

Fig. 29. Field strength dependence of $(S^b)_2$. Numerals on each curve represent the value of R.

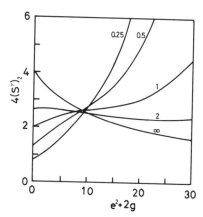

Fig. 30. Field strength dependence of $(S^r)_2$. Numerals are the same as Fig. 29.

the particle, and neglecting interparticle interactions, we may express the normalized birefringence of the decay process at the limiting low field as

$$\Delta n^d(t) = \sum_i \Delta n_i(t) = \frac{2\pi c_v}{n} \sum_i (g_1 - g_2)_i \Phi_i^0 \exp(-6D_i t)$$

$$= \sum_i A_i \exp(-6D_i t) \equiv \sum_i A_i(t) \qquad (5.37a)$$

where

$$A_i = \frac{2\pi c_v}{n} (g_1 - g_2)_i \Phi_i^0 \qquad (5.37b)$$

is the contribution of the particle whose diffusion constant is D_i to the birefringence [cf. Eq. (1.3)]. Hence

$$\Delta n^d(0) = \sum_i \Delta n_i^0 = \sum_i A_i(0)$$

$$(S^d)_2 = \frac{\int_0^\infty \sum_i A_i(t)\, dt}{\Delta n^d(0)} = \frac{\sum_i A_i(0)/6D_i}{\Delta n^d(0)} = \langle \tau^d \rangle \qquad (5.38)$$

Thus the area $(S^d)_2$ seems to correspond with the average of D. We may call this the "birefringence average." In the same way, the buildup process may be expressed as

$$\Delta n^b(t) = \sum_i A_i(0)\left[1 - \frac{3R}{2(R+1)}\exp(-2D_i t) + \frac{R-2}{2(R+1)}\exp(-6D_i t)\right] \qquad (5.39)$$

The area $(S^b)_2$ is then

$$(S^b)_2 = \int_0^\infty \left[1 - \frac{\Delta n^b(t)}{\Delta n^b(\infty)}\right] dt = \frac{4R+1}{6(R+1)}\langle D^{-1} \rangle = \frac{4R+1}{R+1}\langle \tau^d \rangle \qquad (5.40)$$

In exactly the same manner, we have for the areas $(S^r)_2$

$$(S^r)_2 = \int_0^\infty \left[1 - \frac{\Delta n^r(t)}{\Delta n^r(\infty)}\right] dt = \frac{R}{R+1}\langle D^{-1} \rangle = \frac{6R}{R+1}\langle \tau^d \rangle \qquad (5.41)$$

Normally, among the same species, the larger the particle size, the larger the electric dipole moments and the smaller the diffusion constant. Since the contribution to the stationary state birefringence of the particles having larger dipole moments is more prominent in the lower field, that is, Φ_i^0 of a bulky components is larger in the lower field, $\langle \tau^d \rangle$ depends on the field strength, or, more precisely, on the degree of saturation.

As an example, we consider the case of helical polypeptides, such as poly-γ-benzylglutamate in a helicogenic solvent, which may be regarded as a rigid rod with a large permanent dipole moment along the rod axis. In the Kerr region, $\Phi_i^0 \propto \mu_i^2$, where μ_i is the permanent dipole moment of the i component. Since the optical anisotropy factor $(g_1 - g_2)_i$ of the i component is proportional to the volume of the component i, that is proportional to $N_i M_i$, where N_i and M_i are the number and molecular weight of the i component, respectively. Therefore the contribution of the i component to the birefringence is

$$\langle \tau^d \rangle = \frac{\sum_i N_i M_i \mu_i^2 \tau_i^d}{\sum_i N_i M_i \mu_i^2} \tag{5.42}$$

Since $\mu_i \propto M_i$, the average relaxation time is

$$\langle \tau^d \rangle = \frac{\sum_i N_i M_i^3 \tau_i^d}{\sum_i N_i M_i^3} \tag{5.43}$$

This is the $z+1$ average of τ^d. On the other hand, for a sufficiently strong electric field under which the saturation of the orientation of all components is attained, the average relaxation time is given by

$$\langle \tau^d \rangle = \frac{\sum_i N_i M_i \tau_i^d}{\sum_i N_i M_i} \tag{5.44}$$

since the Φ_i^0 is independent of μ_i. This is the weighted average of τ^d.

Incidentally, we shall show that the field-free decay processes of $q_n(\tau)$ obtained by switching off the field at $\tau = a > 0$ in the middle of the buildup process before reaching the stationary state are given in the form of $\exp[-n(n+1)\tau]$, which is similar to the usual decay process after the stationary state, provided that the sample is monodisperse. To this end we return to Eq. (5.2):

$$\frac{dq_n(\tau)}{d\tau} = -n(n+1)q_n(\tau) + e(\tau)w_n'[q_{n-1}(\tau) - q_{n+1}(\tau)]$$

$$+ 2g(\tau)[v_n'q_{n-2}(\tau) + u_nq_n(\tau) - y_n'q_{n+2}(\tau)] \qquad (5.45)$$

with

$$u_n = \frac{n(n+1)}{(2n-1)(2n+3)} \qquad (5.46)$$

Integrating both sides of Eq. (5.45), we have

$$q_n(\tau) = \exp[-n(n+1)\tau]$$

$$\times \left\{ q_n(0) + w_n' \int_0^\tau \exp[n(n+1)\tau']e(\tau')[q_{n-1}(\tau') - q_{n+1}(\tau')] \, d\tau' \right.$$

$$+ 2\int_0^\tau \exp[n(n+1)\tau'] g(\tau')$$

$$\left. \times [v_n'q_{n-2}(\tau') + u_nq_n(\tau') - y_n'q_{n+2}(\tau')] \, d\tau' \right\} \qquad (5.47)$$

We now imagine that a rectangular electric field of height E and width a is applied to the system, which has been free from any orienting field (see Fig. 31). Then,

$$q_n(0) = \delta_{n,0} \qquad (5.48)$$

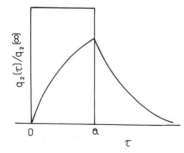

Fig. 31. A decay of the birefringence from the middle of the rise at $\tau = a$.

and

$$q_n(\tau) = q_n^*(a)\exp[-n(n+1)\tau]$$

where

$$q_n^*(a) = w_n'e \int_0^a \exp[n(n+1)\tau'][q_{n-1}(\tau') - q_{n+1}(\tau')]\,d\tau'$$

$$+2g\int_0^\tau \exp[n(n+1)\tau']$$

$$\times [v_n'q_{n-2}(\tau') + u_nq_n(\tau') - y_n'q_{n+2}(\tau')]\,d\tau' \qquad (5.49)$$

For $\tau > a$, $q_n^*(a)$ in Eq. (5.49) is a constant, which shows immediately that $q_n(\tau)$ is proportional to $\exp[-n(n+1)\tau]$.

VI. TRANSIENT AND STEADY-STATE ELECTRIC BIREFRINGENCE IN THE NONLINEAR REGION

A. The Electric Polarization and the Electric Birefringence Caused by Extremely Low Time-Varying Electric Fields

In this subsection we shall obtain analytical expressions for the electric polarization and electric birefringence for time-varying electric fields.[24] To this end we write Eq. (5.2) in the following matrix form, noting that $q_0(\tau) = 1$:

$$\mathbf{Q}(\tau) = \mathbf{D}(\tau)\mathbf{Q}^0 + \int_0^\tau e(\tau)\mathbf{W}(\tau - \tau')\mathbf{Q}(\tau')\,d\tau'$$

$$+2\int_0^\tau g(\tau')\mathbf{G}(\tau - \tau')\mathbf{Q}(\tau')\,d\tau' + \mathbf{R}(\tau) \qquad (6.1)$$

where

$$\mathbf{D}(\tau) = \begin{bmatrix} D_1(\tau) & 0 & 0 & \cdots \\ 0 & D_2(\tau) & 0 & \cdots \\ 0 & 0 & D_3(\tau) & \cdots \\ \cdots & \cdots & \cdots & \cdots \end{bmatrix} \qquad (6.2)$$

$$\mathbf{W}(\tau) = \begin{bmatrix} 0 & -w_1'D_1(\tau) & 0 & 0 & \cdots \\ w_2'D_2(\tau) & 0 & -w_2'D_2(\tau) & 0 & \cdots \\ 0 & w_3'D_3(\tau) & 0 & -w_3'D_3(\tau) & \cdots \\ \cdots & \cdots & \cdots & \cdots & \cdots \end{bmatrix} \qquad (6.3)$$

$\mathbf{G}(\tau)$

$$
= \begin{bmatrix}
u_1 D_1(\tau) & 0 & -y_1 D_1(\tau) & 0 & 0 & \cdots \\
0 & u_2 D_2(\tau) & 0 & -y_2 D_2(\tau) & 0 & \cdots \\
v_3' D_3(\tau) & 0 & u_3 D_3(\tau) & 0 & -y_3 D_3(\tau) & \cdots \\
\cdots & \cdots & \cdots & \cdots & \cdots & \cdots
\end{bmatrix}
$$

$$\tag{6.4}$$

$$
\mathbf{R}(\tau) = \begin{bmatrix}
w_1' \int_0^\tau e(\tau') D_1(\tau - \tau') \, d\tau' \\
2v_2' \int_0^\tau g(\tau') D_2(\tau - \tau') \, d\tau' \\
0 \\
0 \\
\vdots
\end{bmatrix}
\tag{6.5}
$$

in which

$$D_n = \exp[-n(n+1)\tau] \tag{6.6}$$

If we write

$$\mathbf{K}(\tau, \tau') = e(\tau')\mathbf{W}(\tau - \tau') + 2g(\tau')\mathbf{G}(\tau - \tau') \tag{6.7}$$

and

$$\mathbf{P}(\tau) = \mathbf{D}(\tau)\mathbf{Q}^0 + \mathbf{R}(\tau) \tag{6.8}$$

Eq. (6.1) becomes an integral equation

$$\mathbf{Q}(\tau) = \int_0^\tau \mathbf{K}(\tau, \tau')\mathbf{Q}(\tau') \, d\tau' + \mathbf{P}(\tau) \tag{6.9}$$

The solution of Eq. (6.9) is

$$
\mathbf{Q}(\tau) = \int_0^\tau \mathbf{K}(\tau, \tau_1)\mathbf{P}(\tau_1) \, d\tau_1 + \int_0^\tau \int_0^{\tau_1} \mathbf{K}(\tau, \tau_1)\mathbf{K}(\tau_1, \tau_2)\mathbf{P}(\tau_2) \, d\tau_1 \, d\tau_2
$$
$$
+ \int_0^\tau \int_0^{\tau_1} \int_0^{\tau_2} \mathbf{K}(\tau, \tau_1)\mathbf{K}(\tau_1, \tau_2)\mathbf{K}(\tau_2, \tau_3)\mathbf{P}(\tau_3) \, d\tau_1 \, d\tau_2 \, d\tau_3 + \cdots
$$

$$\tag{6.10}$$

It should be noted here that if $e(\tau)$ and $g(\tau)$ are time-independent, Eq. (6.7) gives $K(\tau, \tau') = K(\tau - \tau')$, which enables us to write Eq. (6.10) after taking the Laplace transforms as

$$\tilde{Q}(s) = \left[\tilde{K}(s) + \tilde{K}^2(s) + \tilde{K}^3(s) + \cdots\right]\tilde{P}(s) \tag{6.11}$$

where

$$\tilde{Q}(s) = \mathscr{L}[Q(\tau)] \tag{6.12a}$$

$$\tilde{K}(s) = \mathscr{L}[K(\tau)] \tag{6.12b}$$

and

$$\tilde{P}(s) = \mathscr{L}[P(\tau)] \tag{6.12c}$$

For extremely low values of $E(\tau)$, by taking terms up to the orders of $E(\tau)$ and $E^2(\tau)$, respectively, for $q_1(\tau)$ and $q_2(\tau)$, we find that Eq. (6.10) gives the following expression:

$$q_1(\tau) = q_1(0)\exp(-2\tau) + w_1'\int_0^\tau \exp[-2(\tau - \tau_1)]e(\tau_1)\,d\tau_1 \tag{6.13}$$

$$
\begin{aligned}
q_2(\tau) &= q_2(0)\exp(-6\tau)\left[1 + \frac{2}{7}\int_0^\tau g(\tau_1)\,d\tau_1\right] \\
&+ q_1(0)w_2'\int_0^\tau \exp[-6(\tau - \tau_1)]\exp(-2\tau_1)e(\tau_1)\,d\tau_1 \\
&+ w_1'w_2'\int_0^\tau\int_0^{\tau_1}\exp[-6(\tau - \tau_1)]\exp[-2(\tau_1 - \tau_2)]e(\tau_1)e(\tau_2)\,d\tau_1\,d\tau_2 \\
&+ 2v_2'\int_0^\tau \exp[-6(\tau - \tau_1)]g(\tau_1)\,d\tau_1
\end{aligned} \tag{6.14}
$$

We shall analytically calculate the integrals in Eqs. (6.13) and (6.14) for three types of time-varying fields: (1) continuous reversing square pulses, (2) exponentially rising and decaying fields, and (3) a sinusoidal field. These are particularly interesting from the experimental point of view.

1. Continuous Reversing Square Pulses with a Period of 2T

In order to integrate Eq. (6.13), it is useful to calculate the integral

$$L_m(\tau) = \int_0^\tau e^{\alpha t}U(t)\,dt \tag{6.15}$$

where $U(t)$ is a unit of continuous reversing square pulses with a period $2T$ given in Fig. 32. Since we can always assign an integer m to τ in such a way

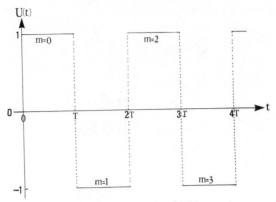

Fig. 32. Definition of $U(t)$.

that

$$m\tau' \leq \tau \leq (m+1)\tau' \qquad (m = 0,1,2,\ldots) \tag{6.16}$$

we see that

$$L_m(\tau) = \sum_{j=0}^{m-1} (-1)^j \int_{j\tau'}^{(j+1)\tau'} e^{\alpha t} dt + (-1)^m \int_{m\tau'}^{\tau} e^{\alpha t} dt$$

$$= \frac{1}{\alpha} \left[\frac{e^{\alpha\tau'} - 1 - 2(-1)^m e^{\alpha(m+1)\tau'}}{1 + e^{\alpha\tau'}} + (-1)^m e^{\alpha\tau} \right] \tag{6.17}$$

where $\tau' = DT$. It is well known that the Laplace transform of $U(t)$ is given by

$$\mathscr{L}[U(t)] = \int_0^\infty U(t)e^{-st} dt = \frac{1}{s}\tanh\left(\frac{s\tau'}{2}\right) \tag{6.18}$$

It is immediately seen that Eq. (6.17), after putting $\alpha = -s$, $m \to \infty$, and $\tau \to \infty$, can be reduced to Eq. (6.18). Using Eq. (6.17), we find from Eq. (6.13) that

$$\frac{q_1(\tau)}{e/3} = (-1)^m \frac{(1 - e^{-2\tau'})e^{-2m\tau'} - 2(-1)^m}{1 + \exp(-2\tau')} \tag{6.19}$$

where

$$\tau = m\tau' + \Delta\tau \qquad (0 \le \Delta\tau < \tau') \tag{6.20}$$

and

$$e(t) = eU(t) \tag{6.21}$$

For the calculation of $q_2(\tau)$, it is necessary to obtain the integral

$$J_m(\tau) = \int_0^\tau \int_0^{\tau_1} e^{4\tau_1} e^{2\tau_2} U(\tau_1) U(\tau_2) \, d\tau_2 \, d\tau_1$$

$$= \int_0^\tau e^{4\tau_1} U(\tau_1) \left[\int_0^{\tau_1} e^{2\tau_2} U(\tau_2) \, d\tau_2 \right] d\tau_1 \tag{6.22}$$

This may be calculated as before:

$$J_m(\tau) = \sum_{j=0}^m (-1)^j x_j + (-1)^m Y_m(\tau) \tag{6.23}$$

where

$$x_j = \int_{j\tau'}^{(j+1)\tau'} e^{4\tau_1} \left[\int_0^{\tau_1} e^{2\tau_2} U(\tau_2) \, d\tau_2 \right] d\tau_1 \tag{6.24}$$

$$Y_m(\tau) = \int_{m\tau'}^\tau e^{4\tau_1} \left[\int_0^{\tau_1} e^{2\tau_2} U(\tau_2) \, d\tau_2 \right] d\tau_1 \tag{6.25}$$

Integrating Eqs. (6.24) and (6.25) by parts and using the results in Eq. (6.17), we obtain

$$4x_j = e^{4(j+1)\tau'} \int_0^{(j+1)\tau'} e^{2\tau_2} U(\tau_2) \, d\tau_2 - 4 e^{4j\tau'} \int_0^{j\tau'} e^{2\tau_2} U(\tau_2) \, d\tau_2$$

$$- \int_{j\tau'}^{(j+1)\tau'} e^{6\tau_1} U(\tau_1) \, d\tau_1$$

$$= \tfrac{1}{2} e^{4j\tau'} (1 - e^{2\tau'})^2 + \tfrac{1}{3} (-1)^j e^{6j\tau'} (e^{2\tau'} - 1)^3 \tag{6.26}$$

$$4Y_m(\tau) = e^{4\tau} \int_0^\tau e^{2\tau_1} U(\tau_1) \, d\tau_1 - e^{4m\tau'} \int_0^{m\tau'} e^{2\tau_1} U(\tau_1) \, d\tau_1 - \int_{m\tau'}^\tau e^{6\tau_1} U(\tau_2) \, d\tau_1$$

$$= \frac{e^{4\tau}}{2} \frac{1 - e^{-2\tau'} - 2(-1)^m e^{2m\tau'}}{1 + e^{-2\tau'}} + (-1)^m \frac{e^{6\tau}}{3} - \frac{e^{4m\tau'}}{2} \frac{1 - e^{-2\tau'}}{1 + e^{-2\tau'}}$$

$$+ (-1)^m \frac{e^{6m\tau'}}{3} \frac{2 - e^{-2\tau'}}{1 + e^{-2\tau'}} \tag{6.27}$$

Equations (6.22) through (6.27) together with (6.14) and $q_1(0) = q_2(0) = 0$ lead to

$$\frac{q_2(\tau)}{e/15} = \left\{ \frac{3}{2} \frac{(1-e^{-2\tau'})^2}{1+e^{-4\tau'}} e^{-2m\tau'} \left[e^{-4m\tau'} - (-1)^m \right] \right.$$

$$+ \frac{(1-e^{-2\tau'})^2(1-e^{-6m\tau'})}{1+e^{-2\tau'}+e^{-4\tau'}} - \frac{3}{2}(-1)^m \frac{1-e^{-2\tau'}}{1+e^{-2\tau'}} e^{-2m\tau'}$$

$$\left. + \frac{2-e^{-2\tau'}}{1+e^{-2\tau'}} \right\} e^{-6\Delta\tau} + \frac{3}{2} \frac{(-1)^m(1-e^{-2\tau'})e^{-2m\tau'}-2}{1+e^{-2\tau'}} e^{-2\Delta\tau}$$

$$+1 + \frac{2g}{e^2}(1-e^{-6\tau}) \tag{6.28}$$

When the reversing pulses are applied for a long time, the system reaches the steady state, removing all the transient effects. Since this limit corresponds to $m \to \infty$ in Eqs. (6.17) and (6.28), we obtain for the electric polarization

$$\frac{q_1(\tau)}{e/3} = (-1)^m \frac{1+e^{-2\tau'}-2e^{-2\Delta\tau}}{1+e^{-2\tau'}} \tag{6.29}$$

and for the electric birefringence

$$\frac{q_2(\tau)}{q_2^0} = 1 + \frac{3R}{R+1} \frac{e^{-2\Delta\tau}}{1+e^{-2\tau'}} \left(\frac{e^{-4\Delta\tau}}{1+e^{-2\tau'}+e^{-4\tau'}} - 1 \right) \tag{6.30}$$

respectively, where

$$q_2^0 = \tfrac{1}{15}(e_0^2 + 2g_0) \tag{6.31}$$

is the birefringence for $\Delta\tau = 0$, $\tau' = \infty$, that is, the birefringence for the static electric field E_0 of limiting low strength [cf. Eq. (1.12)], and $R = e_0^2/2g_0$ [cf. Eq. (1.22)].

In order to show how the system reaches the steady state, $q_2(\tau)/q_2^0$ vs. $\Delta\tau$ in the case of $g_0 = 0$ and $\tau' = 1$ is plotted in Fig. 33. Numerals on the right of the curves represent the values of m. It is evident in this particular case that $m = 3$ already satisfies the condition for the steady state. For the special case where $m = 0$, Eq. (6.28) gives the following result for the rise transient:

$$\frac{q_2(\tau)}{q_2^0} = 1 - \frac{3R}{2(1+R)} e^{-2\Delta\tau} + \frac{R-2}{2(1+R)} e^{-6\Delta\tau} \tag{6.32}$$

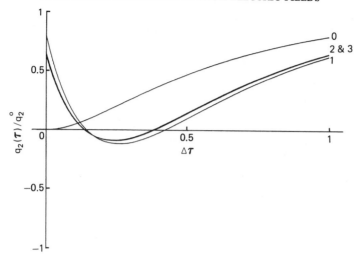

Fig. 33.   Plots of $q_2(\tau)/q_2^0$ obtained from Eq. (6.28) versus $\Delta\tau$ for various values of m, τ', and $g_0 = 0$. The numerals on the right of the curves represent values of m.

This is the condition that $q_2(\tau)/q_2^0$ must satisfy for the case of $m = 0$ [cf. Eq. (2.13)].

In Fig. 34, $q_2(\tau)/q_2^0$ in Eq. (6.30) is plotted against $\Delta\tau/\tau'$ for $g_0 = 0$ and various values of τ'. It is noted in Eq. (6.30) that the induced dipole moment does not contribute to the time dependence of $q_2(\tau)/q_2^0$. The value of R merely affects the level and amplitude of the steady state oscillation of the birefringence. Also, it is seen from this equation that

$$\lim_{\tau' \to \infty} \frac{q_2(\tau)}{q_2^0} = 1 + \frac{3R}{R+1}\left(e^{-6\Delta\tau} - e^{-2\Delta\tau}\right) \qquad (6.33)$$

This corresponds to $q_2(\tau)/q_2^0$ for the rapidly reversing electric field, that is, the birefringence obtained when a constant field $-E_0$ has been applied for a sufficiently long time and suddenly at $\tau = 0$ another electric field of strength E_0 is switched on, provided that the value of E_0 is sufficiently low [cf. Eq. (2.18)]. Putting $\Delta\tau = \tau'$ in Eq. (6.30), we find the same value for $q_2(\tau)/q_2^0$ that is the maximum value of $q_2(\tau)/q_2^0$ and thus represented by $q_2^{\max}$:

$$q_2^{\max} = R + \frac{3R}{R+1}\frac{3e^{-2\tau'}}{1 + e^{-2\tau'} + e^{-4\tau'}} \qquad (6.34)$$

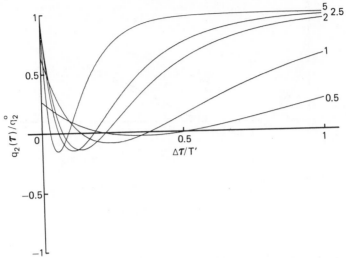

Fig. 34. Plots of $q_2(\tau)/q_2^0$ obtained from Eq. (6.30) versus $\Delta\tau/\tau'$ for various values of τ' and $g_0 = 0$. Numerals on the right of the curves represent values of τ'.

Equation (6.30) also enables us to calculate the time $\Delta\tau_{min}$ at which $q_2(\tau)/q_2^0$ takes the minimum value q_2^{min} by the condition of $dq_2(\Delta\tau)/d(\Delta\tau) = 0$, thus obtaining (cf. Tinoco and Yamaoka[8])

$$\Delta\tau_{min} = \frac{1}{4}\ln\left(\frac{3}{1 + e^{-2\tau'} + e^{-4\tau'}}\right) \qquad (6.35)$$

$$q_2^{min} = 1 - \frac{2}{\sqrt{3}}\frac{R}{R+1}\frac{\left(1 + e^{-2\tau'} + e^{-4\tau'}\right)^{1/2}}{1 + e^{-2\tau'}} \qquad (6.36)$$

In Fig. 35, q_2^{max} and q_2^{min} for several values of R are plotted against τ'. Numerals and the symbol ∞ on the left end of the curves where q_2^{max} and q_2^{min} meet are the values of R. The dashed line is a plot of $\Delta\tau_{min}$. It is seen in Eqs. (6.34) and (6.36) that larger values than $\tau' = 2$ do not contribute appreciably to the maximum and minimum values of the birefringence. In other words, continuous reversing pulses of duration more than about 2 gives the birefringence pattern roughly expressed by Eq. (6.32). It is worthy of special attention in Eq. (6.35) that the time at which $q_2(\tau)/q_2^0$ reaches its minimum is independent of R, that is, independent of the contribution of the induced dipole moment. The depth of the minimum, however, depends on the value of R, as is demonstrated in Fig. 35.

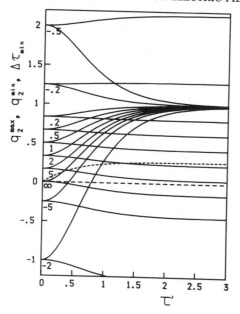

Fig. 35. Plots of q_2^{max} and q_2^{min} obtained from Eqs. (6.34) and (6.36) versus τ'. Dashed line shows $\Delta\tau_{min}$ obtained from Eq. (6.35).

2. Exponentially Rising and Decaying Electric Fields

From the practical point of view, it is not rare that we need to analyze the transients of the electric birefringence for the exponentially rising and decaying electric fields, $E(t) = E_0[1 - \exp(-kt)]$ (field 1) and $E(t) = E_0\exp(-kt)$ (field 2). For these cases integrals in Eq. (6.13) may be readily obtained, leading to

$$q_1(\tau) = q_1(0)e^{-2\tau} - \frac{2}{3}\frac{e_0}{k'-2}(e^{-2\tau} - e^{-k'\tau}) + \frac{1}{3}e_0(1 - e^{-2\tau}) \quad (6.37)$$

and

$$q_1(\tau) = q_1(0)e^{-2\tau} + \frac{2}{3}\frac{e_0}{k'-2}(e^{-2\tau} - e^{-k'\tau}) \quad (6.38)$$

for fields (1) and (2), respectively, where $k' = k/D$.

For the birefringence, we obtain from Eq. (6.14) that

$$
q_2(\tau) = \frac{4}{5} e_0^2 \left\{ \frac{1-e^{-6\tau}}{12} - \frac{(4-k')(e^{-k'\tau}-e^{-6\tau})}{2(2-k')(6-k')} + \frac{k'(e^{-2\tau}-e^{-6\tau})}{8(2-k')} \right.
$$
$$
\left. - \frac{k'[e^{-(2+k')\tau}-e^{-6\tau}]}{2(4-k')(2-k')} + \frac{e^{-2k'\tau}-e^{-6\tau}}{2(2-k')(3-k')} \right\}
$$
$$
+ \frac{4}{5} g_0 \left[\frac{1-e^{-6\tau}}{6} - \frac{2(e^{-k'\tau}-e^{-6\tau})}{6-k'} + \frac{e^{-2k'\tau}-e^{-6\tau}}{2(3-k')} \right] \tag{6.39}
$$

and

$$
q_2(\tau) = q_2(0)e^{-6\tau}\left[1+\frac{2g_0}{7k'}(1-e^{-2k'\tau})\right] + q_1(0)\frac{6e_0}{5(4-k')}\left[e^{-(2+k')\tau}-e^{-6\tau}\right]
$$
$$
+ \frac{4e_0^2}{5}\left[\frac{e^{-2k'\tau}-e^{-6\tau}}{2(2-k')(3-k')} - \frac{e^{-(2+k')\tau}-e^{-6\tau}}{(2-k')(4-k')}\right]
$$
$$
+ \frac{2g_0}{5(3-k')}(e^{-2k'\tau}-e^{-6\tau}) \tag{6.40}
$$

for fields 1 and 2, respectively. It should be noted in Eqs. (6.39) and (6.40) that apparent singularities at $k' = 2$, 3, 4, and 6 may be eliminated using the L'Hospital's theorem. For example, we find for $k' = 2$ that

$$
q_2(\tau) = q_2(0)e^{-6\tau} + q_1(0)\frac{3}{5}e_0(e^{-4\tau}-e^{-6\tau}) + \frac{2}{5}e_0^2\left(\tau e^{-4\tau}+\frac{1}{2}e^{-6\tau}\right)
$$
$$
+ \frac{2g_0}{5}(e^{-4\tau}-e^{-6\tau}) \tag{6.41}
$$

and

$$
q_2(\tau) = \frac{e_0^2}{5}\left(\frac{1}{3}+\tau e^{-4\tau}+\frac{5}{12}e^{-6\tau}\right) + \frac{2g_0}{5}\left(\frac{1}{3}-e^{-2\tau}+e^{-4\tau}-\frac{1}{3}e^{-6\tau}\right) \tag{6.42}
$$

for fields 1 and 2, respectively.

In the case of external field 1, we find from Eq. (6.39) that

$$
\lim_{k' \to 0} q_2(\tau) = 0 \tag{6.43}
$$

and

$$\lim_{k' \to \infty} \frac{q_2(\tau)}{q_2^0} = 1 + \frac{R}{2(R+1)}(e^{-6\tau} - 3e^{-2\tau}) - \frac{1}{R+1}e^{-6\tau} \quad (6.44)$$

Plots of $q_2(\tau)/q_2^0$ vs. τ for various values of k' are shown in Fig. 36.

With respect to external field 2, we shall consider two cases: $(2a)$ A constant field E_0 has been applied for a sufficiently long time and suddenly at $\tau = 0$ we switch on $E(\tau) = E_0 \exp(-k'\tau)$, in which case $q_1(0) = \frac{1}{3}e_0$ and $q_2(0) = \frac{1}{15}(e_0^2 + 2g_0)$, in Eq. (6.40). $(2b)$ We switch on $E(\tau) = E_0 \exp(-k'\tau)$ suddenly at $\tau = 0$ without applying any external field at $\tau < 0$, which corresponds to $q_1(0) = q_2(0) = 0$ in Eq. (6.42).

It is evident in case $2a$ that

$$\lim_{k' \to 0} \frac{q_2(\tau)}{q_2^0} = 1 \quad (6.45)$$

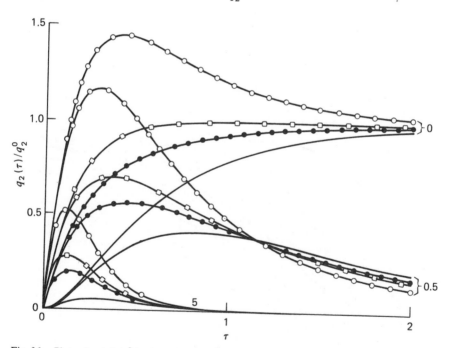

Fig. 36. Plots of $q_2(\tau)/q_2^0$ in the case of field 1 versus τ for various values of R and k'. $R = \infty$ (——); 0.5 (●); −0.5 (○); and 0 (□). Numerals beside the curves represent values of k'.

and

$$\lim_{k' \to \infty} \frac{q_2(\tau)}{q_2^0} = e^{-6\tau} \tag{6.46}$$

while in case $2b$,

$$\lim_{k' \to \infty} \frac{q_2(\tau)}{q_2^0} = 1 + \frac{R}{2(R+1)} \left(e^{-6\tau} - 3e^{-2\tau} \right) - \frac{1}{R+1} e^{-6\tau} \tag{6.47}$$

and

$$\lim_{k' \to 0} \frac{q_2(\tau)}{q_2^0} = 1 \tag{6.48}$$

$q_2(\tau)/q_2^0$ versus τ is plotted in Figs. 37 and 38 for various values of k' and R for cases $2a$ and $2b$, respectively. It should be noted that Eq. (6.46) corresponds to the field-free relaxation, and Eq. (6.47) corresponds to the buildup of the birefringence for the step-up electric field [cf. Eqs. (1.32) and (1.29)].

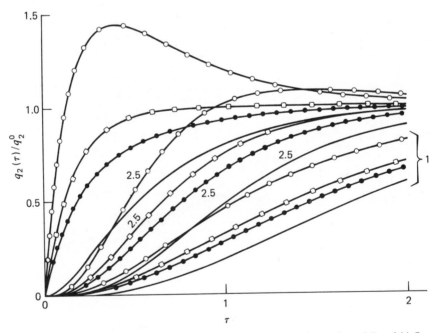

Fig. 37. Plots of $q_2(\tau)/q_2^0$ in the case of field 2a versus τ for various values of R and k'. Symbols and numerals as in Fig. 36.

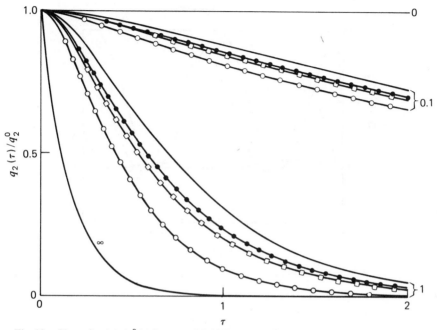

Fig. 38. Plots of $q_2(\tau)/q_2^0$ in the case of field 2b versus τ for various values of R and k'. Symbols and numerals as in Fig. 36.

3. The Sinusoidal Electric Field $E(\tau) = E_0\cos\omega'\tau$

Putting $E(\tau) = E_0\cos\omega'\tau$ in Eqs. (6.13) and (6.14), we obtain, after integration,

$$q_1(\tau) = \frac{2}{3}\frac{e_0}{4+\omega'^2}(2\cos\omega'\tau + \omega'\sin\omega'\tau - 2e^{-2\tau}) \qquad (6.49)$$

and

$$
\begin{aligned}
q_2(\tau) = {}& \frac{4}{5}\frac{e_0^2}{4+\omega'^2} \\
& \times\left\{\frac{1}{6}(1-e^{-6\tau}) - \frac{2}{16+\omega'^2}[4e^{-2\tau}\cos\omega'\tau + \omega'e^{-2\tau}\sin\omega'\tau - 4e^{-6\tau}]\right. \\
& \left. + \frac{1}{36+4\omega'^2}[(6-\omega'^2)\cos 2\omega'\tau + 5\omega'\sin 2\omega'\tau + (\omega'^2-6)e^{-6\tau}]\right\} \\
& + \frac{2g_0}{5}\left\{\frac{1}{6}(1-e^{-6\tau}) + \frac{1}{36+4\omega'^2}[6\cos 2\omega'\tau + 2\omega'\sin 2\omega'\tau - 6e^{-6\tau}]\right\}
\end{aligned}
$$
$$\qquad (6.50)$$

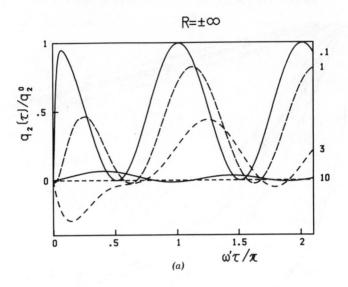

$R = \pm\infty$

(a)

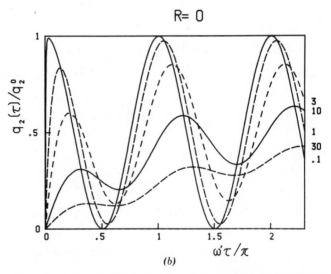

$R = 0$

(b)

Fig. 39. Plots of transient birefringence at the sudden application of the limiting low field $E(\tau) = E_0 \cos \omega' \tau$ obtained from Eq. (6.50) versus $\omega' \tau$ for various values of ω' and R. Numerals at the right are the values of ω'. (a) $R = \pm\infty$. (b) $R = +0$. (c) $R = 0.5$. (d) $R = 2$. (e) $R = -0.5$. (f) $R = -2$.

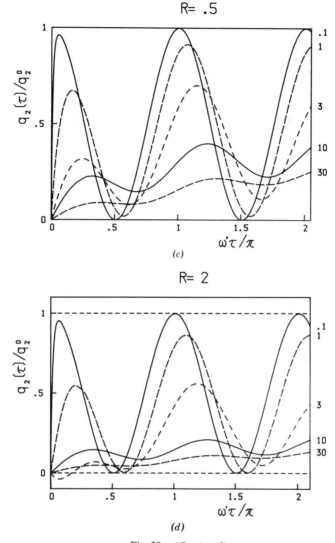

Fig. 39. (*Continued*).

respectively, where $\omega' = \omega/D$ as before. It should be mentioned that Eq. (6.50) is the first complete expression of the transient of the electric birefringence obtained after a sudden application of the sinusoidal electric field $E(\tau) = E_0\cos\omega'\tau$ to a system that has long been free from any orientating field. In Figs. 39a–f we show the transients of the birefringence upon sudden application of the electric field $E(\tau) = E_0\cos\omega'\tau$ for $R = \infty$, $+0$, 0.5, 2,

342   H. WATANABE AND A. MORITA

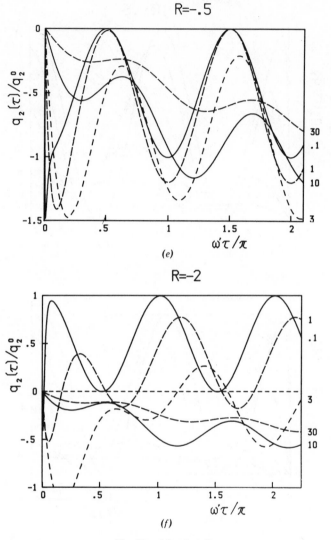

Fig. 39. (*Continued*).

− 0.5, and − 2 and for various values of ω′. On letting τ → ∞ in Eq. (6.50),
we obtain the following expression for the steady state birefringence:

$$q_2(\tau) = \frac{4}{5} \frac{e_0^2}{4+\omega'^2} \left\{ \frac{1}{6} + \frac{1}{36+4\omega'^2} \left[(6-\omega'^2)\cos 2\omega'\tau + 5\omega'\sin 2\omega'\tau \right] \right\}$$
$$+ \frac{2g_0}{5} \left[\frac{1}{6} + \frac{1}{36+4\omega'^2} (6\cos 2\omega'\tau + 2\omega'\sin 2\omega'\tau) \right] \qquad (6.51)$$

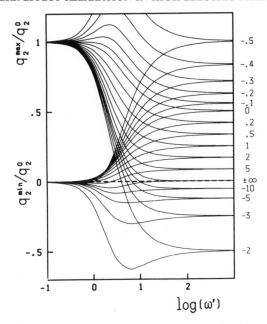

Fig. 40.   Dispersions of the birefringence expressed as the maximum and minimum of the normalized signals versus log(ω') for the limitingly low field. Numerals at the right are the values of R.

This, in fact, fully agrees with the previous results [cf. Eq. (3.12) by Ogawa and Oka[13] and Eqs. (3.15)–(3.21) by Thurston and Bowling[14]]. In Fig. 40, dispersion of the electric birefringence expressed as the top and the bottom of the signals versus log ω' for various values of R is demonstrated. The numerals on the right of the curves are the values of R. It is noted that the dispersion takes place for ω' in the range $0.1 > \omega' > 10$, irrespective of the value of R.

B.   Analytical Expression of the Nonlinear Kerr Effect for the Sinusoidal Electric Field

We shall now obtain an analytical expression for the nonlinear electric birefringence following a sudden application of the sinusoidal electric field $E(\tau) = E_0 \cos \omega' \tau$.[25] To this end, taking terms up to the order of $E^4(\tau)$, we find that Eq. (6.10) gives

$$q_2(\tau) = 2v_2' I_1(\tau) + w_1' w_2' I_2(\tau) - 2y_2' w_1' w_2' I_3(\tau) - w_1'^2 w_2'^2 I_4(\tau)$$
$$- 2w_2' w_3' v_2' I_5(\tau) - w_1' w_2'^2 w_3' I_6(\tau) \qquad (6.52)$$

In Eq. (6.52),

$$I_1(\tau) = \int_0^\tau D_2(\tau, \tau_1) g(\tau_1) \, d\tau_1 \tag{6.53}$$

$$I_2(\tau) = \iint\limits_{\tau \geq \tau_1 \geq \tau_2 \geq 0} D_2(\tau, \tau_1) D_1(\tau_1, \tau_2) e(\tau_1) e(\tau_2) \, d\tau_1 \, d\tau_2 \tag{6.54}$$

$$I_3(\tau) = \iiint\limits_{\tau \geq \tau_1 \geq \tau_2 \geq \tau_3 \geq 0} D_2(\tau, \tau_1) D_1(\tau_1, \tau_2) D_2(\tau_2, \tau_3)$$
$$\times e(\tau_1) e(\tau_2) g(\tau_3) \, d\tau_1 \, d\tau_2 \, d\tau_3 \tag{6.55}$$

$$I_4(\tau) = \iiiint\limits_{\tau \geq \tau_1 \geq \tau_2 \geq \tau_3 \geq \tau_4 \geq 0} D_2(\tau, \tau_1) D_1(\tau_1, \tau_2) D_2(\tau_2, \tau_3) D_1(\tau_3, \tau_4)$$
$$\times e(\tau_1) e(\tau_2) e(\tau_3) e(\tau_4) \, d\tau_1 \, d\tau_2 \, d\tau_3 \, d\tau_4 \tag{6.56}$$

$$I_5(\tau) = \iiint\limits_{\tau \geq \tau_1 \geq \tau_2 \geq \tau_3 \geq 0} D_2(\tau, \tau_1) D_3(\tau_1, \tau_2) D_2(\tau_2, \tau_3)$$
$$\times e(\tau_1) e(\tau_2) g(\tau_3) \, d\tau_1 \, d\tau_2 \, d\tau_3 \tag{6.57}$$

$$I_6(\tau) = \iiiint\limits_{\tau \geq \tau_1 \geq \tau_2 \geq \tau_3 \geq \tau_4 \geq 0} D_2(\tau, \tau_1) D_3(\tau_1, \tau_2) D_2(\tau_2, \tau_3) D_1(\tau_3, \tau_4) e(\tau_1)$$
$$\times e(\tau_2) e(\tau_3) e(\tau_4) \, d\tau_1 \, d\tau_2 \, d\tau_3 \, d\tau_4 \tag{6.58}$$

where

$$D_n(\tau, \tau') = \exp\left[-\int_{\tau'}^\tau d_n(\tau'') \, d\tau'' \right] \tag{6.59}$$

The direct calculation of the integrals in Eq. (6.52) is tedious and lengthy. However, these difficulties may be reduced by taking the Laplace transform of both sides in Eq. (6.52) and using some fundamental theorems such as (1) the convolution theorem stating that

$$\mathscr{L}\left[\iint\limits_{\tau \geq \tau_1 \geq \cdots \geq \tau_n \geq 0} \cdots \int f_1(\tau - \tau_1) f_2(\tau_1 - \tau_2) \cdots f_n(\tau_n) \, d\tau_1 \, d\tau_2 \cdots d\tau_n \right]$$
$$= F_1(s) F_2(s) \cdots F_n(s) \tag{6.60}$$

where the Laplace transform of $f(t)$ has been expressed as

$$\mathscr{L}[f(\tau)] = \int_0^\infty f(\tau) e^{-s\tau} \, d\tau \tag{6.61}$$

and

$$\mathcal{L}[f_i(\tau)] = F_i(s) \qquad (i=1,2,3,\dots,n) \tag{6.62}$$

and (2)

$$\mathcal{L}[f_i(\tau)e^{-a\tau}] = F(s+a) \tag{6.63}$$

Moreover, it is easy to find the terms contributing to the steady state solution, which we have calculated. The final results are

$$
\begin{aligned}
I_1^\infty(\tau) &= \frac{g_0}{4}\left(\frac{1}{3} + \frac{3\cos 2\omega'\tau + \omega'\sin 2\omega'\tau}{9+\omega'^2}\right) \\
&+ \frac{2}{7}g_0^2\left\{\frac{1}{8}\left(\frac{1}{9} + \frac{(9-\omega'^2)\cos 2\omega'\tau + 6\omega'\sin 2\omega'\tau}{(9+\omega'^2)^2}\right)\right. \\
&+ \frac{\sin 2\omega'\tau}{2\omega'}\left(\frac{1}{4}\right)\left(\frac{1}{3} + \frac{3\cos 2\omega'\tau + \omega'\sin 2\omega'\tau}{9+\omega'^2}\right) \\
&\left.- \frac{1}{2\omega'}\frac{1}{4}\left(\frac{-\omega'\cos 2\omega'\tau + 3\sin 2\omega'\tau}{9+\omega'^2} + \frac{3\sin 4\omega'\tau - 2\omega'\cos 4\omega'\tau}{2(9+4\omega'^2)}\right)\right\}
\end{aligned}
\tag{6.64}
$$

$$
\begin{aligned}
I_2^\infty(\tau) &= \frac{e_0^2}{4}\left(\frac{1}{6(2-i\omega')} + \frac{e^{2i\omega'\tau}}{2(3+i\omega')(2+i\omega')}\right) \\
&+ \frac{1}{10}e_0^2 g_0\left(\frac{1}{6(2-i\omega')^2} + \frac{e^{2i\omega'\tau}}{2(3+i\omega')(2+i\omega')^2}\right) \\
&+ \frac{e_0^2 g_0}{40i\omega'}\left(\frac{1}{3(2-i\omega')} + \frac{e^{2i\omega'\tau}}{2(3+i\omega')(2-i\omega')}\right. \\
&+ \frac{e^{4i\omega'\tau}}{2(3+2i\omega')(2+3i\omega')} + \frac{e^{-2i\omega'\tau}}{2(3-i\omega')(2-3i\omega')} \\
&\left.- \frac{e^{4i\omega'\tau}}{2(3+2i\omega')(2+3i\omega')}\right) \\
&+ \frac{e_0^2}{14}g_0\left(\frac{1}{36(2-i\omega')} + \frac{e^{2i\omega'\tau}}{4(3+i\omega')^2(2+i\omega')}\right)
\end{aligned}
$$

$$+ \frac{e_0^2 g_0}{56 i \omega'} \left(\frac{e^{4 i \omega' \tau}}{2(3 + i \omega')(2 + i \omega')} + \frac{1}{2(3 - i \omega')(2 - i \omega')} \right.$$

$$+ \frac{e^{2 i \omega' \tau}}{6(2 - i \omega')} + \frac{e^{2 i \omega' \tau}}{6(2 + i \omega')}$$

$$- \frac{e^{4 i \omega' \tau}}{2(3 + 2 i \omega')(2 + i \omega')} - \frac{e^{2 i \omega' \tau}}{2(3 + i \omega')(2 - i \omega')}$$

$$\left. - \frac{e^{2 i \omega' \tau}}{2(3 + i \omega')(2 + i \omega')} - \frac{1}{6(2 - i \omega')} \right) + \text{C.C.} \qquad (6.65)$$

$$I_3^\infty(\tau) = \frac{e_0^2 g_0}{8} \left(\frac{1}{36} \frac{1}{2 + i \omega'} + \frac{e^{-2 i \omega' \tau}}{12(3 - i \omega')(2 - i \omega')} \right)$$

$$+ \frac{e_0^2 g_0}{16} \left(\frac{e^{-4 i \omega' \tau}}{4(3 - 2 i \omega')(2 - 3 i \omega')(3 - i \omega')} \right.$$

$$+ \frac{e^{-2 i \omega' \tau}}{4(3 - i \omega')^2(2 - i \omega')} + \frac{e^{2 i \omega' \tau}}{4(3 + i \omega')^2(2 + 3 i \omega')}$$

$$\left. + \frac{1}{12(2 + i \omega')(3 + i \omega')} \right) + \text{C.C.} \qquad (6.66)$$

$$I_4^\infty(\tau) = \frac{e_0^4}{16} \left(\frac{e^{-4 i \omega' \tau}}{4(3 - 2 i \omega')(2 - 3 i \omega')(3 - i \omega')(2 - i \omega')} \right.$$

$$+ \frac{e^{-2 i \omega' \tau}}{4(3 - i \omega')^2(2 - i \omega')^2} + \frac{e^{2 i \omega' \tau}}{4(3 + i \omega')^2(2 + 3 i \omega')(2 + i \omega')}$$

$$+ \frac{1}{12(2 + i \omega')^2(3 + i \omega')} + \frac{e^{-2 i \omega' \tau}}{12(3 - i \omega')(2 - i \omega')^2}$$

$$+ \frac{1}{36(4 + \omega'^2)} + \frac{1}{36(2 + i \omega')^2}$$

$$\left. + \frac{e^{-2 i \omega' \tau}}{12(4 + \omega'^2)(3 - i \omega')} \right) + \text{C.C.} \qquad (6.67)$$

$$I_5^\infty(\tau) = \frac{e_0^2 g_0}{8}\left(\frac{1}{36}\frac{1}{12+i\omega'} + \frac{e^{-2i\omega'\tau}}{12(3-i\omega')(12-i\omega')}\right)$$

$$+\frac{e_0^2 g_0}{16}\left(\frac{e^{-4i\omega'\tau}}{12(3-2i\omega')(4-i\omega')(3-i\omega')}\right.$$

$$+\frac{e^{-2i\omega'\tau}}{4(3-i\omega')^2(12-i\omega')} + \frac{e^{2i\omega'\tau}}{12(3+i\omega')(4+i\omega')}$$

$$\left.+\frac{1}{12(12+i\omega')(3+i\omega')}\right)+\text{C.C.} \qquad (6.68)$$

$$I_6^\infty(\tau) = \frac{e_0^4}{16}\left(\frac{e^{-4i\omega'\tau}}{12(3-2i\omega')(4-i\omega')(3-i\omega')(2-i\omega')}\right.$$

$$+\frac{e^{-2i\omega'\tau}}{4(3-i\omega')^2(12-i\omega')(2-i\omega')}$$

$$+\frac{e^{2i\omega'\tau}}{12(3+i\omega')^2(4+i\omega')(2+i\omega')}$$

$$+\frac{1}{12(12+i\omega')(3+i\omega')(2+i\omega')}$$

$$+\frac{e^{-2i\omega'\tau}}{2(3-i\omega')(12-i\omega')(2-i\omega')}$$

$$+\frac{1}{36(12+i\omega')(2-i\omega')}$$

$$+\frac{1}{36(12+i\omega')(2+i\omega')}$$

$$\left.+\frac{e^{-2i\omega'\tau}}{12(3-i\omega')(12-i\omega')(2+i\omega')}\right)+\text{C.C.} \qquad (6.69)$$

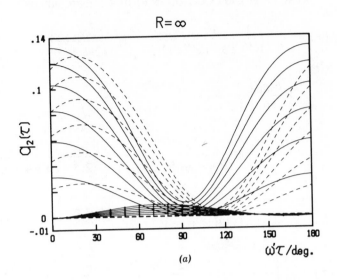

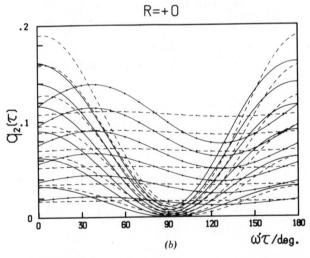

Fig. 41. Field strength dependence of the steady state birefringence obtained from Eq. (6.52) with Eqs. (6.64)–(6.69) versus $\omega'\tau$ for various values of ω' and e^2. From the bottom to top, $e^2 = 0.5$, 1, 1.5, 2, 2.5, and 3. $\omega' = 0.1$ (——); 1 (---); 10 (-●-●-); 100 (-●- -●-); (Curves for $\omega' = 100$ are not shown in (a), because those are essentially equal to zero.) (a) $R = \infty$. (b) $R = +0$. (c) $R = -0$, and the order of the curves is from top to bottom. (d) $R = 0.5$. (e) $R = 2$. (f) $R = -0.5$. (g) $R = -2$, and the order is bottom to top in the upper half and top to bottom in the lower half.

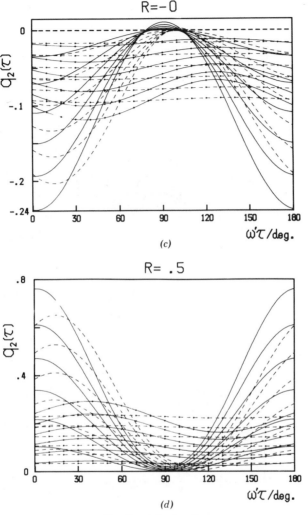

Fig. 41. (*Continued*).

where C.C. represents the complex conjugate terms obtained by replacing $i\omega'$ in the preceding terms by $-i\omega'$. Figures 41a–g shows how the normalized steady-state birefringence $q_2(\tau)/q_2(\infty)$* changes with increasing field strength and angular frequency for $R = \infty$, $+0$, -0, 0.5, 2, -0.5, and -2. The values of R are shown at the top of each part of the figure.

The Laplace transforms of $I_i(t)$ ($i = 1,2,3,4,5,6$) including information on the transient response are collected in Appendix B.

---

*$q_2(\tau)$ is the stationary state birefringence for the static electric field E_0, and is referred as "static" birefringence and its value as "static" value.

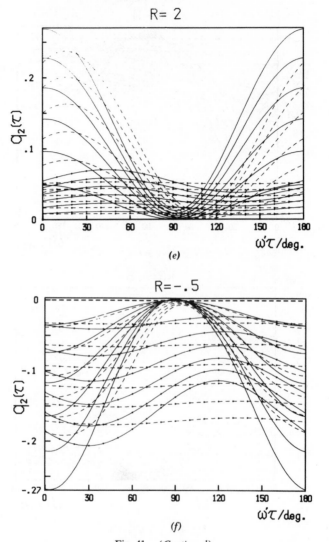

Fig. 41. (*Continued*).

For a sufficiently low angular frequency, letting $\omega' \to 0$, Eq. (6.51) collapses to

$$\frac{q_2(\tau)}{q_2^0} = \frac{1 + \cos 2\omega'\tau}{2} \tag{6.70}$$

Thus the fundamental frequency of the birefringence signal is $2\omega'$. The ana-

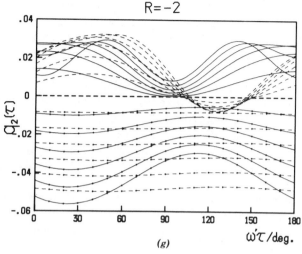

Fig. 41. (*Continued*).

lytical equations (6.52) with (6.64)–(6.69) indicate, however, that the increase of the field strength is responsible for the lowest overtone terms of $\cos 4\omega'\tau$, the amplitude of which is a measure of the nonlinearity of the signals.

C. Numerical Calculations of the Birefringence in High Electric Fields

In order to obtain the values of $q_2(\tau)$ for a large value of $E(\tau)$, the set of the first-order differential equations such as Eq. (5.2) can be solved numerically.[25] Since the previous method of calculating $q_2(\tau)$ for the step-up field, in which we diagonalized the matrix **B** by making use of the eigenvalues and eigenmatrix of **B**, is no longer applicable in this case of the time-varying electric field, the Runge-Kutta-Gill method may be used in the calculation. The results are demonstrated as the normalized birefringence $q_2(\tau)/q_2(\infty)$ in Figs. 42a–f for $R = \infty$, $+0$, 0.5, 2, -0.5, and -2, respectively. These figures show how the transient and steady state birefringence changes with the increase in e_0^2 (from the bottom to the top in each part of the figure) and ω' (from left to right).

Since the electric field $E(\tau) = E_0 \cos \omega'\tau$ is suddenly applied at $\tau = 0$ to a system that has been free from any external field, the value of $q_2(0)$ is initially zero. For the case of $R = \infty$, the signals become very small at high values of ω'; plots are multiplied by a factor of 5 for $\omega' = 10$, 25 for $\omega' = 30$, and 500 for $\omega' = 100$. From the figures we see that the signals of birefringence are positive when $R = \infty$ and 2, whereas they are mainly negative when $R = -0.5$. Note that the static value $q_2(\infty)$ is negative when $R = -0.5$. A complication arises when $R = -2$, for the sign of the static birefringence reverses from positive to negative at $e_0^2 = 10.68$ (see Fig. 1 and Table I). The normalized signals for $e_0^2 = 10$, where the static value is very small, are very large, and the values (multiplied by $1/20$) are presented for the special cases of $e_0^2 = 10$. In the particular case of $R = -2$, the normalized birefringence changes from negative to positive at high values of ω'. The same was seen in the static field (cf. Fig. 21). We note from an overall view of the figures that the contribution of the induced dipole moment to the birefringence is more dominant and the establishment of the steady state is slower, the larger the value of angular frequency. It should, however, be kept in mind that the real time is proportional to τ, so the real time to attain the steady state does not depend much on the angular frequency.

The dependencies of the top and bottom of $q_2(\tau)$ in the steady-state region on the angular frequency ω' are presented in Figs. 43$a-c$ for $R = \infty$ ($g = 0$), 2, and -2, respectively. The values of e_0^2 are presented on the curves.

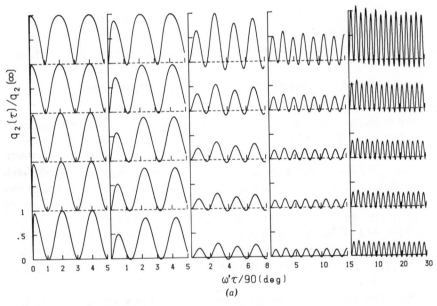

Fig. 42. Normalized transient and steady state birefringence for the sudden application of the strong field $E(\tau) = E_0 \cos \omega'\tau$ versus $\omega'\tau$. From bottom to top: $e_0^2 = 0.01, 1, 3, 10,$ and 30. From left to right: $\omega' = 0.1, 1, 10, 30,$ and 100. (a) $R = \infty$ ($g = 0$). The multiplication factors are: 5 for $\omega' = 10$, 25 for $\omega' = 30$, and 500 for $\omega' = 100$. (b) $R = +0$. The same as (a), but without a multiplication factor. (c) $R = 0.5$. From bottom to top: $e_0^2 = 0.01, 1, 5, 10,$ and 15. Without factor multiplication. (d) $R = 2$. Without factor multiplication. (e) $R = -0.5$. Without factor multiplication. (f) $R = -2$. The multiplication factor is $1/20$ for $e_0^2 = 10$.

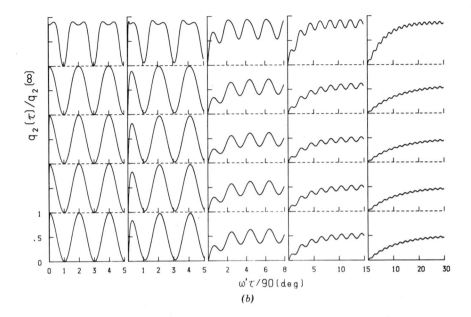

(b)

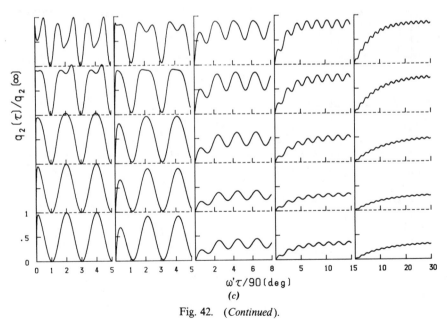

$\omega' \tau / 90$ (deg)

(c)

Fig. 42.   (Continued).

353

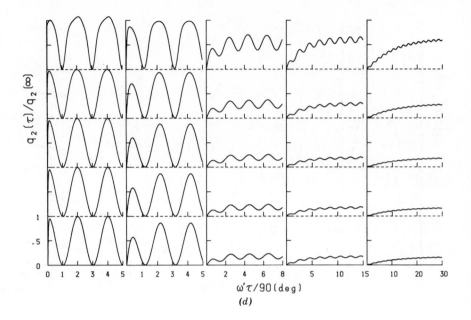

$\omega'\tau/90\,(\text{deg})$

(d)

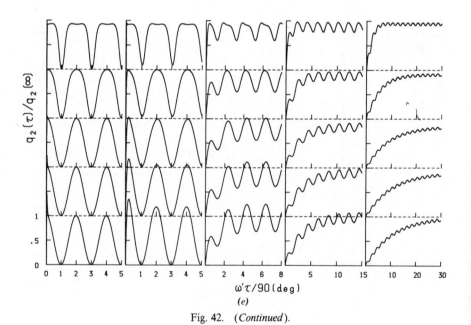

$\omega'\tau/90\,(\text{deg})$

(e)

Fig. 42. *(Continued)*.

354

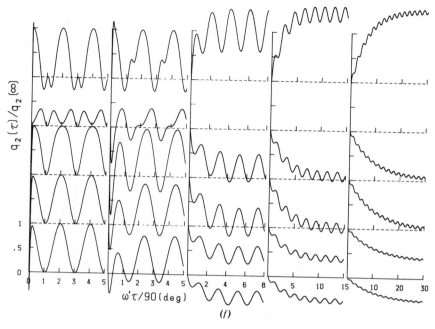

Fig. 42. (*Continued*).

The dependence of the dispersion of the birefringence for pure permanent dipole orientation on the value of e_0 is also demonstrated as normalized dispersion curves in Fig. 44. It is remarkable that the dispersion shifts to the higher value of the angular frequency ω' with increasing values of e_0. This may be understood to be due to the increase of orienting force causing faster reorientation of the particle under the sinusoidal field. The dependence of $(\log \omega')^{1/2}$ at which the normalized birefringence is 0.5 on the value of e_0 is plotted in Fig. 45.

In order to estimate the nonlinearity in birefringence signals for high field strength, contributions of the nonlinear terms expressed as Fourier components

$$a_n = \frac{1}{\pi} \int_0^\pi f(x) \cos nx \, dx \qquad (6.71)$$

where $x = 2\omega'\tau$ and $f(x)$ is the electrooptical signal in the steady state region obtained numerically as demonstrated in Figs. 42 are plotted against e_0^2 in Figs. 46a–c for $R = \infty$ ($g = 0$), 2, and -2. The contributions of the

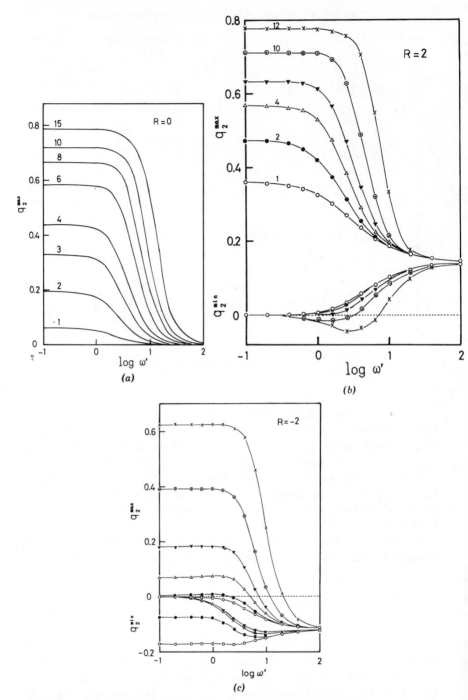

Fig. 43. Field strength dependence of the dispersion of the birefringence expressed as the maximum and minimum of $q_2(\tau)$ in the steady state regions versus $\log \omega'$. (a) $R = 0$ ($e = 0$). (b) $R = 2$. (c) $R = -2$.

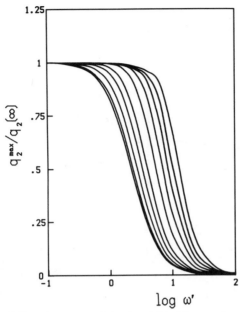

Fig. 44. Normalized dispresion curves of the electric birefringence for pure permanent dipole orientation. From the left to right: $e_0 = 0.001, 1, 2, 3, 4, 6, 8, 10, 12,$ and 15.

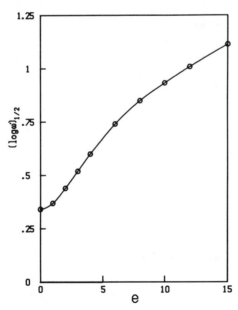

Fig. 45. The dependence of $(\log \omega')_{1/2}$ on the value of e_0 for pure permanent dipole orientation.

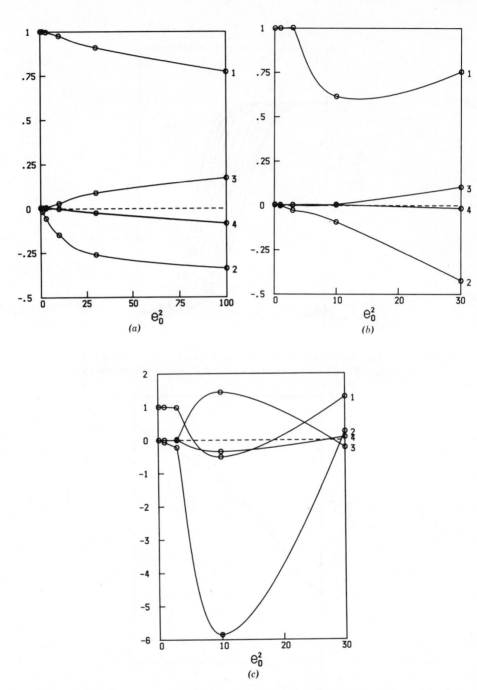

Fig. 46.   Dependence of the nonlinear terms on the value of e_0^2. Numerals at the right are the values of n in Eq. (6.71). (a) $R = \infty$ ($g = 0$). (b) $R = 2$. (c) $R = -2$.

harmonics stay reasonably low as long as the value of e_0^2 is less than 3 but increase noticeably for larger values of e_0^2. Numerals on the right-hand side of each curve are the values of n in Eq. (6.71).

VII.   KERR-EFFECT RELAXATION PROCESSES CAUSED BY A RAPIDLY ROTATING FIELD

The calculations of the electric birefringence in Section V were confined to the cases where (1) $E_0 = 0$ for $t < 0$ and there is a sudden application of E at $t = 0$ (transient rise birefringence) and (2) there is a sudden application of $E = -E_0$ at $t = 0$ after impressing E_0 for a sufficiently long time (the rapidly reversing field). In this section, we shall calculate the electric birefringence for a field where E_0 has been applied for quite a long time and suddenly, at $t = 0$, E is switched on at an angle of 90° with E_0.[26] To this end, we suppose that a rigid symmetrical particle with a permanent dipole moment μ along the symmetrical axis is undergoing rotational Brownian motion and a laboratory coordinate system to describe the motion of the particle is represented by $(Oxyz)$, which is also expressed by the polar coordinate system (r, θ, ϕ). Because the particle is rigid, r, which lies along the symmetrical axis, does not change with time. Then if the electric fields E_0 and E are applied along the directions of the x and z axes, respectively, and a light beam travels along the y axis, as is shown in Fig. 47, the potential energy $V(\theta, \phi)$ arising from the interaction of the permanent dipole and the fields is given by

$$V(\theta, \phi) = -\mu E_0 \sin\theta \cos\phi - \mu E \cos\theta \qquad (7.1)$$

In this case, we suppose that the distribution function $f(\theta, \phi, t)$ satisfies the following Smoluchowski equation obtained by ignoring inertial effects (see

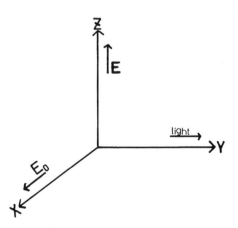

Fig. 47. A configuration of the birefringence measurement in the rotating field.

Section X):

$$\frac{\partial f}{\partial \tau} = \frac{1}{\sin\theta}\left[\frac{\partial}{\partial\theta}\sin\theta\left(\frac{\partial f}{\partial\theta} + \frac{1}{k_B T}\frac{\partial V}{\partial\theta}f\right)\right] + \frac{1}{\sin^2\theta}\frac{\partial}{\partial\phi}\left(\frac{\partial f}{\partial\phi} + \frac{1}{k_B T}\frac{\partial V}{\partial\phi}f\right)$$

(7.2)

The electric field under which the transient electric birefringence shall presently be considered, may be called a rotating field. The potential energy of the particle for the rotating field is then given by

$$V(\theta,\phi) = \begin{cases} -\mu E_0 \sin\theta\cos\phi & (t<0) \\ -\mu E\cos\theta & (t>0) \end{cases}$$

(7.3a)
(7.3b)

The orientation factor, which is responsible for the birefringence, is given by (see Appendix A)

$$\Phi(\tau) = \langle\cos^2\theta\rangle - \langle\sin^2\theta\cos^2\phi\rangle$$
$$= \langle P_2(\cos\theta)\rangle - \tfrac{1}{6}\langle P_2^2(\cos\theta)\cos 2\phi\rangle$$

(7.4)

where $P_n^m(\cos\theta)$ is the associated Legendre polynomial of degree n and order m. Note that the second term on the right of Eq. (7.4) arises from the uneven distribution of the dipole with respect to the angle ϕ due to the application of E_0 along the x axis, which drives the dipole to orient itself along the direction of the x axis. However, this term vanishes for the two conventional experiments (1) and (2), leading to the well-known result [cf. Eq. (1.4)]

$$\Phi(\tau) = \langle P_2(\cos\theta)\rangle$$

(7.5)

For $\tau < 0$, the behavior of the motion of the dipole is described by the Maxwell-Boltzmann distribution function

$$f_0(\theta,\phi) = \frac{\exp(e_0\sin\theta\cos\phi)}{\displaystyle\int_0^\pi \sin\theta\,d\theta\int_0^{2\pi}d\phi\exp(e_0\sin\theta\cos\phi)}$$

(7.6)

where $e_0 = E_0/k_B T$. For $\tau > 0$, Eq. (7.2) with the potential in Eq. (7.3b) governs the random motion of the dipolar particles. Expanding $f(\theta,\phi,\tau)$ as

$$f(\theta,\phi,\tau) = \sum_{n=0}^{\infty}\left[\sum_{m=0}^{n} A_{n,m}(\tau)P_n^m(\cos\theta)\cos m\phi\right.$$
$$\left. + \sum_{m=1}^{n} B_{n,m}(\tau)P_n^m(\cos\theta)\sin m\phi\right]$$

(7.7)

and inserting the result into Eq. (7.2), we find the following recurrence relations for $A_{n,m}(\tau)$:

$$\frac{dA_{n,m}}{d\tau} = -n(n+1)A_{n,m}(\tau)$$

$$+ e\left[\frac{(n+1)(n-m)}{2n-1}A_{n-1,m} - \frac{n(n+m+1)}{2n+3}A_{n+1,m}\right] \quad (7.8)$$

It is immediately seen that on putting $m = 0$ in Eq. (7.8), we obtain the recurrence relations in Eq. (4.1b) for $g = 0$. In view of the relation obtained from Eq. (7.7),

$$C_{n,m}(\tau) = \frac{1}{2(2n+1)}\frac{(n+m)!}{(n-m)!}\frac{A_{n,m}(\tau)}{A_{0,0}}$$

$$= \langle P_n^m(\cos\theta)\cos m\phi\rangle \quad (7.9)$$

we find from Eq. (7.8) that

$$\frac{dC_{n,m}(\tau)}{d\tau} = -n(n+1)C_{n,m}(\tau)$$

$$+ \frac{e}{2n+1}\left[(n+1)(n+m)C_{n-1,m}(\tau)\right.$$

$$\left. - n(n-m+1)C_{n+1,m}(\tau)\right] \quad (7.10)$$

It is obvious from Eqs. (7.4) and (7.9) that

$$\Phi(\tau) = C_{2,0}(\tau) - \tfrac{1}{6}C_{2,2}(\tau) \quad (7.11)$$

Therefore, we need to calculate $C_{2,0}(\tau)$ and $C_{2,2}(\tau)$ for the birefringence. These may be calculated by solving first-order, constant-coefficient, simultaneous differential equations arising from Eqs. (7.10) numerically with reference to the method adopted in Section V. To this end, we should determine the initial conditions for $C_{n,m}(\tau)$ at $\tau = 0$, $C_{n,m}(0)$, which is obtained from Eq. (7.6), leading to

$$C_{n,m}(0) = \frac{\int_0^\pi \sin\theta\, d\theta \int_0^{2\pi} d\phi\, P_n^m(\cos\theta)\cos m\phi f_0(\theta,\phi)}{\int_0^\pi \sin\theta\, d\theta \int_0^{2\pi} d\phi f_0(\theta,\phi)} \quad (7.12)$$

For the calculation of $C_{n,m}(0)$, it is useful to note the relation

$$\exp[\pm ir(\cos\theta\cos\theta' + \sin\theta\sin\theta'\cos\phi)]$$

$$= \sum_{n=0}^{\infty} (\pm i)^n (2n+1) j_n(r)$$

$$\times \left[P_n(\cos\theta) P_n(\cos\theta') \right.$$

$$\left. + 2 \sum_{m=1}^{n} \frac{(n-m)!}{(n+m)!} P_n^m(\cos\theta) P_n^m(\cos\theta')\cos m\phi \right] \quad (7.13)$$

where $j_n(r)$ is the spherical Bessel function of the first kind. This leads to

$$\frac{\int_0^\pi \sin\theta\, d\theta \int_0^{2\pi} \exp[-ir(\cos\theta\cos\theta' + \sin\theta\sin\theta'\cos\phi)] P_n^m(\cos\theta)\cos m\phi\, d\phi}{\int_0^\pi \sin\theta\, d\theta \int_0^{2\pi} \exp[-ir(\cos\theta\cos\theta' + \sin\theta\sin\theta'\cos\phi)]\, d\phi}$$

$$= (-i)^n P_n^m(\cos\theta') \frac{j_n(r)}{j_0(r)} \quad (7.14)$$

On putting $\theta' = \pi/2$ and $r = ie_0$ in Eq. (7.14) and using Eqs. (7.6) and (7.12), we find that

$$C_{n,m}(0) = (-i)^n P_n^m(0) \frac{j_n(ie_0)}{j_0(ie_0)} \quad (7.15)$$

where

$$P_{n+m}^m(0) = \begin{cases} 0 & (n\text{ odd}) & (7.16a) \\ (-1)^{n/2} \dfrac{(n+2m-1)!!}{n!!} & (n\text{ even}) & (7.16b) \end{cases}$$

As $\tau \to \infty$, it is expected that since the distribution function f reaches the equilibrium state described by the Maxwell-Boltzmann distribution function, we require

$$C_{n,m}(\infty) = \lim_{\tau\to\infty} C_{n,m}(\tau)$$

$$= \frac{\int_0^\pi \sin\theta\, d\theta \int_0^{2\pi} \exp(e\cos\theta) P_n^m(\cos\theta)\cos m\phi\, d\phi}{\int_0^\pi \sin\theta\, d\theta \int_0^{2\pi} \exp(e\cos\theta)\, d\phi} \quad (7.17)$$

which, in view of Eq. (7.14) after putting $\theta' = 0$ and $r = ie$, becomes

$$C_{n,m}(\infty) = (-i)^n P_n^m(1) \frac{j_n(ie)}{j_0(ie)} \tag{7.18}$$

In this equation, we note that

$$P_n^m(1) = \delta_{m,0} \tag{7.19}$$

where $\delta_{m,n}$ is the Kronecker delta. Hence, it follows from Eqs. (7.15), (7.18), and (7.19) that if $e_0 = e$, then

$$C_{n,m}(0) = P_n^m(0) C_{n,0}(\infty) \tag{7.20}$$

Thus, our problem is now to calculate

$$C_{n,0}(\infty) = \frac{\int_0^\pi \exp(e\cos\theta) P_n(\cos\theta)\sin\theta\, d\theta}{\int_0^\pi \exp(e\cos\theta)\sin\theta\, d\theta} = (-i)^n \frac{j_n(ie)}{j_0(ie)} \tag{7.21}$$

In view of the relation

$$j_n(ix) = \sqrt{\frac{\pi}{2x}}\, I_{n+1/2}(x) \tag{7.22}$$

we obtain

$$C_{n,0}(\infty) = \frac{I_{n+1/2}(e)}{I_{1/2}(e)} \tag{7.23}$$

where $I_m(z)$ is the modified Bessel function. So, it is appropriate to investigate some useful properties of $I_{n+1/2}(x)$ for the calculation of $C_{n,0}(\infty)$. We note that $I_{n+1/2}(x)$ in fact is given by a closed form. For example,

$$I_{1/2}(x) = \sqrt{\frac{2}{\pi x}}\, \sinh x \tag{7.24a}$$

$$I_{3/2}(x) = \sqrt{\frac{2}{\pi x}}\left(\cosh x - \frac{1}{x}\sinh x\right) \tag{7.24b}$$

and $I_{n+1/2}(x)$ for a large value of n may be calculated by the recurrence relation

$$\frac{2n+1}{x} I_{n+1/2}(x) = I_{n-1/2}(x) - I_{n+3/2}(x) \tag{7.25}$$

By putting

$$\frac{dC_{n,0}(\tau)}{d\tau} = 0 \tag{7.26}$$

in Eq. (7.10), we obtain the relation for $C_{n,0}(\infty)$ in the equilibrium state

$$C_{n,0}(\infty) = \frac{e}{2n+1} \left[C_{n-1,0}(\infty) - C_{n+1,0}(\infty) \right] \tag{7.27}$$

It is seen from Eqs. (7.22)–(7.25) that Eq. (7.27) is also satisfied by $C_{n,0}(\infty)$ in Eq. (7.23). However, it is not convenient for the numerical computation of $C_{n,0}(\infty)$ to use Eq. (7.27) directly, since the difference

$$C_{n-1,0}(\infty) - \frac{2n+1}{e} C_{n,0}(\infty)$$

sometimes leads to unreliable values, canceling significant digits. To avoid this difficulty, we express $C_{n,0}(\infty)/C_{n-1,0}(\infty)$ by the continued fraction

$$G_{n-1} = \frac{C_{n,0}(\infty)}{C_{n-1,0}(\infty)} = \frac{e}{2n+1+eG_n} \tag{7.28a}$$

$$= \cfrac{e}{2n+1+\cfrac{e^2}{2n+3+\cfrac{e^2}{2n+5+\cdots}}} \tag{7.28b}$$

Once we calculate G_{n-1} by the infinite continued fraction in Eq. (7.28b), $G_{n-2}, G_{n-3}, \ldots, G_0$ can be determined successively from the recurrence relation in Eq. (7.28a). Then by using the relation

$$C_{k,0}(\infty) = G_{k-1} G_{k-2} \cdots G_0 \qquad (1 \le k \le n) \tag{7.29}$$

we find $C_{n,0}(\infty)$.

For $e \ll 1$, we find from Eq. (7.10) that

$$\frac{\Phi(\tau)}{\Phi(\infty)} = 1 - \frac{3}{2}\exp(-2\tau) - \frac{1}{2}\left(\frac{e_0}{e}\right)^2\exp(-6\tau) \qquad (7.30)$$

In particular, for $e_0 = e$, we obtain

$$\frac{\Phi(\tau)}{\Phi(\infty)} = 1 - \tfrac{3}{2}\exp(-2\tau) - \tfrac{1}{2}\exp(-6\tau) \qquad (7.31)$$

This may be compared with $\Phi(\tau)/\Phi(\infty)$ for experimental methods 1 and 2:

$$\frac{\Phi(\tau)}{\Phi(\infty)} = 1 - \tfrac{3}{2}\exp(-2\tau) + \tfrac{1}{2}\exp(-6\tau) \qquad (7.32)$$

[cf. Eq. (2.13)], and

$$\frac{\Phi(\tau)}{\Phi(\infty)} = 1 + 3[\exp(-6\tau) - \exp(-2\tau)] \qquad (7.33)$$

[cf. Eq. (2.18)].

In Fig. 48, the contributions from $C_{2,0}(\tau)$ and $C_{2,2}(\tau)$ to $\Phi(\tau)$ for $e = 1$, 5, and 10 in the case of $e_0 = e$ are shown. And in Fig. 49, $\Phi(\tau)/\Phi(\infty)$ is plotted versus τ for various values of e. It is seen that our present method gives a greater difference between $\Phi(\infty)$ and the minimum value of $\Phi(\tau)$ than

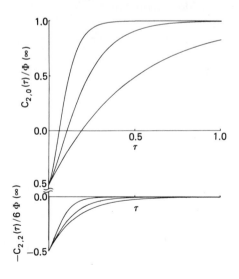

Fig. 48. Plots of $C_{2,0}(\tau)/\Phi(\infty)$ and $-C_{2,2}(\tau)/6\Phi(\infty)$ versus τ for the rotating field method. $e = 1$, 5, and 10 from bottom to top.

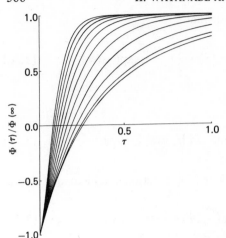

Fig. 49. Plots of $\Phi(\tau)/\Phi(\infty)$ versus τ for the rotating field method. $e = 0.01$, 1, 2, 3, 4, 5, 6, 7, 8, 9, and 10 from bottom to top.

the two conventional methods 1 and 2 (cf. Figs. 16a and b). This certainly is an advantage in using the present method as a technique for transient birefringence measurements.

Moreover, we have extended the calculation of the transient electric birefringence to the case where the angular distribution of the dipoles is not even with respect to the angle ϕ. If we apply a continuously rotating field such as a couple of continuously reversing fields of the same duration, for both of which the duty ratio is unity, at right angles with the phase difference $\pi/2$, as illustrated in Fig. 50, we would observe an oscillating normalized birefringence between -1 and 1. The dispersion of the birefringence due to the increase of frequency of rotation may provide us more information about the relaxation time of the rotational diffusion. The result in Eq. (7.23) has been previously given,[27] and it is quite useful to express $C_{n,0}(\infty)$ by the modified Bessel function $I_{n+1/2}(x)$. However, if we take into account the contribution from the induced dipole moment to the potential in Eq. (7.1), the prob-

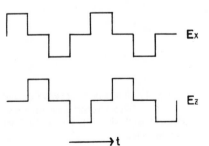

Fig. 50. An example of the continuously rotating field.

lem seems to become quite difficult but may be treated numerically by starting from the Smoluchowski equation, Eq. (7.2).

VIII. THEORY OF TRANSLATIONAL BROWNIAN MOTION

As we have seen, the transient electric birefringence is calculated to obtain the ensemble average $\langle P_2(\cos\theta)\rangle - \langle P_2^2(\cos\theta)\cos 2\phi\rangle/6$ for the uneven distribution about ϕ, or $\langle P_2(\cos\theta)\rangle$ for the even distribution about ϕ [cf. Eqs. (1.4) and (7.4)], with respect to the angular distribution function. The angular distribution function was obtained by solving the Smoluchowski equations, Eq. (1.6) or Eq. (7.2), derived on the assumption that the inertial effects are insignificant.[28] The theory of rotational Brownian motion is based on that of translational motion. It is worthwhile reviewing the outline of the latter theory.

We consider the translational Brownian motion of a particle in a fluid by using the Langevin equation,

$$m\frac{dv(t)}{dt} + m\beta v(t) = mA(t) + F(x,t) \qquad (8.1)$$

where

$$\frac{dx(t)}{dt} = v(t) \qquad (8.2)$$

is the velocity of the particle, $x(t)$ is the position of the particle at time t, m is the mass of the particle, $m\beta$ is the friction constant, $mA(t)$ is the random force arising from the collision between the particle and the surrounding molecules, and $F(x,t)$ is the mechanical force. We confine ourselves in this section to one-dimensional Brownian motion for the sake of convenience, because generalization to the three-dimensional case is straightforward. Equations such as Eqs. (8.1) and (8.2) are called stochastic differential equations, since the random acceleration $A(t)$, whose explicit time dependence is not known except for some statistical properties, is included and this $A(t)$ affects not only $v(t)$ but also $x(t)$. It is evident that when effects due to the collisions are ignored, i.e., $\beta = 0$, $A(t) = 0$, Eq. (8.1) recovers Newton's equation of motion. Thus, the theory of Brownian motion is regarded as a modification of Newton's dynamics in that the former takes into account effects due to collisions. It is usually assumed for $A(t)$ that

1. $A(t)$ is a centered Gaussian random variable.
2. $A(t)$ correlates only with itself, having the correlation function given by

$$\langle A(t_1)A(t_2)\rangle = \sigma^2\delta(t_1 - t_2) \qquad (8.3)$$

where angular brackets represent the ensemble average, σ^2 is a constant to be determined later, and $\delta(t)$ is the Dirac delta function.

Some important properties concerning statistical aspects of a random variable are found in Appendix C.

In the case of $F(x, t) = 0$, we obtain the formal solution of Eqs. (8.1) and (8.2)

$$v(t) = v_0 e^{-\beta t} + e^{-\beta t} \int_0^t e^{\beta t'} A(t') \, dt' \tag{8.4}$$

$$x(t) = x_0 + v_0 \frac{1 - e^{-\beta t}}{\beta} + \frac{1}{\beta} \int_0^t [1 - e^{-\beta(t-t')}] A(t') \, dt' \tag{8.5}$$

It is clear from Eqs. (8.4) and (8.5) that since a linear combination of Gaussian random variables is also a Gaussian random variable, $v(t)$ and $x(t)$ are also Gaussian random variables. Therefore, the distribution function can be written down immediately, once second moments have been calculated (see Appendix D). Those moments, in fact, can be calculated readily by using assumptions 1 and 2:

$$\langle X^2 \rangle = \frac{\sigma^2}{\beta} (1 - e^{-2\beta t}) \tag{8.6}$$

$$\langle XY \rangle = \frac{\sigma^2}{\beta^2} (1 - e^{-\beta t})^2 \tag{8.7}$$

$$\langle Y^2 \rangle = \frac{\sigma^2}{\beta^3} [2\beta t - (1 - e^{-\beta t})(3 - e^{-\beta t})] \tag{8.8}$$

where

$$X = \int_0^t e^{-\beta(t-t')} A(t') \, dt' \tag{8.9}$$

$$Y = \frac{1}{\beta} \int_0^t [1 - e^{-\beta(t-t')}] A(t') \, dt' \tag{8.10}$$

On using the relation

$$\langle v^2(t) \rangle = \langle v_0^2 \rangle e^{-2\beta t} + \langle X^2 \rangle \tag{8.11}$$

and requiring the equipartition law

$$\langle v^2(t) \rangle = \langle v_0^2 \rangle = \frac{k_B T}{m} \tag{8.12}$$

we find that

$$\beta = \sigma^2 \frac{m}{k_B T} \tag{8.13}$$

which is the result from the fluctuation-dissipation theorem relating the friction constant $m\beta$ to the random acceleration $A(t)$.[29] Therefore, it follows that if there are no collisions between the particle and the surrounding molecules, i.e., $A(t) = 0$, $\sigma^2 = 0$, then $\beta = 0$.

The same distribution function $P(x, v, t)$ as the one obtained in the above procedure can be also calculated based on the Fokker-Planck-Kramers equation, which we shall derive in the following. Since

$$\int_{-\infty}^{\infty} \int_{-\infty}^{\infty} P(x, v, t) \, dx \, dv \tag{8.14}$$

must be conserved over time, we require that $P(x, v, t)$ should satisfy the equation of continuity,

$$\frac{\partial P}{\partial t} + \frac{\partial}{\partial x}(\dot{x}P) + \frac{\partial}{\partial v}(\dot{v}P) = 0 \tag{8.15}$$

In view of the equations of motion in Eqs. (8.1) and (8.2), Eq. (8.15) can be written as

$$\frac{\partial P}{\partial t} + v\frac{\partial P}{\partial x} - \beta\frac{\partial}{\partial v}(vP) + \frac{F(x, t)}{m}\frac{\partial P}{\partial v} + A(t)\frac{\partial P}{\partial v} = 0 \tag{8.16}$$

It is assumed in developing the theory of Brownian motion that in Δt, $A(t)$ changes its magnitude quite quickly and randomly, while other quantities such as x, v, and $P(x, v, t)$ change very little. Therefore, from Eq. (8.16), we may write

$$
\begin{aligned}
P(x, v, t + \Delta t) &= G(x, v)P(x, v, t) - \int_t^{t+\Delta t} A(t_1)\frac{\partial P(x, v, t_1)}{\partial v} \, dt_1 \\
&= \left[1 - \int_t^{t+\Delta t} A(t_1) \, dt_1 \frac{\partial}{\partial v} + \int_t^{t+\Delta t}\int_t^{t_1} A(t_1)A(t_2) \, dt_1 \, dt_2 \frac{\partial^2}{\partial v^2} \right. \\
&\quad \left. - \int_t^{t+\Delta t}\int_t^{t_1}\int_t^{t_2} A(t_1)A(t_2)A(t_3) \, dt_1 \, dt_2 \, dt_3 \frac{\partial^3}{\partial v^3} + \cdots \right] \\
&\quad \times G(x, v)P(x, v, t) \\
&= \exp\left[-\int_t^{t+\Delta t} A(t') \, dt' \frac{\partial}{\partial v} \right] G(x, v)P(x, v, t) \tag{8.17}
\end{aligned}
$$

where the operator $G(x, v)$ is

$$G(x, v) = 1 - \left(v \frac{\partial}{\partial x} - \beta \frac{\partial}{\partial v} v + \frac{F(x, t)}{m} \frac{\partial}{\partial v} \right) \Delta t \qquad (8.18)$$

On taking averages of both sides in Eq. (8.17) with respect to $A(t)$, which satisfies assumptions 1 and 2, we deduce that

$$P(x, v, t + \Delta t) = \exp\left(\frac{1}{2} \int_t^{t+\Delta t} \int_t^{t+\Delta t} \langle A(t_1) A(t_2) \rangle \, dt_1 \, dt_2 \frac{\partial^2}{\partial v^2} \right)$$
$$\times G(x, v) P(x, v, t)$$
$$= \exp\left(\sigma^2 \Delta t \frac{\partial^2}{\partial v^2} \right) G(x, v) P(x, v, t), \qquad (8.19)$$

where Eq. (8.3) has been used. Therefore, it follows by keeping terms up to the order of Δt that

$$\frac{\partial P}{\partial t} + v \frac{\partial P}{\partial x} - \beta \frac{\partial}{\partial v} (vP) + \frac{F(x, t)}{m} \frac{\partial P}{\partial v} + \sigma^2 \frac{\partial^2 P}{\partial v^2} = 0 \qquad (8.20)$$

This is the Fokker-Planck-Kramers equation. It should be noted that Eq. (8.20) has been derived in eliminating $A(t)$ by using its statistical properties of the random and rapid change. To solve a stochastic equation is to eliminate a random variable by using its statistical properties and find the probability density function, if possible, or otherwise the average values of slowly changing variables. The requirement that the equilibrium distribution function P_{eq} be given by that of the Maxwell-Boltzmann type

$$P_{eq}(x, v) = C \exp\left(-\frac{mv^2}{k_B T} \right) \exp\left(-\frac{V(x, t)}{k_B T} \right) \qquad (8.21)$$

where C is a constant to be determined by the normalization condition,

$$\int_{-\infty}^{\infty} \int_{-\infty}^{\infty} P_{eq}(x, v, t) \, dx \, dv = 1 \qquad (8.22)$$

and the potential $V(x, t)$ satisfies the relation

$$F(x, t) = -\frac{\partial V}{\partial x} \qquad (8.23)$$

lead to the previous result,

$$\sigma^2 = \beta \frac{k_B T}{m} \tag{8.24}$$

It can be shown that $P(x, v, t)$ calculated from Eq. (8.20) together with Eq. (8.24) in the case of $F(x, t) = 0$, in fact, agree fully with that obtained directly using Eqs. (8.4) and (8.5). Thus we have seen that the same problem may be handled by two approaches, the direct method from a stochastic differential equation and a method based on the Fokker-Planck-Kramers equation.

As time passes, the distribution of v may become almost of the Maxwell-Boltzmann type, while that of x is still away from it. In other words, an order exists for reaching the equilibrium state: first the velocity and then the position. Therefore, in this case, from Eq. (8.1), we write

$$v(t) = \frac{dx}{dt} = \frac{1}{\beta} A(t) + \frac{F(x, t)}{m\beta} \tag{8.25}$$

This corresponds to neglect of the inertial term $dv(t)/dt$, that is,

$$\frac{dv}{dt} \simeq 0 \tag{8.26}$$

It should be noted that even though Eq. (8.26) implies $v(t)$ is constant, Eq. (8.25) should be regarded as the equation of motion governing the evolution of $x(t)$.

In the case of $F(x, t) = 0$, Eq. (8.25) can be easily integrated to give

$$x(t) = x_0 + \frac{1}{\beta} \int_0^t A(t') \, dt' \tag{8.27}$$

from which it follows that

$$\langle [x(t) - x_0]^2 \rangle = \frac{2\sigma^2}{\beta^2} t = \frac{2k_B T}{m\beta} t \tag{8.28}$$

This is precisely the result obtained by Einstein in 1905.[30] As before, we can write the probability density function, noting from Eq. (8.27) that $x(t)$ is a centered Gaussian random variable,

$$\left(\frac{1}{4\pi Dt} \right)^{1/2} \exp\left\{ -\frac{[x(t) - x_0]^2}{4Dt} \right\} \tag{8.29}$$

where $D = \sigma^2/\beta = k_B T/m\beta$ is the diffusion constant. It is interesting to note that if we put $\alpha l = 2D$ and $x_0 = 0$ in Eq. (A3.21), these two equations are identical.

We can derive an equation for the probability density function similar to Eq. (8.20) starting from Eq. (8.25). As before, we use the equation of continuity for $\rho(x, t)$,

$$\frac{\partial f(x, t)}{\partial t} + \frac{\partial}{\partial x}(\dot{x}f) = 0 \tag{8.30}$$

which gives

$$f(x, t + \Delta t) = f(x, t) - \frac{\partial}{\partial x} \frac{F(x, t)}{m\beta} f(x, t)\, \Delta t$$
$$- \frac{1}{\beta} \int_t^{t+\Delta t} A(t') \left[\frac{\partial}{\partial x} f(x, t') \right] dt' \tag{8.31}$$

Applying similar procedures for the derivation of the Fokker-Planck-Kramers equation, we find the following Smoluchowski equation:

$$\frac{\partial f}{\partial t} = \frac{\sigma^2}{\beta^2} \frac{\partial}{\partial x} \left[\frac{\partial f}{\partial x} - \frac{\beta}{\sigma^2 m} F(x, t) f(x, t) \right] \tag{8.32}$$

By putting the Maxwell-Boltzmann distribution function in Eq. (8.32), we find again the relation

$$\sigma^2 = \beta \frac{k_B T}{m} \tag{8.33}$$

It should be noted that the Smoluchowski equation obtained after neglecting the inertial effect [cf. Eq. (8.25)] is an approximated case of the Fokker-Planck-Kramers equation. To demonstrate this point, we shall derive the Smoluchowski equation from the Fokker-Planck-Kramers equation, following Sack's method.[31] Introducing the function,

$$\Psi(u, x, t) = \exp\left(\frac{k_B T}{2m} u^2 \right) \int_{-\infty}^{\infty} P(x, v, t) e^{-iuv}\, dv \tag{8.34}$$

we can transform the Fokker-Planck-Kramers equation, Eq. (8.20), into

$$\frac{\partial \Psi}{\partial t} + i \frac{\partial^2 \Psi}{\partial x\, \partial u} - iu \frac{k_B T}{m} \frac{\partial \Psi}{\partial x} + \frac{iu}{m} F(x, t)\Psi = -\beta u \frac{\partial \Psi}{\partial u} \tag{8.35}$$

By expanding $\Psi(x, u, t)$ in a power series of u,

$$\Psi(x, u, t) = \sum_{n=0}^{\infty} a_n u^n \qquad (8.36)$$

we find that

$$\left(\frac{\partial}{\partial t} + n\beta\right) a_n(x, t) + i(n+1) \frac{\partial a_{n+1}}{\partial x}$$

$$+ i\left(\frac{F(x, t)}{m} - \frac{k_B T}{m} \frac{\partial}{\partial x}\right) a_{n-1} = 0 \qquad (n = 0, 1, 2, \ldots) \quad (8.37)$$

It should be noted that since $\Psi(x, u, t)$ becomes independent of u when v reaches the equilibrium state, higher order terms in Eq. (8.36) do not contribute significantly. Therefore, neglecting $a_2, a_3, a_4, \ldots$, and putting $\partial a_1 / \partial t = 0$, we obtain the same equation as Eq. (8.32) after noting

$$a_0(x, t) = \int_{-\infty}^{\infty} P(x, v, t) \, dv = f(x, t) \qquad (8.38)$$

IX. ROTATIONAL BROWNIAN MOTION OF A PLANE ROTATOR

In order to understand some fundamental properties of the theory of rotational Brownian motion, we take a simple example of a plane rotator in this section. Then we consider the three-dimensional rotational Brownian motion of a rigid rotator in Section X.

The rotational motion of the plane rotator in a fluid can be described by the stochastic differential equations

$$I\frac{d\omega}{dt} + I\beta\omega = IA(t) + N(\phi, t) \qquad (9.1)$$

$$\frac{d\phi}{dt} = \omega \qquad (9.2)$$

where I is the moment of inertia of the rotator, ω is the angular velocity, $IA(t)$ is the random torque, $N(\phi, t)$ is the mechanical torque, and ϕ is an angle between a particular direction of the rotator and a direction fixed in space (see Fig. 51).

It is immediately evident that Eqs. (9.1) and (9.2) have the same forms as Eqs. (8.1) and (8.2), respectively. Thus some of the results in Section VIII can also be used in discussing the rotational Brownian motion of the plane

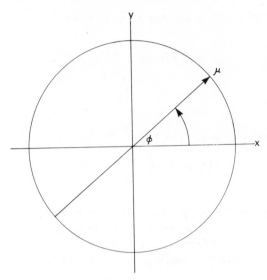

Fig. 51.   Rigid plane rotator with a permanent dipole μ.

rotator. The difference is that $x(t)$ in Section VIII is defined in the range $-\infty < x(t) < \infty$, whereas $0 \leq \phi \leq 2\pi$ (or $-\pi \leq \phi \leq \pi$).

So we obtain the Fokker-Planck-Kramers equation by replacing x, v, and m in Eq. (8.10) by ϕ, ω, and I, respectively:

$$\frac{\partial P}{\partial t} + \omega \frac{\partial P}{\partial \phi} + \frac{N(\phi, t)}{I} \frac{\partial P}{\partial \omega} - \beta \frac{\partial}{\partial \omega}\left(\omega P + \frac{k_B T}{I} \frac{\partial P}{\partial \omega}\right) = 0 \qquad (9.3)$$

And if inertial effects are ignored, Eq. (9.1) gives

$$\frac{d\phi}{dt} = \omega(t) = \frac{1}{\beta} A(t) + \frac{N(\phi, t)}{I\beta} \qquad (9.4)$$

which leads to the Smoluchowski equation

$$\frac{\partial f}{\partial t} = \frac{k_B T}{I\beta} \frac{\partial}{\partial \phi}\left[\frac{\partial \rho}{\partial \phi} - \frac{N(\phi, t)}{k_B T} f(\phi, t)\right] \qquad (9.5)$$

In the case where a permanent dipole lies on the plane rotator and there exist the polarizabilities α_1 and α_2 along the direction parallel and perpendicular to the dipole, respectively, the quantity responsible for the electric birefringence is $\langle \cos 2\phi \rangle$ and the potential energy function V is given by

$$V = -\mu E \cos \phi - \tfrac{1}{2}(\alpha_1 - \alpha_2) E^2 \cos 2\phi \qquad (9.6)$$

where E is an electric field applied along the x axis. If the second term on the left of Eq. (9.6) is neglected, the rotational motion of the dipole interacting with the electric field corresponds to that of a pendulum swinging under the influence of gravity, $\phi = 0$ being the position of the pendulum at rest.

In the particular case where Brownian motion is ignored so that $\beta = A(t) = 0$, the total energy (the Hamiltonian) H is constant over time, and we find

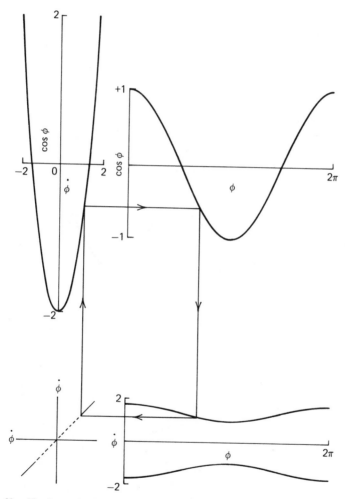

Fig. 52. The figure shows how the phase diagram of the deterministic motion of the plane rotator on the right of the bottom diagrams may be obtained from two simple diagrams; Upper left, a plot of $\cos \phi$ versus $\dot{\phi}$ of the function $\cos \phi = \dot{\phi}^2 - 2$; upper right, a plot of $\cos \phi$ versus ϕ.

by assuming $\Delta\alpha = (\alpha_1 - \alpha_2) = 0$ that

$$\cos\phi = \frac{I}{2\mu E}\dot{\phi}^2 - \frac{H}{\mu E} \tag{9.7}$$

Although the change of ϕ as a function of time can be expressed in terms of an elliptical function, we see from Figs. 52 and 53 (the phase diagrams) how ϕ changes as a function of $\dot{\phi}$, whereas the Brownian motion makes the phase diagrams complicated as seen from Figs. 54–57, which are obtained by computer simulation (see Appendix E).

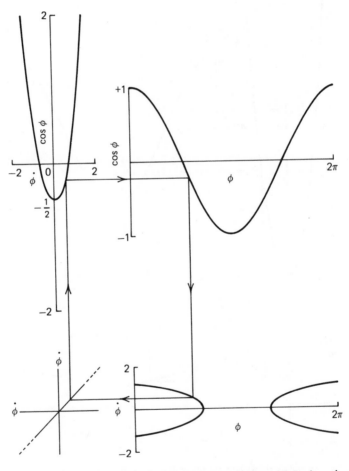

Fig. 53. The same as Fig. 52, except that the diagram on the upper left where the function $\cos\phi = \dot{\phi}^2 - 1/2$ is plotted and the resulting phase diagram.

Fig. 54. The phase diagram taking the rotational Brownian motion into account. $\beta = 1$, $\phi(0) = \pi/2$, $\dot{\phi}(0) = 0.1$, $(\mu E/I) = 1/2$, $\Delta\alpha = \alpha_{\parallel} - \alpha_{\perp} = 0$, $k_B T/I = 1$.

Now we shall investigate how the Fokker-Planck-Kramers equation, Eq. (9.3), may be used to calculate reorientation factor of the birefringence for the plane rotator $\langle \cos 2\phi \rangle$. This equation is transformed by introducing

$$\Psi(u_p, \phi, t) = \exp\left(\frac{u^2}{2I}\right) \int_{-\infty}^{\infty} P(\omega, \phi, t) e^{-i u_p \omega} \, d\omega \qquad (9.8)$$

into

$$\frac{\partial \Psi}{\partial t} + i \frac{\partial^2 \Psi}{\partial \phi \, \partial u_p} - i u_p \frac{k_B T}{I} \frac{\partial \Psi}{\partial \phi} + \frac{i u_p}{I} N(\phi, t) \Psi = -\beta u_p \frac{\partial \Psi}{\partial u_p} \qquad (9.9)$$

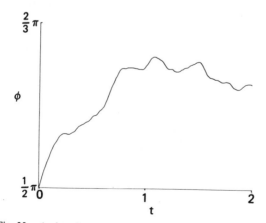

Fig. 55. A plot of ϕ versus t. Parameter values as in Fig. 54.

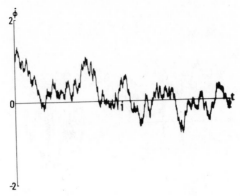

Fig. 56. A plot of $\dot{\phi}$ versus t. Parameter values as Fig. 54.

On expanding $\Psi(u_p, \phi, t)$ in a Fourier series,

$$\Psi(u_p, \phi, t) = \sum_{n=-\infty}^{\infty} a_n(u_p, t) e^{-in\phi} \qquad (9.10)$$

where

$$a_n(u_p, t) = \frac{1}{2\pi} \int_0^{2\pi} \Psi(u_p, \phi, t) e^{in\phi} \, d\phi \qquad (9.11)$$

after putting V of Eq. (9.6) in Eq. (9.9), we find the relation

$$\frac{\partial a_n}{\partial t} + \beta u_p \frac{\partial a_n}{\partial u_p} + n \frac{\partial a_n}{\partial u_p} - \frac{k_B T}{I}(nu_p) a_n$$

$$- \frac{u_p}{2I}\left[\mu E(a_{n+1} - a_{n-1}) + \Delta\alpha E^2(a_{n+2} - a_{n-2})\right] = 0 \qquad (9.12)$$

Particularly when E is not applied for $t > 0$, Eq. (9.12) can be solved exactly, leading to

$$a_n(u_p, t) = a_n\left[e^{-\beta t}\left(u_p + \frac{n}{\beta}\right) - \frac{n}{\beta}, 0\right]$$

$$\times \exp\left\{n\frac{k_B T}{I\beta}\left[\left(u_p + \frac{n}{\beta}\right)(1 - e^{-\beta t}) - nt\right]\right\} \qquad (9.13)$$

where $a_n(u_p, 0)$ is the value of $a_n(u_p, t)$ at $t = 0$, which should be specified

Fig. 57. A plot of $B(\Delta t)$ whose definition is given in Appendix E versus t. Parameter values as in Fig. 54.

as the initial condition. It follows from Eqs. (9.8), (9.11), and (9.13) that

$$\langle e^{in\phi(t)} \rangle = \frac{a_n(0, t)}{a_0(0, t)} = \langle e^{in\phi(0)} \rangle \exp\left[n^2 \frac{k_B T}{I\beta^2} (1 - e^{-\beta t} - \beta t) \right] \quad (9.14)$$

which immediately gives rise to

$$\begin{bmatrix} \langle \cos n\phi(t) \rangle \\ \langle \sin n\phi(t) \rangle \end{bmatrix} = \begin{bmatrix} \langle \cos n\phi(0) \rangle \\ \langle \sin n\phi(0) \rangle \end{bmatrix} \exp\left[n^2 \frac{k_B T}{I\beta^2} (1 - e^{-\beta t} - \beta t) \right] \quad (9.15)$$

It therefore becomes evident that the electric birefringence for the sudden removal of the field at $t = 0$ decays, when it is normalized, in accordance with the function

$$\exp\left[\frac{4k_B T}{I\beta^2} (1 - e^{-\beta t} - \beta t) \right] \quad (9.16)$$

in which two parameters, $D = k_B T/I\beta$ (the rotational diffusion constant) and β, should be specified (see Fig. 58). Equation (9.16) leads to the two limiting values. As $\beta \to 0$, when Brownian motion is not considered.

$$\frac{\langle \cos n\phi(t) \rangle}{\langle \cos n\phi(0) \rangle} = \exp\left(-n^2 \frac{k_B T}{2I} t^2 \right) \quad (9.17)$$

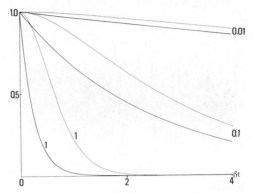

Fig. 58.   Plots of ($\cdots$) the function in Eq. (9.16) and (——) $\exp(-4Dt)$ obtained neglecting inertial effects for the decay birefringence versus βt. Numerals on the curves represent the values for $k_B T/I\beta^2$.

whereas when β' is very large we find that

$$\frac{\langle \cos n\phi(t) \rangle}{\langle \cos n\phi(0) \rangle} = \exp(-n^2 Dt) \tag{9.18}$$

This result can also be obtained from the Smoluchowski equation in Eq. (9.5) with $N(\phi, t) = 0$. The results in Eqs. (9.17) and (9.18) show directly that Eqs. (9.1) and (9.2) include the deterministic and stochastic limits, respectively.[32]
     When the applied electric field E is taken into account, it no longer seems possible to solve Eq. (9.12) analytically. On expanding $a_n(u_p, t)$ in a power series of u_p,

$$a_n(u_p, t) = \sum_{m=0}^{\infty} b_{m,n}(t) u_p^m \tag{9.19}$$

and putting this into Eq. (9.13), we find that

$$\frac{db_{m,n}(t)}{dt} + m\beta b_{m,n} + n(m+1)b_{m+1,n} - \frac{k_B T}{I} n b_{m-1,n}$$

$$- \frac{1}{2I}\left[\mu E(b_{m-1,n+1} - b_{m-1,n-1}) + \Delta\alpha E^2(b_{m-1,n+2} - b_{m-1,n-2})\right] = 0 \tag{9.20}$$

If we assume $b_{2,n} = b_{3,n} = \cdots = 0$ and $(db_{1,n}/dt) = 0$ (the same assumption used in Section VIII when the Smoluchowski equation was derived from the

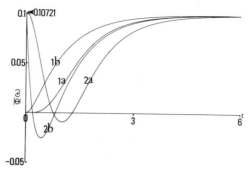

Fig. 59. Plots of $\Phi(t) = \langle \cos 2\phi \rangle$ for the buildup (1a and 1b) and reversing (2a and 2b) fields with (1a and 2a) and without (1b and 2b) inertial effects versus t. [$\beta = 1$, $k_B T / I\beta^2 = 1$ ($\mu E / k_B T) = 1$ and $\Delta\alpha = 0$.]

Fokker-Planck-Kramers equation), we find that

$$\frac{db_{0,n}}{dt} + n\left\{ \frac{k_B T}{I\beta} nb_{0,n} + \frac{1}{2I\beta}\left[\mu E(b_{0,n+1} - b_{0,n-1}) \right.\right.$$

$$\left.\left. + \Delta\alpha E^2(b_{0,n+2} - b_{0,n-2})\right]\right\} = 0 \qquad (9.21)$$

We have calculated $\langle \cos 2\phi \rangle$ numerically from Eq. (9.21) and show the results in Figs. 59–61.

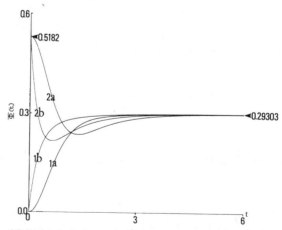

Fig. 60. Plots of $\Phi(t)$ for the buildup (1a and 1b) and the reversing (2a and 2b) fields with (1a and 2a) and without (1b and 2b) inertial effects versus t. [$\beta = 1$, $(\mu E / k_B T) = 1$, and $\Delta\alpha = 1$.]

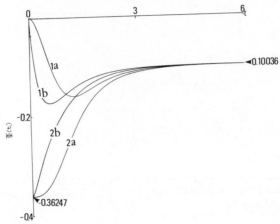

Fig. 61.   Plots of $\Phi(t)$ for the buildup (1a and 1b) and reversing (2a and 2b) fields with (1a and 2a) and without (1b and 2b) inertial effects versus t. [$\beta = 1, (\mu E/k_B T) = 1$, and $\Delta\alpha = -1$.]

X. ROTATIONAL BROWNIAN MOTION OF A RIGID BODY

Since a molecule in a fluid does not rotate around a fixed axis in space except in very limited cases, we must treat the rotational Brownian motion of a molecule fixed in a space point to describe the Kerr-effect relaxation. We assume that separation of translational and rotational motion is possible and that the molecule is rigid with an arbitrary shape. Although flexibility of electrons attached to the molecule is allowed, we assume that electron motion does not contribute to the rotational motion as a whole. We shall therefore use the following equations of motion, which are written using Euler's equations with terms arising from Brownian motion:

$$\frac{d}{dt}\omega_x(t) - \frac{I_y - I_z}{I_x}\omega_y(t)\omega_z(t) = -\beta_x\omega_x(t) + A_x(t) + \frac{1}{I_x}N_x(\theta, \phi, \psi)$$

$$(10.1)$$

$$\frac{d}{dt}\omega_y(t) - \frac{I_z - I_x}{I_y}\omega_z(t)\omega_x(t) = -\beta_y\omega_y(t) + A_y(t) + \frac{1}{I_y}N_y(\theta, \phi, \psi)$$

$$(10.2)$$

$$\frac{d}{dt}\omega_z(t) - \frac{I_x - I_y}{I_z}\omega_x(t)\omega_y(t) = -\beta_z\omega_z(t) + A_z(t) + \frac{1}{I_z}N_z(\theta, \phi, \psi)$$

$$(10.3)$$

$$\frac{d}{dt}\theta(t) = \omega_x \cos\psi - \omega_y \sin\psi \tag{10.4}$$

$$\frac{d}{dt}\phi(t) = \frac{1}{\sin\theta}\left(\omega_x \sin\psi + \omega_y \cos\psi\right) \tag{10.5}$$

$$\frac{d}{dt}\psi(t) = \omega_z - \cot\theta\left(\omega_x \sin\psi + \omega_y \cos\psi\right) \tag{10.6}$$

where $\omega_i(t)$ ($i = x$, y, or z) is the angular velocity about the principal axis labeled i, I_i is the moment of inertia, $I_i\beta_i\omega_i(t)$ is the angular velocity-dependent damping torque, $N_i(\theta, \phi, \psi)$ is the external torque, $I_iA_i(t)$ is the random torque, and θ, ϕ, and ψ are the Eulerian angles specifying the orientation of the body (see for example, Ref. 33). Instead of Eqs. (10.4)–(10.6), we may use the differential equations for three independent directional cosines, M_x, M_y, and M_z:

$$\frac{d}{dt}M_x = \omega_z M_y - \omega_y M_z \tag{10.7}$$

$$\frac{d}{dt}M_y = \omega_x M_z - \omega_z M_x \tag{10.8}$$

$$\frac{d}{dt}M_z = \omega_y M_x - \omega_x M_y \tag{10.9}$$

M_x, M_y, and M_z are the directional cosines along three orthogonal axes of the coordinate system fixed in space when the unit vector of one of the three principal axes is projected. So, depending on which unit vector is projected, we have three cases:

CASE 1

$$M_x = \cos\psi\cos\phi - \cos\theta\sin\phi\sin\psi \tag{10.10a}$$
$$M_y = -\sin\psi\cos\phi - \cos\theta\sin\phi\cos\psi \tag{10.10b}$$
$$M_z = \sin\theta\sin\phi, \tag{10.10c}$$

CASE 2

$$M_x = \cos\psi\sin\phi + \cos\theta\cos\phi\sin\psi \tag{10.11a}$$
$$M_y = -\sin\psi\sin\phi + \cos\theta\cos\phi\cos\psi \tag{10.11b}$$
$$M_z = -\sin\theta\cos\phi, \tag{10.11c}$$

CASE 3

$$M_x = \sin\theta\sin\psi \tag{10.12a}$$
$$M_y = \sin\theta\cos\psi \tag{10.12b}$$
$$M_z = \cos\theta \tag{10.12c}$$

It should be noted that M_x, M_y, and M_z in case (3) do not contain ϕ explicitly. The Fokker-Planck-Kramers equation for the probability density function P can be derived from either the continuity equation,

$$\frac{\partial P}{\partial t} + \sum_{i=x,y,z} \frac{\partial}{\partial \omega_i}(\dot{\omega}_i P) + \frac{\partial}{\partial \theta}(\dot{\theta}P) + \frac{\partial}{\partial \phi}(\dot{\phi}P) + \frac{\partial}{\partial \psi}(\dot{\psi}P) = 0$$

(10.13)

or

$$\frac{\partial P}{\partial t} + \sum_{i=x,y,z} \left[\frac{\partial}{\partial \omega_i}(\dot{\omega}_i P) + \frac{\partial}{\partial M_i}(\dot{M}_i P) \right] = 0 \qquad (10.14)$$

Using Eq. (10.13) or (10.14), the assumptions for $A_i(t)$, (1) $A_i(t)$ is a centered Gaussian random variable and (2) $A_i(t)$ correlates only with itself, and

$$\langle A_i(t_1) A_j(t_2) \rangle = \sigma_i^2 \delta_{ij} \delta(t_1 - t_2) \qquad (10.15)$$

where δ_{ij} is the Kronecker delta, and the requirement that in equilibrium P becomes the Maxwell-Boltzmann distribution function, we find after taking the procedure in Section VIII that the Fokker-Planck-Kramers equation is given by

$$\frac{\partial P}{\partial t} + \sum_{j=x,y,z} \left[i\omega_j L_j P - \beta_j \frac{\partial}{\partial \omega_j} \left(\omega_j P + \frac{k_B T}{I_j} \frac{\partial P}{\partial \omega_j} \right) \right]$$
$$+ \sum_{i,j,k} \frac{1}{I_i} \left[\omega_j \omega_k (I_j - I_k) + N_i \right] \frac{\partial P}{\partial \omega_i} = 0 \qquad (10.16)$$

where i, j, and k are cyclic indices, and

$$L_x = \frac{1}{i} \left(M_z \frac{\partial}{\partial M_y} - M_y \frac{\partial}{\partial M_z} \right) = \frac{1}{i} \left(\cos\psi \frac{\partial}{\partial \theta} + \frac{\sin\psi}{\sin\theta} \frac{\partial}{\partial \phi} - \cot\theta \sin\psi \frac{\partial}{\partial \psi} \right)$$

(10.17a)

$$L_y = \frac{1}{i} \left(M_x \frac{\partial}{\partial M_z} - M_z \frac{\partial}{\partial M_x} \right) = \frac{1}{i} \left(-\sin\psi \frac{\partial}{\partial \theta} + \frac{\cos\psi}{\sin\theta} \frac{\partial}{\partial \phi} - \cot\theta \cos\psi \frac{\partial}{\partial \psi} \right)$$

(10.17b)

$$L_z = \frac{1}{i} \left(M_y \frac{\partial}{\partial M_x} - M_x \frac{\partial}{\partial M_y} \right) = \frac{1}{i} \frac{\partial}{\partial \psi} \qquad (10.17c)$$

The reason the quantum-mechanical angular operators L_x, L_y, and L_z appear in Eqs. (10.17) becomes clear from Eqs. (10.7)–(10.12). At this stage, it is worth noting that since

$$N = M \times F, \tag{10.18}$$

and

$$F_i = -\frac{\partial}{\partial M_i} V(\theta, \phi, \psi) \tag{10.19}$$

where $V(\theta, \phi, \psi)$ is the potential energy and F is the force,

$$N_i = -\sqrt{-1}\, L_i V(\theta, \phi, \psi) \tag{10.20}$$

The Smoluchowski equation for the probability density function w can be found by using the following stochastic differential equations obtained after neglecting inertial effects in Eqs. (10.1)–(10.3):

$$\omega_x(t) = \dot{\phi}\sin\theta\sin\psi + \dot{\theta}\cos\psi = \frac{1}{\beta_x} A_x(t) + \frac{1}{I_x\beta_x} N_x(\theta, \phi, \psi) \tag{10.21}$$

$$\omega_y(t) = \dot{\phi}\sin\theta\cos\psi - \dot{\theta}\sin\psi = \frac{1}{\beta_y} A_y(t) + \frac{1}{I_y\beta_y} N_y(\theta, \phi, \psi) \tag{10.22}$$

$$\omega_z(t) = \dot{\phi}\cos\theta + \dot{\psi} = \frac{1}{\beta_z} A_z(t) + \frac{1}{I_z\beta_z} N_z(\theta, \phi, \psi) \tag{10.23}$$

and the equation of continuity, either

$$\frac{\partial w}{\partial t} + \frac{\partial}{\partial \theta}(\dot{\theta}w) + \frac{\partial}{\partial \phi}(\dot{\phi}\omega) + \frac{\partial}{\partial \psi}(\dot{\psi}w) = 0 \tag{10.24}$$

or

$$\frac{\partial w}{\partial t} + \sum_{i=x,y,z} \frac{\partial}{\partial M_i}(\dot{M}_i w) = 0 \tag{10.25}$$

The Smoluchowski equation is

$$\frac{\partial w}{\partial t} + \sum_{j=x,y,z} \left[\frac{k_B T}{I_j \beta_j} L_j^2 w + \frac{1}{I_j \beta_j}(iL_j)(N_j w) \right] = 0 \tag{10.26}$$

For symmetric-top molecules where $I_x = I_y$ and $\beta_x = \beta_y$, Eq. (10.26) becomes

$$\frac{\partial w}{\partial t} = \frac{k_B T}{I_x \beta_x} \left(\frac{\partial^2 w}{\partial \theta^2} + \cot \theta \frac{\partial w}{\partial \theta} + \frac{1}{\sin^2 \theta} \frac{\partial^2}{\partial \phi^2} + \frac{1}{\sin^2 \theta} \frac{\partial^2 w}{\partial \psi^2} - 2 \frac{\cos \theta}{\sin^2 \theta} \frac{\partial^2 w}{\partial \phi \, \partial \psi} \right)$$

$$+ \frac{k_B T}{I_z \beta_z} \frac{\partial^2 w}{\partial \psi^2} - \frac{1}{I_x \beta_x} \left[\cos \psi \frac{\partial}{\partial \theta} (N_x w) - \sin \psi \frac{\partial}{\partial \theta} (N_y w) \right.$$

$$+ \frac{1}{\sin \theta} \left(\sin \psi \frac{\partial}{\partial \phi} (N_x w) + \cos \psi \frac{\partial}{\partial \phi} (N_y w) \right)$$

$$\left. - \cot \theta \left(\sin \psi \frac{\partial}{\partial \psi} (N_x w) + \cos \psi \frac{\partial}{\partial \psi} (N_y w) \right) \right]$$

$$- \frac{1}{I_z \beta_z} \frac{\partial}{\partial \psi} (N_z w) \tag{10.27}$$

Particularly when $V(\theta, \phi, \psi)$ is a function of θ only, Eq. (10.27) can be reduced to

$$\frac{\partial f(\theta, t)}{\partial t} = \frac{1}{\sin \theta} \frac{\partial}{\partial \theta} \left[\sin \theta \left(\frac{k_B T}{I_x \beta_x} \frac{\partial f}{\partial \theta} + \frac{1}{I_x \beta_x} \frac{\partial V(\theta)}{\partial \theta} f \right) \right] \tag{10.28}$$

where

$$f(\theta, t) = \int_0^{2\pi} \int_0^{2\pi} w(\theta, \phi, \psi, t) \, d\phi \, d\psi \tag{10.29}$$

In this article, we have often used this Smoluchowski equation for investigating Kerr-effect relaxation in high fields. We must bear in mind that this equation is valid for symmetric-top molecules, ignoring inertial effects.

The first analytical attempt to treat the rotational Brownian motion of a nonspherical rigid body immersed in a fluid with inertial effects has been carried out by one of the authors.[34] The exact solution for the conditional probability function

$$P_\omega(\omega_x, \omega_y, \omega_z, t) = \int_0^\pi \int_0^{2\pi} \int_0^{2\pi} P(\omega_x, \omega_y, \omega_z, \theta, \phi, \psi, t) \sin \theta \, d\theta \, d\phi \, d\psi$$

in the case of $I_x = I_y$, $\beta_x = \beta_y$, and $N_i(\theta, \phi, \psi) = 0$ has been calculated. The angular velocity correlation functions are given by

$$\langle \omega_x(0)\omega_y(t)\rangle = \langle \omega_y(0)\omega_y(t)\rangle$$

$$= \frac{k_B T}{I_x}e^{-\beta_x t}\exp\left[-\frac{k_B T}{I_z}\left(\frac{a}{\beta_z}\right)^2(\beta_x t + e^{-\beta_z t}-1)\right]$$

$$\langle \omega_z(0)\omega_z(t)\rangle = \frac{k_B T}{I_z}e^{-\beta_z t}$$

$$\langle \omega_i(t_1)\omega_j(t_2)\rangle = 0 \qquad (i\neq j)$$

where
$$a = 1 - \frac{I_z}{I_x}$$

The Laplace transform of $\langle\cos\theta(t)\rangle$ after the sudden removal of the static field along the z axis is also calculated. For symmetric molecules, $\mathscr{L}[\langle\cos\theta(t)\rangle]$ is expressed as

$$\mathscr{L}[\langle\cos\theta(t)\rangle]$$

$$= \frac{\mu^2 E_0/3k_B T}{s + \cfrac{2k_B T/I_x}{s + \beta_x + \cfrac{2k_B T/I_x}{s + 2\beta_x + \cfrac{4k_B T/I_x}{s + 3\beta_x}}} + \cfrac{(k_B T/I_x)(I_z/I_x)}{s + \beta_x + \beta_z + \cfrac{(2k_B T/I_x)(I_z/I_x)}{s + \beta_x + 2\beta_z + \cfrac{2k_B T/I_x}{s + 2\beta_z}}}}$$

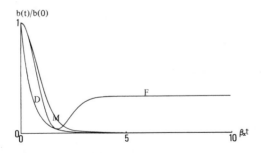

Fig. 62.   Plots of $b(t)/b(0)$ for $k_B T/I_x\beta_x^2 = 1$ and $I_x = I_y = I_z$. (F) Curve obtained using the free rotational model. (D) Curve obtained the single exponential decay function. (M) Curve obtained by Morita's method.

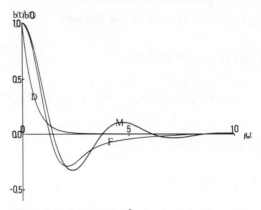

Fig. 63.   Plots of $b(t)/b(0)$ for $k_B T/I_x \beta_x^2 = 1$, where $I_x = I_y$ and $I_z = 0$. (F) Curve obtained using the free rotational model. (D) Curve obtained using the single exponential decay function. (M) Curve obtained by Morita's method.

Plots of $b(t) = \langle \cos \theta(t) \rangle$ versus $\beta_x t$ are shown in Figs. 62–65. It is noted that inertial effects affect the behavior of the dynamic processes considerably, especially at short times where the deterministic motion is predominant.

It becomes difficult to treat the rotational Brownian motion of an asymmetrical body immersed in a fluid. The full detailed description is found in a paper of Morita.[35] Related problems are discussed in the recent books by McConnell[36] and by Evans et al.[37]

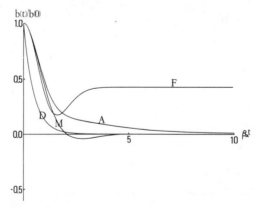

Fig. 64.   Plots of $b(t)/b(0)$ for $k_B T/I_x \beta_x^2 = 1$, where $I_x = I_y$ and $I_z/I_x = 1.9$. (F) Curve obtained using the free rotational model. (D) Curve obtained using the single exponential decay function. (M) Curve obtained by Morita's method with $\beta_z/\beta_x = 10$. (A) Curve obtained by Morita's method with $\beta_z/\beta_x = 0.1$.

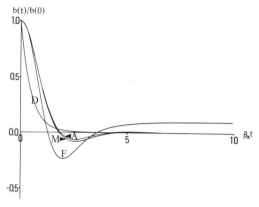

Fig. 65.   Plots of $b(t)/b(0)$ for $k_BT/I_x\beta_x^2 = 1$, where $I_x = I_y$ and $I_z/I_x = 0.1$. (F) Curve obtained using the free rotational model. (D) Curve obtained using the single exponential decay function. (M) Curve obtained by Morita's method with $\beta_z/\beta_x = 10$. (A) Curve obtained by Morita's method with $\beta_z/\beta_x = 0.1$.

XI.   ELECTRIC BIREFRINGENCE IN INFINITELY HIGH, TIME-VARYING ELECTRIC FIELDS

Since for an infinitely large value of E, N_i also becomes very large, the random torque $A_i(t)$ may be neglected. This means that N_i is so large that $A_i(t)$ arising from collisions between the rigid body and surrounding molecules may not be significant. Therefore, Eqs. (10.21)–(10.23) should be replaced by

$$\omega_x(t) = \frac{1}{I_x\beta_x}N_x(\theta, \phi, \psi) \tag{11.1}$$

$$\omega_y(t) = \frac{1}{I_y\beta_y}N_y(\theta, \phi, \psi) \tag{11.2}$$

$$\omega_z(t) = \frac{1}{I_z\beta_z}N_z(\theta, \phi, \psi) \tag{11.3}$$

Equations (10.7)–(10.9) after Eqs. (11.1)–(11.3) are put where $M_x = \sin\theta\sin\phi$, $M_y = \sin\theta\cos\phi$, and $M_z = \cos\theta$ describe the orientation of the body in a very high field when inertial effects are ignored. It should be noted that since the stochastic variable $A_i(t)$ is missing in Eqs. (11.1)–(11.3), Eqs. (10.7)–(10.9) give the nonstochastic (deterministic) solutions, which do not contain the effect due to collisions.

The continuity equation for the distribution function f,

$$\frac{\partial f}{\partial t} + \frac{\partial}{\partial M_x}(\dot{M}_x f) + \frac{\partial}{\partial M_y}(\dot{M}_y f) + \frac{\partial}{\partial M_z}(\dot{M}_z f) = 0 \qquad (11.4)$$

gives rise to

$$\frac{\partial f}{\partial t} + i \sum_{j=x,y,z} L_j \frac{N_j f}{I_j \beta_j} = 0 \qquad (11.5)$$

It is seen by comparing this equation with Eq. (10.26) that the diffusion term

$$\sum_{j=x,y,z} D_j L_j^2 f \qquad (11.6)$$

in the Smoluchowski equation, Eq. (10.26), is absent from Eq. (11.5). Equation (11.5) is a first-order partial differential equation, which is sometimes easy to solve. It is worth noting that the diffusion term results from the random torque.

For the problem at hand where the body is symmetric ($I_x = I_y$ and $\beta_x = \beta_y$) and the potential energy V, which is related to the mechanical torque N_i by Eq. (10.20), is given by Eq. (1.7), Eq. (11.5) becomes

$$\frac{\partial f}{\partial \tau} = -\frac{\partial}{\partial u}\left\{\left[e(\tau)(1-u^2)+2g(\tau)(1-u^2)u\right]f\right\} \qquad (11.7)$$

where $u = M_z = \cos\theta$, while Eqs. (10.7)–(10.9) together with Eqs. (10.12) lead to

$$\frac{du}{d\tau} = e(\tau)(1-u^2)+2g(\tau)(1-u^2)u \qquad (11.8)$$

It is seen that Eq. (11.7) is just the continuity equation for $f(u, t)$ with the velocity given by Eq. (11.8).

Both Eqs. (11.7) and (11.8) can be solved analytically for the special cases (1) $e \neq 0$, $g = 0$ (the pure permanent dipole orientation) and (2) $e \neq 0$, $g = 0$ (the pure induced dipole orientation). The solutions for Eq. (11.7) with the initial conditions

$$f(u,0) = C_1 \exp(e_0 u) \qquad (11.9)$$

for case 1 and

$$f(u,0) = C_2 \exp(g_0 u^2) \qquad (11.10)$$

for case 2 are given by

$$f(u,\tau)=C_1\frac{4x(\tau)}{[(1-u)x(\tau)+1+u]^2}\exp(e_0)\exp\left[\frac{2e_0(u-1)x(\tau)}{1+u+(1-u)x(\tau)}\right]$$

$$(11.11)$$

and

$$f(u,\tau)=C_2\frac{y(\tau)}{[u^2+(1-u^2)y(\tau)]^{3/2}}\exp\left[g_0\frac{u^2}{u^2+(1-u^2)y(\tau)}\right]$$

$$(11.12)$$

respectively, where C_1 and C_2 are constants to be determined from the normalization condition for $f(u,0)$,

$$x(\tau)=\exp\left[2\int_0^\tau e(\tau')\,d\tau'\right]$$

$$(11.13)$$

and

$$y(\tau)=\exp\left[4\int_0^\tau g(\tau')\,d\tau'\right]$$

$$(11.14)$$

Equations (11.11) and (11.12), after lengthy and tedious calculations, lead to

$$\langle u^2\rangle=\frac{2e_0x(\tau)}{\sinh(e_0)}[1-x(\tau)]^{-2}\{2P\sinh(e_0)+2\cosh(e_0)$$

$$+[e_0(P^2-1)+2P]\exp(e_0P)I\}+P^2$$

$$(11.15)$$

for case 1 and

$$\langle u^2\rangle=\frac{y(\tau)}{y(\tau)-1}\left[1-\frac{\int_0^{\sqrt{g_0}}\dfrac{\exp(z^2)}{\{[y(\tau)-1]/g_0\}z^2+1}\,dz}{\int_0^{\sqrt{g_0}}\exp(z^2)\,dz}\right]$$

$$(11.16)$$

for case 2, where

$$P(\tau)=\frac{1+x(\tau)}{1-x(\tau)}$$

$$(11.17)$$

and

$$I = \int_{(P+1)e_0}^{(P-1)e_0} z^{-1} e^{-z} \, dz \tag{11.18}$$

Particularly when the electric field e_0 has not been applied for $t \leq 0$, Eq. (11.15) after expanding $\exp(-z)$ in the integrand of Eq. (11.18) and taking the limit of $e_0 \to 0$, we find that

$$\lim_{e_0 \to 0} \langle u^2 \rangle = 1 + \frac{8x(\tau)}{(x(\tau)-1)^2} - \frac{8x(\tau)(x(\tau)+1)}{(x(\tau)-1)^3} \int_0^\tau e(\tau') \, d\tau' \tag{11.19}$$

whereas if we expand $\exp(z^2)$ in the integrands of both the nominator and denominator on the right-hand side of Eq. (11.16) in a power series of z and take the limit of $g_0 \to 0$, we obtain

$$\lim_{g_0 \to 0} \langle u^2 \rangle = \frac{y(\tau)}{y(\tau)-1} \left(1 - \frac{\tan^{-1}\sqrt{y(\tau)-1}}{\sqrt{y(\tau)-1}} \right) \tag{11.20}$$

The solutions for Eq. (11.8) for cases 1 and 2 are

$$u(\tau) = 1 - \frac{2(1-u_0)}{1 + x(\tau) + [x(\tau)-1]u_0} \tag{11.21}$$

and

$$u^2(\tau) = 1 - \frac{1-u_0^2}{1 + [y(\tau)-1]u_0^2} \tag{11.22}$$

respectively, where u_0 is $u(\tau) = \cos\theta(\tau)$ at $\tau = 0$. Equations (11.21) and (11.22) determine $u(\tau)$ once u_0 is specified. In our problem where the system consists of a large number of molecules, however, we do not have detailed knowledge concerning the u_0 of each molecule, except that we know the system is in an equilibrium state for $\tau \leq 0$, where the molecules take quite random values of u_0 whose distribution is described by $f(u,0)$ in Eq. (11.9) or (11.10). In other words, u_0 can be treated as a stochastic random variable, because it is very difficult, if not impossible, to find the precise position of each molecule that undergoes quite irregular collisions, whereas it has been assumed for $\tau > 0$ that torques due to such collisions can be neglected in comparison with those due to the very strong electric field. This assumption leads to the smooth curve of $u(\tau)$ as given by Eqs. (11.21) and (11.22). Therefore, in calculating $\langle u^2 \rangle$, u^2 arising from Eqs. (11.21) and (11.22) should be averaged out with respect to the initial value of u_0:

$$\langle u^2(\tau) \rangle = \frac{\int_{-1}^{+1} f(u_0,0)\, u^2(\tau)\, du_0}{\int_{-1}^{+1} f(u_0,0)\, du_0} \tag{11.23}$$

which, in fact, leads to the same results in Eqs. (11.15) and (11.16).
On imposing the initial condition

$$f(u,0) = \delta(u - u_0) \tag{11.24}$$

instead of the one in Eq. (11.9) for case 1 and solving Eq. (11.7) we find that

$$f(u,\tau) = \frac{1 - u'^2}{1 - u^2}\delta(u' - u_0) \tag{11.25}$$

where

$$u' = \frac{1 + u - x(\tau)(1 - u)}{1 + u + x(\tau)(1 - u)} \tag{11.26}$$

This shows immediately that since $f(u, \tau)$ is still given by the delta function for $\tau > 0$, which is zero except in the case of $u' = u_0$, there is no chance for u to fluctuate; it must take the deterministic value that satisfies $u' = u_0$. In fact, putting $u' = u_0$, we obtain the previous result in Eq. (11.21). In view of the fact that $f(u, \tau)$ in Eq. (11.25) is the conditional probability density function, we can recover $f(u, \tau)$ in Eq. (11.11) by calculating the integral

$$\frac{1 - u'^2}{1 - u^2}\int_{-1}^{+1}\delta(u' - u_0)\big[C_1\exp(e_0 u_0)\big]\, du_0 \tag{11.27}$$

A similar approach can be taken when the deterministic nature of u for $\tau > 0$ is considered for case 2.

It follows from Eqs. (11.11) and (11.12) that for time-dependent $e(\tau)$ and $g(\tau)$, the distribution function f in the equilibrium state at

$$\lim_{\tau \to \infty} f(u,\tau) = C_1\delta(u - 1) \tag{11.28}$$

and

$$\lim_{\tau \to \infty} f(u,\tau) = C_2\delta(u^2 - 1) \tag{11.29}$$

respectively. As expected from the deterministic behavior for $\tau > 0$, these are not the Maxwell-Boltzmann distribution functions and ensure that as $\tau \to \infty$, every molecule must have $u = 1$ and $u = \pm 1$ for cases 1 and 2, respectively. Of course, this is in agreement with the behavior expected in infinitely high fields and also follows by puting $[du(\tau)/d\tau] = 0$ in Eq. (11.8).

For case 2 with time-independent g and $e_0 = 0$ and for case 1 with time-independent e and $g_0 = 0$, $f(u, t)$ has been calculated first by Schwarz[9] and later by O'Konski, Yoshioka, and Oruttung,[4] who used Schwarz's approach. Schwarz' method is different from ours, and its physical significance is not so persuasive. The equation of motion used by both Schwarz and O'Konski et al. agree with Eq. (11.8) when it is written for $\dot{\theta}$. However, it is not clear whether they consider the motion for the plane rotator or the symmetric body, because for both cases the equation for $\dot{\theta}$ happens to be the same (see Section IX). So even though the solution is the same, the final results for $\langle u^2 \rangle$ are different, because for the plane rotator,

$$\langle u^2 \rangle = \frac{\int_0^{2\pi} u^2(\tau) f(\theta(0),0) \, d\theta(0)}{\int_0^{2\pi} f(\theta(0),0) \, d\theta(0)} \tag{11.30}$$

and for the symmetric body $\langle u^2 \rangle$ should be calculated from Eq. (11.23).

In the case when $e(\tau)$ and $g(\tau)$ are considered simultaneously, it is difficult to treat the problem in the above analytical approach. We will publish the results elsewhere.

APPENDIX A.   THE QUANTITY OBSERVED IN KERR-EFFECT EXPERIMENTS

As stated in the text, the electric birefringence $\Delta n = n_{\parallel} - n_{\perp}$ is observed to arise from the anisotropy produced by an application of an electric field. Here we show how Δn is related to a quantity that provides information on the orientation of particles in fluids.

We introduce two coordinate systems, the laboratory coordinate system, $Oxyz$, and the coordinate system fixed on the particle frame, $Ox'y'z'$. The laboratory coordinate system is needed when the direction of the applied electric field, which is fixed in space, is specified, while the $Ox'y'z'$ system changes direction continuously with respect to the laboratory frame as the particles rotate. It is always possible that a vector in the $Ox'y'z'$ system can be expressed in terms of three components along the x, y, and z axes through a transformation matrix $\mathbf{A}$, as given by

$$\mathbf{A} = \begin{bmatrix} \cos\psi\cos\phi - \cos\theta\sin\phi\sin\psi & -\sin\phi\cos\phi - \cos\theta\sin\phi\cos\psi & \sin\theta\sin\phi \\ \cos\psi\sin\phi + \cos\theta\cos\phi\sin\psi & -\sin\psi\sin\phi + \cos\theta\cos\phi\cos\psi & -\sin\theta\cos\phi \\ \sin\theta\sin\psi & \sin\theta\cos\psi & \cos\theta \end{bmatrix}$$

$$(A1.1)$$

where θ, ϕ, and ψ are Euler's angles (see Ref. 33).

Thus the induced dipole moments $\mathbf{m}$ and $\mathbf{m}'$ and the electric fields $\mathbf{E}$ and $\mathbf{E}'$ with respect to the $Oxyz$ and $Ox'y'z'$ systems are given by

$$\mathbf{m} = \mathbf{A}\mathbf{m}' \qquad (A1.2)$$

$$\mathbf{E} = \mathbf{A}\mathbf{E}' \qquad (A1.3)$$

The permanent dipole moment $\boldsymbol{\mu}'$ and the polarizability tensor $\boldsymbol{\alpha}'$, both of which have characteristic values inherent to the molecules, give rise to the induced dipole $\mathbf{m}'$:

$$\mathbf{m}' = \boldsymbol{\mu}' + \boldsymbol{\alpha}'\mathbf{E}' \qquad (A1.4)$$

On writing

$$\mathbf{m} = \boldsymbol{\mu} + \boldsymbol{\alpha}\mathbf{E} \qquad (A1.5)$$

we find from Eqs. (A1.2)–(A1.5) that

$$\boldsymbol{\mu} = \mathbf{A}\boldsymbol{\mu}'$$
$$\boldsymbol{\alpha} = \mathbf{A}\boldsymbol{\alpha}'\mathbf{A}^{-1} \qquad (A1.6)$$

It should be noted that both $\boldsymbol{\mu}$ and $\boldsymbol{\alpha}$ are related through $\mathbf{A}$ and $\mathbf{A}^{-1}$ to the quantities $\boldsymbol{\mu}'$ and $\boldsymbol{\alpha}'$, which are entirely determined by the geometrical configuration and rigidity of electrons on the particle with which we are concerned. In order to emphasize the dependence of $\mathbf{m}$, $\boldsymbol{\mu}$, and $\boldsymbol{\alpha}$ on $\mathbf{E}$, we write Eq. (A1.5) as

$$\mathbf{m}(\mathbf{E}) = \boldsymbol{\mu}(\mathbf{E}) + \boldsymbol{\alpha}(\mathbf{E})\mathbf{E} \qquad (A1.7)$$

For an optically isotropic system prior to the application of $\mathbf{E}$, such as liquids and solutions, the change in the direction of $\mathbf{E}$ to $-\mathbf{E}$ with the same field strength $|\mathbf{E}|$ leads to

$$\langle \mathbf{m}(\mathbf{E}) \rangle = -\langle \mathbf{m}(-\mathbf{E}) \rangle \qquad (A1.8)$$

where angular brackets represent the ensemble average.

It immediately follows from Eqs. (A1.7) and (A1.8) that

$$\langle \mu(\mathbf{E}) \rangle = -\langle \mu(-\mathbf{E}) \rangle \tag{A1.9}$$

$$\langle \alpha(\mathbf{E}) \rangle = \langle \alpha(-\mathbf{E}) \rangle \tag{A1.10}$$

These show explicitly that μ and α are odd and even functions of $\mathbf{E}$, respectively. Since μ' and α' are determined by particle properties, they are independent of the field $\mathbf{E}$ and Euler's angle. Thus we find from Eqs. (A1.6) and (A1.9) that

$$\langle \mu(\mathbf{E}) \rangle = \langle \mathbf{A}(\mathbf{E}) \rangle \mu' \tag{A1.11}$$

$$\langle \mathbf{A}(\mathbf{E}) \rangle = -\langle \mathbf{A}(-\mathbf{E}) \rangle \tag{A1.12}$$

In view of the relations

$$\mathbf{A}(\mathbf{E})\mathbf{A}^{-1}(\mathbf{E}) = \mathbf{A}^{-1}(\mathbf{E})\mathbf{A}(\mathbf{E}) = \mathbf{A}(-\mathbf{E})\mathbf{A}^{-1}(-\mathbf{E}) = \mathbf{A}^{-1}(-\mathbf{E})\mathbf{A}(-\mathbf{E}) = \mathbf{I} \tag{A1.13}$$

where $\mathbf{I}$ is the unit matrix, we have

$$\langle \mathbf{A}^{-1}(\mathbf{E}) \rangle = -\langle \mathbf{A}^{-1}(-\mathbf{E}) \rangle \tag{A1.14}$$

It is useful to note that both $\langle \mathbf{A} \rangle$ and $\langle \mathbf{A}^{-1} \rangle$ are odd functions of $\mathbf{E}$. We have now shown that

$$\langle \alpha(\mathbf{E}) \rangle = \alpha_2 E^2 + \alpha_4 E^4 + \cdots \tag{A1.15}$$

This result evidently depends on the validity of Eq. (A1.8).

Now if an electric field $\mathbf{E}^e$ is applied from the z axis and a plane-polarized light beam oriented at an angle 45° with respect to the z axis is put along the x axis, the induced dipole moment $\mathbf{m}^0$ due to the electric field of the light beam denoted by $\mathbf{E}^0$ becomes

$$\mathbf{m}^0 = \alpha^0 \mathbf{E}^0 \tag{A1.16}$$

Here,

$$\alpha^0 = \mathbf{A}\alpha'^{(0)}\mathbf{A}^{-1} \tag{A1.17}$$

where $\alpha'^{(0)}$ is the molecular polarizability tensor at an optical frequency due to $\mathbf{E}^0$. It has been assumed that this frequency is large enough to allow us to neglect the induced dipole moment arising from the orientational motion of a permanent dipole moment. We therefore see from Eq. (A1.16) that two components of $\mathbf{m}^0$ parallel and perpendicular to $\mathbf{E}^e$, denoted by m_y^0 and m_z^0, respectively, are given by

$$m_y^0 = \left(\alpha_{yy}^0 + \alpha_{yz}^0 \right) \frac{E^0}{\sqrt{2}} \tag{A1.18}$$

$$m_z^0 = \left(\alpha_{zy}^0 + \alpha_{zz}^0 \right) \frac{E^0}{\sqrt{2}} \tag{A1.19}$$

In view of the general relation obtained, neglecting the internal field,

$$\left(n_i^2 - 1 \right) E_i^0 = 4\pi N \langle m_i^0 \rangle \tag{A1.20}$$

where E_i^0 ($i = x$, y, or z) is the component of $\mathbf{E}^0$ along the i axis, n_i is the refractive index for the i axis, and N is the number of particles in a unit volume, putting $E_y^0 = E_z^0 = E^0/\sqrt{2}$ in Eq. (A1.19), we find that

$$n_y^2 - 1 = 4\pi N \langle \alpha_{yy}^0 + \alpha_{yz}^0 \rangle \tag{A1.21}$$

$$n_z^2 - 1 = 4\pi N \langle \alpha_{zy}^0 + \alpha_{zz}^0 \rangle \tag{A1.22}$$

Therefore, it follows that

$$\Delta n = n_z - n_y = \frac{4\pi N}{n_z + n_y} \langle \alpha_{zz}^0 - \alpha_{yy}^0 \rangle$$

$$\simeq \frac{2\pi N}{n} \langle \alpha_{zz}^0 - \alpha_{yy}^0 \rangle \tag{A1.23}$$

where $n_y \simeq n_z \simeq n$ and $\langle \alpha_{zy}^0 \rangle = \langle \alpha_{yz}^0 \rangle$ have been assumed.

Since both $\langle \mathbf{A} \rangle$ and $\langle \mathbf{A}^{-1} \rangle$ are odd functions of $\mathbf{E}^e$, we see from Eq. (A1.17) that the electric birefringence Δn is an even function of E^e.

On writing the matrix $\mathbf{A}$ in Eq. (A1.1) as

$$\mathbf{A} = \begin{bmatrix} a_{11} & a_{12} & a_{13} \\ a_{21} & a_{22} & a_{23} \\ a_{31} & a_{32} & a_{33} \end{bmatrix} \tag{A1.24}$$

and using the relation that holds for the orthogonal transformation,

$$\mathbf{A}^{-1} = \tilde{\mathbf{A}} \tag{A1.25}$$

where $\tilde{\mathbf{A}}$ is the transpose of $\mathbf{A}$, we find that

$$
\begin{aligned}
\langle \alpha_{zz}^0 - \alpha_{yy}^0 \rangle = \Big\langle & \big(a_{31}^2 - a_{21}^2 \big) \alpha_{xx}'^{(0)} + \big(a_{32}a_{31} - a_{22}a_{21} \big) \alpha_{yx}'^{(0)} \\
& + \big(a_{33}a_{31} - a_{23}a_{21} \big) \alpha_{zx}'^{(0)} + \big(a_{31}a_{32} - a_{21}a_{22} \big) \alpha_{xy}'^{(0)} \\
& + \big(a_{32}^2 - a_{22}^2 \big) \alpha_{yy}'^{(0)} + \big(a_{33}a_{32} - a_{23}a_{22} \big) \alpha_{zy}'^{(0)} \\
& + \big(a_{31}a_{33} - a_{21}a_{23} \big) \alpha_{xz}'^{(0)} + \big(a_{32}a_{33} - a_{22}a_{23} \big) \alpha_{yz}'^{(0)} \\
& + \big(a_{33}^2 - a_{23}^2 \big) \alpha_{zz}'^{(0)} \Big\rangle
\end{aligned}
\tag{A1.26}
$$

The potential energy V due to the interaction between the induced dipole moment $\mathbf{m}^e$ and the applied electric field $\mathbf{E}^e$ is given by

$$
\begin{aligned}
V = & -\mathbf{m}^e \cdot \mathbf{E}^e \\
= & -m_z^e E^e \\
= & -\big(a_{31}\mu_x'^{(e)} + a_{32}\mu_y'^{(e)} + a_{33}\mu_z'^{(e)} \big) E^e \\
& -\Big[\big(a_{31}\alpha_{xx}'^{(e)} + a_{32}\alpha_{yx}'^{(e)} + a_{33}\alpha_{zx}'^{(e)} \big) a_{31} + \big(a_{31}\alpha_{xy}'^{(e)} + \alpha_{32}\alpha_{yy}'^{(e)} + a_{33}\alpha_{zy}'^{(e)} \big) a_{32} \\
& + \big(a_{31}\alpha_{xz}'^{(e)} + a_{32}\alpha_{yz}'^{(e)} + a_{33}\alpha_{zz}'^{(e)} \big) a_{33} \Big] (E^e)^2.
\end{aligned}
\tag{A1.27}
$$

If there is symmetry for α', Eqs. (A1.26) and (A1.27) can be simplified. Holcomb and Tinoco[33] considered the case where the principal axis of the optical and electric polarizability are coincident and the off-diagonal elements in α' are zero. In this case, we have

$$
\langle \alpha_{zz}^0 - \alpha_{yy}^0 \rangle = \big\langle \big(a_{31}^2 - a_{21}^2 \big) \alpha_{xx}'^{(0)} + \big(a_{32}^2 - a_{22}^2 \big) \alpha_{yy}'^{(0)} + \big(a_{33}^2 - a_{23}^2 \big) \alpha_{zz}'^{(0)} \big\rangle
\tag{A1.28}
$$

and

$$
\begin{aligned}
V = & -\big(a_{31}\mu_x'^{(e)} + a_{32}\mu_y'^{(e)} + a_{33}\mu_z'^{(e)} \big) E^e \\
& -\big(a_{31}^2 \alpha_{xx}'^{(e)} + a_{32}^2 \alpha_{yy}'^{(e)} + a_{33}^2 \alpha_{zz}'^{(e)} \big) (E^e)^2
\end{aligned}
\tag{A1.29}
$$

Further simplification is often made in the literature for the case of

$$
\alpha_{xx}^{(0)} = \alpha_{yy}^{(0)} = g_2, \quad \alpha_{zz}^{(0)} = g_1
\tag{A1.30}
$$

$$
\alpha_{xx}^{(e)} = \alpha_{yy}^{(e)} = \alpha_2, \quad \alpha_{zz}^{(e)} = \alpha_1
\tag{A1.31}
$$

and

$$\mu_x'^{(e)} = \mu_y'^{(e)} = 0, \quad \mu_z'^{(e)} = \mu \tag{A1.32}$$

which enable us to write

$$
\begin{aligned}
\langle \alpha_{zz}^0 - \alpha_{yy}^0 \rangle &= \langle (g_1 - g_2)(\cos^2\theta - \sin^2\theta\cos^2\phi) \rangle \\
&= \langle (g_1 - g_2)[\langle P_2(\cos\theta) \rangle - \tfrac{1}{6}\langle P_2^2(\cos\theta)\cos 2\phi \rangle \tag{A1.33}
\end{aligned}
$$

$$V = -\mu E^e \cos\theta - (\alpha_1 - \alpha_2)(E^e)^2 \cos^2\theta - \alpha_2(E^e)^2 \tag{A1.34}$$

In this article, we are mainly concerned with the case where assumptions in Eqs. (A1.30)–(A1.32) are used. With the potential in Eq. (A1.34), which is independent of ϕ, the second term on the right in Eq. (A1.33) vanishes, giving

$$\langle \alpha_{zz}^0 - \alpha_{yy}^0 \rangle = (g_1 - g_2)\langle P_2(\cos\theta) \rangle$$

However, when V is a function of ϕ as in the case of Section VII, $\langle P_2^2(\cos\theta)\cos 2\phi \rangle$ contributes to the electric birefringence.

The electric birefringence at the equilibrium state can be calculated using the Maxwell-Boltzmann distribution function f, which is given by

$$f = C\exp\left(-\frac{V}{k_B T}\right) \tag{A1.35}$$

where C is a constant to be determined from the normalization requirement. The analytical expression of Δn for the most general case, where both the electrical and optical polarizability have no symmetry and the permanent dipole moment may have any orientation, is solved by Holcomb and Tinoco.[37]

Equations (A1.26) and (A1.27) are useful when the dynamics are considered and the principal axes are taken for the moments of inertia of a rigid particle for neither $\alpha'^{(0)}$ nor $\alpha'^{(e)}$. In this case, equations of motion describing the evolution of the orientation of the particle as a function of time become simple.

APPENDIX B. CALCULATION OF THE LAPLACE TRANSFORM OF EQS. (6.57) TO (6.62)

$$
\mathcal{L}[I_1(\tau)] = \frac{G(s)}{s+6} + \frac{2}{7}g_0\frac{G(s)}{(s+6)^2} + \frac{1}{4i\omega'}\left[\frac{G(s-2i\omega')}{s+6-2i\omega'} - \frac{G(s+2i\omega')}{s+6+2i\omega'}\right]
$$

$$
- \frac{1}{4i\omega'}\frac{1}{s+6}[G(s-2i\omega') - G(s+2i\omega')]
$$

where
$$G(s) = \mathcal{L}[g(\tau)] = \frac{g_0}{2}\left[\frac{1}{s} + \frac{s}{s^2 + 4\omega'^2}\right]$$

$$\mathcal{L}[I_2(\tau)] = \frac{e_0^2}{4}\left[\frac{1}{(s+6)(s+2-i\omega')(s-2i\omega')} + \frac{1}{(s+6)(s+2-i\omega')s}\right]$$

$$+ \frac{e_0^2}{10}g_0\left[\frac{1}{(s+6)(s+2-i\omega')^2(s-2i\omega')} + \frac{1}{(s+6)(s+2-i\omega')^2 s}\right]$$

$$+ \frac{e_0^2}{40i\omega'}g_0\frac{1}{s+6}\left[\frac{1}{(s+2-3i\omega')(s-4i\omega')}\right.$$

$$+ \frac{2}{(s+2-i\omega')s} - \frac{1}{(s+2-i\omega')(s-4i\omega')}$$

$$+ \frac{1}{(s+2-3i\omega')(s-2i\omega')}$$

$$\left.+ \frac{1}{(s+2-i\omega')(s+2i\omega')}\right]$$

$$+ \frac{e_0^2}{14}g_0\frac{1}{(s+6)^2}\left[\frac{1}{(s+2-i\omega')(s-2i\omega')} + \frac{1}{(s+2-i\omega')s}\right]$$

$$+ \frac{e_0^2}{56i\omega'}g_0\left[\frac{1}{(s+6-2i\omega')(s+2-3i\omega')(s-4i\omega')}\right.$$

$$+ \frac{1}{(s+6-2i\omega')(s+2-i\omega')s}$$

$$+ \frac{1}{(s+6-2i\omega')(s+2-3i\omega')(s-2i\omega')}$$

$$+ \frac{1}{(s+6-2i\omega')(s+2-i\omega')(s-2i\omega')}$$

$$- \frac{1}{(s+6)(s+2-3i\omega')(s-4i\omega')}$$

$$- \frac{1}{(s+6)(s+2-3i\omega')(s-2i\omega')}$$

$$- \frac{1}{(s+6)(s+2-i\omega')(s-2i\omega')}$$

$$\left.- \frac{1}{(s+6)(s+2-i\omega')s}\right] + \text{C.C.}$$

$$\mathscr{L}[I_3(\tau)] = \frac{e_0^2}{8} g_0 \left[\frac{1}{(s+6)(s+2+i\omega')(s+6)s} \right.$$

$$+ \left. \frac{1}{(s+6)(s+2+i\omega')(s+6+2i\omega')(s+2i\omega')} \right]$$

$$+ \frac{e_0^2}{16} g_0 \left[\frac{1}{(s+6)(s+2+i\omega')(s+6+2i\omega')(s+4i\omega')} \right.$$

$$+ \frac{1}{(s+6)(s+2+i\omega')(s+6)(s+2i\omega')}$$

$$+ \frac{1}{(s+6)(s+2+i\omega')(s+6)(s-2i\omega')}$$

$$+ \left. \frac{1}{(s+6)(s+2+i\omega')(s+6+2i\omega')s} \right] + \text{C.C.}$$

$$\mathscr{L}[I_4(\tau)] = \frac{e_0^4}{16} \left[\frac{1}{(s+6)(s+2+i\omega')(s+6+2i\omega')(s+2+3i\omega')(s+4i\omega')} \right.$$

$$+ \frac{1}{(s+6)(s+2+i\omega')(s+6)(s+2+i\omega')(s+2i\omega')}$$

$$+ \frac{1}{(s+6)(s+2+i\omega')(s+6)(s+2-i\omega')(s-2i\omega')}$$

$$+ \frac{1}{(s+6)(s+2+i\omega')(s+6+2i\omega')(s+2+i\omega')s}$$

$$+ \frac{1}{(s+6)(s+2+i\omega')(s+6+2i\omega')(s+2+i\omega')(s+2i\omega')}$$

$$+ \frac{1}{(s+6)(s+2+i\omega')(s+6)(s+2-i\omega')s}$$

$$+ \frac{1}{(s+6)(s+2+i\omega')(s+6)(s+2+i\omega')s}$$

$$+ \left. \frac{1}{(s+6)(s+2+i\omega')(s+6+2i\omega')(s+2+3i\omega')(s+2i\omega')} \right]$$

$$+ \text{C.C.}$$

$$\mathcal{L}[I_5(\tau)] = \frac{e_0^2}{8} g_0 \left[\frac{1}{(s+6)(s+12+i\omega')(s+6)s} \right.$$

$$+ \left. \frac{1}{(s+6)(s+12+i\omega')(s+6+2i\omega')(s+2i\omega')} \right]$$

$$+ \frac{e_0^2}{16} g_0 \left[\frac{1}{(s+6)(s+12+i\omega')(s+6+2i\omega')(s+4i\omega')} \right.$$

$$+ \frac{1}{(s+6)(s+12+i\omega')(s+6)(s+2i\omega')}$$

$$+ \frac{1}{(s+6)(s+12+i\omega')(s+6)(s-2i\omega')}$$

$$+ \left. \frac{1}{(s+6)(s+12+i\omega')(s+6+2i\omega')s} \right] + \text{C.C.}$$

$$\mathcal{L}[I_6(\tau)] = \frac{e_0^4}{16} \left[\frac{1}{(s+6)(s+12+i\omega')(s+6+2i\omega')(s+2+3i\omega')(s+4i\omega')} \right.$$

$$+ \frac{1}{(s+6)(s+12+i\omega')(s+6)(s+2+i\omega')(s+2i\omega')}$$

$$+ \frac{1}{(s+6)(s+12+i\omega')(s+6)(s+2-i\omega')(s-2i\omega')}$$

$$+ \frac{1}{(s+6)(s+12+i\omega')(s+6+2i\omega')(s+2+i\omega')s}$$

$$+ \frac{1}{(s+6)(s+12+i\omega')(s+6+2i\omega')(s+2+i\omega')(s+2i\omega')}$$

$$+ \frac{1}{(s+6)(s+12+i\omega')(s+6)(s+2-i\omega')s}$$

$$+ \frac{1}{(s+6)(s+12+i\omega')(s+6)(s+2+i\omega')s}$$

$$+ \left. \frac{1}{(s+6)(s+12+i\omega')(s+6+2i\omega')(s+2+3i\omega')(s+2i\omega')} \right]$$

$$+ \text{C.C.}$$

APPENDIX C. RANDOM WALK

In order to clarify the significance of the Gaussian distribution, we consider the problem of the random walk. The Bernoulli distribution $f(n, N)$ arises when we consider the following problem: If we put N identical and noninteracting particles into two boxes, A and B, and furthermore, if the probabilities of the particles entering boxes A and B are p and q, respectively, the probability of finding n particles in box A is given by

$$f(n, N) = \frac{N!}{n!(N-n)!} p^n q^{N-n} \qquad (A3.1)$$

where

$$p + q = 1 \qquad (A3.2)$$

The average value of n, $\langle n \rangle$, and the variance $\sigma^2 = \langle n^2 \rangle - \langle n \rangle^2$ are obtained from the relation

$$\langle n^k \rangle = \sum_{n=0}^{N} n^k f(n, N) \qquad (A3.3)$$

where $k = 1, 2, 3, \ldots,$. In order to calculate the average value in Eq. (A3.3), it is convenient to introduce a generating function

$$F(z, n, N) = \sum_{n=0}^{N} z^n f(n, N) = (pz + q)^N \qquad (A3.4)$$

It is evident that

$$\langle n \rangle = \left. \frac{\partial F}{\partial z} \right|_{z=1} = Np \qquad (A3.5)$$

$$\langle n^2 \rangle - \langle n \rangle = \left. \frac{\partial^2 F}{\partial z^2} \right|_{z=1} = N(N-1)p^2. \qquad (A3.6)$$

Therefore, Eqs. (A3.5) and (A3.6) lead to

$$\sigma^2 = \langle (n - \langle n \rangle)^2 \rangle = \langle n^2 \rangle - \langle n \rangle^2 = Npq \qquad (A3.7)$$

Now we shall derive the Gaussian distribution by employing some approximation in Eq. (A3.1). In doing so, we define the function

$$\Phi(n, N) = \ln f(n, N) \qquad (A3.8)$$

On using Stirling's formula, which is valid for large values of N,

$$\ln N! = N \ln N - N$$

we find that

$$\Phi(n, N) \simeq N \ln N - n \ln n - (N-n) \ln(N-n) + n \ln p + (N-n) \ln q$$
(A3.9)

The value of n that maximizes $\Phi(n, N)$ or $f(n, N)$ represented by n^* can be obtained from the condition of

$$\frac{\partial \Phi(n, N)}{\partial n} = 0,$$
(A3.10)

as

$$n^* = Np = \langle n \rangle$$
(A3.11)

By expanding $\Phi(n, N)$ with respect to n^*, we find that

$$\Phi(n, N) = \Phi(n^*, N) + \frac{1}{2!} \Phi^{(2)}_{n=n^*} (n - n^*)^2 + \frac{1}{3!} \Phi^{(3)}_{n=n^*} (n - n^*)^2 + \cdots$$
(A3.12)

where $\Phi^{(k)}_{n=n^*}$ stands for $\partial^k \Phi / \partial n^k$ at $n = n^*$, which can be calculated easily to give

$$\frac{\partial^k \Phi}{\partial n^k} = (k-2)! (-1)^k \left[(n-N)^{1-k} - n^{1-k} \right]$$
(A3.13)

In view of Eq. (A3.7), it is seen that since

$$|n - n^*| \simeq \sqrt{Npq}$$
(A3.14)

$\Phi^{(k)}_{n=n^*}/k!$ is on the order of $N^{1-(k/2)}$, enabling us to neglect higher order terms than $\Phi^{(3)}_{n=n^*}/3!$ when N is very large. Therefore, we find that

$$\frac{f(n)}{f(n^*)} = \exp\left[-\frac{1}{2} \frac{(n-n^*)^2}{Npq} \right]$$
(A3.15)

It has been shown that the Bernoulli distribution function becomes Gaussian when N is so large that we can use the approximation in Eq. (A3.9).

With the above result, we shall now consider the random walk problem. We assume that a random walker takes a step either forward or backward with the same probability, $p = q = \frac{1}{2}$, and ask what the probability $W(m, N)$ is of finding the walker at m, which is one of the positions $-N, -(N-1), \ldots, -1, 0, 1, \ldots, (N-1), N$, after taking N steps. This problem may be treated by representing the numbers of steps taken forward and backward by M_f and M_b, respectively. Then we have

$$M_f + M_b = N \tag{A3.16a}$$

$$M_f - M_b = m \tag{A3.16b}$$

which lead to

$$M_f = \tfrac{1}{2}(N + m) \tag{A3.17a}$$

$$M_b = \tfrac{1}{2}(N - m) \tag{A3.17b}$$

It is noted from Eqs. (A3.17) that m should be an even or odd integer, depending on whether N is even or odd, respectively, because M_f or M_b is an integer. Now, on putting $n = M_f$ and $p = q = \frac{1}{2}$ in Eq. (A3.1), we find that

$$W(m, N) = \frac{N!}{M_f! M_b!} \left(\frac{1}{2}\right)^N \tag{A3.18}$$

When N is very large, it follows from Eq. (A3.15) that

$$\frac{W(m, N)}{W(0, N)} = \exp\left(-\frac{1}{2}\frac{m^2}{N}\right) \tag{A3.19}$$

where $W(0, N)$ for very large N is given by

$$W(0, N) = \sqrt{\frac{2}{\pi N}} \tag{A3.20}$$

If the displacement of a step is l and the position of the walker is x, so that $x = lm$ and if $Nl = \alpha t$, where α is the velocity of the walker, then Eqs. (A3.19) and (A3.20) leads to

$$W(m, N) = \sqrt{\frac{2l}{\pi \alpha t}} \exp\left(-\frac{1}{2}\frac{x^2}{\alpha l t}\right) \tag{A3.21}$$

This is the Gaussian distribution function for the position x of the random walker.

It should be pointed out that the Bernoulli distribution function in Eq. (A3.18) leads to the Gaussian in Eq. (A3.21) under the condition of the large value of N. From this example, we have seen that a Gaussian random variable whose average is zero (the centered random variable) behaves just like x.

APPENDIX D. GAUSSIAN RANDOM VARIABLES

Random variables $x_1, x_2, \ldots, x_M$, where x_i $(i = 1, 2, 3, \ldots, M)$ is defined in the range between $-\infty$ and ∞, are called Gaussian when they obey the distribution function

$$g(x_1, x_2, \ldots, x_M) = (2\pi)^{-M/2} |\mathbf{M}|^{-1/2} \exp\left(-\tfrac{1}{2} \tilde{\mathbf{x}} \mathbf{M}^{-1} \mathbf{x}\right) \quad \text{(A4.1)}$$

where

$$\mathbf{M} = \begin{bmatrix} \langle x_1 x_1 \rangle & \langle x_1 x_2 \rangle & \cdots & \langle x_1 x_M \rangle \\ \langle x_2 x_1 \rangle & \langle x_2 x_2 \rangle & \cdots & \langle x_2 x_M \rangle \\ \cdots\cdots\cdots\cdots\cdots\cdots\cdots\cdots\cdots\cdots\cdots \\ \langle x_M x_1 \rangle & \langle x_M x_2 \rangle & \cdots & \langle x_M x_M \rangle \end{bmatrix} \quad \text{(A4.2)}$$

and $\mathbf{x}$ is the column matrix whose transposed matrix is $\tilde{\mathbf{x}}$, given by

$$\tilde{\mathbf{x}} = [x_1, x_2, \ldots, x_M] \quad \text{(A4.3)}$$

The exponent in Eq. (A4.1) may be also written by noting

$$\tilde{\mathbf{x}} \mathbf{M}^{-1} \mathbf{x} = \sum_{m=1}^{M} \sum_{n=1}^{M} \frac{M_{m,n}}{|\mathbf{M}|} x_m x_n \quad \text{(A4.4)}$$

where $M_{m,n}$ is the cofactor of $\langle x_m x_n \rangle$ in $|\mathbf{M}|$. Therefore it is seen that for Gaussian random variables the distribution function in Eq. (A4.1) can be written down readily once the second moment $\langle x_i x_j \rangle$ is obtained.

In particular, for two Gaussian random variables x and y, $g(x, y)$ is

$$g(x, y) = \frac{\left(\langle x^2 \rangle \langle y^2 \rangle - \langle xy \rangle^2\right)^{-1/2}}{2\pi} \exp\left[\frac{-\langle y^2 \rangle x^2 - \langle x^2 \rangle y^2 + 2\langle xy \rangle xy}{2(\langle x^2 \rangle \langle y^2 \rangle - \langle xy \rangle^2)}\right]$$

$$\text{(A4.5)}$$

The Gaussian distribution function g has another nice property when it is used to express the characteristic function $G(u_1, u_2, \ldots, u_M)$,

$$
\begin{aligned}
G(u_1, u_2, \ldots, u_M) &= \langle \exp(-iu_1 x_1 - iu_2 x_2 - \cdots - iu_M x_M) \rangle \\
&= \int_{-\infty}^{\infty} dx_1 \int_{-\infty}^{\infty} dx_2 \cdots \int_{-\infty}^{\infty} dx_M \, g(x_1, x_2, \ldots, x_M) \\
&\quad \times \exp(-iu_1 x_1 - iu_2 x_2 - \cdots - iu_M x_M) \\
&= \exp\left(-\tfrac{1}{2} \sum_{m=1}^{M} \sum_{n=1}^{M} \langle x_m x_n \rangle u_m u_n\right)
\end{aligned}
\tag{A4.6}
$$

It follows that the sum of Gaussian random variables is also a Gaussian random variable.

APPENDIX E.   COMPUTER SIMULATION OF LANGEVIN'S EQUATION

To obtain the phase diagram in Fig. 54 by computer simulation, we start with Langevin's equation

$$
\frac{d\omega(t)}{dt} + \beta\omega(t) = A(t) - \frac{\mu E}{I} \sin\phi
$$

This may be put into

$$
\omega(t + \Delta t) = \omega(t) - \beta\omega(t)\Delta t - \frac{\mu E}{I}\Delta t \sin\phi + B(\Delta t)
$$

where

$$
B(\Delta t) = \int_{t}^{t + \Delta t} A(t') \, dt'
$$

It is assumed as before that in Δt, $A(t)$ changes quite randomly and swiftly, whereas $\omega(t)$ and $\phi(t)$ do not. Since $A(t)$ is a Gaussian random variable, $B(\Delta t)$ is also a Gaussian random variable. The variance of $B(\Delta t)$, $\langle B^2(\Delta t) \rangle$ is calculated to be

$$
\begin{aligned}
\langle B^2(\Delta t) \rangle &= \int_{t}^{t + \Delta t} \int_{t}^{t + \Delta t} \langle A(t_1) A(t_2) \rangle \, dt_1 \, dt_2 \\
&= 2 \int_{t}^{t + \Delta t} dt_1 \int_{t}^{t_1} \langle A(t_1) A(t_2) \rangle \, dt_1 \, dt_2 \\
&= 2\sigma^2 \Delta t
\end{aligned}
$$

We can generate the Gaussian random variable X, if the variance $\langle X^2 \rangle$ and the average $\langle X \rangle$ are given, from the uniform random number Y_i which is readily available in most computers, by using the theorem stating that

$$Z = \sum_{i=1}^{12} Y_i$$

satisfies

$$X = (Z - 6)\sqrt{\langle X^2 \rangle} + \langle X \rangle$$

REFERENCES

1. See the following books:
   E. Fredericq and C. Houssier, "Electric Dichroism and Electric Birefringence" Clarendon Press, Oxford (1973);
   "Molecular Electro-Optics" Parts 1 and 2, C. T. O'Konski Ed., Marcel Dekker, Inc., New York (1976);
   "Electro-Optics and Dielectrics of Macromolecules and Colloids" B. R. Jennings Ed., Plenum Press, New York (1979);
   "Molecular Electro-Optics- Electro-Optic Properties of Macromolecules and Colloids in Solution-" S. Krause Ed., Plenum Press, New York (1981).
2. A. Peterlin and H. A. Stuart, *Hand und Jahrbuch der Chemischen Physik*, Vol. 8, Akademische Verlagses, Leipzig, 1943.
3. A. D. Buckingham and J. A. Pople, *Proc. Phys. Soc.* **A68**, 905 (1955).
4. C. T. O'Konski, K. Yoshioka, and W. H. Oruttung, *J. Phys. Chem.* **63**, 1558 (1959).
5. M. J. Shah, J. Phys. Chem. **67**, 2215 (1963)
6. W. B. Jones and W. J. Thron, *Continued Fractions*, Addison-Wesley, Reading, MA, 1980.
7. H. Benoit, *Ann. Phys.* **6**, 561 (1951).
8. I. Tinoco and K. Yamamoka, *J. Phys. Chem.* **63**, 423 (1959).
9. G. Schwarz, *Z. Phys.* **145**, 563 (1956).
10. K. Nishinari and K. Yoshioka, *Kolloid-Z. Z. Polymere* **235**, 1189 (1969).
11. M. Matsumoto, H. Watanabe, and K. Yoshioka, *J. Phys. Chem.* **74**, 2182 (1970).
12. G. Koopmans, J. de Bore, and J. Greve, *Electro-optics and Dielectrics of Macromolecules and Colloids*, B. R. Jennings, Ed., Plenum, New York, 1979.
13. S. Ogawa and S. Oka, *J. Phys. Soc. Japan* **15**, 658 (1960).
14. G. B. Thurston and D. L. Bowling, *J. Colloid Interface Sci.* **30**, 34 (1969).
15. H. H. Käs and R. Brückner, *Z. Angew. Phys.* **36**, 368 (1969).
16. A. Morita, *J. Phys. D* **11**, 1357 (1978).
17. A. Morita and H. Watanabe, *J. Chem. Phys.* **70**, 4708 (1979).
18. H. S. Wall, *Analytical Theory of Continued Fractions*, Chelsea Publishing, New York, 1967.
19. M. Abramowitz and I. A. Stegun, (Eds.), *Handbook of Mathematical Functions*, Dover, New York, 1970.
20. H. Watanabe and A. Morita, *J. Chem. Phys.* **73**, 5884 (1980).

21. H. Watanabe and A. Morita, *J. Chem. Phys.* **75**, 379 (1981).

22. K. Yoshioka and H. Watanabe, *Nippon Kagaku Zasshi* (*J. Chem. Soc. Japan*) **84**, 626 (1964).

23. K. Yoshioka and H. Watanabe, in *Physical Principles and Techniques of Protein Chemistry*, Part A, S. J. Leach, Ed. Academic Press, New York, 1969.

24. A. Morita and H. Watanabe, *J. Chem. Phys.* **75**, 1320 (1981).

25. H. Watanabe and A. Morita, *J. Chem. Phys.* **78**, 5311 (1983).

26. A. Morita and H. Watanabe, *J. Chem. Phys.* **77**, 1193 (1982).

27. A. Morita, *J. Phys* **A12**, 991 (1977)

28. S. Chandrasekhar, *Rev. Mod. Phys.* **15**, 1 (1943).

29. R. Kubo, in *1965 Tokyo Summer Lectures in Theoretical Physics*, R. Kubo, Ed., Shokabo Tokyo, and Benjamin, New York, 1966.

30. A. Einstein, *Investigations on the Theory of the Brownian Movement by Albert Einstein*, *Ph.D.* Edited by Fruth and translated by A. D. Coper, Dover, New York, 1956.

31. R. A. Sack, *Proc. Phys. Soc. London Sect. B* **70**, 402 (1957).

32. B. K. P. Scaife, *Complex Permittivity*, English Univ. Press, London, 1971.

33. H. Goldstein, *Classical Mechanics*, Addison-Wesley, Reading, MA, 1980.

34. A. Morita, Colloquium on Rotational Brownian Motion held at the School of Theoretical Physics, Dublin Institute for Advanced Studies, 24 August 1976.

35. A. Morita, *J. Chem. Phys.* **76**, 3198 (1982).

36. J. McConnell, *Rotational Brownian Motion and Dielectric Theory*, Academic Press, London, 1980.

37. M. W. Evans, G. J. Evans, W. T. Coffey, and P. Grigolini, *Molecular Dynamics*, Wiley-Interscience, New York, 1982.

38. D. N. Holcomb and I. Tinoco, *J. Phys. Chem.* **67**, 2691 (1963).

THE INTERNAL FIELD PROBLEM IN DEPOLARIZED LIGHT SCATTERING

THOMAS KEYES

Department of Chemistry
Boston University
Boston, MA 02215

BRANKA M. LADANYI

Department of Chemistry
Colorado State University
Fort Collins, CO 80523

CONTENTS

I. INTRODUCTION

A. Ground Rules and Definitions

This article is a review of the developments of roughly the last ten years concerning the importance of the internal field problem (IFP) in depolarized light scattering; that importance is considerable. Although we want to keep to this specific focus as much as possible, the task is not easy. Discussion of the IFP is impossible without some background in the basic theory of light scattering. Furthermore, much of the confusion that has surrounded the IFP over the years seems to be due to a surprising misunderstanding of these ideas in the scientific community. Presenting them is therefore more than offering background; they are actually part of our topic.

In view of this, after a brief collection of basic facts and definitions, we will present our view of the essentials of light scattering theory. We hope that this initial discussion will clarify the review of the IFP in depolarized scattering that occupies the remaining, and major part, of the article.

A light scattering experiment is shown in Fig. 1. The incident light is assumed polarized along the z, or vertical (v), axis and propagates in the xy plane. Scattered light propagates in the xy, or horizontal (h), plane and is collected with either z polarization for polarized (vv) scattering, or h polarization for depolarized (vh) or "anisotropic" scattering. The angle between the incident and scattered beams is denoted θ. A wavevector $\mathbf{k}_{i,s}$ is associ-

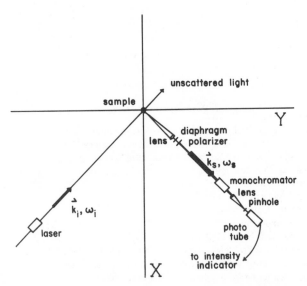

Fig. 1.   Schematic representation of a typical light scattering experiment.

ated with the incident and scattered light,

$$\mathbf{k}_{i,s} = \frac{2\pi}{\lambda_{i,s}} \hat{e}_{i,s} \tag{1.1}$$

where $\hat{e}$ is the direction of propagation and λ the wavelength; for quasi-elastic scattering, we may let $\lambda_s = \lambda_i = \lambda$ when evaluating $\mathbf{k}$. Of course, $\mathbf{k}$ is related to the angular frequency ω by the relation

$$\omega = \frac{|k|c}{n} \tag{1.2}$$

and c/n is the velocity of light in a medium of refractive index n.

The incident beam is usually focused to a point in the sample, which is also the focal point of a lens in the collection optics. Thus, a "scattering volume" V is roughly defined, being that part of the sample from which scattered light originates and is collected. We shall soon see that great care must be used with this concept.

If the scattered light is frequency-analyzed before detection, the spectrum $S(\omega)$ is obtained. In the absence of a monochromator, the intensity I is determined, and $I = \int d\omega \, S(\omega)$. The intensities I^{vv} and I^{vh} are related according to

$$I^{vv} = I^{vh} + I^{iso} \tag{1.3}$$

where the isotropic scattering I^{iso} arises due to optically isotropic fluctuations in the sample (in contrast to anisotropic scattering). Our use of this terminology should become clear later on.

The phenomena we are to discuss have virtually no θ dependence. Thus, we will be thinking mostly of a $\theta = 90°$ experiment, and we will assume that I and S still have their zero-angle forms at $90°$. Finally, we will consider just two classes of molecule here: rigid, optically symmetric molecules and flexible chain molecules.

B. Basic Theory

The light scattering experiment cannot be interpreted without an understanding of the physics involved. The basic idea is very simple.[1] The applied field induces a dipole moment μ_i in the ith molecule in the sample according to the relation

$$\mu_i = \underset{\sim}{\alpha}_i \cdot \mathbf{E}_i^L \tag{1.4}$$

where $\underset{\sim}{\alpha}_i$ is the molecular polarizability tensor in the laboratory frame and $\mathbf{E}_i^L$ is the instantaneous field at the molecule. Since $\mathbf{E}$ is an optical field, μ is an oscillating dipole, which radiates according to classical electrodynamics.[2] The radiation from all the particles in the sample is the scattered field.

More formally, one must solve Maxwell's equations in the presence of a polarizable sample, which is furthermore undergoing thermal fluctuations in its optical properties. The polarization of the sample by the incident light creates a source, or inhomogeneous, term in the equations that gives rise to the scattered electric field. The equations, after some simple manipulations, become

$$\nabla^2 \mathbf{E} + \left(k_0^2 - \nabla \nabla \cdot \right) \mathbf{E} = - k_0^2 \mathbf{P} \tag{1.5}$$

where $\mathbf{k}_0$ is the wavevector in the absence of the sample ($n = 1$), and we have assumed that the time dependence of all quantities is $e^{i\omega_0 t}$. The homogeneous (left-hand side) part of this equation is just that for electric field propagation in vacuo, while the inhomogeneity $\mathbf{P}$ on the right-hand side is the polarization, or induced dipole moment, density; the solution is[2]

$$\mathbf{E} = \mathbf{E}^0 + k_0^2 (4\pi D)^{-1} e^{ik_0 D} \int_V d\mathbf{r} \, \mathbf{P}(\mathbf{r}) \exp[- i\mathbf{k}_{0,s} \cdot \mathbf{r}] \tag{1.6}$$

where D and $\mathbf{r}$ are the distances from the center of the scattering volume to the detector ($D \gg \lambda$) and to a point in V, and $\mathbf{E}^0$ is the field in the absence of sample. It is straightforward to use Eq. (1.6) to construct the scattered intensity and to perform an average ($\langle \ \rangle$) over thermal fluctuations with the result

$$I^\gamma = \int_V d\mathbf{r} \, d\mathbf{r}' \exp[i\mathbf{q}_0 \cdot (\mathbf{r} - \mathbf{r}')] \langle P^\gamma(\mathbf{r}) P^\gamma(\mathbf{r}') \rangle$$

$$(\gamma = z, vv; \ \gamma \text{ in } xy \text{ plane}, vh) \tag{1.7}$$

where I is now in arbitrary units,

$$\mathbf{q}_0 = \mathbf{k}_{0,i} - \mathbf{k}_{0,s} \tag{1.8}$$

and

$$|q_0| = \frac{4\pi}{\lambda} \sin\left(\frac{\Theta}{2}\right) \tag{1.9}$$

Of course, we have assumed $\mathbf{E}^0 = 0$ at the detector.

If we divide **P** into its average and its fluctuation,

$$\mathbf{P} = \langle \mathbf{P} \rangle + \delta \mathbf{P} \tag{1.10}$$

then, since $\langle \mathbf{P} \rangle$ is a constant, the integral in Eq. (1.7) requires that $\langle \mathbf{P} \rangle$ contribute only if $q_0 = 0$, or $\theta = 0$, that is, for forward scattering, which we do not wish to discuss. Thus, we find the important result that only fluctuations contribute to light scattering. For the problems to be discussed here, the q dependence of the integral over the fluctuation correlation function is negligible, and we finally obtain what is the starting point of many papers,

$$I^{\gamma} = \int_{V} d\mathbf{r} \, d\mathbf{r}' \langle \delta P^{\gamma}(\mathbf{r}) \delta P^{\gamma}(\mathbf{r}') \rangle \tag{1.11}$$

According to Eq. (1.11), all we need for a light scattering theory is to evaluate $\delta \mathbf{P}$ from Eq. (1.4) and the definition of P as the dipole moment density and take the average. This statement is correct. It is also the cause of almost all of the confusion that has historically plagued the theory of light scattering.

The key point is that the integral over the "scattering volume" in Eqs. (1.6) and (1.7) is a pure hoax. If real samples consisted of tiny globules floating in the laser beam, all would be well. In reality, however, the sample is large. The incident field passes through a converging lens, through the cell-air interface, and through considerable sample before reaching minimum width. The point of minimum width is arranged as the focal point of a lens along the y axis, which (for 90° scattering) channels the scattered light into the detection system. Scattering originates from *every part* of the cell—from the directly illuminated parts and from the other regions via multiple scattering. How, then, is one supposed to express all this as a contribution from a simple, small scattering volume? One might be tempted to just draw a small sphere around the focal point, which cannot be justified, as we will now see.

A paper by Fixman[3] provides the answers to these questions. We now outline the paper. It is simple to calculate what happens to light upon entering and leaving the cell. So it makes sense to set up our light scattering problem in a universe filled with sample—the interior of the cell—and deal with everything else by macroscopic dielectric theory. In this formulation, $\mathbf{E}^{0}$, the field in the absence of sample, is not a natural quantity, and it seems reasonable to write the scattering as the correction to a different zero-order, or unperturbed, field. The field that immediately suggests itself is the averaged Maxwell field in the sample, $\langle \mathbf{E} \rangle$, which is related to $\langle \mathbf{P} \rangle$ via the dielectric constitutive relation,

$$\langle \mathbf{P} \rangle = \frac{\langle \varepsilon \rangle - 1}{4\pi} \langle \mathbf{E} \rangle \tag{1.12}$$

and $\langle \varepsilon \rangle$ is the averaged, usual, optical dielectric constant.

These ideas are formally implemented[3] by simply adding a factor of $k_0^2[(\langle \varepsilon \rangle - 1)/4\pi]\mathbf{E}$ to both sides of Eq. (1.5), with the result

$$\nabla^2 \mathbf{E} + (k^2 - \nabla\nabla\cdot)\mathbf{E} = -k_0^2 \left[\mathbf{P} - \frac{\langle \varepsilon \rangle - 1}{4\pi} \mathbf{E} \right] \tag{1.13}$$

The inhomogeneous term on the right-hand side of Eq. (1.13) will be denoted $\mathbf{S}$, for source, and

$$\mathbf{S} = \mathbf{P} - \frac{\langle \varepsilon \rangle - 1}{4\pi} \mathbf{E} \tag{1.14}$$

Note that $\delta \mathbf{S} = \mathbf{S}$, that is, $\langle \mathbf{S} \rangle = 0$. The right-hand side of Eq. (1.13) is zero if the fields have their averaged values, and the solution of the homogeneous equation must be $\langle \mathbf{E} \rangle$. We then obtain the desired result by solving Eq. (1.13):

$$\mathbf{E} = \langle \mathbf{E} \rangle + k_0^2 (4\pi D)^{-1} e^{ikD} \int_V d\mathbf{r} \, \mathbf{S}(\mathbf{r}) \exp(i\mathbf{k}_s \cdot \mathbf{r}) \tag{1.15}$$

where we still retain the ill-defined concept of scattering volume.

It is important to realize that both Eqs. (1.15) and (1.6) are correct; they just represent different choices of a zero-order system. In Eq. (1.6), scattering is the correction to $\mathbf{E}^0$, the in vacuo result; the source for this expansion is $\delta\mathbf{P}$, and the in vacuo wave vector appears.

In Eq. (1.15), scattering is the correction to the field, $\langle \mathbf{E} \rangle$, in a medium with nonfluctuating optical properties; the source is $\mathbf{S}$, and the "in-medium" wavevector $\mathbf{k}$ appears.

Which of these formulations should be used in the construction of a theory of the light scattering experiment? It seems logical to choose the zero-order approximation, which best describes the scattering medium. And, optically, a liquid resembles a nonfluctuating dielectric (dielectric continuum) far more closely than it does a vacuum. For dilute gases, the formulations become equivalent, since $\langle \varepsilon \rangle \to 1$.

A much more substantive argument may be made about the desirability of $\mathbf{S}$ as the source. Let us return to the problem of describing the real geometry and optics of a scattering experiment. Fixman proposed the following hypothetical system: Suppose the cell is filled with dielectric continuum everywhere except in a tiny sphere around the focal point of the collecting lens, which is filled with real, fluctuating sample (the "scattering volume"). While an idealization, this clearly represents an improvement over considering *only* the tiny sphere and ignoring the rest of the cell.

In the continuum, even though **E** fluctuates due to scattering from the real sample, the constitutive relation holds, and **S** = 0. Thus, we can forget the rest of the cell and concentrate on a simple, small "scattering volume," as is essential if a simple theory is going to be possible. The resulting expression for I is, again neglecting q dependence,

$$I^\gamma = \int_V d\mathbf{r}\, d\mathbf{r}' \langle S^\gamma(\mathbf{r}) S^\gamma(\mathbf{r}') \rangle \qquad (1.16)$$

Recall, now, that use of Eq. (1.6) and the assumption that only a small volume need be considered gave Eq. (1.11), which would be obtained for the Fixman idealization as well. Equations (1.11) and (1.16) cannot both be correct. They both arise from correct equations by ignoring all but the scattering volume. However, **S** is in fact zero outside, while $\delta\mathbf{P}$ is not. One concludes that a theory based upon $\delta\mathbf{P}$ must consider more than a simple scattering volume; it must treat the entire cell, a complication that is highly undesirable. We therefore come to the main conclusion of this section:

Light scattering theories should be formulated in terms of the fluctuations in **S**. Theories using $\delta\mathbf{P}$ are wrong unless:

1. Scattering from the entire cell is considered.
2. The density is very low (dilute gas) and $\mathbf{S} = \delta\mathbf{P}$.
3. Due to luck, the consequences of using the wrong source are unimportant.

Now, many of the modern papers on light scattering still use $\delta\mathbf{P}$ as the source. Of course, we urge an end to this practice, but what are the actual consequences? Light scattering research is of two main types: spectral studies, where the shape of $S(\omega)$ is considered with no regard to its magnitude, and studies of I. For the former case, the use of **P** usually does little damage. It is virtually impossible, however, to construct a theory of I in this way, which is one of the main themes of this article.

II. THE INTERNAL FIELD PROBLEM AND THE LOW-DENSITY LIMIT

We have reduced the light scattering problem to evaluation of the mean-square fluctuations in **S**; it is here that the IFP arises. Note that the IFP also arises in evaluation of $\langle \delta\mathbf{P}\delta\mathbf{P} \rangle$; however, since $\delta\mathbf{P}$ is not **S**, even a correct evaluation of this quantity would not allow, without considerable extra work, determination of I. Within the $\mathbf{P}$ approach, this extra work would look like part of the IFP. Thus, from that point of view, using S automatically incorporates part of the IFP.

As a first step, consider the limit of zero density (ρ). Since $(\langle \varepsilon \rangle - 1) \propto \rho$ as $\rho \to 0$, $\mathbf{S} = \delta \mathbf{P}$ here, and both formulations are equivalent. According to Eq. (1.5) and the definition of $\mathbf{P}$ as dipole moment density,

$$\mathbf{P}(\mathbf{r}) = \sum_i \delta(\mathbf{r} - \mathbf{r}_i) \underset{\sim}{\alpha}_i \cdot \mathbf{E}_i^L \tag{2.1}$$

where the sum is over all molecules. The local field $\mathbf{E}_i^L$ is the field "seen" by molecule i. In general, $\mathbf{E}_i^L$ is different for different i and is a complicated function of the arrangement of the neighbor molecules, unlike the other fields, $\mathbf{E}^0$ and $\langle \mathbf{E} \rangle$. However, in the $\mathbf{S}$ approach, $\langle \mathbf{E} \rangle$ is assumed to be under control, so we want $I(\langle \mathbf{E} \rangle)$, which requires $\mathbf{E}^L(\langle \mathbf{E} \rangle)$; to a large extent, finding $\mathbf{E}^L(\langle \mathbf{E} \rangle)$ is the IFP.

As $\rho \to 0$, $\mathbf{E}_i^L$ obviously becomes equal to $\mathbf{E}^0 + O(\rho)$; the perturbation by neighboring molecules of the field that a molecule "sees" vanishes as the number of neighbors vanish. For the same reason, $\langle \mathbf{E} \rangle \to \mathbf{E}^0$. If we define the Rayleigh ratio R by

$$R = \frac{I}{|\langle E \rangle|^2} \tag{2.2}$$

we then have, for the case of 90° scattering

$$R^\gamma = \int_V d\mathbf{r}\, d\mathbf{r}' \left\langle \delta\left(\sum_i \delta(\mathbf{r} - \mathbf{r}_i) \alpha_i^{z\gamma} \right) \delta\left(\sum_j \delta(\mathbf{r}' - \mathbf{r}_j) \alpha_j^{z\gamma} \right) \right\rangle \tag{2.3}$$

At this point, the two types of molecules under discussion show quite different behavior. For chain molecules, $\underset{\sim}{\alpha}$ depends on the full chain configuration and on solution of an intramolecular IFP, where each part of the molecule sees the remainder as a "solvent." So even letting ρ approach zero does not banish the IFP, and we defer treatment of this case until Section VI.

For rigid, axially (z) symmetric molecules, $\underset{\sim}{\alpha}$ is conveniently divided[1] into a trace and a traceless part,

$$\underset{\sim}{\alpha} = \alpha_0 \mathbf{1} + \Delta\alpha \mathbf{Q} \tag{2.4}$$

where

$$\alpha_0 = \tfrac{1}{3} \mathrm{Tr}\, \underset{\sim}{\alpha} \tag{2.5}$$

$$\Delta\alpha = \alpha_{zz} - \alpha_{xx}$$

1 is the unit tensor,

$$Q_i = \hat{u}_i \hat{u}_i - \tfrac{1}{3} 1 \tag{2.6}$$

where $\hat{u}_i$ is a unit vector along the symmetry axis of molecule i. Combination of Eqs. (2.3)–(2.5) gives

$$R^\gamma = \delta_{\gamma z} \alpha_0^2 \int_V d\mathbf{r}\, d\mathbf{r}' \langle \delta n(\mathbf{r}) \delta n(\mathbf{r}') \rangle + (\Delta \alpha)^2 \int_v d\mathbf{r}\, d\mathbf{r}' \langle Q^{z\gamma}(\mathbf{r}) Q^{z\gamma}(\mathbf{r}') \rangle \tag{2.7}$$

where the density $n(\mathbf{r})$ and the orientational order parameter $Q(\mathbf{r})$ are given by the relations

$$n(\mathbf{r}) = \sum_i \delta(\mathbf{r} - \mathbf{r}_i) \tag{2.8}$$

$$Q(\mathbf{r}) = \sum_i Q_i \delta(\mathbf{r} - \mathbf{r}_i) \tag{2.9}$$

(Note $\langle Q \rangle = 0$.)

Equation (2.7) displays the basic features of scattering from nonspherical molecules. As mentioned earlier, scattering arises from fluctuations in S, and those fluctuations arise from fluctuations in more physically conceivable variables, such as the density. So, Eq. (2.7) contains contributions from both density (denoted R^ρ) and orientational (R^ϕ) fluctuations. This makes perfect sense; the optical properties of the scattering volume change due to changes in the number of molecules present and due to reorientation of molecules already present via the Q term in Eq. (2.4). Density fluctuations produce polarized or isotropic scattering involving only the trace or isotropic part of α, and orientational fluctuations produce polarized and depolarized anisotropic scattering involving the polarizability anisotropy $\Delta \alpha$.

This mean square fluctuation in Eqs. (2.7) is related[4] to the fluid structure by the relations

$$\int_V d\mathbf{r}\, d\mathbf{r}' \langle \delta n(\mathbf{r}) \delta n(\mathbf{r}') \rangle = \rho V (1 + \rho \hat{h}(o)) \tag{2.10}$$

$$\int_V d\mathbf{r}\, d\mathbf{r}' \langle Q^{z\gamma}(\mathbf{r}) Q^{z\gamma}(\mathbf{r}') \rangle = \rho V \left(1 + \tfrac{1}{3} \delta_{\gamma z}\right)(1 + \rho f) \tag{2.11}$$

where $\hat{h}(0)$ is the integral over the pair correlation function, and f is a measure of the orientational pair correlations in the system:

$$f = V \frac{\langle Q_i^{xz} Q_j^{xz} \rangle}{\langle |Q_i^{xz}| \rangle^2} \tag{2.12}$$

Fluid structure thus enters the anisotropic scattering via a factor of $(1 + \rho f) \equiv g^{(2)}$. These expressions hold for liquids as well as for dilute gases, and one might be tempted to simply plug them into Eq. (2.7) and obtain a liquid state theory. Such a procedure would hold out the prospect of obtaining the valuable quantities $\hat{h}(0)$ and f. This would be wrong, however, since Eq. (2.7) ignored the IFP and is valid only as $\rho \to 0$. The only conclusion which we can draw with confidence so far is

$$\lim_{\rho \to 0} \frac{R^{\gamma}}{V} = \delta_{\gamma z} \alpha_0^2 \rho + \left(1 + \delta_{\gamma z} \frac{1}{3}\right)(\Delta \alpha)^2 \rho \qquad (2.13)$$

It is not difficult to calculate R as $\rho \to 0$; conversely, the content, or available information, of a light scattering experiment is clear. A far more difficult problem is the subject of this article: construction of a theory valid at liquid density. Of course, such a theory is necessary if light scattering is to be used as a probe (e.g., to determine f) of liquids.

III. HISTORICAL SURVEY AND THE EINSTEIN-SMOLUCHOWSKI THEORY

With the foregoing as background, let us now consider the history of theories of light scattering by dense systems. Early in this century, Einstein[4] and Smoluchowski[5] gave a successful theory of isotropic scattering from liquids. The ES theory was notably simple and may be summarized, in the notation of the previous section, as

$$\mathbf{S}(\mathbf{r}) = \frac{1}{4\pi} \delta\varepsilon(\mathbf{r})\langle \mathbf{E} \rangle \qquad (3.1)$$

$\delta\varepsilon$ being the fluctuation in the dielectric constant. Furthermore, the fluctuations in ε were assumed to follow the local fluctuations in the thermodynamic state variables n and T (temperature) such that

$$\delta\varepsilon(\mathbf{r}) = \left(\frac{\partial\langle\varepsilon\rangle}{\partial\langle n\rangle}\right)_T \delta n(\mathbf{r}) + \left(\frac{\partial\langle\varepsilon\rangle}{\partial\langle T\rangle}\right)_n \delta T(\mathbf{r}) \qquad (3.2)$$

where the derivatives are ordinary thermodynamic derivatives defined on equilibrium states only. It has always been found that $\partial\langle\varepsilon\rangle/\partial\langle T\rangle$ is negligible. Use of Eqs. (3.1) and (3.2) plus a hydrodynamic calculation of the time-dependent density correlation function to calculate $S(k, \omega)$ gives[6,1e,1f] the famous Rayleigh-Brillouin spectrum. A calculation of R^{ρ} gives the zero-density result Eq. (2.13), with α replaced by $\partial\langle\varepsilon\rangle/\partial\langle n\rangle$; within the ES theory, the IFP is entirely incorporated in this replacement. Thus, solution of the scattering problem has been reduced to finding a good equilibrium theory for $\langle \varepsilon \rangle$.

An accurate expression for $\langle \varepsilon \rangle$ is that of Clausius and Mossotti,

$$\frac{\langle \varepsilon \rangle - 1}{\langle \varepsilon \rangle + 2} = \frac{4\pi}{3} \alpha_0 \langle n \rangle \tag{3.3}$$

Equations (3.1)–(3.3) give

$$R(ES) = \left(\frac{\langle \varepsilon \rangle + 2}{3} \right)^4 R \text{ (no IFP)} \tag{3.4}$$

that is, the "internal field correction" is a fourth power of the Lorentz factor, $L \equiv (\langle \varepsilon \rangle + 2)/3$. Equation (3.4) is[7] in excellent agreement with experiment, after years of exhaustive testing. For liquids, where $\langle \varepsilon \rangle \sim 2$, the L^4 factor is large and a sensitive function of $\langle \varepsilon \rangle$, so the agreement is not trivial. Other local field factors have been proposed[8] to explain the small discrepancies that do exist between L^4 and experiment, but these have almost always been based upon equations for $\langle \varepsilon \rangle$ ($\langle n \rangle$) supposedly superior to Eq. (3.3). The ES idea that the problems of light scattering can be solved given a good theory of $\langle \varepsilon \rangle$ and a knowledge of the important fluctuations has been widely accepted. Note that focusing on "important fluctuations" makes it possible to be much more physical and intuitive than if one just wrote down **S** microscopically and proceeded. The phase-space variables describing the important fluctuations have been referred to[9] as "primary variables."

Nevertheless, the ES theory has generated controversy, for reasons which a reading of the last section should make easy to understand. If the equilibrium quantities in the dielectric constitutive relation, Eq. (1.12), are allowed to have small fluctuations, one obtains

$$\delta \mathbf{P} = \frac{1}{4\pi} \delta \varepsilon \langle \mathbf{E} \rangle + \frac{\langle \varepsilon \rangle - 1}{4\pi} \delta \mathbf{E} \tag{3.5}$$

So, if one believes that $\delta \mathbf{P}$ is the appropriate source for light scattering, the right-hand side of Eq. (3.5) contains the ES source plus an extra term containing E that would seem to have been ignored by ES. This problem is immediately cleared up by a fluctuation analysis of the correct source **S** [Eq. (1.14)]; the $[(\langle \varepsilon \rangle - 1)/4\pi]\mathbf{E}$ term gives rise to $[-(\langle \varepsilon \rangle - 1)/4\pi]\delta \mathbf{E}$, which cancels the "extra" part of Eq. (3.5) (note that $\langle \varepsilon \rangle$ here is, by construction, truly $\langle \varepsilon \rangle$ and cannot conceivably fluctuate). Thus,

$$\delta \mathbf{S} = \frac{1}{4\pi} \delta \varepsilon \langle \mathbf{E} \rangle \tag{3.6}$$

and so a theory based upon **S** is in perfect agreement with the ES theory.

In sum, a simple, accurate theory for the IFP in light scattering by thermodynamic (basically density) fluctuations has been available for a long time. Such is far from the case, however, for other types of scattering, such as that due to orientational fluctuations. The main difficulty is that, being a "thermodynamic" quantity, $\langle \varepsilon \rangle$ can depend only on $\langle n \rangle$ and $\langle T \rangle$, so the effect of other fluctuations on the scattering cannot be immediately determined. And it is not completely obvious what other fluctuations are important, although the low-density results make it clear that orientational fluctuations must be included.

For these reasons, a good theory of depolarized scattering was not developed until recently. We see no point in giving a detailed review of the theories[10] in disagreement with the principles outlined so far that arose in the interim. The basic conclusion was that the scattering was due to orientational fluctuations and that the "local field factor" for scattering was L^2, in marked contrast to the L^4 behavior of isotropic scattering. Broadly speaking, the theories erred in use of $\delta \mathbf{P}$ as the source.

In addition, however, it seems likely that experimental data on R predisposed theorists to favor an "L^2 law." Of course, there is no reason why the internal field correction should simply involve multiplication by L raised to any power, $I \sim L^x (\Delta \alpha)^2 \rho g^{(2)}$, but the great success of the ES theory suggests this as an attractive possibility. The data in Table I clearly show that if R were to contain L^4 it would be very much larger than its experimental value unless $g^{(2)}$ were considerably less than unity, which would require that neighbor molecules in a liquid have a strong preference to "pack" with their long axes perpendicular. Although this idea was considered,[11,12] it did not, even before $g^{(2)}$ was available independently for some simple liquids, seem terribly sensible. On the other hand, an L^2 factor does quite a good job of making $g^{(2)} \lesssim 1$ for a wide range of liquids, a reasonable assumption. This is no doubt responsible in large part, for the wide acceptance gained by the L^2 law.

TABLE I
Depolarized Rayleigh Ratios and Lorentz Factors

Liquid	$R^\phi / R^\phi (\varepsilon = 0)$	L^2	L^4
Benzene	1.19	2.02	4.1
Chloroform	1.35	1.86	3.46
Carbon disulfide	2.57	2.43	5.89
Acetonitrile	1.13	1.61	2.60

Source: Reference 14d.

Several authors[13] presented more careful approaches. Their theories were too idealized to treat scattering from real liquids, but they were fundamentally sound and definitively established that an L^2 law made no sense. Then a large amount of experimental and theoretical work was carried out[14] by Alms, Burnham, Gierke, and Flygare. The experiments of this group were particularly important, both because of their high quality and because, for the first time, they actually measured $R(\langle \varepsilon \rangle)$ versus $\langle \varepsilon \rangle$. This may sound strange, but consider: measurements of $R(\langle \varepsilon \rangle)$ on pure liquids vary $\langle \varepsilon \rangle$ and *everything else* in passing from liquid to liquid. Such measurements do not provide an ideal way to obtain the functional form of $R(\langle \varepsilon \rangle)$. Gierke and Flygare[14a] and Alms, Gierke, and Flygare[14b] measured R for dilute solutions of liquid-crystal-forming molecules in various solvents. Since the solute molecules are immensely strong scatterers, scattering by the solvent alone can be ignored, and since the solute is dilute, $g^{(2)} = 1$. Thus, varying the solvent varies $\langle \varepsilon \rangle$ and, to a good approximation, leaves everything else constant.

These authors found that even though an L^2 law might give a reasonable R for a single $\langle \varepsilon \rangle$, $R(\langle \varepsilon \rangle)$ *was a sharply decreasing function of* $\langle \varepsilon \rangle$; a plot of R versus $\langle \varepsilon \rangle$ for the nematogens MBA and EBA is given in Fig. 2. In our view, this constitutes the most powerful empirical argument against an L^2 law, while also proving that no power law is correct. Burnham, Alms, and Flygare[14c] then collected and analyzed considerable data with a phenomenological theory relying heavily on dielectric theory of nonspherical molecules. The main conclusion is that R contains a factor of L^4 *but* the polarizability anisotropy in Eq. (2.13), $\Delta \alpha$, must also be replaced by an effective anisot-

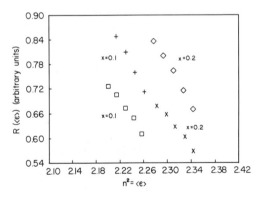

Fig. 2. Experimental demonstration that the depolarized Rayleigh ratio, R^ϕ, is a decreasing function of the optical dielectric constant, $\langle \varepsilon \rangle = n^2$, for MBA and EBA. Taken from Ref. 14a.

ropy, $\gamma(\langle\varepsilon\rangle)$. The form of γ varies with the dielectric theory used but is a decreasing function of $\langle\varepsilon\rangle$. This conclusion is broadly correct, although the theory is only partly correct.

The early errors and later partial successes of theories for R contrast sharply with the short, simple story of the ES theory. As mentioned earlier, we believe this is due to the difficulty of differentiating $\langle\varepsilon\rangle$ with respect to nonthermodynamic variables, leaving one with recourse to either relatively difficult, fully "microscopic" theories or to phenomenology. We will now show that a "generalized ES" theory is really not very difficult and is, unsurprisingly, successful.

IV.   EINSTEIN-SMOLUCHOWSKI THEORY FOR DEPOLARIZED SCATTERING

A.   General Considerations

To circumvent the problems just discussed, we consider[15] an ensemble, denoted $^-$, where $\overline{Q}_i \neq 0$. Such ensembles exist in nature, as in the case of liquid crystals, but this is really irrelevant. We can construct the $^-$ system using response theory; $\overline{\varepsilon}$ will then depend on $\overline{Q}$, and the ES procedure can be carried out. Since the fluctuations of interest are those about equilibrium, the derivative must be evaluated at $\overline{Q} = 0$. The point is, of course, that we must retain the possibility of a nonzero $\overline{Q}$ until after the derivative is taken. This approach is absolutely equivalent to the ES theory for thermodynamic fluctuations; there is nothing fundamentally different about fluctuations about zero and nonzero average, but in the former case $\langle\varepsilon\rangle$ does not contain the information needed to evaluate the intensity.

If the k (angle) dependence of R is unimportant, it is sufficient to consider an ensemble characterized by $\overline{Q}$ where, for any variable X,

$$X = \int d\mathbf{r}\, X(\mathbf{r}) \tag{4.1}$$

The ES approach then gives

$$\delta S^\gamma = \frac{1}{4\pi}\left(\frac{\partial\overline{\varepsilon}^{\gamma z}}{\partial\overline{Q}^{\gamma z}}\right)Q^{\gamma z}\langle|E|\rangle \tag{4.2}$$

and the depolarized scattering due to orientational fluctuations, R^ϕ, is

$$R^\phi = \left(\frac{\partial\overline{\varepsilon}^{\gamma z}}{\partial\overline{Q}^{\gamma z}}\right)^2 (1+\rho f) \tag{4.3}$$

where we have used Eq. (2.11) and $\gamma \neq z$. Since we are interested mainly in corrections to R's derived ignoring the IFP [Eq. (2.7)], we now introduce new quantities, R^*, the ratio of R to its gas-phase form. Thus,

$$R^{*\phi} = \left[\frac{4\pi}{\Delta\alpha} \left(\frac{\partial\bar{\varepsilon}^{\gamma z}}{\partial\overline{Q}^{\gamma z}} \right) \right]^2 \qquad (4.4)$$

Additional arguments are needed, however, before Eq. (4.4) can be useful. The legitimacy of assuming that R^ϕ is basically the same thing as the depolarized intensity for liquids must be established. This equation is tied up with the very nature of an intensity measurement. There is no such thing as a pure intensity measurement. Most experiments that do not use monochromators use broad filters to eliminate Raman scattering, and the response of the phototubes gives an ultimate frequency cutoff. In other words, an intensity measurement is in reality a measurement of the spectrum over a broad, but finite, window centered at the laser frequency.

Thus, the questions of the nature of the important fluctuations and of their time dependence become linked. The most slowly varying fluctuations will dominate R, as long as the frequencies sampled are not "too large," *irrespective* of the importance of rapidly varying fluctuations. Furthermore, only for slowly varying variables x will $\partial\bar{\varepsilon}/\partial\bar{x}$ be relevant for fluctuations. In this regard, of course, the ES theory is ideal, since the thermodynamic variables are the most slowly varying of all, and there is no doubt that the fluctuations of S due to n and T are determined by $\partial\langle\varepsilon\rangle/\langle\partial n\rangle$ and $\partial\langle\varepsilon\rangle/\partial\langle T\rangle$. It is still true, however, that the ES theory will fail if the intensity measurement includes frequencies that are too high, for example, Raman scattering. The theory will only describe an experiment if the phototube as filter "cuts off" *before* scattering by more rapidly varying fluctuations becomes important.

In applying the ES local equilibrium theory to Q fluctuations, we are stretching things a bit. Q has nothing like the status of n and T as a "slow" variable. Even so, an impressive amount of evidence indicates[16] that a good local-equilibrium description of liquids composed of nonspherical molecules can be obtained by adding Q to the thermodynamic variables. It is still necessary that an I or R measurement cut off frequencies where scattering due to faster fluctuations is important. If all that is true, so is Eq. (4.2), and furthermore, $R \sim R^\phi$. We are left with the problem of evaluating $\partial\bar{\varepsilon}/\partial\overline{Q}$.

Any version of dielectric theory can be used in the calculation. We consider those due to Kirkwood[17] and Yvon.[18] Since this is not a review of dielectric theory, we merely state the results. The most general dielectric relation, of which Eq. (1.12) is a special case, is

$$\bar{P}(r) = \frac{1}{4\pi} \int dr' \left[\underset{\sim}{\bar{\varepsilon}}(r,r') - 1\delta(r-r') \right] \cdot \bar{E}(r') \qquad (4.5)$$

where $\bar{P}$ and $\bar{E}$ are averages in an arbitrary ensemble and $\underset{\sim}{\varepsilon}$ is the nonlocal dielectric tensor kernel. For translationally invariant systems, $\underset{\sim}{\varepsilon}$ will be a function of $r-r'$ only, and for isotropic systems $\underset{\sim}{\bar{\varepsilon}}$ will be a multiple of the unit tensor. A local dielectric relation is obtained by noting that $\underset{\sim}{\bar{\varepsilon}}(r-r')$ is short-ranged, so

$$\int dr' \underset{\sim}{\bar{\varepsilon}}(r,r') \cdot \bar{E}(r') \sim \int dr' \underset{\sim}{\bar{\varepsilon}}(r,r') \cdot \bar{E}(r) = \underset{\sim}{\bar{\varepsilon}}(r) \cdot \bar{E}(r)$$

that is, $\underset{\sim}{\bar{\varepsilon}}(r)$, the position-dependent dielectric tensor, denotes the integral. For systems with translational invariance, the only ones we consider here, $\underset{\sim}{\bar{\varepsilon}}$ is independent of r, so we drop the argument and write $\underset{\sim}{\bar{\varepsilon}}$ from here on.

With an obvious operator notation in r space, Eq. (4.5) becomes

$$\bar{P} = \frac{1}{4\pi} (\mathscr{E} - I) \bar{E} \qquad (4.6)$$

where

$$(\mathscr{E})_{r,r'} = \underset{\sim}{\bar{\varepsilon}}(r,r') \qquad (4.6a)$$

and

$$(I)_{r,r'} = 1\delta(r-r'). \qquad (4.6b)$$

In the same notation, both Kirkwood's and Yvon's theories are based upon the relation

$$\left(\frac{\mathscr{E} - I}{4\pi} \right) \left(\frac{\mathscr{E} + 2I}{3} \right)^{-1} = \overline{\left| \mathscr{T} + \overline{\left((I - A\mathscr{T})^{-1} A \right)^{-1}} \right|}^{-1} \qquad (4.7)$$

where $\mathscr{T}$ is the dipole tensor operator

$$(\mathscr{T})_{r,r'} = T'(r-r') \qquad (4.8)$$

T' being the "cutout" dipole tensor,

$$T'(\mathbf{r}) = \begin{cases} r^{-3}(3\hat{r}\hat{r} - 1) & r > a \\ 0 & r < a \end{cases} \qquad a \to 0 \qquad (4.9)$$

and A is the polarizability tensor operator,

$$(A)_{r,r'} = \left[\sum_{i=1}^{N} \underset{\sim}{\alpha_i} \delta(\mathbf{r} - \mathbf{r}_i) \right] \delta(\mathbf{r} - \mathbf{r}') \qquad (4.10)$$

Equation (4.7) is an exact expression for the Clausius-Mossotti operator in the dipole-induced dipole (DID) approximation. The correctness of the DID approximation, where molecules are assumed to possess point induced dipoles determined by their polarizability tensor and the local electric field (which is itself determined by the applied field and induced dipoles), has been actively discussed.[19] While the approximation presents no problems for the response of an isolated molecule to a spatially smooth field, the finite size and quantum nature of molecules can show up[19a] in response to the extremely rapidly varying field from a polarized near neighbor. Less subtly, for the chain molecules, the DID approach must be applied to appropriate chosen segments, not to entire molecules. The DID approximation may lead to some small errors in the IFP for nonspherical molecules, but the evidence so far[19a] is that it is remarkably accurate, and we will employ it in the following.

To use Eq. (4.7), some approximation must be made. The most obvious is to expand in the number of $\mathscr{T}$'s, which, roughly speaking, corresponds to the "number" of interactions. Kirkwood's theory is based upon a truncation of that expansion; the most important aspects of DLS are contained in the first two terms,

$$\left(\frac{\mathscr{E} - I}{4\pi} \right) \left(\frac{\mathscr{E} + 2I}{3} \right)^{-1} = \bar{A} + \overline{\delta A \mathscr{T} \delta A} + \cdots \qquad (4.11)$$

where $\delta x = x - \bar{x}$. On the other hand, Yvon's approach suggests truncation of an expansion of the inverse of the Clausius-Mossotti function, and

$$\left(\frac{\mathscr{E} + 2I}{3} \right) \left(\frac{\mathscr{E} - I}{4\pi} \right)^{-1} = \bar{A}^{-1} - \bar{A}^{-1} \overline{\delta A \mathscr{T} \delta A} \, \bar{A}^{-1} + \cdots \qquad (4.12)$$

The definitions of the operators may now be used to obtain expressions for $\underset{\sim}{\bar{\epsilon}}$:

KIRKWOOD:

$$\frac{3}{4\pi}(\bar{\underset{\sim}{\varepsilon}}-1)(\bar{\underset{\sim}{\varepsilon}}+21)^{-1}=\bar{\underset{\sim}{\alpha}}(r)+\int d\mathbf{r}'\,\overline{\delta\underset{\sim}{\alpha}(\mathbf{r})\cdot\mathbf{T}'(\mathbf{r}-\mathbf{r}')\cdot\delta\underset{\sim}{\alpha}(\mathbf{r}')} \quad (4.13)$$

YVON:

$$\frac{3}{4\pi}(\bar{\underset{\sim}{\varepsilon}}-1)(\bar{\underset{\sim}{\varepsilon}}+21)^{-1}$$

$$=\left(\left[\overline{\underset{\sim}{\alpha}(\mathbf{r})}\right]^{-1}\left\{\left[\bar{\underset{\sim}{\alpha}}(\mathbf{r})\right]^{-1}\int d\mathbf{r}'\,\overline{\delta\underset{\sim}{\alpha}(\mathbf{r})\cdot\mathbf{T}'(\mathbf{r}-\mathbf{r}')\cdot\delta\underset{\sim}{\alpha}(\mathbf{r}')}\left[\bar{\underset{\sim}{\alpha}}(\mathbf{r}')\right]^{-1}\right\}\right)^{-1}$$

$$(4.14)$$

Equations (4.13) and (4.14) are just typical examples of standard dielectric theory. The new feature present here is that we are not going to have the overbar represent a thermal equilibrium average ($\langle\,\rangle$) but rather an orientationally ordered system. Assuming that $\overline{Q^{\gamma z}}$ is known, then the local equilibrium ensemble constructed via response theory gives

$$\bar{g}=\langle g\rangle+\langle gQ^{\gamma z}\rangle\langle Q^{\gamma z}Q^{\gamma z}\rangle^{-1}\overline{Q}^{\gamma z} \quad (4.15)$$

where g is any phase-space function. Combination of this result with any dielectric theory will then yield

$$\bar{\varepsilon}^{\gamma z}=\langle\varepsilon\rangle\delta_{\gamma z}+\varepsilon^{\gamma z}_{(1)}\overline{Q}^{\gamma z} \quad (4.16)$$

and $\underset{\sim}{\varepsilon}_{(1)}$ is the desired derivative.

Thus, we use Eq. (4.15) on the right-hand side of both the Kirkwood [Eq. (4.13)] and Yvon [Eq. (4.14)] theories, and on the left we divide $\bar{\underset{\sim}{\varepsilon}}$ into equilibrium and first-order parts, as in Eq. (4.16). The results are[20] (for $\gamma\neq z$)

KIRKWOOD:

$$V\varepsilon^{\gamma z}_{(1)}=4\pi\left(\frac{\langle\varepsilon\rangle+2}{3}\right)^{2}(\Delta\alpha+\mathscr{L}) \quad (4.17)$$

YVON:

$$V\varepsilon^{\gamma z}_{(1)}=4\pi\left(\frac{\langle\varepsilon\rangle+2}{3}\right)^{2}\left(1-\frac{\kappa}{\alpha_0\rho}\right)^{-2}\left[\left(1-\frac{\kappa}{\alpha_0\rho}\right)\Delta\alpha+\mathscr{L}\right] \quad (4.18)$$

where

$$\kappa=\int d\mathbf{r}'\langle\left[\delta\underset{\sim}{\alpha}(\mathbf{r})\cdot\mathbf{T}'(\mathbf{r}-\mathbf{r}')\cdot\delta\underset{\sim}{\alpha}(\mathbf{r}')\right]^{zz}\rangle \quad (4.19)$$

and

$$\mathscr{L} = \frac{V \langle Q^{yz} \int d\mathbf{r}' \left(\delta\underline{\alpha}(\mathbf{r}) \cdot \mathbf{T}'(\mathbf{r}-\mathbf{r}') \cdot \delta\underline{\alpha}(\mathbf{r}') \right)^{yz} \rangle}{\langle Q^{yz} Q^{yz} \rangle} \tag{4.20}$$

Of course, these averages do not depend on $\mathbf{r}$.

Equations (4.17) and (4.18) show that if κ, $\mathscr{L}$ are set to zero, the local field correction for R^ϕ is identical to that obtained in the ES theory with the Clausius-Mossotti equation. As mentioned earlier, such a correction, and nothing else, is in striking disagreement with experiment. There is good evidence that $\kappa/\alpha_0\rho$ is small, since in the current approximation[20]

$$\frac{\langle \varepsilon \rangle - 1}{\langle \varepsilon \rangle + 2} = \frac{4\pi}{3} \alpha_0 \rho \times \begin{cases} \left(1 + \dfrac{\kappa}{\alpha_0\rho} \right) & \text{(Kirkwood)} \\[2ex] \left(1 - \dfrac{\kappa}{\alpha_0\rho} \right)^{-1} & \text{(Yvon)} \end{cases} \tag{4.21}$$

and it is an experimental fact that the corrections to the Clausius-Mossotti relation are very small. If the ES theory for R^ϕ is to be successful, then, we must look to $\mathscr{L}$ as the key quantity. With this in mind, we define

$$R^\phi = \left(\frac{\langle \varepsilon \rangle + 2}{3} \right)^4 \left(\frac{\Delta\alpha}{4\pi} \right)^2 \left(\frac{\gamma}{\Delta\alpha} \right)^2 (1 + \rho f) \tag{4.22a}$$

or

$$R^{*\phi} = L^4 \left(\frac{\gamma}{\Delta\alpha} \right)^2 \tag{4.22b}$$

and γ is an effective polarizability anisotropy for liquids. Equations (4.17) and (4.18) gives

$$\frac{\gamma}{\Delta\alpha} = 1 + \frac{\mathscr{L}}{\Delta\alpha} \qquad \text{(Kirkwood)} \tag{4.23a}$$

$$\frac{\gamma}{\Delta\alpha} = \left(1 - \frac{\kappa}{\alpha_0\rho} \right)^{-2} \left[\left(1 - \frac{2\kappa}{\alpha_0\rho} \right) + \frac{\mathscr{L}}{\Delta\alpha} \right] \qquad \text{(Yvon)} \tag{4.23b}$$

Obviously, $\mathscr{L}/\Delta\alpha$ must be substantial and negative if the theory is to be successful. Perhaps this might seem unlikely, since it constitutes the first correction, in powers of $\mathbf{T}'$, to $\varepsilon_{(1)}$, and the analogous correction to $\langle \varepsilon \rangle$ is very small. Nevertheless, Eqs. (4.23) allow construction of a theory in good agreement with experiment.

B. Formal Evaluation of $\mathscr{L}$

Evaluation of $\mathscr{L}$ is a straightforward, formal problem in equilibrium statistical mechanics. While working out the details of these calculations, we will always try to emphasize interpretations of $\mathscr{L}$ in terms of physical ideas, such as the local field.

Define the quantity t',

$$\mathscr{L} = \frac{t'}{\langle Q^{yz}Q^{yz}\rangle} \tag{4.24}$$

We wish to express t' in terms of averages over the molecular coordinates. This is complicated by the presence of $\delta\alpha$, rather than α, in Eqs. (4.20). When the averages are expressed as integrals over distribution functions, however, the presence of $\delta\alpha$ merely corresponds to removing the asymptotic, long-range part of the distribution functions. Thus, we consider the auxiliary quantity t,

$$t = \int d\mathbf{r}\, d\mathbf{r}'\, d\mathbf{r}'' \langle Q^{yz}(\mathbf{r})[\underset{\sim}{\alpha}(\mathbf{r}')\cdot\mathbf{T}'(\mathbf{r}'-\mathbf{r}'')\cdot\underset{\sim}{\alpha}(\mathbf{r}'')]^{yz}\rangle \tag{4.25}$$

which, with the above remarks in mind, is equivalent to t'; here we have used Eq. (4.1) and noted that $\int d\mathbf{r}\, d\mathbf{r}''$ is independent of $\mathbf{r}'$, so that the integration just produces a factor of V.

Inserting the definitions of $\mathbf{Q}(\mathbf{r})$ and $\underset{\sim}{\alpha}(\mathbf{r})$, we immediately see[20] that t is a sum of two- and three-body terms,

$$t = t^{(2)} + t^{(3)} \tag{4.26}$$

where

$$t^{(2)} = N(N-1)\left[\langle Q_1^{yz}(\underset{\sim}{\alpha}_1\cdot\mathbf{T}_{12}'\cdot\underset{\sim}{\alpha}_2)\rangle^{yz} + \langle Q_2^{yz}(\underset{\sim}{\alpha}_1\cdot\mathbf{T}_{12}'\cdot\underset{\sim}{\alpha}_2)\rangle^{yz}\right] \tag{4.27}$$

$$t^{(3)} = N(N-1)(N-2)\langle Q_3^{yz}(\underset{\sim}{\alpha}_1\cdot\mathbf{T}_{12}'\cdot\underset{\sim}{\alpha}_2')\rangle^{yz} \tag{4.28}$$

and $\mathbf{T}_{12}'$ denotes $\mathbf{T}'(\mathbf{r}_1-\mathbf{r}_2)$, etc. These averages are most conveniently expressed in terms of the pair and three-particle distribution functions,

$$t^{(2)} = \frac{2\rho^2}{(8\pi^2)^2}\int d\mathbf{R}_1\, d\mathbf{R}_2\, g^{(2)}(\mathbf{R}_1,\mathbf{R}_2)$$

$$\times \left\{Q^{yz}(\Omega_1)[\underset{\sim}{\alpha}(\Omega_1)\cdot\mathbf{T}'(r_1-r_2)\cdot\underset{\sim}{\alpha}(\Omega_2)]^{yz}\right\} \tag{4.29}$$

$$t^{(3)} = \frac{\rho^3}{(8\pi^2)^3}\int d\mathbf{R}_1\, d\mathbf{R}_2\, d\mathbf{R}_3\, g^{(3)}(\mathbf{R}_1,\mathbf{R}_2,\mathbf{R}_3)$$

$$\times \left\{Q^{yz}(\Omega_3)[\underset{\sim}{\alpha}(\Omega_1)\cdot\mathbf{T}'(r_1-r_2)\cdot\underset{\sim}{\alpha}(\Omega_2)]^{yz}\right\} \tag{4.30}$$

where $\mathbf{R}_i = (\mathbf{r}_i, \Omega_i)$ denotes the center-of-mass position $\mathbf{r}_i$ and the Euler angles Ω_i, giving the position and orientation of the ith molecule. The object we really want, t', is obtained by replacing $g^{(2)}$ by $h^{(2)}$, the pair correlation function, in Eq. (4.29), where

$$h^{(2)} = g^{(2)} - 1 \qquad (4.31)$$

and by replacing $g^{(3)}$ with $h^{(3)}$ in Eq. (4.30)

$$h^{(3)}(\mathbf{R}_1, \mathbf{R}_2, \mathbf{R}_3) = g^{(3)}(\mathbf{R}_1, \mathbf{R}_2, \mathbf{R}_3) - g^{(2)}(\mathbf{R}_1, \mathbf{R}_2)$$
$$- g^{(2)}(\mathbf{R}_1, \mathbf{R}_3) - g^{(3)}(\mathbf{R}_2, \mathbf{R}_3) + 2 \qquad (4.32)$$

The position- and angle-dependent g's for nonspherical molecules are very much more complicated than the well-understood $g^{(2)}(r)$ for atoms. It is only a recent development[21] that some information about $g^{(2)}$ is available, and virtually nothing is known about $g^{(3)}$. Thus, we approximate $g^{(3)}$ by the Kirkwood superposition approximation,[20]

$$g^{(3)}(\mathbf{R}_1, \mathbf{R}_2, \mathbf{R}_3) = g^{(2)}(\mathbf{R}_1, \mathbf{R}_2) g^{(2)}(\mathbf{R}_1, \mathbf{R}_3) g^{(2)}(\mathbf{R}_2, \mathbf{R}_3) \qquad (4.33)$$

or

$$h^{(3)}(\mathbf{R}_1, \mathbf{R}_2, \mathbf{R}_3) = h^{(2)}(\mathbf{R}_1, \mathbf{R}_2) h^{(2)}(\mathbf{R}_1, \mathbf{R}_3) + h^{(2)}(\mathbf{R}_1, \mathbf{R}_2) h^{(2)}(\mathbf{R}_2, \mathbf{R}_3)$$
$$+ h^{(2)}(\mathbf{R}_1, \mathbf{R}_3) h^{(2)}(\mathbf{R}_2, \mathbf{R}_3)$$
$$+ h^{(2)}(\mathbf{R}_1, \mathbf{R}_2) h^{(2)}(\mathbf{R}_1, \mathbf{R}_3) h^{(2)}(\mathbf{R}_2, \mathbf{R}_3) \qquad (4.34)$$

We may now determine $\mathscr{L}$ in terms of $h^{(2)}$. Similar manipulations give[20]

$$\kappa = \frac{\rho^2}{(8\pi^2)^2} \frac{1}{V} \int d\mathbf{R}_1 d\mathbf{R}_2 \, h^{(2)}(\mathbf{R}_1, \mathbf{R}_2) \left[\underset{\approx}{\alpha}(\Omega_1) \cdot \mathbf{T}'(\mathbf{r}_1 - \mathbf{r}_2) \cdot \underset{\approx}{\alpha}(\Omega_2) \right]^{zz}$$

$$(4.35)$$

and

$$f = \frac{1}{V^2 (8\pi^2)^2 \langle Q_i^z Q_i^z \rangle} \int d\mathbf{R}_1 d\mathbf{R}_2 \, h^{(2)}(\mathbf{R}_1, \mathbf{R}_2) Q^{yz}(\Omega_1) Q^{yz}(\Omega_2)$$

$$(4.36)$$

It is difficult to obtain, or even think about, a function of so many variables as $h^{(2)}$; fortunately, this is unnecessary. An arbitrary function,

$X(\mathbf{R}_1, \mathbf{R}_2)$, may be expanded[1a] in an infinite sum of angular functions and radial coefficients,

$$X(\mathbf{R}_1, \mathbf{R}_2) = \sum_{N(1)} \sum_{N(2)} X_{N(1), N(2)}(r_{12}) D_{N(1)}(\Omega_1^{12}) D_{N(2)}(\Omega_2^{12}) \quad (4.37)$$

where N denotes the three indices of the Wigner rotation functions $D_{m\mu}^j$, and Ω_i^{12} are the angles specifying the orientation of molecule i with respect to a coordinate frame whose z axis is parallel to $\mathbf{r}_{12}$. For $\mu = 0$, the D's reduce to the spherical harmonics Y_{jm}.

When Eq. (4.37) is used for $h^{(2)}$, many averages are seen, due to the orthogonality properties of the D's, to involve only a few $X_{N(1), N(2)}$. Thus, we can concentrate on those X's and try to calculate them or find them in a computer simulation, irrespective of the rapidity of convergence of Eq. (4.37). In this way, the messy averages arising for nonspherical molecules are finally reduced to simple radial integrals, as is the case for atoms.

For simplicity in illustration, we now specialize to symmetrical top molecules. None of the basic physics of the IFP is lost here; the extension to molecules of arbitrary shapes has been given by Gierke.[22] The details and algebra are given in Ref. 20. We find

$$f = \frac{1}{25} \sum_{R=-2}^{2} (-)^R \int dr\, 4\pi r^2 g_{2R0, 2-R0}(r) \quad (4.38)$$

$$\kappa = \frac{2\rho^2\alpha_0}{15}\left[2\,\Delta\alpha\, \tau_{20} - \frac{(\Delta\alpha)^2}{15\alpha_0}\tau_{22} \right] \quad (4.39)$$

and

$$t^{(2)} = \frac{2N\rho}{25}\left[\alpha_0^2\tau_{20} + \frac{\alpha_0\Delta\alpha}{3}\left(\tau_{20} - \tfrac{1}{5}\tau_{22} \right) + \left(\frac{\Delta\alpha}{9} \right)^2\left(2\tau_{20} - \tfrac{1}{5}\tau_{22} \right) \right] \quad (4.40)$$

where

$$\tau_{20} = \int dr\, \frac{4\pi}{r} g_{200,000} \quad (4.41)$$

and

$$\tau_{22} = \int dr\, \frac{4\pi}{r}\left[2g_{220, 2-20}(r) + g_{210, 2-10}(r) - g_{200, 200}(r) \right] \quad (4.42)$$

We are now writing g instead of h, since the two differ only in that part of the expansion with no angular dependence, $N(1)$, $N(2) = 000$, and this does not contribute to any of our results.

Evaluation of $t^{(3)}$ is more difficult. Note that the integrand giving $t^{(3)}$ depends on $\mathbf{r}_{12}$ through T', over and above whatever comes from $h^{(3)}$. Thus, referring to Eq. (4.34) for $h^{(3)}$ in the superposition approximation, it is natural to classify the terms according to whether they introduce one (first two on the right) or two interparticle separations in addition to r_{12}; the contributions to $t^{(3)}$ are denoted $t_p^{(3)}$ for product and $t_c^{(3)}$, for connected, respectively. The reason for this terminology is that the product integrals factor into products of two two-body integrals, and we find the suggestive result,

$$t_p^{(3)} = \rho f t^{(2)} \qquad (4.43)$$

We have not evaluated $t_c^{(3)}$. No simple answer is possible, and we originally assumed that it was small due to its high "connectedness." This point was later pursued by Frankel and McTague,[23] who evaluated the averages via direct computer simulation; they concluded that our approximation is questionable for liquid diatomics. Of course, the approximation is in no way fundamental. If it is wrong, three-body h's just have to be kept. For simplicity, and to illustrate our original theory, we now assume

$$t \sim t^{(2)} + t_p^{(3)} = (1 + \rho f)t^{(2)} \qquad (4.44)$$

Equation (4.44) is striking, since it means that $\mathscr{L}$ [Eq. (4.24)] contains no explicit factor of $\langle Q^{\gamma z}Q^{\gamma z}\rangle^{-1}$, this being canceled by the $(1 + \rho f)$. In fact, we now have

$$\mathscr{L} = \frac{15}{N} t^{(2)} \qquad (4.45)$$

The linear response ensemble, given in Eq. (4.15), directly introduces $\langle Q_0^{\gamma z}Q_0^{\gamma z}\rangle^{-1}$ into $\partial\bar{\varepsilon}/\partial\bar{Q}$, but the zero$-T'$ and one$-T'$ terms in $\bar{\varepsilon}$ themselves contain factors to cancel the inverse, although the presence of these factors, especially for $\mathscr{L}$, is not immediately obvious. Perhaps some general principle exists to explain this, but we are not aware of it.

The effective anisotropy γ is now expressed in terms of α_0, $\Delta\alpha$, ρ, and two parameters, τ_{22} and $\bar{\tau}_{22}$, involving the detailed pair structure of the fluid as it affects the dipolar interaction. Let us discuss the physical meaning of these parameters. The average field on an atom due to an induced dipole on a neighbor vanishes, due to the random orientation of the interpair vector. The average field on a nonspherical molecule with a *given orientation* due to a neighbor does not vanish, for two reasons. First, the hard core of the molecule makes a nonspherical volume of space inaccessible, and the interpair vector is not randomly oriented at small r_{12}. Second, the orientation of the neighbor is correlated with its position. In particular, at small r_{12}, the only orientations possible are those with r_{12} perpendicular to the long axis of the neighbor. So even if r_{12} were randomly oriented, the dipole induced in molecule 2 would vary, on average, with r_{12} as the preferred orientation, and thus the polarizability tensor, of molecule 2 changes. The consequence, again, is a nonzero average field.

All this means that an orientational fluctuation (molecule with fixed orientation) "turns on" a contribution from its neighbors to the local field seen by that molecule. Since orientational fluctuations cause depolarized light scattering, the extra local field will, broadly speaking, affect the scattering intensity. The effect of the orientation of a molecule on the possible center-of-mass positions of a neighbor is described by τ_{20}. This is determined by $g_{N(1), N(2)}$ with $j_1 = 2$ (orientation dependence) and $j_2 = 0$ (no orientation; center of mass only). The role of pure orientation-orientation correlations comes from τ_{22}. We will see that τ_{20} dominates $\mathscr{L}$. Its orientation-position correlations exist for an atom polarizing a molecule and even for a continuum. If a nonspherical molecule simply excludes a continuum from a nonspherical cavity, the spatial distribution of the continuum is determined by the orientation of the molecule.

C. The Depolarized Intensity

The parameters τ_{20} and τ_{22} can be obtained from a computer simulation[21a,b] or from an approximate theory[21c] for $g^{(2)}$. Streett and Tildesley have obtained the necessary $g_{N(1), N(2)}(r)$ in simulations of model diatomic liquids[21a] and of a hard triatomic model[21b] of CS_2, and we calculated the τ's. We also studied[24] CS_2 with structural information derived from the equation of Hsu et al.[21c] Theory and experiment cannot really be compared for the diatomics,[20] because the simulation was not designed to study any real molecule, and also because orientational fluctuations may not be regarded as slow (see p. 000) for diatomics. Therefore, we use those studies only to establish general trends. Assuming, as is typical, that $\Delta\alpha = \alpha_0/2$, we plot R versus α_0/σ^3 (σ is the atom-atom separation of the diatomic) in Fig. 3. This figure demonstrates what we regard as the most important features of depolarized light scattering in liquids. Even though L^4 increases sharply with α_0 ($\langle\varepsilon\rangle$ increases with α_0), the negative $\mathscr{L}$ causes the effective anisotropy γ to decrease even more strongly, and R^ϕ is a decreasing function of α_0.

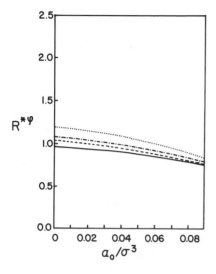

Fig. 3. Dependence of the depolarized Rayleigh ratio, R^{ϕ}, on α_0 according to the Kirkwood theory, Eq. (4.23a), (dashed line) and the Yvon theory, Eq. (4.23b) (solid line). Liquid structure is taken from a computer simulation of a diatomic liquid (Ref. 20) at two typical thermodynamic states, ρ^* (reduced density) T^* (reduced temperature) of (0.6, 1.166), and (0.5, 0.98); $\Delta\alpha = \alpha_0/2$.

Of course, one cannot, in reality, vary $\langle \varepsilon \rangle$ and leave all else constant, but Fig. 3 exposes the basic facts: Depending on the range of $\langle \varepsilon \rangle$, or α_0, in a class of experiments, the shape of $I(\langle \varepsilon \rangle)$ will differ drastically. Thus, no "power law" L^{α} can ever give a good local field correction to R^{ϕ}. For most liquids, $\langle \varepsilon \rangle$ is such that $R(\langle \varepsilon \rangle)$ is in the decreasing regime. This is why L^4 predicts too much intensity. The solution is not, however, to invoke L^2, which is still an increasing function. This can be seen most clearly by studies of scattering from large solute molecules, where $\langle \varepsilon \rangle$ can be varied, with everything else more or less constant, by varying the solvent. Two-component systems are discussed in Section IV.E.

Frenkel and McTague[23] carried out similar studies, without involving the superposition approximation and neglecting $t_c^{(3)}$. They found larger values of $(\gamma/\Delta\alpha)$ than did we for a fixed $\langle \varepsilon \rangle$, but the trends remain the same. The CS_2 simulation[21b] was explicitly carried out to compare[24] theory and experiment. Two versions of the theory were used, the one just given and an extended version[25] where CS_2 has three-point isotropic polarizabilities and induced dipoles, one at the center of each atom and interacting via the DID mechanism. The DID interactions cause the total molecular polarizability to be anisotropic. The aim was to determine whether the resulting spreading out of the polarizability and dipole might give a better description of local field effects involving close neighbors. No adjustable parameters were used in either version of the theory; the atomic polarizabilities in the three-point polarizability model were determined by insisting that an isolated molecule show the measured α_0 and $\Delta\alpha$.

We actually calculate the ratio $\delta \equiv (\alpha/\Delta\alpha)/(\alpha_0^{\text{eff}}/\alpha_0)$, where α_0^{eff} is the analogue of γ for isotropic scattering; δ is particularly accessible experimentally. However, since $(\alpha_0^{\text{eff}}/\alpha_0) \sim 1$, δ is basically $\gamma/\Delta\alpha$. The one-point dipole, three-point dipole, and experimental values for δ are 0.51, 0.66, and 0.71, respectively.[24] Thus, the three-point-dipole theory is in excellent agreement with experiment. To our knowledge, Ref. 24 contains the first completely microscopic calculation of the anisotropic scattering intensity of a pure liquid; the good quality of the result argues in favor of the ES theory.

Why is the $\mathscr{L}$ term so important for depolarized scattering? We have calculated the analogous term for isotropic scattering, and it is completely negligible. This is obviously true from an empirical point of view, since L^4 is an excellent local field correction for isotropic scattering. A more physical understanding is possible also. As was previously discussed, $\mathscr{L}$ is related to the effect of an orientational fluctuation on the average local field generated by the neighbors; the light-scattering fluctuation "turns on" some new local fields. Such *correlations* determine $\mathscr{L}$, while the L^4 factor involves no correlations.

Since isotropic scattering arises from density fluctuations, the analogous effect must involve the way in which a density fluctuation changes the averaged local field. But a long-range density fluctuation does not break the local symmetry, which is the most important determinant of the local field. This is one explanation of why internal field effects are more complicated for fluctuations which, if "frozen," most effectively turn on the average dipolar fields from the neighbors. Prime candidates for this are fluctuations that break angular isotropy.

D. Dielectric Cavity Models

While some intuitive understanding of the results of the last section was obtained, contact was not made with the area where internal field problems have been most studied: dielectric theory. Of course, the questions are different. While light scattering is a fluctuation phenomenon, in dielectrics one needs to know the average (over neighbor configurations) field seen by a fixed molecule. On the other hand, the two problems are linked in the ES approach. Much effort has been expended in obtaining the factors that relate the averaged local field at a molecule to the fully averaged field $\langle \mathbf{E} \rangle$ ($\langle \mathbf{E} \rangle$ contains contributions from points in free space as well as in molecules and is evaluated with no molecule fixed in position and orientation). It would be very desirable if these factors, a_{ii}, could be incorporated into a theory of light scattering. Their definition is as follows: Given that the principal values of the gas-phase polarizability tensor are α_{ii}, then, in a fluid,

$$\mu_i = \underset{\sim}{\alpha}^e \cdot \langle \mathbf{E} \rangle \tag{4.46}$$

and the principal values of the effective polarizability tensor $\underset{\sim}{\alpha}^e$ are $a_{ii}\alpha_{ii}$. The a_{ii} are equal to $\langle E_L \rangle_0 / \langle E \rangle$ when $\langle E \rangle$ is parallel to the ith axis.

The best-known expressions for the a_{ii} come from the Lorentz-Raman-Krishnan (LRK)[26] and Onsager-Scholte[27] dielectric theories. In the LRK theory, the local field is assumed to be that inside an empty cavity with the molecular shape in a dielectric continuum of average dielectric constant $\langle \varepsilon \rangle$. The result for ellipsoids is

$$a_{ii}^L = 1 + (\varepsilon - 1) S_{ii} \tag{4.47}$$

where S_{ii} is the "demagnetizing factor" for the ith axis, a measure of the molecular shape. In all cases, $S > 0$,

$$\sum_i S_{ii} = 1 \tag{4.48}$$

and so, for symmetric tops, the only additional information needed is the "shape anisotropy," ΔS,

$$\Delta S = S_{zz} - S_{xx} \tag{4.49}$$

where z is the unique axis. Note that for spherical molecules, $S_{ii} = \frac{1}{3}$ and $a_{ii} = L$. In the other extreme case, that of a "needle," $\Delta S = -1$.

The Onsager-Scholte theory adds the effect of the reaction field, whereby the field in the empty cavity is modified by the presence of the induced molecular dipole, whose field changes the polarization of the continuum, which in turn changes the local field. Thus,

$$a_{ii}^{OS} = \left\{ 1 - \left(\frac{\varepsilon - 1}{\varepsilon} \right) S_i \left[1 + 3(1 - S_i) \alpha_{ii} (r_1 r_2 r_3)^{-1} \right] \right\}^{-1} \tag{4.50}$$

where the r_i are the semiaxes of the cavity. The common feature of the LRK and Onsager-Scholte theories is that the neighbor molecules are approximated as a continuum. A whole host of molecular theories, such as those of Kirkwood and Yvon, can also be used to find a_{ii}. Since a molecular theory is difficult, such theories must also make approximations, such as the truncation of the $\mathscr{T}$ series discussed earlier.

The concept of an effective polarizability defined in Eq. (4.46) is appealing. What does it have to do with light scattering? We now consider this question for the LRK model.[28]

If one believed, wrongly, that scattering intensities could be calculated from the mean-square polarization fluctuations, one would be strongly tempted to apply the gas-phase theory to liquids by merely replacing α with α^e or, for DLS, $\Delta\alpha$ with $\Delta\alpha^e$. The results of such a procedure are instructive. The Lorentz expression for $\langle\varepsilon\rangle$, based upon Eq. (4.47), may be written

$$\frac{\langle\varepsilon\rangle-1}{\langle\varepsilon\rangle+2} = \frac{4\pi}{3}\alpha_0\rho\frac{1}{1-(8\pi/9)\rho\Delta\alpha\Delta S} \qquad (4.51)$$

which reduces to the Clausius-Mossotti relation for a sphere ($\Delta S = 0$). Using Eqs. (4.47) and (4.51), we can construct $\Delta\alpha^e$ and eliminate $\langle\varepsilon\rangle$, α_{ii}, and S_{ii} in favor of α_0, $\Delta\alpha$, ρ, and ΔS, with the result

$$\Delta\alpha^e = \frac{\Delta\alpha + 4\pi\alpha_0\rho\left(\alpha_0 + \frac{1}{3}\Delta\alpha - \frac{2}{9}(\Delta\alpha)^2/\alpha_0\right)\Delta S}{1 - \frac{4}{3}\pi\alpha_0\rho - \frac{8}{9}\pi\Delta\alpha\rho\Delta S} \qquad (4.52)$$

Further use of Eq. (4.51) shows that the denominator is proportional to $\langle\varepsilon\rangle - 1$, and[28]

$$\Delta\alpha^e = \left(\frac{\langle\varepsilon\rangle-1}{4\pi\alpha_0\rho}\right)\left[\Delta\alpha + 4\pi\alpha_0\rho\left(\alpha_0 + \frac{1}{3}\Delta\alpha + \frac{2}{9}(\Delta\alpha)^2/\alpha_0\right)\Delta S\right] \qquad (4.53)$$

Now, referring to the molecular expression [Eq. (4.23b)] for γ in the Yvon theory, and Eq. (4.21) for $\langle\varepsilon\rangle$ in that theory to the same order of approximation, it is seen that the factor of $(1-\kappa/\alpha_0\rho)^{-2}$ in γ^2 combines with L^4 to produce a factor of $[(\langle\varepsilon\rangle-1)/4\pi\alpha_0\rho]^4$. So a more natural definition of γ in the Yvon version of the ES theory might be

$$R^{*\phi} \equiv \left(\frac{\langle\varepsilon\rangle-1}{4\pi\alpha_0\rho}\right)^4\left(\frac{\gamma^y}{\Delta\alpha}^2\right) \equiv (L^y)^4\left(\frac{\gamma^y}{\Delta\alpha}\right)^2 \qquad (4.54)$$

Using this definition for L^y, it is seen that simple replacement of $\Delta\alpha$ in the gas-phase formula with $\Delta\alpha^e$ leads to an R^* proportional to $(L^y)^2$ and to the square of the quantity in square brackets in Eq. (4.53), which is very reminiscent of γ [Eq. (4.23)]. Thus, the naive replacement *roughly* seems to predict $R^* \propto L^2(\gamma/\Delta\alpha)^2$ an expression for R^* not seen so far. The quantity in square brackets in Eq. (4.53) has the correct essential properties. Since ΔS is negative for a long principal axis, the second term will subtract from $\Delta\alpha$ in a way that is proportional to $\alpha_0\rho$, as we found in the molecular theories. Within the dielectric model, it is easy to understand why this is so. According to Eq.

(4.47), the local field is enhanced above $\langle E \rangle$ in all directions, but more so along the short axis, since S is larger for the short axis. This larger enhancement partly compensates for the (usually) smaller polarizability along the short axis, making the effective polarizability more isotropic and leading to less orientational scattering. The reason for this effect is that the continuum that can approach closest to the molecular polarizability and that is most important for the local field is to be found by going perpendicular to the long axis at the center of the molecule. Adding more optically responsive material here must make the combined molecule-plus-continuum more isotropic (as opposed to adding continuum at the ends of the long axis). Note that this argument has a flavor similar to that given about the τ_{20} terms in the molecular theory, and the two are basically equivalent.

The question remains, What relation does any of this have to reality? If $\Delta \alpha^e / L^y$ is replaced by $\Delta \alpha$, we find the "L^2 law." Since the different power laws, by definition, hold when the only aspect of the IFP taken into account is L to some power, it follows that $\underset{\sim}{\alpha} \rightarrow \underset{\sim}{\alpha}^e$ alone produces an L^2 law. This fits perfectly with what we argued earlier—namely, that treating $\delta \mathbf{P}$ as the source, in which case $\underset{\sim}{\alpha} \rightarrow \underset{\sim}{\alpha}^e$ alone should be correct, also gives the erroneous L^2 law.

Burnham, Alms, and Flygare[14c] (BAF) realized that any approach implying an L^2 law was wrong but also wanted to use the concept of $\underset{\sim}{\alpha}^e$ to construct a theory for $R^{*\phi}$. Thus, very logically, they suggested

$$(\alpha^e_{BAF})_{ii} = a^2_{ii} \alpha_{ii} \tag{4.55}$$

which gives a multiplicative L^4 factor. BAF interpreted a large amount of data using Eq. (4.55). Since the molecular shape is the key quantity in the dielectric cavity models, they tried to use depolarized light scattering to determine molecular shapes. Overall, they found shapes that seemed too isotropic.

The ES theory can unambiguously answer the question of how to use the a_{ii}. Again, consider the Lorentz model. Equation (4.51) is an expression for the inverse of the Clausius-Mossotti function, which is of zero- and first-order in the polarizability; that is, it has precisely the same polarizability dependence as the truncated Yvon theory for $\langle \varepsilon \rangle$, Eq. (4.21). Thus, the Lorentz model must be a special case of the truncated Yvon theory. Consequently, we can compare the two expressions for $\langle \varepsilon \rangle$ and obtain expressions for the parameters in the molecular theory based upon the model. Upon using Eq. (4.39) for κ, we immediately find

$$\tau^L_{20} = \frac{10\pi}{3} \Delta S \tag{4.56a}$$

$$\tau^L_{22} = 0 \tag{4.56b}$$

Equation (4.56b) is expected, since τ_{22} contains true orientational pair correlations, but the continuum has no orientational degrees as freedom. Equation (4.56a) expresses the orientation-position correlations in the model; very simply, the molecule's nonspherical shape excludes the continuum.

Given these results, we can now consistently construct a Lorentz model for depolarized light scattering from the molecular theory; the result[28] is Eq. (4.54), with

$$\frac{\gamma^{y,L}}{\Delta\alpha} = 1 + 4\pi\alpha_0\rho\left[\frac{\alpha_0}{\Delta\alpha} + \frac{1}{3} - \frac{2}{9}\frac{\Delta\alpha}{\alpha_0}\right]\Delta S \qquad (4.57)$$

Note that this quantity is just $(L^y)^{-1}(\Delta\alpha^e/\Delta\alpha)$.

The situation is now clear. The BAF procedure overcounts powers of $\gamma/\Delta\alpha$ by two, due to the extra a_{ii} in Eq. (4.55), but gives the proper L^4. On the other hand, direct use of $(\Delta\alpha^e)^2$ gives $\gamma/\Delta\alpha$ correctly but misses two powers of L. The correct answer can be obtained only from a complete theory that does not use $\delta\mathbf{P}$ as a source. Although all our analysis has been based upon the Lorentz model, the obvious generalization to other dielectric models is

$$R^{*\phi} = L^2\left(\frac{\Delta\alpha^e}{\Delta\alpha}\right)^2 \qquad (4.58)$$

Burnham[14d] and Burnham and Gierke[29] reanalyzed the data based upon these considerations. The resulting molecular shapes were more realistic (anisotropic), since a large ΔS is needed to get the same R in Eq. (4.58) (two a_{ii}'s) than in Eq. (4.55) (four a_{ii}'s). Some of these results are given in Table II. Equation (4.58) has also been used by Cox, Battaglia, and Madden[30a] in their analysis of scattering CS_2. These authors have shown[30b] how to determine effective anisotropies by combination of Cotton-Mouton effect measurements with depolarized light scattering, thereby determining $g^{(2)}$.

TABLE II
Estimates of Molecular Shape Anisotropy
from Light Scattering

Liquid	Experiment		van der Waals	
	ΔS	$r_\parallel/r_\perp$	ΔS	$r_\parallel/r_\perp$
Benzene	.22	.59	.25	.55
Chloroform	.22	.59	.22	.59
Carbon disulfide	$-.22$	1.89	$-.23$	1.91
Acetonitrile	$-.31$	2.63	$-.20$	1.75

Source: Reference 14d.

In sum, the ES theory not only allows a fully microscopic evaluation of R, if appropriate liquid structural data are available. It is also essential in adapting approximate dielectric models to light scattering calculations.

E. Two-Component Systems

So far we have discussed pure fluids. However, two-component systems are extremely important for depolarized light scattering intensity studies. This is because the clearest distinction between the different theories is usually in the predicted $\langle \varepsilon \rangle$ dependence, but a pure liquid has, practically, a single $\langle \varepsilon \rangle$. When different liquids are compared, other parameters change as well. This experimental difficulty is, in our view, a large part of the reason behind the "L^2 law"; for the single $\langle \varepsilon \rangle$ available in many liquids, L^2 and $L^4(\gamma/\Delta\alpha)^2$ are numerically close, even if their $\langle \varepsilon \rangle$ dependence is wildly different.

On the other hand, consider a two-component system where one component (A) is far more anisotropic than the other (B). If several relatively isotropic B components are available with a range of $\langle \varepsilon \rangle$'s, and if changing B changes $\langle \varepsilon \rangle$ and nothing else, then the intensity of scattering from A might be obtained as a function of $\langle \varepsilon \rangle$. This idea has been employed in a very attractive approach, the "dilution experiment," to determination of $g^{(2)}$. The data provide a clear elucidation of the $\langle \varepsilon \rangle$ dependence of R^ϕ, which is seen to be decreasing sharply.

In order to obtain $g^{(2)}$ from R^ϕ, one must know L^4, which is easy, and $(\gamma/\Delta\alpha)^2$, which is not. Of course, $\gamma/\Delta\alpha$ can now be calculated with reasonable confidence, but it would be nice to have a method that required no calculation. One tries to achieve this in the dilution experiment by measuring $R^{*\phi}$ for a dilute solution of an A solute in several B solvents. The data are fit to an empirical function, $\tilde{R}(\langle \varepsilon \rangle)$ such as a polynomial. A-A pair correlations will be negligible if the solution is dilute enough, but, clearly, different solvents could affect the scattering power of A in ways not described by $\langle \varepsilon \rangle$ alone. It is assumed, however, that the dilute solution fit, $\tilde{R}$, is in fact the full $\langle \varepsilon \rangle$ dependence of R, and

$$g_A^{(2)} = \frac{R_A^{*\phi}}{\tilde{R}(\langle \varepsilon \rangle_A)} \tag{4.59}$$

where A denotes pure A. Some values of $g^{(2)}$ so obtained are given in Table III. In another application, the technique was used[14a] at different temperatures by Gierke and Flygare to find the T dependence of $g^{(2)}$ above a liquid crystal transition temperature T_c, where it diverges. They found $g^{(2)} \propto (T - T^*)^{-1}$ for $T > T_c$, where T^* is an apparent transition temperature that cannot be reached since $T_c > T^*$. The data in Fig. 2 come from this experiment.

TABLE III
Estimates of $g^{(2)}$ from Light Scattering

Liquid	$g^{(2)}$
Benzene	0.95
Cumene	0.95
Fluorobenzene	1.00
Toluene	1.05
Acetophenone	1.40
Benzaldehyde	1.50
Benzoylchloride	2.10
Nitrobenzene	2.77

Source: Reference 31.

Equation (4.59) not only assumes that different B-type solvents affect A via $\langle \varepsilon \rangle$ alone, but that $\tilde{R}$ so obtained also works for A-type solvents (pure A). These assumptions can be checked via the ES theory. It is easy[32] to carry out the calculations in Section IV.B for a solution of A molecules in an isotropic solvent of B molecules, with mole fractions X_A, X_B, and total density ρ. The results already obtained hold, as long as we use the new expressions,

$$\mathscr{L} = \tfrac{6}{5}\rho\left(X_A\left\{ \left[\alpha_{0_A}^2 + \tfrac{1}{3}\alpha_{0_A}\Delta\alpha + \tfrac{2}{9}(\Delta\alpha)^2 \right] \tau_{20}^{AA} - \tfrac{1}{15}\left(\alpha_{0_A}\Delta\alpha + \tfrac{1}{3}(\Delta\alpha)^2 \right)\tau_{22}^{AA} \right\} \right.$$
$$\left. + X_B\left[\alpha_{0_A}\alpha_{0_B} + \tfrac{1}{3}\alpha_{0_B}\Delta\alpha\tau_{20}^{AB} \right] \right) \tag{4.60}$$

$$\kappa = \frac{4X_A}{15\left(X_A\alpha_{0_A} + X_B\alpha_{0_B} \right)}\left\{ X_a\left(\alpha_{0_A}\Delta\alpha\tau_{20}^{AA} - \tfrac{1}{30}(\Delta\alpha)^2\tau_{22}^{AA} \right) + X_B\alpha_{0_B}\Delta\alpha\,\tau_{20}^{AB} \right\} \tag{4.61}$$

with obvious notation; of course, $g^{(2)}$ refers now to an AA pair. We will use the truncated Kirkwood theory [Eq. (4.23a)] in the following.

First consider the limit $X_A \to 0$. Since $\langle \varepsilon \rangle$ will be $\langle \varepsilon \rangle_B$ here, we use the Clausius-Mossotti relation to eliminate α_{0_B} and find a theoretical expression for the quantity R,

$$\lim_{X_A \to 0} R^{*\phi} = \left(\frac{\langle \varepsilon \rangle + 2}{3} \right)^4\left\{ 1 + \frac{9}{10\pi}\left(\frac{\langle \varepsilon \rangle - 1}{\langle \varepsilon \rangle + 2} \right)\left(\frac{\alpha_{0_A}}{\Delta\alpha} + \frac{1}{3} \right)\tau_{20}^{AB} \right\}^2 \tag{4.62}$$

note the absence of τ_{22} and $(\Delta\alpha)^2$ terms, both of which vanish when one member of the interacting pair is spherical.

From our viewpoint, then, the dilution experiment first determines a fit to Eq. (4.62). The assumption that the right-hand side is a function of $\langle \varepsilon \rangle$ only for a given A is not true, due to the presence of τ_{20}^{AB}, which could vary with B. However, recall that in the Lorentz model, $\tau_{20} = \frac{10}{3}\pi\Delta S$, and ΔS is a property of A only. So, if a continuum model is valid, for which a large A molecule is indicated, fitting $\tilde{R}$ as a function of $\langle \varepsilon \rangle$ only makes sense. For A molecules comparable in size to B molecules, one must argue that τ_{20}^{AB} is fairly constant with B. This will be true if the arrangement of B around A is determined mainly by exclusion of B from the hard core of A, which in fact seems true in non-hydrogen-bonded liquids. Thus, there is some reason to believe that $\tilde{R}$ is a function of $\langle \varepsilon \rangle$ only given A, and Eq. (4.62) then tells us precisely what is determined.

Turning now to the second step of the "theory of the experiment," we have, at $X_A = 1$,

$$\lim_{X_A \to 1} R^{*\phi} = \left(\frac{\langle\varepsilon\rangle+2}{3}\right)^2 \left\{ 1 + \frac{9}{10\pi}\left(\frac{\langle\varepsilon\rangle-1}{\langle\varepsilon\rangle-2}\right)\left[\left(\frac{\alpha_{0_A}}{\Delta\alpha} + \frac{1}{3} + \frac{2}{9}\frac{\Delta\alpha}{\alpha_{0_A}}\right)\tau_{20}^{AA}\right.\right.$$
$$\left.\left. - \frac{1}{15}\left(1 + \frac{1}{3}\frac{\Delta\alpha}{\alpha_{0_A}}\right)\tau_{22}^{AA}\right]\right\}$$

(4.63)

Equation (4.62) must give a good approximation to Eq. (4.63) if $g^{(2)}$ is to be obtained from Eq. (4.59). For this to be true, first the arrangement of the center of mass of a nonspherical A molecule around an A, as measured by τ_{20}^{AA}, must be about the same as for an isotropic B molecule (τ_{20}^{AB}); this is plausible. Next, τ_{22} must be negligible, since it is absent from $\tilde{R}$. In computer simulations, we have found that τ_{22} is in fact negligible for small linear molecules. Finally, $\frac{2}{9}\Delta\alpha/\alpha_{0_A}$ should be much less than $\alpha_{0_A}/\Delta\alpha + \frac{1}{3}$; this is true for small diatomics but not for CS_2 and several benzene derivatives.

Based on these considerations, we suggested a slightly different approach to the dilution experiment if the polarizabilities are known. Use the low-density experiments to determine τ_{20}. Then, estimate $R^{*\phi}$ for pure A, using that τ_{20} including the $\frac{2}{9}\Delta\alpha/\alpha_{0_A}$ term, ignoring τ_{22}. It seems to us that this is the best way available to get $g^{(2)}$ without an actual calculation of the τ's.

It should also be noted that, given τ_{20}, Eq. (4.62) can be used to extract the ratio $\Delta\alpha/\alpha_{0_A}$ from $\tilde{R}$. In most cases, the polarizabilities are far better known than τ_{20}, so this might seem unrewarding. Such is not the case, however, for large molecules. Accurate determinations of $\Delta\alpha$ and α_0 are best made in the gas phase, and dilute gases of large molecules are usually unavailable. For large rigid molecules, estimates of τ_{20} based on shape alone should be accurate. So, from a knowledge of the shape, $\Delta\alpha/\alpha_0$ can be estimated in a situation where it is usually unaccessible. We[32] analyzed the data of Gierke and Flygare[14a] this way and found $R(\langle\varepsilon\rangle)$ to depend sensitively on $\Delta\alpha/\alpha_0$. Under the assumption that MBBA is needle-shaped, we concluded that $\Delta\alpha/\alpha_0 \sim 0.8$. We are not aware of any other estimates of this sort for such complex molecules. The technique seems appealing for cases where gas-phase data are unavailable. Of course, attempts to extract similar information from solution-phase measurements without a good treatment of the internal field problem are meaningless.

V. MICROSCOPIC BASIS OF THE ES THEORY

Having applied the ES theory extensively, it is time to further discuss its underpinnings. There is little reason to doubt that the ES theory is correct for slow primary variables; in the presence of a slow fluctuation, the source S, or ε, ought to respond as in local equilibrium. However, a full understanding of the ES theory and/or a treatment of the influence of rapid fluctuations, can be obtained only by direct calculation of the time-dependent source correlation function whose Fourier transform gives $S(\omega)$. The spectrum for liquids composed of nonspherical molecules is[16d] now well characterized; the narrow reorientational line sits atop a broad "Rayleigh wing" with a width of perhaps 20 cm^{-1}, and the integrated intensities of the two structures are comparable. These features should be derivable from a fully microscopic theory, and in particular the intensity of the narrow line should be given by the ES expression. Various attempts have been made[16d] along these lines. The current consensus is that the Rayleigh wing arises from rapidly varying internal field fluctuations. (And, of course, there is no argument about the sharp line.) The rapidity of the internal field fluctuations is believed due to the extreme sensitivity of the factors of $1/R_{ij}$ in the dipole tensor to shifts in intermolecular spacing.

The accepted view of the broad line might seem to pose a paradox: How can we both calculate an internal field correction to the slow orientational scattering and argue that internal field fluctuations are fast? We will now show that a unified picture of all the important facts and ideas can be constructed.

The ES theory can be employed with any dielectric theory. In Sections III and IV, we chose theories that would allow simple approximate calculations of $\gamma/\Delta\alpha$. Here, however, top priority is that we be able to relate expressions from scattering theory to those from dielectric theory. This is particularly easy to do with an approach[33] of Felderhof's, where the medium is regarded, to zero order, as having no fluctuations; the zero-order system does not scatter and has the Clausius-Mossotti dielectric constant, denoted $\langle \varepsilon \rangle^0$. The scattering and the corrections to $\langle \varepsilon \rangle^0$ are developed in closely related expansions ordered by the number of fluctuations present. One slight consequent drawback is that Felderhof's light scattering theory is best adapted to treating a scattering volume containing real fluid surrounded by continuum with $\langle \varepsilon \rangle^0$, not $\langle \varepsilon \rangle$. The difference is numerically negligible, and we shall use $\langle \varepsilon \rangle$ and $\langle \varepsilon \rangle^0$ interchangeably in the following.

Calculation of $\partial \bar{\varepsilon} / \partial \overline{Q}$ is straightforward, using our generalization of Felderhof's work to nonspherical molecules.[34] The result is

$$\frac{\partial \bar{\varepsilon}^{\alpha\beta}}{\partial \overline{Q}^{\alpha\beta}} = \left(\frac{\langle \varepsilon \rangle + 2}{3} \right)^2 \left(\Delta\alpha + \frac{\langle Q^{\alpha\beta} F^{\alpha\beta} \rangle}{\langle Q^{\alpha\beta} Q^{\alpha\beta} \rangle} \right) \tag{5.1}$$

$Q^{\alpha\beta}$ is the integral given in Eq. (4.1), and $F^{\alpha\beta}$ is a double integral,

$$F^{\alpha\beta} = \int d\mathbf{r} \, d\mathbf{r}' \, (\mathscr{F})_{r,r'} \tag{5.2}$$

of the matrix element of the operator $\mathscr{F}$ defined in the sense discussed in the early part of Section IV. The equations that give $\mathscr{F}$ are

$$\mathscr{F} \equiv N - \delta A \tag{5.3}$$

$$N = (I - \Delta \delta A U)^{-1} \delta A \tag{5.4}$$

$$U = \mathscr{T} R = \mathscr{T} (I - \alpha_0 \rho \mathscr{T})^{-1} \tag{5.5}$$

$\mathscr{T}$ and δA are given in Eqs. (4.8) and (4.10), and Δ is the operator that subtracts from anything its equilibrium average. Equation (5.1) is formally exact and thus completely equivalent to the Kirkwood theory, Eq. (4.7). Expansion of the right-hand side in "powers" of $\mathscr{T}$ immediately gives the truncated Kirkwood theory, Eq. (4.11), since to lowest order, $U = \mathscr{T}$, and

$$N = \delta A + \delta A \mathscr{T} \delta A + \cdots$$

The whole point of using Felderhof's theory, however, is that we do not want to make such expansions; in this case, use of formal expressions proves fruitful.

Turning to the source itself, the theory gives

$$\mathbf{S} = \left(\frac{\langle \varepsilon \rangle + 2}{3} \right)^2 [\delta A + F] \cdot \langle \mathbf{E} \rangle \tag{5.6}$$

Equation (5.6) makes clear why we are using this formulation; all the interesting effects are contained in the same operator F, in both the ES and fully microscopic approaches. Thus,

$$S(\omega) = \left(\frac{\langle \varepsilon \rangle + 2}{3}\right)^4 (\Delta\alpha)^2$$

$$\times \int dt\, e^{i\omega t}\left[\langle Q^{\alpha\beta}(t)Q^{\alpha\beta}\rangle + \frac{1}{\Delta\alpha}(\langle Q^{\alpha\beta}(t)F^{\alpha\beta}\rangle + \langle F^{\alpha\beta}(t)Q^{\alpha\beta}\rangle)\right.$$

$$\left. + (\Delta\alpha)^{-2}\langle F^{\alpha\beta}(t)F^{\alpha\beta}\rangle\right]\langle |E|\rangle^2 \qquad (\alpha \neq \beta) \qquad (5.7)$$

where we have noted that $\delta A^{\alpha\beta} = \Delta\alpha Q^{\alpha\beta}$.

We now come to a crucial point in relating the two approaches. No further progress can be made without a theory for the time correlation functions in Eq. (5.7). Since F (every term of which contains one or more dipole tensors) is considered to vary rapidly while Q is slowly varying, the most obvious way to reproduce sharp and broad lines is to discard the cross correlations. The sharp line is then of pure Q "character" and comes from the Q autocorrelation, and similarly for the broad line and F. This simple recipe has indeed been tried, but it presents difficulties. First, it gives the coefficient of the sharp line intensity (the Q autocorrelation function) as a pure "fourth-power law." Second, it predicts a relative broad/sharp intensity of $\langle FF\rangle/(\Delta\alpha)^2\langle QQ\rangle$, which, when evaluated, is much too small. Of course, it is wrong to just neglect the cross correlations. In fact, they are essential to finding the ES answer, as we now discuss.

Keyes, Kivelson, and McTague[35] studied the problem of the coupled dynamics of a slow and a fast variable via the Mori formalism. Their most important conclusion was that the time dependence of the cross correlations is almost entirely the slow time dependence. Also, the "fast" variable picks up a bit of slow character, and vice versa, through the coupling. When all the slowly varying parts of Eq. (5.7) are added up, we obtain[34]

$$R^{*\phi}(\text{slow}) = \left(\frac{\langle \varepsilon \rangle + 2}{3}\right)^4\left(\frac{\partial\bar{\varepsilon}^{\alpha\beta}}{\partial\bar{Q}^{\alpha\beta}}\right)^2 \qquad (5.8)$$

with the derivative given by Eq. (5.1). Thus, the fully microscopic theory and the ES theory are in perfect agreement for the sharp line.

In going from Eq. (5.7) to Eq. (5.8), the Q autocorrelations, of course, give the $(\Delta\alpha)^2$ term in $(\partial\varepsilon/\partial Q)^2$. The dominant slow part of the cross time correlations give the cross terms in $(\partial\varepsilon/\partial Q)^2$. The small slow part of the F autocorrelation function, which automatically comes out of the Mori approach, gives the remaining component of $(\partial\varepsilon/\partial Q)^2$. All the results for the contribution of the different correlation functions to the sharp line depend on the separation of time scales, and corrections to Eq. (5.8) exist of order

τ(short)/τ(long). To the same order of approximation, the intensity of the sharp line is just $L^4 \langle F^{\alpha\beta} F^{\alpha\beta} \rangle$, that is, there is no change from the most naive estimate.

This simple calculation merits some discussion. To a large extent, it constitutes a derivation of the local equilibrium hypothesis from first principles. Suppose internal field effects seem to be, in some crude sense, "fast." Fluctuations in slow variables then determine, to a good approximation, the *static* environment that the fast variable sees, which in turn affects its dynamics. Thus the dynamics of the "fast" variable can never be completely fast, since its response to the slow fluctuation is superimposed on any other time dependence. This is why internal fields can contribute to the sharp line. In a calculation of the source correlation function, these ideas manifest themselves formally in the slow time dependence of correlation functions that, naively, appear fast. The argument can be reversed to explain why the broad line intensity is so much simpler to evaluate. Consequently, the broad/sharp intensity ratio is greatly reduced from the estimate discussed just preceding Eq. (5.8), in agreement with experiment ($\gamma/\Delta\alpha < 1$).

We can now give some general guidelines for constructing theories with several primary variables, which has been done[16d] by several authors. If the variables are slow with respect to F, writing

$$S = \sum_i \left(\frac{\partial \varepsilon}{\partial X_i} \right) X_i \tag{5.9}$$

followed by evaluation of the source correlation function, is correct. On the other hand, Eq. (5.9) cannot be used with both slow and fast X's, since (1) ε will not be in local equilibrium with a fast X and (2) the response of the fast X's (X^F) to the slow X's (X^s) will be an essential element of ($\partial\varepsilon/\delta X^s$), so writing Eq. (5.9), and then evaluating $\langle X^F(t) X^s \rangle$ and $\langle X^F(t) X^F \rangle$, which, as we just saw, have large slow parts, will overcount this response. To repeat, the ES approach is apt for slow fluctuations only. Once any direct treatment of fast fluctuations is desired, the theory must work with the source itself, without the aid of thermodynamic derivatives.

VI. THE INTERNAL FIELD PROBLEM IN LIGHT SCATTERING FROM FLEXIBLE MOLECULES

In our terminology, "flexible" molecules are molecules that possess internal flexibility due to torsional degrees of freedom. Chain polymers are the most important class of these molecules, but many others, such as cyclohexane, exist. The polarizability of most of these molecules changes due to conformational rearrangements, so that light scattering is an important technique for the study of the internal structure and the dynamics of these systems. Since changes in the internal molecular structure are an interesting property of flexible molecules, experiments on dilute solutions of dilute gases are quite common. The data on dilute gases are much easier to interpret, but in most cases of interest, namely chain polymers, these data are experimentally accessible only for relatively small oligomers.

In the case of flexible molecules, the solution of the internal field problem is necessary even when one is considering light scattering from a dilute gas where the intermolecular interactions can safely be neglected. This additional internal field problem arises from the fact that dipoles induced in one part of the molecule interact with dipoles induced in the other parts. This IFP exists in rigid polyatomic molecules as well, but it is much more important in the present case, where the polarizability fluctuations due to conformational changes are of primary interest in light scattering experiments. As was pointed out earlier, in a dilute gas the sum of the mean square fluctuations in the molecular induced moments is responsible for light scattering. Our ability to relate the light scattering data to the fundamental properties of these molecules, such as the torsional potential parameters and the relaxation rates for internal motion, depends on an accurate solution of the intramolecular internal field problem. Even though light scattering has been used for a number of years to measure the properties mentioned above, especially in dilute polymer solutions, the internal field problem in these systems has received very little attention. The main reason for this is, in our opinion, the fact that the solution results, which contain both the intra- and intermolecular internal field effects, have not been critically analyzed with the goal of separating these effects. The experiments designed to isolate the intermolecular internal field problem, analogous to those carried out by Flygare and coworkers[14] on fluids of rigid molecules discussed in Section IV, have never been carried out on solutions of flexible molecules. A model, which we call the *bond additive approximation* (BAA), and which is also called the valence optical scheme, based on the picture of a polymer as a collection of independent scatterers, is almost universally used in interpreting the solution data. As will become clear from the following development, the deviations from this model are solvent dependent and may be quite large in some solvents. We start by describing first light scattering from dilute gases, where only the intramolecular internal field effects need to be considered.

A. Light Scattering from Dilute Gases of Flexible Molecules

The total dipole moment μ_M induced in a flexible molecule has contributions from several sites on the molecule and from intersite interactions. The simplest model that approximately describes this situation is the *interacting atom model* (IAM), based on the early work of Gray[36] and Silberstein[37] and developed by Applequist and coworkers.[38,39] In this model, an ideal dipole μ_i is induced in the ith atom. The atoms are assumed to be optically isotropic, so that the ith atom has the polarizability α_i. The molecular polarizability $\underset{\sim}{\alpha}_M$ is defined as the tensor relating the total dipole moment μ_M induced in the molecule to the applied field E_0:

$$\mu_M = \underset{\sim}{\alpha}_M \cdot E_0 \qquad (6.1)$$

where, for an m-atomic molecule

$$\mu_M = \sum_{i=1}^{m} \mu_i \qquad (6.2)$$

and μ_i is given by

$$\mu_i = \alpha_i \left[E_0 + \sum_{j \neq i}^{m} T_{ij} \cdot \underset{\sim}{\mu}_j \right] \qquad (6.3)$$

where $T_{ij} = T(r_i - r_j)$ is the dipole tensor defined by Eq. (4.9). Equation (6.3) may be written as a matrix equation

$$\tilde{A}\tilde{\mu} = \tilde{E}_0 \qquad (6.4)$$

where $\tilde{A}$ is a $3m \times 3m$ matrix and $\tilde{\mu}$ and $\tilde{E}_0$ are $3m$-dimensional column vectors. If m is not too large, Eq. (6.4) may be inverted to give

$$\tilde{\mu} = \tilde{B}\tilde{E}_0 \qquad (6.5)$$

where B is called the "relay tensor," since the 3×3 submatrix B_{ij} expresses the dipole moment relayed to atom i by the field and applied to atom j. In many applications of the IAM, for example, the contribution to light scattering from intermolecular interactions, all elements of the relay tensor are needed. If, however, we are interested in the total molecular polarizability, they need not all be evaluated. It suffices to solve the linear system of Eqs. (6.4) for $\tilde{\mu}$ as a function of $\tilde{E}_0$. In terms of the relay tensor elements, $\underset{\sim}{\alpha}_M$ is given by

$$\underset{\sim}{\alpha}_M = \sum_{i=1}^{m} \sum_{j=1}^{m} B_{ij} \qquad (6.6)$$

Due to its dependence on the B_{ij}'s, $\underset{\sim}{\alpha}_M$ varies with the changes in the molecular conformation. In light scattering, one measures the average over the molecular conformations of the mean squared isotropic or anisotropic polarizability. Thus in isotropic scattering the Rayleigh ratio per molecule for light scattering from an ideal gas of flexible molecules is given by

$$\hat{R}_{\text{iso}} = \langle \alpha_0^2 \rangle \tag{6.7}$$

and

$$\alpha_0 = \tfrac{1}{3} \text{Tr} \, \mathbf{\alpha}_M \tag{6.8}$$

The depolarized Rayleigh ratio per molecule is given by

$$\hat{R}_{\text{dep}} = \frac{\langle \gamma^2 \rangle}{15} \tag{6.9}$$

where the anisotropic polarizability γ is defined as

$$\gamma^2 = \tfrac{3}{2} \text{Tr} \big[(\mathbf{\alpha}_M - \alpha_0 1) \cdot (\mathbf{\alpha}_M - \alpha_0 1) \big] \tag{6.10}$$

Equations (6.8) and (6.10) represent the generalization of the Eqs. (2.5) to asymmetric molecules. A quantity often measured is the depolarization ratio ρ, which is

$$\rho_u = \frac{5 \langle \gamma^2 \rangle}{45 \langle \alpha_0^2 \rangle + 7 \langle \gamma^2 \rangle} \tag{6.11}$$

for unpolarized incident light, and

$$\rho_p = \frac{\langle \gamma^2 \rangle}{15 \langle \alpha_0^2 \rangle} \tag{6.12}$$

for polarized incident light.

A prototypical system for the study of properties of flexible molecules is the set of n-alkane chains. The chain-length dependence of a variety of their properties has been measured. The IAM has been applied to the calculation of the depolarization ratio ρ_u of light scattering from a series of dilute gases of short n-alkane molecules.[40] It has successfully explained the observed chain-length dependence of ρ_u, something that the still widely used BAA has failed to do. Light scattering from dilute n-alkane gases was measured by several groups in the 1960s and 1970s.[41,42] Aval et al.[42] summarized most of the experimental and early theoretical results, based on the BAA and on several models for the polymer conformational statistics. They found that the BAA always fell below the observed ρ_u versus $N = n - 1$ even if one assumed that the molecules were frozen in their all-*trans* conformation, something that is not likely to occur at room temperature, where the experiments were carried out.

We will briefly describe the BAA and its predictions in the case of chain polymers. The BAA was proposed by Denbigh[43] in 1940 and applied to depolarized light scattering from polymers soon after.[44] In the BAA, an axially symmetric polarizability is assigned to each bond. The molecular polarizability is obtained by simply summing the bond polarizabilities. Thus, for a molecule containing N_b bonds, the molecular polarizability is

$$\underset{\sim}{\alpha}_M^{BAA} = \sum_{i=1}^{N_b} \underset{\sim}{\alpha}_i^{(b)} \qquad (6.13)$$

where the bond polarizability $\underset{\sim}{\alpha}_i^{(b)}$ is defined by Eq. (2.4), with Q_i being a function of $\hat{u}_i$, the unit vector along the bond. In the BAA the anisotropic polarizability γ is a function of the relative orientations of bonds i and j,

$$\gamma^2 = \sum_{i=1}^{N_b} \Delta\alpha_i^{(b)} \left[\Delta\alpha_i^{(b)} + \sum_{j \neq i}^{N_b} \Delta\alpha_j^{(b)} P_2(\hat{u}_i \cdot \hat{u}_j) \right] \qquad (6.14)$$

while the isotropic polarizability is independent of the internal molecular coordinates

$$\alpha_0 = \sum_{i=1}^{N_b} \alpha_{0i}^{(b)} \qquad (6.15)$$

In this model $\langle \gamma^2 \rangle$ and ρ_u measure the intramolecular orientational pair correlations. In the commonly used models of polymer conformational statistics, if the excluded volume effects are ignored, quantities such as $\langle P_2(\hat{u}_i \cdot \hat{u}_j) \rangle$ depend only on the relative location of the ith and jth bonds along the polymer backbone, not on the distance between them. The IAM is fundamentally different in that the inverse powers of interatomic distances, arising from interactions between nonbonded atoms, play a crucial role in both $\langle \alpha_0^2 \rangle$ and $\langle \gamma^2 \rangle$. The most important consequence of this for light scattering is the fact that $\langle \gamma^2 \rangle$ in the IAM has a much stronger dependence on the chain length than it does in the BAA. The chain-length dependence of $\langle \alpha_0^2 \rangle$ is almost the same in the two models, namely $\langle \alpha_0^2 \rangle \propto N^2$. In the case of *n*-alkanes, the

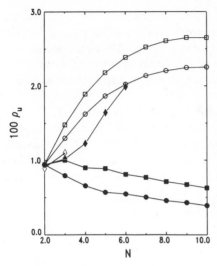

Fig. 4. Depolarization ratio ρ_u versus N, the number of C—C bonds for n-alkane gases at room temperature. Experimental results are those of Ref. 41a (solid diamonds) and Ref. 42 (open diamonds). Theoretical results (from Ref. 40) are for the rotational isomeric state model with $E_g = 475$ cal/mol (squares) and $E_g = 0$ (circles), both with infinite g^+g^- energies. The IAM is represented by open symbols, the BAA by solid symbols.

consequences of this difference are dramatic. If the results for ρ_u in two models, evaluated using the rotational isomeric state model of the polymer conformation,[46] are compared, the IAM predicts ρ_u to be an *increasing* function of N, while the BAA predicts it to *decrease* with increasing N. The predictions of the two models as well as the experimental results for ρ_u are shown in Fig. 4.[47] The *gauche* (g) states are assumed to be 475 cal/mol higher in energy than the *trans* state, and the g^+g^- and g^-g^+ sequences are assumed to have infinite energies. Note that the experimental data show an increase of ρ_u with increasing N, in agreement with the IAM. Thus it is seen that the intramolecular internal field problem has to be solved in order to explain the gas-phase experimental data.

While the IAM is quite accurate in describing the optical response of n-alkanes, it is sometimes inaccurate, especially in modeling that of molecules containing double or triple bonds. The main deficiency of the IAM is in its treatment of the interactions between atoms directly bonded to each other. Given their proximity and the electron density contained in the bond, these atoms are not likely to actually interact as ideal dipoles. The simple IAM tends to overestimate the polarizability parallel to the bond axis. Two very simple modifications of the original model have recently been proposed to deal with this problem. The first one is due to Birge[48] and consists in replacing the isotropic atomic polarizability α_i by an anisotropic polarizability $\underset{\sim}{\alpha}_i$. The anisotropy in $\underset{\sim}{\alpha}_i$, which is due to the chemical environment, is assumed to be a function of the electronic repulsion integrals g_{ij} between atom i and all the other atoms j in the molecule. Each g_{ij} is raised to the same power κ, the repulsion exponent, which is an additional global parameter for the whole molecule. The other modification, due to Thole,[47] retains the isotropic atomic polarizabilities but modifies the DID interactions, so that they are no longer given by T_{ij}, but by a modified interaction tensor t_{ij}, which reduces to T_{ij} when r_{ij} becomes larger than $a(\alpha_i\alpha_j)^{1/6}$, where a is a number of order unity and depends on the functional form of the short-range interaction. The quantity $(\alpha_i\alpha_j)^{1/6}$ sets the distance scale. The interaction functions are chosen starting from simple functions describing the charge density around an atom. Thole found that he could improve substantially upon the accuracy of the polarizabilities calculated by Applequist, Carl, and Fung[48] by using a single, relatively simple, interaction function with fewer atomic polarizability parameters. Both of the modified IAMs result in a decrease of the polarizability along bonds and an increase of the polarizability perpendicular to bonds relative to the original IAM. Neither one of these modified IAMs have yet been applied to light scattering from flexible molecules. Birge and coworkers have, however, treated quite successfully a closely related problem of the Raman intensities for the ring puckering[49] and torsional[50] vibrations of several molecules. Before closing the discussion of the intramolecular internal field problem, we note that a model similar to the IAM has also been proposed, with bonds instead of atoms as the fundamental polarizable units.[51] This model has not yet been applied to the internal field problem in flexible molecules. Its application to the polarizabilities and the hyperpolarizabilities of halomethanes has shown that it would also lead to results similar to those of the IAM.[52]

B. Light Scattering from Solutions of Flexible Molecules

Most of the light scattering experiments on flexible molecules are done on dilute solutions of these molecules in optically isotropic solvents, although in the case of alkanes many experiments on neat liquids have been done as well. As should be clear from the discussion of the previous sections, the quantities observed in these experiments are effective polarizabilities. The situation here parallels the one described in the case of rigid molecules. However, due to the increased difficulties associated with dealing with internal molecular degrees of freedom, the theoretical understanding of depolarized light scattering from these systems in the liquid phase is considerably less advanced.

For reasons similar to the ones already discussed, isotropic scattering from flexible molecules in the liquid state has for a long time been well understood in the context of the ES theory. In the case of neat liquids of flexible molecules, the ES theory discussed in Section III applies. If the Clausius-

Mossotti equation is used as well, one has to assume that α_0 fluctuates to a negligible extent, namely that

$$\langle \alpha_0 \rangle^2 = \langle \alpha_0^2 \rangle \tag{6.16}$$

In the BAA, Eq. (6.16) holds exactly, since α_0 is independent of the chain conformation. In the IAM this relation is approximate, but at least for n-alkanes it turns out to be an excellent approximation, since α_0 is only weakly conformation-dependent.[40,51] As is true in the case of rigid molecules, the effective isotropic polarizability of flexible molecules in the liquid phase is almost identical to its gas-phase counterpart. In the case of dilute solutions, the ES theory applies as well. Here the solute concentration c is an additional thermodynamic variable. In the limit of infinite dilution, the excess scattering due to the solute is proportional to $c(\partial \varepsilon / \partial c)^2$,[52] where ε is the optical dielectric constant of the solution. An expression for $\partial \varepsilon / \partial c$ in terms of the isotropic solute polarizability may be derived by using a Clausius-Mossotti equation generalized to mixtures. This generalization of the ES theory to dilute solutions of flexible molecules is due to Debye.[53] Isotropic scattering from polymers in solution has been a very valuable technique for measuring the overall chain dimensions and its molecular weight.[54,55]

The generalized ES theory for anisotropic scattering discussed in Section IV is not straightforwardly applicable to all flexible molecules. If we ignore the rapid local field fluctuations, which always contribute a broad background to the spectrum, two types of dynamical processes still remain. One of them is the overall rotation of the molecule, and the other consists of internal structural rearrangements. In the case of small, flexible molecules, the depolarized light scattering spectrum is dominated by molecular reorientation; in the case of long-chain molecules, by local structural rearrangements. The spectrum is usually a composite of several relaxation rates. Even in small molecules, where the overall rotation is the only process that need be considered, conformers of different shapes reorient at different rates. In longer molecules, different modes of internal rotation will have different relaxation rates. In spite of all this, the experimental spectrum appears to be surprisingly simple in many cases,[55-58] so that there is a hope that a modified form of the generalized ES theory might be applicable. Again, we may cite the n-alkanes as a typical example. Depolarized light scattering spectra of neat liquids near room temperature have recently been measured for the values of n between 5 and 71.[56,57] The spectra can all be fit by single Lorentzian line shapes, indicating that a single relaxation mechanism dominates. The relaxation times τ_{vh} obtained in this way may be fit to the expression

$$\tau_{\mathrm{vh}} = \frac{A\eta}{T} + B \tag{6.17}$$

where η is the shear viscosity, A depends on the size and the shape of the molecule, and B is the intercept believed to represent the inertial limit for the reorientation of the molecule at temperature T. When τ_{vh} is plotted against η/T, the slope A increases with increasing n for n between 5 and 22. This indicates that the overall molecular rotation is the dominant relaxation mechanism, since A increases with both the size and elongation of the molecule. For larger n-alkanes, the opposite trend is seen. This means that the local structural rearrangements are the important mechanism of the optical anisotropy relaxation. Analogous results have also been observed in depolarized light scattering experiments on polystyrene solutions.[55]

Different models for the effective molecular polarizability weigh differently the importance of contributions from various relaxation mechanisms to the depolarized light scattering spectrum. Thus here the internal field problem, especially its intramolecular part, affects not only the intensity but the spectrum as well. This point was illustrated by Keyes et al.,[59] who calculated the relaxation rate of the effective anisotropic polarizability of a series of short chain n-alkanes ($5 \leq n \leq 11$) in solution. The chain dynamics was modeled using a Brownian dynamics simulation of a chain having independent torsional potentials. Two models for the effective polarizability were compared. One of them was the BAA, described above. The other was the first-order perturbation expansion of the IAM; namely, $\underset{\sim}{\alpha}_{M,\text{eff}}$ was expressed as

$$\underset{\sim}{\alpha}_{M,\text{eff}} = \sum_{i=1}^{m} \alpha_i \left[1 + \sum_{\substack{j \neq i}}^{m} \alpha_j T_{ij} \right] \tag{6.18}$$

The rationale for the use of this expansion, which would drastically underestimate $\underset{\sim}{\alpha}_M$, will be much more apparent when the depolarized light scattering intensity of flexible molecules in the liquid phase is discussed in detail. For the present, it suffices to say that the solvent appears to greatly weaken the intramolecular internal field effects in these systems, possibly to the point where they might be treated perturbatively. The results of these calculations for the relaxation time as a function of $N = n - 1$ are shown in Fig. 5. The relaxation times are expressed in reduced units such that $\tau^* = \tau k_B T / 3\pi \eta r_{\text{CC}}^2 d_C$, where $r_{\text{CC}} = 1.54$ Å is the carbon–carbon bond length and $d_C = 3.40$ Å is the methylene group covalent diameter. The figure depicts $\tau T / \eta$ (at constant T), which roughly corresponds to the quantity A in Eq. (6.17). In addition to the polarizability anisotropy relaxation times in the two models, two relaxation times of second-rank variables τ_E^* and τ_D^* are depicted. τ_E^* corresponds to the relaxation of $\mathbf{EE} - (1/3)E^2$, where $\mathbf{E}$ is the chain end-to-end vector, and τ_D^* corresponds to the relaxation of $\mathbf{D}_1\mathbf{D}_1 - (1/3)D_1^2$, where

D_1 is the difference vector between the first pair of backbone bond vectors b_1 and b_2, that is, $D_1 = (b_1 - b_2)$. τ_E^* and τ_D^* are, respectively, the longest and shortest relaxation times for the second-rank tensors of a given chain. The figure also shows the experimental results for τ_{vh}^* for neat liquids of n-alkanes. The figure shows that $\tau_{IAM, P}^*$ [P denotes use of Eq. (6.18)] is always longer than τ_{BAA}^*, with the difference between the two increasing with increasing chain length. $\tau_{IAM, P}^*$ stays close to τ_E^*, while τ_{BAA}^* is closer to the local relaxation time τ_D^*. Through the DID interactions between parts of the chain distant along the backbone, the effective polarizability in the IAM acquires a substantially larger contribution from the overall molecular rotation than its BAA counterpart does. Scattering appears, roughly, to be a collective property in the IAM and a local property in the BAA. The figure also shows that τ_{vh}^* is larger than even τ_E^*. This may seem at first glance to be a paradoxical result. There are, however, three important differences between the experiment and the calculation that account for it. First, since τ_{vh}^* is measured in a neat liquid, it contains single-molecule and pair contributions, while the simulation contains only a single-molecule contribution. The effect of static second-rank pair correlations is to increase τ_{vh}^* above the corresponding single-molecule relaxation time. Also, for short chains these correlations will increase with increasing chain length. The simulations did not include the coupling between adjacent torsional angles through the "pentane effect." As a result, the actual n-alkanes are stiffer than the molecules represented by the model. Finally, had the higher order terms in the IAM been included in the representation of $\alpha_{M, eff}$, τ_{IAM}^* would be closer to τ_E^*. The most important point illustrated by the figure is that it is impossible to

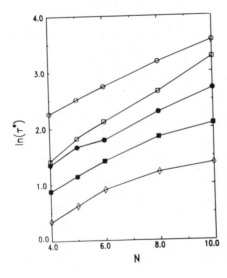

Fig. 5. Logarithms of reduced relaxation times τ^* for second-rank variables versus N, the number of C—C bonds for n-alkanes in the liquid phase. Brownian dynamics results of Ref. 59 for τ_E^* (open squares), $\tau_{IAM, P}^*$ (solid circles), τ_{BAA}^* (solid squares), and $\tau_{D_1}^*$ (open diamonds) are compared to experimental depolarized light scattering relaxation times of Ref. 58 (open circles).

explain the condensed-phase depolarized light scattering spectra without taking the long-range intramolecular internal field effects into account. In particular, the figure shows that τ^*_{BAA} starts to level off by $n = 11$, while τ^*_{vh} and $\tau^*_{IAM, P}$ are still increasing. In fact, as was mentioned earlier, τ^*_{vh} is still increasing for $n = 22$.

As we have mentioned already, most of the depolarized light scattering intensity measurements involving chain molecules in the liquid phase are carried out on dilute solutions. Optically isotropic solvents are normally used. In addition to that, the scattering intensity from the pure solvent, sometimes multiplied by the ratio of the isotropic local field factors for the pure solvent and the solution, is subtracted from the measured Rayleigh ratio. Using the Lorentz-Lorenz expression for the local field, the Rayleigh ratio for the excess scattering due to the polymers is then given by

$$R^{(p)}_{vh} = R_{vh} - \phi_s \left(\frac{\varepsilon + 2}{\varepsilon_s + 2} \right)^4 R^{(s)}_{vh} \qquad (6.19)$$

where ε_s is the solvent optical dielectric constant and ϕ_s is the solvent volume fraction.[60] The effective anisotropic polarizability of flexible molecules in solution may be defined in a way analogous to the effective optical anisotropy of rigid molecules in dilute solution, namely,

$$\langle \gamma^2_{eff} \rangle = \frac{15}{\rho_p} \left(\frac{3}{\varepsilon + 2} \right)^4 R^{(p)}_{vh} \qquad (6.20)$$

where ρ_p is the solute number density. From our discussion of both the molecular and dielectric cavity theories of depolarized light scattering from liquids of rigid molecules, we expect $\langle \gamma^2_{eff} \rangle$ to be smaller than $\langle \gamma^2 \rangle$ and solvent dependent. This first expectation is indeed realized, and the second will most likely be realized also once the experiments designed to test it are carried out. To our knowledge, the solvent dependence of $\langle \gamma^2_{eff} \rangle$ has never been measured. Given the fact that $\langle \gamma^2_{eff} \rangle \ll \langle \gamma^2 \rangle$ for the solvents having refractive indices in the typical liquid range, one would expect that $\langle \gamma^2_{eff} \rangle$ would decrease with the increasing solvent refractive index. We would further anticipate that the solvent effects on $\langle \gamma^2_{eff} \rangle$ would be more and more pronounced for more elongated, and therefore more optically anisotropic, molecules. This expectation is indeed borne out by experiment. For example, in the measurement of $\langle \gamma^2_{eff} \rangle$ for a series of n-alkanes dissolved in carbon tetrachloride, the difference between $\langle \gamma^2 \rangle$ and $\langle \gamma^2_{eff} \rangle$ becomes more and more pronounced as n increases. Figure 6 shows the experimentally determined $\langle \gamma^2 \rangle$ and $\langle \gamma^2_{eff} \rangle$ in CCl_4 solvent[60,61] and in the neat liquid.[58] The neat liquid

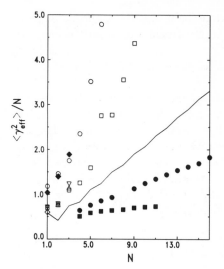

Fig. 6. $\langle\gamma_{\text{eff}}^2\rangle/N$ versus N, the number of C—C bonds for n-alkanes in gas and liquid phases. Gas-phase data are from Refs. 41a (open circles), 41b (open inverted triangles), 41c (open triangles), 41d (open diamonds), 41e (open squares), and 42 (solid diamonds). Neat liquid data are from Ref. 58 (filled circles) and solution (in CCl_4) data from Ref. 61 (filled circles). Solid line is the BAA result for all *trans* chains.

results have not been corrected for the orientational pair correlations, which are expected to be small (but increasing with increasing n) for the short chains depicted here. The optical dielectric constant of carbon tetrachloride is equal to 2.14, and that of the liquid n-alkanes ranges from 1.83 for n-pentane to 1.99 for n-dodecane. The increasing optical dielectric constant of the pure liquids as n increases will counteract to some extent the effects of the orientational pair correlations on the displayed $\langle\gamma_{\text{eff}}^2\rangle$. The mean squared effective anisotropic polarizabilities exhibit the trends expected from the known behavior of rigid molecules in solution. The solvent effects are seen to be very large, especially for longer chains.

Even though the qualitative trends in the behavior of $\langle\gamma_{\text{eff}}^2\rangle$ of flexible molecules in solution are readily predicted on the basis of the generalized ES theory for rigid molecules, the actual quantitative prediction of $\langle\gamma_{\text{eff}}^2\rangle$ for a series of chain polymers even as simple as n-alkanes has not yet been achieved, although progress has been made toward this goal. This statement may come as a surprise to readers familiar with the literature on depolarized light scattering in polymer solutions. In the vast majority of papers on this subject, the BAA is used as a model for $\langle\gamma_{\text{eff}}^2\rangle$. While the BAA is a dismal failure when applied to $\langle\gamma^2\rangle$, it has been relatively much more successful in modeling the optical response of polymers in solution. For example, $\langle\gamma_{\text{eff}}^2\rangle$ of n-alkanes in CCl_4 is well described by the BAA with realistic values of the torsional potential parameters.[60] The difficulties with the BAA arise if one is trying to describe the solvent effects on depolarized light scattering from chain molecules. There is absolutely no mechanism for incorporating these effects into the BAA. This means that the three sets of results represented on Fig. 6 could never be accounted for by the BAA. From the available evidence, it appears that the BAA is a reasonable model for $\langle \gamma_{\text{eff}}^2 \rangle$ for chain molecules dissolved in solvents of molecules that are more highly polarizable than the monomeric units of the chain molecule. In this case the polymer-solvent DID interactions are relatively stronger than the intramolecular DID interactions of the polymer molecule. As a result of this, the interactions between distant parts of the polymer become highly screened by the solvent and can safely be neglected.

A reasonable procedure for developing the theory of depolarized light scattering from solutions of flexible molecules is to start with small flexible molecules, for which the theory of light scattering from rigid molecules in solution is applicable with only minor modifications. The insights gained from the results of this work can then be applied to depolarized light scattering from larger chain molecules, where it becomes highly impractical to take into account the details of the solvent structure and of the polymer-solvent interactions. For small-chain molecules, such as small oligomers of chain polymers, we may assume that the solution consists of an equilibrium mixture of different conformers. The anisotropic polarizability of each conformer plays the role analogous to that of Q_i in the generalized ES theory. The molecular polarizability as well as the solute-solvent interactions may be modeled by using the IAM for the polymer optical response. Adopting this procedure for a dilute solution of flexible molecules in an optically isotropic solvent with molecular polarizability $\alpha^{(s)}$, we find the following "Kirkwood" expansion for the xz component of the polymer's effective anisotropic polarizability[62]:

$$
\alpha_{M,\text{eff}}^{xz} = \alpha_M^{xz} \left\{ 1 + \frac{15 \rho_s \alpha^{(s)}}{\langle \gamma^2 \rangle} \left\langle \alpha_M^{xz} \int d\{\mathbf{r}_{i\beta}\} \left[\left(2 h_{ps}(\{\mathbf{r}_{j\beta}\}) \right. \right. \right. \right.
$$
$$
\left. \left. \left. \times \sum_{j=1}^m \sum_{k=1}^m \left[\mathbf{B}_{jk} \cdot \mathbf{T}'(\mathbf{r}_{k\beta}) \right]^{xz} \right) + \alpha_{ps}^{(s)} \int d\mathbf{r}_{\mu\nu}\, h_{pss'}^{(3)}(\{\mathbf{r}_{i\beta}, \mathbf{r}_{\mu\nu}\}) T'^{xz}(\mathbf{r}_{\mu\nu}) \right] \right\rangle_c \right\}
$$

$$(6.21)$$

where ρ_s is the solvent number density, h_{ps} is the pair correlation function for a single conformer of the flexible molecule (p) and a solvent molecule (s). $h_{pss'}^{(3)}$ is the three-molecule correlation between a conformer of the flexible molecule and two solvent molecules (s and s'). The latin subscripts denote the sites on the polymer molecule (total of m sites), and the Greek subscripts the solvent molecules. The notation $\{\mathbf{r}_{i\beta}\}$ means the set of distance vectors between the sites on the polymer and the center of the βth solvent molecule and $\langle\ldots\rangle_c$, the average over conformations. The effective polarizability is seen to depend on the DID interactions between the relay tensor elements and the solvent molecules. $\langle\gamma_{\text{eff}}^2\rangle$ is given by

$$\langle\gamma_{\text{eff}}^2\rangle = 15\langle(\alpha_{M,\text{eff}}^{xz})^2\rangle_c \qquad (6.22)$$

If the "product" approximation discussed in Section IV is used, the three-molecule correlations disappear altogether, and Eq. (6.21) becomes

$$\alpha_{M,\text{eff}}^{xz} = \alpha_M^{xz}\left[1 + \frac{30\rho_s\alpha^{(s)}}{\langle\gamma^2\rangle}\left\langle\int d\{\mathbf{r}_{i\beta}\}\, h_{ps}(\{\mathbf{r}_{j\beta}\})\right.\right.$$
$$\left.\left.\times \alpha_M^{xz}\sum_{j=1}^{m}\sum_{k=1}^{m}\left[\mathbf{B}_{jk}\cdot\mathbf{T}'(\mathbf{r}_{k\beta})\right]^{xz}\right\rangle_c\right] \qquad (6.23)$$

This expression appears only slightly more complicated than the effective polarizability expressions obtained earlier for the dilute solutions of rigid molecules. We may point out, parenthetically, that the relay tensor formalism for intermolecular DID interactions is useful also in representing the internal field effects on light scattering from liquids of rigid polyatomic molecules. As has been shown in the case of liquid CS_2, a more realistic representation of the intermolecular interactions is obtained if one represents the molecules as containing three polarizable sites instead of one.[21,24]

In order to develop some insight into the solvent effects described by Eq. (6.23), it is instructive to consider a simple but nontrivial example, that of n-butane in dilute solution. Since n-butane is a small molecule, we assume that the intermolecular interaction term in Eq. (6.23) may be approximated by a DID interaction between a single polarizable site whose induced dipole is proportional to $\boldsymbol{\alpha}_M$ of a given conformer, and a solvent molecule. Furthermore, we approximate the solvent distribution as that corresponding to a uniform density of solvent outside the ellipsoidal cavity occupied by the n-butane molecule. n-Butane has two *gauche* conformers, which have the same ellipsoidal shape and polarizability components, and one *trans* conformer. Using a reasonable set of shape parameters, the IAM polarizability components, and Gierke's work on depolarized light scattering from anisometric molecules,[22] the following expression for $\langle\gamma_{\text{eff}}^2\rangle$ of n-butane is found[62]:

$$\langle\gamma_{\text{eff}}^2\rangle = \langle\gamma^2\rangle\left[1 - \left(\frac{\varepsilon_s-1}{\varepsilon_s+2}\right)\frac{19.81+13.88\sigma}{11.28+10.67\sigma}\right] \qquad (6.24)$$

$\sigma = \exp(-E_g/RT)$ is the Boltzmann factor associated with the *gauche* states. The estimates of *trans-gauche* energy difference E_g range from 300 to 700 cal/mol. In Fig. 7 we plot $\langle\gamma_{\text{eff}}^2\rangle$ for two different values of σ as a function

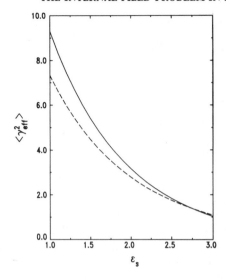

Fig. 7. $\langle \gamma_{\text{eff}}^2 \rangle$ of n-butane versus solvent optical dielectric constant ε_s calculated in Ref. 62 using the dielectric cavity approximation.

of the solvent optical dielectric constant ε_s. Since the *trans* conformer is more elongated and more optically anisotropic than the *gauche* ones, $\langle \gamma^2 \rangle$ is larger when $\sigma = 0.25$ than when $\sigma = 1.0$. However, the solvent effects are more pronounced in the first case, so that the less anisotropic molecule will actually have a larger $\langle \gamma_{\text{eff}}^2 \rangle$ for $\varepsilon_s \geq 2.7$. It is also seen that the solvent effects are quite substantial and much more pronounced for CCl_4 ($\varepsilon_s = 2.14$ near $T = 300$ K) than for n-pentane ($\varepsilon_s = 1.83$ near $T = 300$ K).

In principle, Eq. (6.23) can be solved for a flexible molecule of any size or complexity, but in practice this is at present not feasible. The dielectric cavity approach, while simpler than the evaluation of a realistic h_{ps}, also becomes difficult to deal with when it is no longer reasonable to use the ellipsoidal cavity approximation. Thus, for more complex flexible molecules it is necessary to use cruder models for $\langle \gamma_{\text{eff}}^2 \rangle$. An essential component of such models is the requirement that they result in the screening by the solvent of intramolecular DID interactions. We have proposed several such models, all based on the IAM of the polymer polarizability. One of them, the perturbation expansion of the IAM, has already been described in the discussion of n-alkane depolarized light scattering spectra. The other two use the full IAM with either the site polarizabilities or the intramolecular DID interactions scaled down to model the solvent screening. Thus, in the first one, the gas-phase atom polarizability α_i is replaced by $\alpha_{i,\text{eff}}$, defined as

$$\alpha_{i,\text{eff}} = f\alpha_i \qquad (6.25)$$

where f is a number in the range $0 \leq f \leq 1$. In the second one, the atom polarizabilities retain their gas-phase values, but the intramolecular interactions $T(r)$ between atoms more than a distance r_0 apart are replaced by

$$T_{\text{eff}}(\mathbf{r}) = \frac{1}{\varepsilon_s} T(\mathbf{r}). \qquad (6.26)$$

The results obtained using these two models for $\langle \gamma_{\text{eff}}^2 \rangle$ and $\langle \alpha_{0,\text{eff}}^2 \rangle$ of a series of n-alkanes are described in Ref. 51. Both models show changes in the n dependence of $\langle \gamma_{\text{eff}}^2 \rangle$ and of the depolarization ratio as the extent of screening is increased. For a very large amount of screening, their behavior becomes analogous to that of the BAA.

None of these simple models for $\langle \gamma_{\text{eff}}^2 \rangle$ has been derived from a molecular optical response theory. It would certainly be desirable to formulate them as well-defined approximations based on a systematic approach. The model defined by Eq. (6.26) has, however, been tested against the dielectric cavity theory calculation for n-butane, described above. The resulting $\langle \gamma_{\text{eff}}^2 \rangle$ versus $1 - f$ for the range $0.84 \leq f \leq 1.0$ resembles Fig. 6 very closely.[62] This suggests that Eq. (6.25) might be a reasonable model for the solvent screening of the intramolecular DID interactions.

A different approach to the internal field problem in dilute solutions of large flexible molecules is to treat both the polymer molecule and the solvent as continuum dielectric media. This point of view has been used to describe the "shape" anisotropy of polymer molecules observed in various birefringence experiments, most notably in flow birefringence[63] and in permanent birefringence of polymer crystals.[64] In these applications, the "shape" assumed in order to solve the electrostatic problem has been that of an infinite cylinder. While this may be a reasonable approximation in the case of relatively rigid macromolecules, such as biopolymers in their helical states, it becomes less reasonable in describing many other polymer molecules, which are substantially more flexible. In order to develop a continuum model appropriate for the optical response of flexible polymers, we have considered the wormlike chain model[65] of molecular conformation.[54] The molecular shape in this model is that of a flexible cylinder of constant cross section. The optical response has been obtained approximately for a dilute solution of polymer molecules having a scalar internal optical dielectric constant ε_p in a solvent with a scalar optical dielectric constant ε_s. It was found that taking into account the molecular flexibility is important and that the "shape" anisotropy deviates substantially from that of an infinite cylinder if the molecules are long as well as flexible. In this model, the solvent screening manifests itself in a dramatic way. It is found that $R_{\text{vh}}^{(p)}$ is proportional to the fourth power of $y = (\varepsilon_p - \varepsilon_s)/\varepsilon_s$. Thus the excess scattering disappears when $y = 0$ or $\varepsilon_p = \varepsilon_s$. It turns out that this model is somewhat unrealistic in that the permanent optical anisotropy, which has been neglected, of the polymer chain contributes importantly to $R_{vh}^{(p)}$ as well. However, work recently completed on a model where the chain has an anisotropic optical dielectric constant indicates that the "shape" anisotropy contribution remains important, except in the vicinity of $y = 0$ (which is now defined in terms of the isotropic part of ε_p).[66]

VII. CONCLUSION

Except for those cases described by the original ES theory, the calculation of light scattering intensities and spectra is difficult, due to the internal field problem. Quantitatively accurate theories are now available for depolarized scattering by rigid nonspherical molecules in liquids and by chain molecules in the gas phase. An understanding of the qualitative features of scattering by chain molecules has been achieved, but more work is needed here. Even less advanced is evaluation[67] of the role of the internal field problem for "collision-induced" phenomena; we did not discuss this subject here, because it would have made the article unreasonably long, but it may be quite important.

In our view, the problems discussed in this article are interesting for two reasons: First, they are fundamental to the interaction of light and matter, and second, they must be solved before light scattering can be used as a probe of molecular and fluid structural and dynamical properties. With respect to the second reason, we feel that the outstanding current problem is development of a theory that will allow precise interpretation of experiments on chain molecules in solution. Assertions of the authors to the contrary, we do not believe that the conclusions found in most existing papers on this topic are terribly meaningful.

REFERENCES

1. (a) R. Pecora and W. A. Steele, *J. Chem. Phys.* **42**, 1872 (1965). (b) W. A. Steele, *J. Chem. Phys.* **39**, 3197 (1963). (c) A. Ben Reuven and N. Gershon, *J. Chem. Phys.* **51**, 893 (1967). (d) A. D. Buckingham and M. Stephen, *Trans. Faraday Soc.* **53**, 88 (1957). (e) B. J. Berne and R. Pecora, *Dynamic Light Scattering*, Wiley, New York, 1976. (f) I. Fabelinskii, *Molecular Scattering of Light*, Plenum, New York, 1968.

2. L. Landau and E. M. Lipfshitz, *Electrodynamics of Continuous Media*, Addison-Wesley, 1960.

3. M. Fixman, *J. Chem. Phys.* **23**, 2074 (1955).

4. A. Einstein, *Ann. Phys.* **33**, 1275 (1910).

5. M. Smoluchowski, *Ann. Phys.* **25**, 205 (1908).

6. R. D. Mountain, *Rev. Mod. Phys.* **38**, 205 (1966).

7. D. Beysens, *J. Chem. Phys.* **64**, 2579 (1976).

8.   Y. Rocard, *Ann. Phys. (Paris)* **10**, 116 (1928); M. Vuks, *Opt. Spec.* **25**, 479 (1968).

9.   T. Keyes and D. Kivelson, *J. Chem. Phys.* **56**, 1057 (1972).

10.   D. Coumou, *Trans. Faraday Soc.* **65**, 2654 (1969); S. Kielich, *J. Chem. Phys.* **46**, 4090 (1967); see also p. 42, Ref. 1f.

11.   N. Gershon, E. Zamir, and A. Ben Reuven, *Ber. Bunsengen. Phys. Chem.* **75**, 316 (1971).

12.   P. Schoen, P. Cheung, D. Jackson, and J. Powles, *Mol. Phys.* **29**, 41 (1975); E. Zamir, N. Gershon, and A. Ben Reuven, *J. Chem. Phys.* **55**, 3397 (1971).

13.   M. S. Malmberg and E. R. Lippincott, *J. Coll. Int. Sci.* **27**, 591 (1961); R. Fulton, *J. Chem. Phys.* **61**, 4141 (1974); G. D. Patterson, *J. Chem. Phys.* **63**, 4032 (1975).

14.   (a) T. D. Gierke and W. H. Flygare, *J. Chem. Phys.* **61**, 2231 (1974). (b) G. R. Alms, T. D. Gierke, and W. H. Flygare, *ibid.* **61**, 4083 (1974). (c) A. K. Burnham, G. R. Alms, and W. H. Flygare, *J. Chem. Phys.* **62**, 3289 (1975). (d) A. K. Burnham, Ph.D. thesis, University of Illinois, 1977.

15.   T. Keyes, *J. Chem. Phys.* **63**, 815 (1975).

16.   (a) T. Keyes, Ph.D. thesis, U.C.L.A., 1971. (b) T. Keyes and D. Kivelson, *J. Chem. Phys.* **54**, 1786 (1971). (c) N. Gershon and I. Oppenheim, *Physica* **64**, 297 (1973). (d) D. Kivelson and P. A. Madden, *Ann. Rec. Phys. Chem.* **31**, 523 (1980). This references other work by Kivelson et al. in this area.

17.   J. Kirkwood, *J. Chem. Phys.* **4**, 592 (1936).

18.   J. Yvon, *Actualities Scientifiques et Industrielles*, Nos. 542, 543, Hermann, Paris.

19.   (a) K. L. Clarke, P. A. Madden, and A. D. Buckingham, *Mol. Phys.* **36**, 301 (1978); J. van Krenendonk and J. Sipe, *ibid.*, **35**, 1579 (1978). (b) L. Frommhold, *Adv. Chem. Phys.* **46**, 1 (1981); many other papers on the DID approximation are referenced here.

20.   B. M. Ladanyi and T. Keyes, *Mol. Phys.* **33**, 1063 (1977).

21.   (a) W. B. Streett and S. Tildesley, *Proc. Roy. Soc. Lond.* **A348**, 485 (1976); **A355**, 239 (1977). (b) B. M. Ladanyi, T. Keyes, D. Tildesley, and W. B. Strett, *Mol. Phys.* **39**, 645 (1980). (c) C. S. Hsu, D. Chandler, and L. S. Lowden, *Chem. Phys.* **14**, 213 (1976).

22.   T. D. Gierke, *Mol. Phys.*, submitted.

23.   D. Frenkel and J. P. McTague, *J. Chem. Phys.* **72**, 2801 (1980).

24.   B. M. Ladanyi and T. Keyes, *J. Chem. Phys.* **68**, 3217 (1978).

25.   T. Keyes and B. M. Ladanyi, *Mol. Phys.* **33**, 1271 (1977).

26.   C. Raman and K. Krishnan, *Phil. Mag.* **5**, 498 (1928).

27.   T. Scholte, *Physica* **15**, 437 (1949).

28.   T. Keyes and B. M. Ladanyi, *Mol. Phys.* **34**, 765 (1977).

29.   A. K. Burnham and T. D. Gierke, *J. Chem. Phys.* **73**, 4822 (1980).

30.   (a) T. I. Cox, M. A. Battaglia, and P. A. Madden, *Mol. Phys.* **38**, 1539 (1979). (b) M. A. Battaglia, T. I. Cox, and P. A. Madden, *ibid.* **37**, 1413 (1979).

31.   S. Bertucci, A. Burnham, G. Alms, and W. H. Flygare, *J. Chem. Phys.* **66**, 605 (1976).

32.   T. Keyes and B. M. Ladanyi, *Mol. Phys.* **38**, 605 (1979).

33.   B. U. Felderhof, *Physica* **76**, 486 (1975).

34.   T. Keyes and B. M. Ladanyi, *Mol. Phys.* **33**, 1099 (1977).

35.   T. Keyes, D. Kivelson, and J. P. McTague, *J. Chem. Phys.* **55**, 4096 (1971).

36.   F. Gray, *Phys. Rev.* **7**, 472 (1916).

37.   L. Silberstein, *Phil. Mag.* **33**, 92 (1917).

38.   J. Applequist, J. R. Carl, and K. K. Fung, *J. Am. Chem. Soc.* **94**, 2952 (1972).

39. J. Applequist, *Acct. Chem. Res.* **10**, 79 (1977).

40. B. M. Ladanyi and T. Keyes, *Mol. Phys.* **37**, 1809 (1979).

41. (a) S. Powers, D. A. Keedy, and R. S. Stein, *J. Chem. Phys.* **35**, 376 (1961). (b) F. R. Dintzis and R. S. Stein, *ibid.* **40**, 1459 (1964). (c) F. T. Gucker, S. Basu, A. A. Pulido, and G. Chin, *ibid.* **50**, 2526 (1969). (d) N. J. Bridge and A. D. Buckingham, *Proc. Roy. Soc. Lond. A* **293** (1966). (e) A. Massoulier, *C. R. Hebd. Seanc. Acad. Sci., Paris* **267**, 132 (1968).

42. G. M. Aval, R. S. Rowell, and J. J. Barrett, *J. Chem. Phys.* **57**, 3104 (1972).

43. K. G. Denbigh, *Trans. Faraday Soc.* **36**, 936 (1940).

44. W. Kuhn and F. Grun, *Kolloid Z.* **101**, 248 (1942); R. Kubo, *J. Phys. Soc. Japan* **2**, 47 (1947); **3**, 85, 119 (1947–48); **4**, 120 (1949).

45. P. J. Flory, *Statistical Mechanics of Chain Molecules*, Interscience, New York, 1969.

46. R. R. Birge, *J. Chem. Phys.* **72**, 5312 (1980).

47. B. T. Thole, *Chem. Phys.* **59**, 341 (1981).

48. D. F. Bocian, G. A. Schick, and R. R. Birge, *J. Chem. Phys.* **74**, 3660 (1981); **75**, 2626, 3215 (1981);

49. D. F. Bocian, G. A. Schick, J. K. Hurd, and R. R. Birge, *J. Chem. Phys.* **76**, 4828, 6454 (1982).

50. R. L. Rowell and R. S. Stein, *J. Chem. Phys.* **47**, 2985 (1967); R. P. Smith and E. M. Mortensen, *ibid.* **35**, 714 (1961); J. J. Teixeira-Dias and J. N. Murrell, *Mol. Phys.* **19**, 329 (1970); C. K. Miller, B. J. Orr, and J. F. Ward, *J. Chem. Phys.* **54**, 2109 (1977).

51. B. M. Ladanyi and T. Keyes, *J. Chem. Phys.* **76**, 2047 (1982).

52. C. Strazielle, in *Light Scattering from Polymer Solutions*, M. B. Huglin, Ed., Academic Press, New York, 1972.

53. P. Debye, *J. Appl. Phys.* **15**, 338 (1944); R. H. Ewart, C. P. Roe, P. Debye, and J. P. McCartney, *J. Chem. Phys.* **14**, 687 (1946); P. Debye, *J. Phys. Colloid. Chem.* **51**, 18 (1947).

54. H. Yamakawa, *Modern Theory of Polymer Solutions*, Harper & Row, New York, 1971.

55. D. R. Bauer, J. R. Brauman, and R. Pecora, *Macromol.* **8**, 443 (1975).

56. G. D. Patterson, C. P. Lindsay, and G. R. Alms, *J. Chem. Phys.* **69**, 3250 (1978).

57. G. D. Patterson and P. J. Carroll, *J. Chem. Phys.* **76**, 4356 (1982).

58. J. V. Champion, A. Dandridge, and G. H. Meeten, *Faraday Discuss.* **66**, 266 (1979).

59. T. Keyes, G. T. Evans, and B. M. Ladanyi, *Faraday Discuss.* **74**, 3779 (1981).

60. G. D. Patterson and P. J. Flory, *J. Chem. Soc. Faraday Trans. II* **68**, 1098 (1972).

61. P. Bothorel and G. Fourche, *J. Chem. Soc. Faraday Trans II* **69**, 441 (1973).

62. B. M. Ladanyi, *J. Chem. Phys.* **76**, 4303 (1982).

63. M. Tricot and C. Houssier, in *Polyelectrolytes*, K. C. Frisch, D. Klempner, and A. V. Patsis, Eds., Technomic, Westport, CT, 1976.

64. R. S. Stein, *J. Polymer Sci. Pt. A-2* **7**, 1021 (1969); S. D. Hong, C. Chang, and R. S. Stein, *J. Polymer Sci.: Polymer Phys.* **13**, 1447 (1975).

65. B. M. Ladanyi and T. Keyes, *Mol. Phys.* **42**, 501 (1981).

66. M. H. Dung and B. M. Ladanyi, to be submitted.

67. T. Keyes, B. Ladanyi, and P. Madden, *Chem. Phys. Lett.* **64**, 479 (1779); T. Keyes, *J. Chem. Phys.* **70**, 5438 (1979); B. Ladanyi and T. Keyes, *Can. J. Phys.* **59**, 1421 (1981); T. Keyes and P. Madden, *Can. J. Phys.* **59**, 1560 (1981); H. Kildal and S. Brueck, *J. Chem. Phys.* **73**, 4951 (1980).

A CONSISTENT MOLECULAR TREATMENT OF DIELECTRIC PHENOMENA*

PAUL MADDEN[†]

Theoretical Physics Section
Royal Signals and Radar Establishment
Great Malvern, Worcs, United Kingdom WR 14 3PS

DANIEL KIVELSON

Department of Chemistry, University of California
Los Angeles, CA 90024

CONTENTS

*Supported in part by a NATO research grant and by the National Science Foundation Grant NSF CHE77-15387.
†Advanced Research Fellow of SERC.

I. INTRODUCTION

In this article, we present a consistent molecular theory of frequency-dependent dielectric permittivity, $\varepsilon(\omega)$, for simple one-component molecular liquids. The material is in part review, in that many of the results have been obtained previously by others. Our principal concern is to provide a consistent molecular basis for an understanding of dielectric relaxation, so that dynamical information obtained from this source may be compared readily with that from other experiments, such as magnetic resonance and light scattering. The dielectric relaxation problem is inextricably linked to an understanding of the equilibrium structure of the fluid, so we include a treatment of the dielectric constant (i.e., the permittivity at zero frequency).

By a "molecular theory," we mean a theory that treats intermolecular interactions *explicitly* at all stages and avoids the introduction of the cavities and continuum models that occur in conventional discussions of dielectrics (for reviews, see Refs. 1–3). Such constructs as cavities in dielectric continua were introduced at an early stage into theories of the dielectric constant [i.e., $\varepsilon(\omega = 0)$] because they enabled exact treatment of the long-range dipole-dipole interactions in a polar fluid; an exact treatment of these long-range effects is a minimum requirement for a correct theory. We feel, however, that the introduction of these constructs is a barrier to the understanding of dielectric data on a molecular basis and to their appreciation by workers in other areas of liquid study. Since the dielectric constant (or zero-frequency permittivity) is an equilibrium property, we are able to describe its dependence on the equilibrium structure of the liquid by recourse to the molecular distribution function (g). More important, although the cavity constructs deal correctly with the long-range interactions in the equilibrium problem, they have been far less successful in providing useful relationships between the frequency-dependent permittivity and molecular quantities. Consequently, in the interpretation of dielectric relaxation data, much confusion persists concerning the "correct" local field to be used and the role of such phenomena as dipole-dipole correlations and dielectric friction. A clear specification of such concepts and an understanding of how they affect the dynamical behavior of dielectrics can be obtained by tracing their origins to fundamental molecular interactions.

In recent years, following the work of Nienhuis and Deutch,[4] there has been a renaissance of interest in molecular theories of the dielectric constant. Progress has been made on the development of rigorous theories that show the existence of $\varepsilon(0)$ as an intensive material property and introduce systematic approximations leading to computationally tractable evaluation procedures. This work has been reviewed comprehensively.[5-7] The present work does not belong in this category. We *assume* that $\varepsilon(\omega)$ is intensive, so our theory is phenomenologically based. Furthermore, in discussing the static problem, we describe the fluid structure through the readily understood distribution function g rather than by means of the direct correlation function[7] c or similar graphically defined objects. The direct correlation function c appears in many theories because its properties are more readily deduced from the pair potential than are those of g, and because systematic approximations leading to numerical results are more readily carried through with c. However, arguments with intuitive appeal concerning the form of g that lead to correct results for ε can be made, and we feel that the procedures we adopt give insight into the full development.

The connection between time-dependent molecular properties and the dielectric permittivity is made through linear response theory.[8] We make this connection by calculating the response of the molecules to the "vacuum field" (i.e., the field present in the absence of the sample), rather than to the "local field." The vacuum field is the one to which response theory is most straightforwardly applied (see also Ref. 9) and most readily visualized, since the perturbing field is independent of the system. Because of this choice of fields, our treatment of polarizable molecules assumes a different form from that of Felderhof,[10] Mazur,[11] and others.[12] Still other fields have been discussed by Sullivan.[13]

The formal framework of our theory, which relates the macroscopic quantity $\varepsilon(\omega)$ to molecular correlation functions, follows a procedure outlined by Fulton[14]; it is given in Sections II and III. Although the various molecular correlation functions and the distribution function depend upon sample shape, because of the long-range character of the dipole-dipole interactions, any sample can be used to calculate $\varepsilon(\omega)$ since it is intensive. We have used an infinite sample, as suggested by Fulton. In the succeeding sections, we evaluate the correlation functions, first for rigid dipolar molecules (Sections IV–VII), then for polarizable nonpolar molecules (Section VIII), and finally for the polar-polarizable case (Section IX and X). Throughout, we devote considerable attention to expressing the correlation functions in terms of "short-range" properties of the fluid, that is, in terms of averages that involve explicitly groups of molecules separated only by microscopic (molecular) distances. There are several reasons for this concern. First, we wish to make contact with the classical theories of the dielectric constant, in which $\varepsilon(0)$ is related to the average dipole moment of a small region of the sample.[1-3] Second, at finite frequency, the objective of theory is to relate the time dependence of the collective, many-molecule correlation functions that enter the expression for $\varepsilon(\omega)$ to the motion of individual molecules; this is possible only if the correlation function are short-range. Many aspects of our approach to the frequency-dependent problem are similar to the work of Sullivan.[15]

The major thrust of our work, described in Sections V–VII, X, and XI, is toward the dynamical or frequency-dependent behavior of $\varepsilon(\omega)$. Our approach is based on Mori's continued-fraction expansion, which enables us to relate $\varepsilon(\omega)$ to a hierarchy of Fourier-transformed memory functions.[16] In requiring that the frequency dependence of the permittivity be expressible through relaxation parameters that are short-range, we impose the same criterion on the first-order memory function as Sullivan and Deutch did.[15]

In Section VII, we formulate the dielectric problem in terms of a correlation function that is very closely related to the autocorrelation function of the intermolecular torques. This appears to be an extremely useful starting point for theories of dielectric relaxation, since it enables us to concentrate on the physically meaningful intermolecular torques rather than merely on the stochastic behavior of molecular reorientations. In Section XI, we also examine a number of specific dynamical models and study their strengths and shortcomings. We also show how the existence of dipolarons is a natural consequence of these models. In Section XII, we look briefly at the extension of our procedures to problems of dynamical light scattering.

II. RELATIONSHIP OF CALCULABLE RESPONSE FUNCTIONS TO PHENOMENOLOGICAL ONES

In theories of dielectric response, we are interested in the polarization induced in a material by a weak, spatially and temporally varying electric field (**E**). If the electric field is weak and the material is at equilibrium and, on average, isotropic and uniform, the polarization may be written in terms of a response function or susceptibility ($\underset{\sim}{\chi}$)

$$\mathbf{P}(\mathbf{r}, t) = \int d\mathbf{r}' \int_{-\infty}^{t} dt' \, \varepsilon_0 \underset{\sim}{\chi}(\mathbf{r} - \mathbf{r}', t - t') \cdot \mathbf{E}(\mathbf{r}', t') \tag{2.1}$$

where χ is a property of the material and is independent of $\mathbf{E}$, and ε_0 is the permittivity of free space. This relationship assumes a simpler form when we take its space and time transform:

$$P(\mathbf{k}, \omega) = \varepsilon_0 \chi(\mathbf{k}, \omega) \cdot E(\mathbf{k}, \omega) \tag{2.2}$$

where

$$\mathbf{P}(\mathbf{k}, \omega) = \int d\mathbf{r}\, e^{i\mathbf{k}\cdot\mathbf{r}} \int_{-\infty}^{\infty} dt\, e^{i\omega t} \mathbf{P}(\mathbf{r}, t) \tag{2.3a}$$

$$\mathbf{E}(\mathbf{k}, \omega) = \int d\mathbf{r}\, e^{i\mathbf{k}\cdot\mathbf{r}} \int_{-\infty}^{\infty} dt\, e^{i\omega t} \mathbf{E}(\mathbf{r}, t) \tag{2.3b}$$

$$\chi(\mathbf{k}, \omega) = \int d\mathbf{r}\, e^{i\mathbf{k}\cdot\mathbf{r}} \int_{0}^{\infty} dt\, e^{i\omega t} \chi(\mathbf{r}, t) \tag{2.3c}$$

Note that $\mathbf{E}(\mathbf{k}, \omega)$ and $\mathbf{P}(\mathbf{k}, \omega)$ are full transforms over time, whereas $\chi(\mathbf{k}, \omega)$ is a half-transform. The dielectric permittivity tensor ε relates the electric displacement $\mathbf{D}$ to the electric field:

$$\mathbf{D}(\mathbf{k}, \omega) = \varepsilon_0 \varepsilon(\mathbf{k}, \omega) \cdot \mathbf{E}(\mathbf{k}, \omega) \tag{2.4}$$

and, since $\mathbf{D}$ and $\mathbf{E}$ determine the polarization through

$$\mathbf{P}(\mathbf{k}, \omega) = \mathbf{D}(\mathbf{k}, \omega) - \varepsilon_0 \mathbf{E}(\mathbf{k}, \omega) \tag{2.5}$$

we see that χ and ε are closely related:

$$\chi(\mathbf{k}, \omega) = \varepsilon(\mathbf{k}, \omega) - I \tag{2.6}$$

Furthermore, χ and ε are known to be intensive properties; if an experiment is performed to measure ε for a given material, the results will be the same irrespective of the shape and size of the sample. The intensive character of χ (or ε) implies that χ (ε) depends only on correlations between molecules that are short-range (with characteristic length r_c) compared to the macroscopic lengths L that characterize experimental sample sizes (i.e., $L \gg r_c$). Alternatively, we can say that the integrand in Eq. (2.1) is significant only for $|\mathbf{r} - \mathbf{r}'| < r_c$. Since the nonuniform fields in which we shall be interested are electromagnetic waves with propagation vector $\mathbf{k}$, even at optical frequencies their nonuniformity is on a scale large compared to molecular lengths; that is, $kr_c \ll 1$. Equation (2.1) then shows that the response $\mathbf{P}(\mathbf{r})$ is insensitive to such large-scale nonuniformities, so that in this low $|\mathbf{k}|$ limit the susceptibility must be independent of both the magnitude and direction of $\mathbf{k}$. These considerations find their formal expression in the equation

$$\lim_{k \to 0} \underset{\sim}{\chi}(\mathbf{k}, \omega) = \chi(\omega)\mathbf{l} \tag{2.7}$$

In this article, we obtain molecular expressions for $\underset{\sim}{\varepsilon}$, but we assume throughout that the intensive nature of $\underset{\sim}{\varepsilon}$ is an experimentally verified fact that enables us to impose Eq. (2.7) as a condition on our molecular expressions. This makes our theory a phenomenological one. To prove from molecular considerations that $\underset{\sim}{\varepsilon}$ is intensive is a difficult task, which has been addressed by a number of workers.[4-7]

The tensor $\underset{\sim}{\chi}(\mathbf{r}, t)$ appears in Eq. (2.1) as the response function for the "Maxwell field" ($\mathbf{E}$), but the application of response theory to this field is not straightforward, since it is not a true "external" perturbing field, but one whose value depends on the state of the material. A more convenient field for our purposes is $\mathbf{E}^0$, which is the field that would be present in the space (vacuum) occupied by the sample if the sample were absent, that is, the field caused by the external charges. Instead of writing $\mathbf{P}(\mathbf{r}, t)$ as a response to $\mathbf{E}(\mathbf{r}, t)$, as in Eq. (2.1), we can express it equally well as a response to $\mathbf{E}^0(\mathbf{r}, t)$. Or equivalently, we can rewrite Eq. (2.2) as

$$\mathbf{P}(\mathbf{k}, \omega) = \varepsilon_0 \underset{\sim}{\chi}^0(\mathbf{k}, \omega) \cdot \mathbf{E}^0(\mathbf{k}, \omega) \tag{2.8}$$

and in the linear response regime of interest to us, $\underset{\sim}{\chi}^0(\mathbf{k}, \omega)$ is independent of $\mathbf{E}^0(\mathbf{k}, \omega)$.

The property $\underset{\sim}{\chi}(\mathbf{k}, \omega)$ cannot be expressed easily in terms of molecular quantities, whereas $\underset{\sim}{\chi}^0(\mathbf{k}, \omega)$ can be. Thus, $\underset{\sim}{\chi}^0(\mathbf{k}, \omega)$ is the *basic quantity* that we wish to calculate in a molecular theory, and we must obtain a relationship between $\underset{\sim}{\chi}$ and $\underset{\sim}{\chi}^0$ in order to obtain a molecular expression for χ. To obtain the relationship between $\underset{\sim}{\chi}$ and $\underset{\sim}{\chi}^0$, we require a relationship between $\mathbf{E}$ and $\mathbf{E}^0$; this is given by[11,17]:

$$\mathbf{E}(\mathbf{r}, \omega) = \mathbf{E}^0(\mathbf{r}, \omega) + \int d\mathbf{r}' \, \mathsf{F}(\mathbf{r} - \mathbf{r}', \omega) \cdot \mathbf{P}(\mathbf{r}', \omega) \tag{2.9}$$

The propagator F is given by

$$\mathsf{F}(\mathbf{r}, \omega) = \begin{cases} (4\pi\varepsilon_0)^{-1} \left[\nabla \nabla + \left(\dfrac{\omega^2}{c^2} \right) \mathbf{l} \right] \dfrac{e^{ir\omega/c}}{r} & |r| \geq \sigma \\[4mm] -\dfrac{1}{3\varepsilon_0} \delta(\mathbf{r})\mathbf{l} & |r| < \sigma \end{cases} \tag{2.10}$$

and the limit, $\sigma \to 0$ is to be taken in evaluating the integral in Eq. (2.9).

At large $\mathbf{r}$, F is a slowly decaying function of $\mathbf{r}$, so that the value of the integral in Eq. (2.9) depends upon the shape of the sample volume. Since the relationship between $\mathbf{E}$ and $\mathbf{E}^0$ depends upon sample shape, and since χ and ε are intensive, it follows from Eqs. (2.1) and (2.8) that χ^0, as well as the relationship between χ and χ^0, are sample-shape-dependent. However, if both χ^0 and its relation to χ are calculated correctly for a specific sample geometry, then for that geometry χ can be related to molecular quantities and, since χ is intensive, the expression so derived for χ may be used to interpret data on χ or ε gathered on a sample of arbitrary shape. We follow Fulton[14] in choosing an infinite sample; for this sample, we need not worry about the effect of boundary conditions on molecular correlation functions. In the remainder of this section, we obtain the relationship between χ and χ^0 for an infinite sample, and the rest of the article will be concerned with the calculation of molecular expressions for χ^0.

For sufficiently small $\mathbf{r}$, $r \ll c/\omega$, the ω^2/c^2 term in Eq. (2.10) can be neglected; this truncated form of $F(\mathbf{r}, \omega)$ is the static dipole operator. The ω/c term arises from the retardation of the fields due to a time-dependent polarization and gives an electromagnetic wave character to the fields at large distances. We propose to drop these terms. The justification for this step has been considered by Fulton,[18] and we discuss it further in Appendix D. For the present, we simply note that in the expressions for χ, F always appears in conjunction with the molecular distribution function, which ensures that the effective range of $\mathbf{r}$ is of the order of r_c, the molecular correlation length. For χ to be intensive, $r_c \ll c/\omega$ even up to optical frequencies; far from critical points, this must surely be true for isotropic fluids. Consequently, our expressions for χ must be independent of the retardation contributions.

With this simplification, the relationship between $\mathbf{E}$ and $\mathbf{E}^0$ may be obtained from Eq. (2.9):

$$\mathbf{E}(\mathbf{k}, \omega) = \mathbf{E}^0(\mathbf{k}, \omega) + \mathsf{T}(\mathbf{k}) \cdot \mathbf{P}(\mathbf{k}, \omega) \qquad (2.11)$$

where $\mathsf{T}(\mathbf{k})$ is the Fourier transform of the static dipole operator; that is, of $F(\mathbf{r}, \omega)$ with the ω/c term missing. As shown in Appendix C, Eq. (C.9),

$$\mathsf{T}(\mathbf{k}) = -\varepsilon_0^{-1} \hat{\mathbf{k}} \hat{\mathbf{k}} \qquad (2.12)$$

where $\hat{\mathbf{k}}$ is a unit vector parallel to $\mathbf{k}$. It is convenient to choose a coordinate system in which $\mathbf{k}$ defines the z direction. We may now combine Eqs. (2.6), (2.8), (2.11), and (2.12) to obtain

$$\lim_{k \to 0} \chi_{zz}^0(\mathbf{k}, \omega) = \chi(\omega)[1 + \chi(\omega)]^{-1} = \frac{\varepsilon(\omega) - 1}{\varepsilon(\omega)} \qquad (2.13a)$$

$$\lim_{k \to 0} \chi_{xx}^0(\mathbf{k}, \omega) = \lim_{k \to 0} \chi_{yy}^0(\mathbf{k}, \omega) = \chi(\omega) = \varepsilon(\omega) - 1 \qquad (2.13b)$$

$$\lim_{k \to 0} \chi_{ab}^0(\mathbf{k}, \omega) = 0 \qquad (a \neq b) \qquad (2.13c)$$

where χ^0_{zz} is the longitudinal and χ^0_{xx} and χ^0_{yy} are the transverse quasi-susceptibilities. The susceptibility χ is isotropic, reflecting the symmetry of the medium, whereas the quasi-susceptibility χ^0 is clearly not isotropic even in isotropic media; the symmetry of χ^0 is lower than that of χ. Furthermore, as we have seen, although $\underset{\sim}{\varepsilon}$ and $\underset{\sim}{\chi}$ are independent of $\mathbf{k}$ at low k, χ^0 is not.

Here, we have shown how χ^0 can be related to $\underset{\sim}{\chi}$ and $\underset{\sim}{\varepsilon}$; in the next section we show how χ^0 may be related to molecular properties. However, if we merely recognize that χ^0 is closely related to molecular dipoles, permanent and induced, even before obtaining detailed expressions of χ^0 we can explain why there is anisotropy in χ^0 despite the fact that the medium is isotropic. To understand this, we note that because dipolar interactions are long-range, as we move along a line that we specify as the $\mathbf{k}$ direction, the distribution and dynamic behavior of end-to-end or longitudinal dipole components may be quite different from those of the side-to-side or transverse components; at the end of Section III we give a qualitative discussion of why $\chi^0_{zz}(k,0)$ is less than $\chi^0_{xx}(k,0)$.

III. MOLECULAR EXPRESSIONS FOR THE RESPONSE FUNCTION

In the last section, we showed how the permittivity $\varepsilon(\omega)$ could be related to the quasi-susceptibility (χ^0), which specifies the polarization as a response to the vacuum field $\mathbf{E}^0$ [see Eqs. (2.13)]. In this section, we show how the polarization may be expressed in terms of molecular properties in the presence of $\mathbf{E}^0$, thereby opening a route through which the permittivity may be expressed in molecular terms. We may conveniently relate the polarization of the material to molecular quantities by examining the change in the internal energy (U_{E^0}) of the material in the presence of $\mathbf{E}^0$. The terms in the Hamiltonian that are linear and quadratic in $\mathbf{E}^0$ are[19]:

$$\mathscr{H}_{\mathbf{E}^0} = -\sum_j \mathbf{m}^j \cdot \mathbf{E}^0(\mathbf{r}^j) - \frac{1}{2}\sum_{jn} \mathbf{E}^0(\mathbf{r}^j) \cdot \mathbf{a}_n^j \cdot \mathbf{E}^0(\mathbf{r}^n) \tag{3.1}$$

where $\mathbf{m}^j$ is the dipole moment of molecule j in the absence of $\mathbf{E}^0$, and $\mathbf{a}_n^j \cdot \mathbf{E}^0(\mathbf{r}^n)$ is the dipole induced in molecule j by the effect of $\mathbf{E}^0$ on molecule n. In writing down this expression, we have assumed that the field $\mathbf{E}^0$ is sufficiently uniform over molecular dimensions for the interaction of the higher order molecular multipoles with the gradients of $\mathbf{E}^0$ to be neglected.[20] The change in the internal energy of the molecules due to the application of $\mathbf{E}^0$ is

$$U_{E^0} = \int_0^{\mathbf{E}_0} d\mathscr{E}^0 \cdot \nabla_{\mathscr{E}^0} \langle \mathscr{H}_{\mathscr{E}^0} \rangle_{\mathscr{E}^0} \tag{3.2}$$

where $\langle\ \rangle_{\mathscr{E}^0}$ denotes a statistical average in the presence of $\mathscr{E}^0$, and $\nabla_{\mathscr{E}^0}$ represents the functional derivative with respect to $\mathscr{E}^0(\mathbf{r})$. The average of the Hamiltonian that enters Eq. (3.2) may be written conveniently as

$$\langle\mathscr{H}_{\mathscr{E}^0}\rangle = -\int d\mathbf{r}\left\langle \sum_j \mathbf{m}^j\delta(\mathbf{r}-\mathbf{r}^j)\right\rangle_{\mathscr{E}^0}\cdot\mathscr{E}^0(\mathbf{r})$$

$$-\frac{1}{2}\int d\mathbf{r}\int d\mathbf{r}'\,\mathscr{E}^0(\mathbf{r})\cdot\left\langle \sum_{jn}\delta(\mathbf{r}-\mathbf{r}^j)\underset{\sim}{\mathbf{a}}_n^j\delta(\mathbf{r}'-\mathbf{r}^n)\right\rangle_{\mathscr{E}^0}\cdot\mathscr{E}^0(\mathbf{r}')\quad(3.3)$$

On the other hand, macroscopic considerations[1] show that, for a slowly spatially varying field, U_{E^0} is given by

$$U_{E^0} = -\frac{1}{2}\int d\mathbf{r}\,\mathbf{P}(\mathbf{r})\cdot\mathbf{E}^0(\mathbf{r})\qquad(3.4)$$

so that $\mathbf{P}(\mathbf{r})$ may be found from the functional derivative of Eqs. (3.4) and (3.2) with respect to $\mathbf{E}^0$. We wish to retain only the lowest-order terms in $\mathbf{E}^0$, and in order to evaluate $\mathbf{P}(\mathbf{r}, t)$ in this limit, we expand the average $\langle\ \rangle_{\mathscr{E}^0}$:

$$\langle\ \rangle_{\mathscr{E}^0} = \langle\ \rangle + [\nabla_{\mathscr{E}^0}\langle\ \rangle_{\mathscr{E}^0}]_{\mathscr{E}^0=0}\cdot\mathscr{E}^0 + \cdots\qquad(3.5)$$

where $\langle\ \rangle$ indicates a statistical average in the absence of $\mathbf{E}^0$. If the first term on the right in Eq. (3.3) is expanded as indicated in Eq. (3.5), then the first term in the expansion is $\langle\mathbf{m}^j\delta(\mathbf{r}-\mathbf{r}^j)\rangle$, and it vanishes. Consequently, comparison of Eqs. (3.2)–(3.5) leads to

$$\mathbf{P}(\mathbf{r}, t) = \left\langle\sum_j \mathbf{m}^j(t)\delta(\mathbf{r}-\mathbf{r}^j(t))\right\rangle_{\mathbf{E}^0} + \left\langle\sum_{jn}\underset{\sim}{\mathbf{a}}_n^j\delta(\mathbf{r}-\mathbf{r}^j(t))\right\rangle_{\mathbf{E}^0}\cdot\mathbf{E}^0(\mathbf{r}^j, t)$$

$$(3.6)$$

Although the averages that appear in this expression are equilibrium averages, they may depend upon time through the time-dependence of $\mathbf{E}^0$ and may also depend upon the past history of the field $\mathbf{E}^0$.

The effective permanent moment $\mathbf{m}^j$ of the jth molecule, which enters into the Hamiltonian in Eq. (3.1), is given by

$$\mathbf{m}^j = \boldsymbol{\mu}^j + \underset{\sim}{\boldsymbol{\alpha}}^j\cdot\sum_n \mathbf{T}^{jn}\cdot\boldsymbol{\mu}^n + \underset{\sim}{\boldsymbol{\alpha}}^j\cdot\sum_{nl}\mathbf{T}^{jn}\cdot\underset{\sim}{\boldsymbol{\alpha}}^n\cdot\mathbf{T}^{nl}\cdot\boldsymbol{\mu}^l + \cdots\qquad(3.7)$$

where $\boldsymbol{\mu}^j$ and $\underset{\sim}{\boldsymbol{\alpha}}^j$ are the (gas-phase) dipole moment and polarizability, respectively, of the jth molecule, and T^{jn} is the dipole-dipole interaction tensor between a pair of molecules [viz. Eq. (2.10)]:

$$\mathsf{T}^{jl} = \mathsf{F}(\mathbf{r}^j - \mathbf{r}^l, 0) \tag{3.8a}$$

$$\mathsf{T}^{jj} = 0 \tag{3.8b}$$

In Eqs. (3.7) and (3.8), we see that the field at molecule j at time t due to a dipole on molecule l is expressed as $[\mathsf{T}^{jl} \cdot \boldsymbol{\mu}^l(t)]$, whereas the correct (retarded) value of this field is

$$\int_{-\infty}^{t} dt' \int d\omega \, e^{i\omega(t-t')} \mathsf{F}(\mathbf{r}^j - \mathbf{r}^l, \omega) \cdot \boldsymbol{\mu}^l(t') \tag{3.9}$$

However, as we discussed briefly in Section II, only short-range correlations between dipoles, both induced and permanent, can enter our final expressions for χ, so that consistent omission of the additional long-range terms that result from using the correctly retarded propagator can have no effect on our expressions for χ. This is discussed in Appendix D.

Equation (3.7) expresses the effective permanent moment of a molecule as the sum of a gas-phase permanent moment and all possible permanent dipole-induced (DID) terms. Similarly, the second term in Eq. (3.6) contains $\mathsf{a}_n^j \cdot \mathbf{E}^0(\mathbf{r}^n)$, which is the effective induced moment and is given by

$$\mathsf{a}_n^j \cdot \mathbf{E}^0(\mathbf{r}^n) = \left\{ \underset{\sim}{\alpha}^j \delta_{jn} + \underset{\sim}{\alpha}^j \cdot \mathsf{T}^{jn} \cdot \alpha^n + \sum_l \underset{\sim}{\alpha}^j \cdot \mathsf{T}^{jl} \cdot \underset{\sim}{\alpha}^l \cdot \mathsf{T}^{ln} \cdot \underset{\sim}{\alpha}^n + \cdots \right\} \cdot \mathbf{E}^0(\mathbf{r}^n) \tag{3.10}$$

In using the expressions in Eqs. (3.7) and (3.10) for the effective permanent and induced moments, we have assumed that the molecules are geometrically *rigid*, that electron overlap makes no contribution to the moments induced by intermolecular interactions, that the molecular polarizability is concentrated at a point (so that there are no field-gradient-induced dipoles), and that the higher order multipoles induced by field gradients do not contribute to the dielectric response. Experimental studies in the gas phase establish that this is a reasonably sound framework on which to model the dielectric response.[21]

It is convenient to rearrange the expressions for $\mathbf{m}^j$ and a_n^j in order to separate those terms in which the chain of dipole-induced dipole events begins on the jth molecule. We obtain

$$\mathbf{m}^j = \boldsymbol{\mu}^j + \sum_m \underset{\sim}{\alpha}^j \cdot \mathsf{T}^{jm} \cdot \underset{\sim}{\alpha}^m \cdot \underset{\sim}{\mathscr{T}}^{mj} \cdot \boldsymbol{\mu}^j + \sum_l \underset{\sim}{\alpha}^j \cdot \underset{\sim}{\mathscr{T}}^{jl} \cdot \boldsymbol{\mu}^l \tag{3.11}$$

where

$$\mathcal{T}^{jl} = \mathsf{T}^{jl} + \sum_m \mathsf{T}^{jm} \cdot \underset{\sim}{\alpha}^m \cdot \mathsf{T}^{ml} + \cdots \qquad j \neq l \qquad (3.12)$$

and

$$\mathsf{a}_n^j = \left(\underset{\sim}{\alpha}^j + \underset{\sim}{\alpha}^j \cdot \sum_l \mathsf{T}^{jl} \cdot \underset{\sim}{\alpha}^l \cdot \mathcal{T}^{lj} \cdot \underset{\sim}{\alpha}^j \right) \delta_{jn} + \underset{\sim}{\alpha}^j \cdot \mathcal{T}^{jn} \cdot \underset{\sim}{\alpha}^n \qquad (3.13)$$

In both Eqs. (3.11) and (3.13), the first term is a gas-phase molecular property, and the third term gives the moment induced by the fields from all other molecules in the sample. These terms are analogous to those that appear in the classical Onsager theory[13,22] for the response of a molecule to the "cavity field"; the second term of both equations gives the "reaction field" induced dipole.[13,22]

To obtain an expression for $\mathbf{P}(\mathbf{r}, t)$ involving the molecular properties introduced above, it is convenient to write the nonuniform, time-dependent field $\mathbf{E}^0$ in terms of its spatial Fourier components,

$$\mathbf{E}^0(\mathbf{r}, t) = (2\pi)^{-3} \int d\mathbf{k}' e^{-i\mathbf{k}' \cdot \mathbf{r}} \mathbf{E}^0(\mathbf{k}', t) \qquad (3.14)$$

and to make use of Eq. (3.6) to write

$$\mathbf{P}(\mathbf{r}, t) = \left\langle \sum_j \mathbf{m}^j(t) \delta(\mathbf{r} - \mathbf{r}^j(t)) \right\rangle_{\mathbf{E}^0}$$
$$+ (2\pi)^{-3} \int d\mathbf{k}' \left\langle \sum_{jl} \delta(\mathbf{r} - \mathbf{r}^j(t)) \exp[-i\mathbf{k}' \cdot \mathbf{r}^l(t)] \mathsf{a}_l^j(t) \right\rangle \cdot \mathbf{E}^0(\mathbf{k}', t) \qquad (3.15)$$

Since we are interested only in the contributions to $\mathbf{P}$ that are linear in $\mathbf{E}^0$, the equal-time ensemble average that appears in the second term is the field-free equilibrium average $\langle \cdots \rangle$, which is independent of time. To obtain $\chi^0(\mathbf{k}, \omega)$, we must evaluate $\mathbf{P}(\mathbf{k}, \omega)$, which, in turn, may be calculated by transforming Eq. (3.15):

$$\mathbf{P}(\mathbf{k}, \omega) = \int_{-\infty}^{\infty} dt\, e^{i\omega t} \left\langle \sum_j \mathbf{m}^j(t) \exp[i\mathbf{k} \cdot \mathbf{r}^j(t)] \right\rangle_{\mathbf{E}^0}$$
$$+ (2\pi)^{-3} \int d\mathbf{k}' \left\langle \sum_{jl} [\exp(i\mathbf{k} \cdot \mathbf{r}^j - i\mathbf{k}' \cdot \mathbf{r}^l)] \mathsf{a}_l^j \right\rangle \cdot \mathbf{E}^0(\mathbf{k}', \omega) \qquad (3.16)$$

where $\mathbf{E}^0(\mathbf{k}, \omega)$ is defined in Eqs. (2.3b) and (2.8). If we make use of the fact that the free-field average is translationally invariant, the second term in Eq. (3.16) becomes

$$\rho\langle A(\mathbf{k})\rangle \cdot \mathbf{E}^0(\mathbf{k}, \omega) \tag{3.17}$$

where ρ is the number density and $A(\mathbf{k})$ is an effective molecular polarizability for molecule 1:

$$A(\mathbf{k}) = \sum_j a_j^1 \exp\left[i\mathbf{k}\cdot(\mathbf{r}^1 - \mathbf{r}^j)\right] \tag{3.18}$$

The first term in Eq. (3.16) must be examined more carefully; the $\mathbf{E}^0$-independent term in the expansion of the average in powers of $\mathbf{E}^0$ vanishes, and the second term, which is linear in $\mathbf{E}^0$, can be obtained by linear response theory.[8,9,23] In this manner, taking advantage of space and time translational invariance of the field-free average, we can evaluate the first term in Eq. (3.16):

$$\frac{\rho}{k_B T N_0}\left\{\langle |M(\mathbf{k}, 0)|^2\rangle + i\omega \int_0^\infty dt\, e^{i\omega t}\langle M(\mathbf{k}, t)M(\mathbf{k}, 0)^*\rangle\right\}\cdot \mathbf{E}^0(\mathbf{k}, \omega) \tag{3.19}$$

where $M(\mathbf{k}, t)$ is the transform of the dipolar moment density defined as

$$M(\mathbf{k}, t) = \sum_j \mathbf{m}^j(t)\exp\left[i\mathbf{k}\cdot\mathbf{r}^j(t)\right] \tag{3.20}$$

k_B is the Boltzmann constant, T is the temperature, and N_0 is the number of molecules. At $t = 0$, we write $M(\mathbf{k}, t)$ as $M(\mathbf{k})$; a similar convention will be used for other quantities. We can now evaluate $\mathbf{P}(\mathbf{k}, \omega)$ by substituting the expressions in Eqs. (3.19) and (3.17) into Eq. (3.16); by comparing this expression for $\mathbf{P}(\mathbf{k}, \omega)$ with that in Eq. (2.8), we obtain

$$\underset{\sim}{\chi}^0(\mathbf{k}, \omega) = \frac{\rho}{k_B T N_0 \varepsilon_0}\left\{\langle |M(\mathbf{k})|^2\rangle + i\omega \int_0^\infty dt\, e^{i\omega t}\langle M(\mathbf{k}, t)M(\mathbf{k})^*\rangle\right\}$$
$$+ \left(\frac{\rho}{\varepsilon_0}\right)\langle A(\mathbf{k})\rangle \tag{3.21}$$

This is the expression we sought, since it relates χ^0 to molecular quantities.

The static permittivity or *dielectric constant* $\varepsilon(0)$ is obtained from $\underset{\sim}{\chi}^0(\mathbf{k}, \omega)$ at $\omega = 0$:

$$\underset{\sim}{\chi}^0(\mathbf{k}, 0) = \frac{\rho}{N_0 k_B T \varepsilon_0}\langle M(\mathbf{k}, 0)M(\mathbf{k}, 0)^*\rangle + \frac{\rho}{\varepsilon_0}\langle A(\mathbf{k})\rangle \tag{3.22}$$

At very high frequency, $\underset{\sim}{\chi}^0(\mathbf{k}, \omega)$ reduces to

$$\underset{\sim}{\chi}^0(\mathbf{k}, \infty) = \frac{\rho}{\varepsilon_0}\langle A(\mathbf{k})\rangle \tag{3.23}$$

$\chi^0(\mathbf{k}, \infty)$ determines $\varepsilon(\infty)$ and thereby the refractive index of the material. In the remainder of the article, we evaluate the right-hand side of Eq. (3.21) for various molecular models.

IV. NONPOLARIZABLE (RIGID) DIPOLES: THE DIELECTRIC CONSTANT

In this section, we obtain molecular expressions for the permittivity $\varepsilon(0)$, also called the dielectric constant, for conformationally rigid, nonpolarizable symmetric top molecules. Our principal result is[24,1-3]

$$\frac{[\varepsilon(0)-1][2\varepsilon(0)+1]}{\varepsilon(0)} = \frac{\mu^2\rho}{k_BT\varepsilon_0}g^S \tag{4.1}$$

where μ is the magnitude of the gas-phase molecular dipole and ρ is the number density, while g^S is the Kirkwood g factor, which can be expressed as

$$g^S = 1 + 3\left\langle \sum_{j\neq 1} \cos\theta^1 \cos\theta^j \right\rangle_s \tag{4.2}$$

where θ^j is the polar angle of molecule j with respect to some external direction, and $\langle\ \rangle_s$ indicates that the average is over those molecules within a small macroscopic sphere around molecule 1. (This small sphere is surrounded by and interacts with the remainder of the sample.) $\varepsilon(0)$ is thereby demonstrated to depend only upon short-range intermolecular angular correlations. This result agrees with the one originally obtained by Kirkwood and since then by numerous other workers. The framework of our approach follows closely that suggested by Fulton.[14]

For nonpolarizable molecules, the term $\langle A(\mathbf{k}) \rangle$ in Eq. (3.21) disappears, and the moment $\mathbf{m}^j$ in that equation depends only on the coordinates of a single molecule. For symmetric top molecules, we may write the components of the total moment $\mathbf{M}$ as

$$M_z(\mathbf{k}) = \mu\sum_j C_{10}(\Omega^j)e^{i\mathbf{k}\cdot\mathbf{r}^j} \tag{4.3a}$$

$$M_x(\mathbf{k}) = -\mu 2^{-1/2}\sum_j \left[C_{11}(\Omega^j) - C_{1-1}(\Omega^j)\right]e^{i\mathbf{k}\cdot\mathbf{r}^j} \tag{4.3b}$$

$$M_y(\mathbf{k}) = i\mu 2^{-1/2}\sum_j \left[C_{11}(\Omega^j) + C_{1-1}(\Omega^j)\right]e^{i\mathbf{k}\cdot\mathbf{r}^j} \tag{4.3c}$$

where Ω^j denotes the polar angles subtended by the molecular symmetry axis in the laboratory coordinate system, and the C_{lm} are unnormalized spherical harmonics.[25] As can be seen from Eq. (3.21), we must evaluate the mean square values of the components of $\mathbf{M}(\mathbf{k})$, quantities that involve correlations between the orientations of pairs of molecules. To obtain suitable averages, we need the pair distribution function $g^{(2)}(r^{12}, \Omega^1, \Omega^2)$, which may be expressed in the invariant form[26,27]

$$g^{(2)}(\mathbf{r}^{12}, \Omega^1, \Omega^2) = \sum_{l_1 l_2 l_{12}} i^{l_1 - l_2 - l_{12}} \begin{pmatrix} l_1 & l_2 & l_{12} \\ \nu_1 & \nu_2 & \nu_{12} \end{pmatrix}$$

$$\times g^{l_1 l_2 l_{12}}(r^{12}) C_{l_1 \nu_1}(\Omega^1) C_{l_2 \nu_2}(\Omega^2) C_{l_{12} \nu_{12}}(\hat{\mathbf{r}}^{12}) \quad (4.4)$$

a form that ensures that the pair correlations share the spatial isotropy of the sample. In Eq. (4.4) and throughout this paper, we use the convention that repeated Greek indices are summed. Here, $\hat{\mathbf{r}}^{12}$ represents the angles associated with $\mathbf{r}^{12}$, and $r^{12} = |\mathbf{r}^2 - \mathbf{r}^1|$. We show in Appendix A that

$$\chi_{zz}^0 = (k_B T N_0 \varepsilon_0)^{-1} \rho \langle |M_z(\mathbf{k})|^2 \rangle = \frac{(\mu)^2 \rho}{3 k_B T \varepsilon_0} \left[g^S(\mathbf{k}) + 2 g^L(\mathbf{k}) \right] \quad (4.5a)$$

and

$$\chi_{xx}^0 = \chi_{yy}^0 = (k_B T N_0 \varepsilon_0)^{-1} \rho \langle |M_x(\mathbf{k})|^2 \rangle = \frac{(\mu)^2 \rho}{3 k_B T \varepsilon_0} \left[g^S(\mathbf{k}) - g^L(\mathbf{k}) \right]$$

$$(4.5b)$$

where

$$g^S(\mathbf{k}) = 1 - 4\pi\rho (3)^{-3/2} \int_0^\infty dr \, r^2 j_0(kr) g^{110}(r), \quad (4.6)$$

$$g^L(\mathbf{k}) = \frac{4\pi\rho}{3} (30)^{-1/2} \int_0^\infty dr \, r^2 j_2(kr) g^{112}(r). \quad (4.7)$$

In the dielectric literature,[5] g^{110} and g^{112} are often referred to as h_Δ and h_D, respectively, and $j_l(x)$ is a spherical Bessel function. The presence of long-range dipolar correlations in χ^0 is signaled by the dependence of its elements on the direction of $\mathbf{k}$, even when k^{-1} is a macroscopic length.

We can obtain some insight into the form of the coefficients $(g^{11l'})$ of the invariant expansion in Eq. (4.4) if we study the general form of the pair distribution function $g^{(2)}(1,2)$ given by

$$g^{(2)}(1,2) = Q_N^{-1} \int dq^3 \cdots dq^{N_0} \exp\left(-\beta \sum_{i<j} U^{SR}(i,j)\right)$$

$$\times \exp\left(-\beta \sum_{i<j} U^{DD}(i,j)\right) \qquad (4.8)$$

Here, q^j represents all the coordinates of the jth particle, $\beta = (k_B T)^{-1}$, Q_N is the configurational integral, and U^{SR} and U^{DD} are the short-range and dipole-dipole contributions to the pair potential. The dipole-dipole contributions to the intermolecular potential are separated from the other parts because they play a special role in dielectric theory; they vary as r^{-3}, and this long-range behavior has a crucial effect on $g^{(2)}$. The dipole-dipole interaction takes the form[19,26]

$$U^{DD}(\mathbf{r}^{12}, \Omega^1, \Omega^2) = \boldsymbol{\mu}^1 \cdot \mathbf{T}^{12} \cdot \boldsymbol{\mu}^2$$

$$= -\frac{(\mu)^2 (30)^{1/2}}{4\pi\varepsilon_0 (r^{12})^3} \sum_{m_1 m_2 m_{12}} \begin{pmatrix} 1 & 1 & 2 \\ m_1 & m_2 & m_{12} \end{pmatrix}$$

$$\times C_{1m_1}(\Omega^1) C_{1m_2}(\Omega^2) C_{2m_{12}}(\hat{\mathbf{r}}^{12}) \qquad (4.9)$$

where the dipole-dipole tensor $\mathbf{T}$ is specified by Eqs. (3.8a) and (2.10). By expanding the second exponential factor in Eqs. (4.8), terms may be found in each order of the expansion that will contribute to $g^{(2)}(1,2)$ and give it a long-range character; for large r^{12}, the first exponential factor in Eq. (4.8) may be replaced by unity. The first-order term $-\beta U^{DD}(1,2)$ clearly makes a contribution only to g^{112}, and it varies as $(r^{12})^{-3}$ [compare Eq. (4.4)]. For an *infinite sample*, the second-order term $\{\beta^2 \int dq^3 U^{DD}(1,3) U^{DD}(3,2)\}$, and all similar "chainlike" terms in higher order, such as $\int dq^j \cdots dq^k U^{DD}(1,j)$ $\cdots U^{DD}(k,2)$, may be shown to be proportional to $U^{DD}(1,2)$ at large r^{12}. Hence, only g^{112} has long-range contributions [cf. Eq. (C.16) and the argument of Appendix B]. For finite samples, however, terms of this kind may also contribute to g^{110}, and this long-range contribution to g^{110} is, in part, the source of the "sample-shape" problems that have plagued the theory of dielectrics. For both infinite and finite samples, chainlike terms of the kind mentioned above can contribute to the short-range character of $g^{110}(r)$, but, since the intermediate particles over whose coordinates we integrate may be at long range, even the short-range parts of $g^{110}(r)$ are dependent on sample shape. Arguments of much greater precision than these on the form of $g^{(2)}$ have been given by others, in particular by Nienhuis and Deutch[4] and Høye and Stell.[5,28]

From Eqs. (4.6) and (4.7), we see that two components of the pair correlation function, $g^{110}(r)$ and $g^{112}(r)$, enter into the theory of dielectrics. At large r, $g^{11l}(r)$ varies as r^{-n_l}; as discussed above, for an infinite sample, we expect $n_2 = 3$ and $n_0 > 3$. As discussed in Section II, we are interested in the low k limit of the quasi-susceptibility χ^0, which depends upon the low k limit of $g^L(\mathbf{k})$ and $g^S(\mathbf{k})$, which in turn depend upon the low k values of the integrals in Eqs. (4.6) and (4.7). The integrals in these two equations can be separated into short- and long-range parts;

$$\int_0^\infty dr\, r^2 j_l(kr) g^{11l}(r) = \int_0^{r_c} dr\, r^2 j_l(kr) g^{11l}(r) + \int_{r_c}^\infty dr\, r^2 j_l(kr) g^{11l}(r)$$

(4.10a)

where r_c is a distance that is large compared to the "cutoff" of the short-range correlations. At very small values of kr, the Bessel function $j_l(kr)$ is proportional to $(kr)^l$; if this limiting value of $(kr)^l$ is substituted into the first (short-range) integral in Eq. (4.10a), we see that the short-range integral vanishes in the $k \to 0$ limit for $l = 2$ but contributes in this limit for $l = 0$. The second (long-range) integral in Eq. (4.10a) is proportional to

$$k^{3-n_l} \int_{r_c k}^\infty dx\, x^2 j_l(x) x^{-n_l}$$

(4.10b)

this term vanishes for $k \to 0$ except for $l = 2$, where $n_l = 3$. Thus, we conclude that both g^S and g^L are nonvanishing and *independent of* $\mathbf{k}$ *for low k values*. Furthermore, it is the *short-range correlations* that contribute to g^S and the *long-range dipole-dipole correlations* that contribute to g^L.

The long-range contributions to g^L are the source of the dependence of χ^0 on the direction of $\mathbf{k}$ indicated in Eqs. (4.5ab). The long-range g^L terms may be eliminated from these equations by taking the trace of χ^0:

$$\lim_{k \to 0} \mathrm{Tr}\, \chi^0 = \frac{(\mu)^2 \rho}{k_B T \varepsilon_0} \lim_{k \to 0} g^S(\mathbf{k}) = \frac{[\varepsilon(0)-1][2\varepsilon(0)+1]}{\varepsilon(0)}$$

(4.11)

Since, as we have shown, $g^{110}(r)$ is short-ranged, Eq. (4.6) can be rewritten as

$$g^S \equiv \lim_{k \to 0} g^S(\mathbf{k}) = 1 - 4\pi\rho \int_0^{r_c} dr\, r^2 g^{110}(r)$$

(4.12)

where the Bessel function $j_0(kr) \to 1$, and $g^{110}(r)$ is negligible for $r < r_c$. Since we expect all intermolecular interactions except the dipole-dipole interactions to be very short-ranged, and because $\varepsilon(0)$ is, by hypothesis and experience, intensive, we expect r_c to be not more than several molecular diameters and, at the very least, to be much less than macroscopic distances.

By using the orthogonality of the spherical harmonics and the invariant expression for $g^{(2)}(\mathbf{r}^{12}, \Omega^1, \Omega^2)$ given in Eq. (4.4), we may write Eq. (4.12) in the more readily interpretable form

$$
g^S = 1 + 3\rho \int_0^{r_c} dr^{12} (r^{12})^2 (4\pi)^{-2} \int d\Omega^1 \, d\Omega^2 \, d\hat{\mathbf{r}}^{12}
$$
$$
\times \cos\theta^1 \cos\theta^2 g(\mathbf{r}^{12}, \Omega^1, \Omega^2) \tag{4.13}
$$

This is identical with Eq. (4.2). Combining Eq. (4.13) with Eq. (4.11) leads to the Kirkwood result, Eq. (4.1).

Although g^{110} is short-ranged in that its r dependence falls off rapidly, as discussed above, its form is specific to the infinite sample that we have considered, and the relation that we have drawn between the permittivity and the integral over g^{110} is also specific to this sample. Stell et al.[5] have presented a clear discussion of how the structural parameters calculated in a computer simulation of a polar fluid depend upon the long-range interactions and how they may be related to the permittivity.

The calculation of g^S, and hence of $\varepsilon(0)$, from an intermolecular potential remains an objective in liquid-state structure calculations. The problems caused by the long-range interactions for computer simulations have only recently been sorted out.[29] To date, the only simulations of sufficient size and length for studying dielectric properties have been made for molecules with unrealistic shapes and charge distributions (e.g., for dipolar spheres) in order to test the methodology.[29] Computationally tractable theories that deal successfully with such idealized models have been developed, principally by Patey.[30] The most successful theories of liquid structure that describe the effects of molecular shape are based upon interaction-site models.[3] Although until lately the only available scheme derived from such theories (RISM) gave values of $g^S = 1$ and $g^L = 0$,[31b] a recent article[31c] indicates how to correct for this failure by means of a related (ISM) calculation; predictions based on this modification have yet to be compared with experiment. In his recent compilation of experimentally derived g^S values, Buckingham[32] found that g^S was indeed close to unity, but he also found that g^S could be correlated with molecular shape. Alternative theories that yield nonzero values for $g^S - 1$ are currently being developed.[31c] Although realistic potential models have not as yet been used in simulations of sufficient size for an accurate determination of $\varepsilon(0)$, shorter simulations of realistic models have been made, and the importance of the molecular shape has been emphasized.[33]

The results above also permit us to determine $g^{112}(r)$ at long range. We have already argued that, at long range,

$$g^{112}(r) \to Ar^{-3} \tag{4.14}$$

and that g^L depends entirely upon the long-range contributions of $g^{112}(r)$. If we substitute Eq. (4.14) into Eq. (4.7) and evaluate the integral, we obtain

$$g^L \equiv \lim_{k \to 0} g^L(\mathbf{k}) = A\frac{4\pi\rho}{9}\left(\frac{1}{30}\right)^{1/2} \tag{4.15}$$

Alternatively, we can obtain g^L from Eqs. (4.5), (4.11), and (4.1):

$$g^L = -\frac{\varepsilon(0)}{[2\varepsilon(0)+1]^2}\rho\frac{(\mu g^S)^2}{k_B T \varepsilon_0} \tag{4.16}$$

Comparison of Eqs. (4.15) and (4.16) yields a value of A that can be substituted into Eq. (4.14), and the resultant expression for the asymptotic value of $g^{112}(r)$ can be combined with Eq. (4.4) to yield

$$\lim_{r^{12} \to \infty} g^{(2)}(\mathbf{r}^{12}, \Omega^1, \Omega^2) = 1 + \frac{(30)^{1/2}}{k_B T 4\pi\varepsilon_0\varepsilon(0)}\left(\frac{3\varepsilon(0)\mu g^S}{2\varepsilon(0)+1}\right)^2\left(\frac{1}{m_1}\quad\frac{1}{m_2}\quad\frac{2}{m_{12}}\right)$$

$$\times (r^{12})^{-3}C_{1m_1}(\Omega^1)C_{1m_2}(\Omega^2)C_{2m_{12}}(\hat{\mathbf{r}}^{12}) \tag{4.17}$$

Comparison of Eqs. (4.9) and (4.17) shows that $g(\mathbf{r}^{12}, \Omega^1, \Omega^2)$, at long range, has the same angular and radial form as a dipole-dipole interaction. We can therefore rewrite Eq. (4.17):

$$\lim_{r^{12} \to \infty} g^{(2)}(\mathbf{r}^{12}, \Omega^1, \Omega^2) = 1 - (k_B T \varepsilon(0))^{-1}\tilde{\mu}^2 \cdot \mathbf{T}^{21} \cdot \tilde{\mu}^1 \tag{4.18}$$

where

$$\tilde{\mu}^j = \frac{\mu^j g^S 3\varepsilon(0)}{2\varepsilon(0)+1} \tag{4.19}$$

This result agrees with that obtained (for an infinite sample) by Nienhuis and Deutch[34] by a direct calculation of the structure of a polar fluid. Here, $\tilde{\mu}^1$ is the "apparent" dipole of molecule 1; when viewed at long range in a dielectric medium, g^S can be envisaged as the result of molecule 1 orienting its neighbors at short range, and the factor $3\varepsilon/(2\varepsilon+1)$ as the result of molecule 1 and its near neighbors polarizing the surrounding medium. The factor

$3\varepsilon/(2\varepsilon+1)$ agrees with the enhancement of the dipole in a cavity that would be predicted by a continuum calculation.[35] The field at molecule 2 produced by this dipole is reduced by $\varepsilon(0)^{-1}$ from its value in vacuo.

From the long-range form of $g^{(2)}$ given in Eq. (4.18), we can understand why $\langle|M_x(\mathbf{k})|^2\rangle$ is larger and $\langle|M_z(\mathbf{k})|^2\rangle$ smaller than $N\mu^2/3$, the single-molecule value of both quantities in the absence of dipole–dipole correlations. We focus on a given dipole, $\boldsymbol{\mu}^1$, and examine other dipoles, $\boldsymbol{\mu}^j$, at a large distance, r^{1j}. In the direction of $\boldsymbol{\mu}^1$ the dipoles $\boldsymbol{\mu}^j$ tend to be oriented parallel to $\boldsymbol{\mu}^1$, so that they make a positive contribution to $\langle\boldsymbol{\mu}^1\cdot\mathbf{M}\rangle$; dipoles at the same distance but in the two directions orthogonal to $\boldsymbol{\mu}^1$ tend to be oriented antiparallel to $\boldsymbol{\mu}^1$ with a probability that is half as great as that of the first group, and they therefore make a negative contribution to $\langle\boldsymbol{\mu}^1\cdot\mathbf{M}\rangle$. In $\langle|M_x(\mathbf{k})|^2\rangle$ and $\langle|M_z(\mathbf{k})|^2\rangle$, the contributions from the correlations between dipoles is modified by the phase factor $\exp(i\mathbf{k}\cdot\mathbf{r}^{ij})$; this phase factor has no effect on the dipoles at sites located in the two directions orthogonal to $\mathbf{k}$, but the oscillatory nature of the phase factor diminishes the net contribution from dipoles at sites located in the direction along $\mathbf{k}$. To envisage the "longitudinal phenomenon" we consider $\boldsymbol{\mu}^1$ directed along $\mathbf{k}$, in which case the dipole–dipole correlations that arise principally from dipoles at sites perpendicular to $\boldsymbol{\mu}^1$, are negative, and $\langle|M_z(\mathbf{k})|^2\rangle$ is then less than $N\mu^2/3$. To envisage the "transverse phenomenon" we consider $\boldsymbol{\mu}^1$ directed perpendicularly to $\mathbf{k}$, in which case the phase factor diminishes the negative contributions from dipoles at sites in one of the directions perpendicular to $\boldsymbol{\mu}^1$; of the remaining contributions, the positive one from dipoles at sites in the direction parallel to $\boldsymbol{\mu}^1$ is twice as great as the negative one from dipoles at sites in the direction orthogonal to both $\mathbf{k}$ and $\boldsymbol{\mu}^1$. Thus $\langle|M_x(\mathbf{k})|^2\rangle$ is enhanced by long-range correlations by one half the amount by which $\langle|M_z(\mathbf{k})|^2\rangle$ is reduced. For large $\varepsilon(0)$, $k\to 0$, and $g^S=1$, Eqs. (4.5) and (2.13) lead to $\langle|M_x(\mathbf{k})|^2\rangle=(3/2)(N\mu^2/3)$ and $\langle|M_z(\mathbf{k})|^2\rangle=(3/2\varepsilon(0))(N\mu^2/3)$, so that in the longitudinal case the intermolecular dipole–dipole correlations almost cancel the single molecule terms.

In this section, we have shown by a simple calculation what is known about the dielectric behavior of nonpolarizable polar fluids. The remarkable thing is that the dielectric properties should follow so readily by the simple imposition of the phenomenological requirement that $\lim_{k\to 0}\varepsilon(\mathbf{k})$ be intensive and isotropic.

V. RIGID DIPOLES: FREQUENCY-DEPENDENT PERMITTIVITY

In the preceding section, we showed how the static permittivity for nonpolarizable rigid dipoles could be related to molecular properties and the short-range liquid structural parameter g^S. In this section, we discuss the related time-dependent problem of obtaining expressions for $\lim_{k\to 0}\varepsilon(\mathbf{k},\omega)$ in terms of molecular dynamical properties that, in the infinite sample, are independent of long-range correlations. In the next two sections, we take up the problem of modeling the dynamical properties.

Our starting point is the relationship in Eq. (3.21), which, for nonpolarizable molecules, reduces to

$$\chi_{aa}^0 = \left(k_B T N_0 \varepsilon_0 \right)^{-1} \rho \langle M_a(\mathbf{k},0) M_a(-\mathbf{k},0) \rangle \left[1 + i\omega \Phi_{aa}(\mathbf{k},\omega) \right] \quad (5.1)$$

where the normalized response function Φ_{aa} is defined by

$$\Phi_{aa}(\mathbf{k},\omega) \equiv \int_0^\infty dt\, e^{i\omega t} \frac{\langle M_a(\mathbf{k},t) M_a(-\mathbf{k},0) \rangle}{\langle M_a(\mathbf{k},0) M_a(-\mathbf{k},0) \rangle} \quad (5.2)$$

The collective dipole moment $\mathbf{M(k)}$ is defined in Eq. (4.3), and its components are given by Eq. (4.3). A more revealing expression for χ^0 can be obtained by transforming Eq. (5.2) according to the procedures of Mori,[16]

$$\Phi_{aa}(\mathbf{k},\omega) = \frac{T_a(\mathbf{k},\omega)}{1 - i\omega T_a(\mathbf{k},\omega)} \quad (5.3)$$

This transformation is discussed elsewhere.[15,23] The frequency-dependent relaxation time $T_a(\mathbf{k},\omega)$ is defined by the relation

$$T_a^{-1}(\mathbf{k},\omega) \equiv \int_0^\infty dt\, e^{i\omega t} \frac{\langle \dot{M}_a(\mathbf{k},t_{M_a}) \dot{M}_a(-\mathbf{k},0) \rangle}{\langle M_a(\mathbf{k},0) M_a(-\mathbf{k},0) \rangle} \quad (5.4)$$

in which t_{M_a} indicates that the time dependence is projected orthogonally to $M_a(\mathbf{k},0)$.[16] Combining Eqs. (5.1) and (5.3), we obtain

$$\chi_{aa}^0(\mathbf{k},\omega) = \rho \left(k_B T N_0 \varepsilon_0 \right)^{-1} \frac{\langle M_a(\mathbf{k},0) M_a(-\mathbf{k},0) \rangle}{1 - i\omega T_a(\mathbf{k},\omega)} \quad (5.5)$$

Both $T_a(\mathbf{k},\omega)$ in Eq. (5.4) and the $\omega = 0$ contribution to $\chi^0(\mathbf{k},\omega)$ in Eq. (5.5) are proportional to $\langle |M_a(\mathbf{k})|^2 \rangle$; hence, they are both $\mathbf{k}$-dependent since, as discussed in the last section, $\langle |M_a(\mathbf{k})|^2 \rangle$ contains the long-range quantity g^L. However, in Section IV we showed that for our infinite sample each component of $\langle |M_a|^2 \rangle$ could be expressed as a function of $\varepsilon(0)$. Although the long-range parameter g^L is then suppressed, its presence is reflected in the fact that the functional form of the relation between $\langle |M_a|^2 \rangle$ and $\varepsilon(0)$ depends upon the direction a. In addition, we can envisage that the numerator of Eq. (5.4) will contain other long-range terms whose presence will be signaled by a dependence of the numerator upon a. If we combine Eqs. (2.13) and (5.5), we obtain

$$\frac{\varepsilon(\omega)-1}{\varepsilon(\omega)} = \frac{\varepsilon(0)-1}{\varepsilon(0)} \left\{ 1 - i\omega \mathcal{R}_z(\omega) \frac{[\varepsilon(0)-1]}{\varepsilon(0)} \right\}^{-1} \tag{5.6}$$

and

$$\varepsilon(\omega)-1 = [\varepsilon(0)-1]\{1 - i\omega \mathcal{R}_x(\omega)[\varepsilon(0)-1]\}^{-1} \tag{5.7}$$

where

$$\mathcal{R}_a^{-1}(\omega) \equiv (k_B T N_0 \varepsilon_0)^{-1} \rho \lim_{k \to 0} \int_0^\infty dt\, e^{i\omega t} \langle \dot{M}_a(\mathbf{k}, t_{M_a}) \dot{M}_a(\mathbf{k})^* \rangle \tag{5.8}$$

Because we have demanded that $\lim_{k \to 0} \underline{\varepsilon}(\mathbf{k}, \omega)$ be isotropic, Eqs. (5.6) and (5.7) give two independent expressions for the same function $\varepsilon(\omega)$. Consistency demands that

$$\mathcal{R}_x(\omega) = \mathcal{R}_z(\omega) \equiv \mathcal{R}(\omega) \tag{5.9}$$

This identity shows that the integrand in Eq. (5.8) contains *no direct long-range correlation*, so that $\mathcal{R}(\omega)$ is a convenient quantity in a theory in which $\varepsilon(\mathbf{k}, \omega)$ is demonstrated to be intensive. This important conclusion was first reached by Sullivan and Deutch.[15]

Equations (5.4)–(5.7) show that the longitudinal and transverse components of the polarization relax with distinct rates, as was pointed out by Berne[36] and Fulton.[14] Because of the long-range nature of the dipole-dipole interactions and the different geometrical arrangement of the dipoles at large distances, $\langle |M_x(k)|^2 \rangle$ and $\langle |M_z(k)|^2 \rangle$ differ, and this leads to different relaxation times for the two polarizations. The physical picture of this difference can be understood for large $\varepsilon(0)$ by noting that the "parallel" projections of the molecular dipoles constituting M_z almost cancel each other, so that it requires much less than a 180° molecular rotation on average to change the sign of M_z; thus the reorientation time for M_z is much less than that for a molecular dipole. On the other hand, the "perpendicular" projections of the molecular dipoles constituting M_x reinforce each other slightly, so that it requires somewhat more effort (and time) to change the sign of M_x than that of individual dipoles.

Many relationships have appeared in the literature to relate $\varepsilon(\omega)$ to collective relaxation times, such as $\mathcal{R}$. Most have been concerned with the rotational diffusion limit in which, as we show in Section VI, $\mathcal{R}$ is independent of frequency provided that the dipole lies along a principal axis of the diffusion tensor. In this limit, Eq. (5.7) is simply the Debye equation[37]

$$\frac{\varepsilon(\omega)-1}{\varepsilon(0)-1} = [1 - i\omega\tau_D]^{-1} \qquad (5.10)$$

where the Debye time τ_D is defined as

$$\tau_D \equiv [\varepsilon(0)-1]\mathscr{R}(0) \qquad (5.11)$$

Equation (5.10) was obtained from an analysis of the M_x correlation, that is, from the *transverse* modes in an infinite sample.

The Debye functional of $\varepsilon(\omega)$ is not the only one that relaxes with a single decay time in the rotational diffusion limit. Equation (5.6) shows that $[\varepsilon(\omega) -1]\varepsilon(\omega)^{-1}$ has a single correlation time, albeit a different one, $\tau_D\varepsilon(0)^{-1}$, and this time corresponds to the relaxation of M_z or the *longitudinal* modes in an infinite sample. It can also be readily shown that the "Glarum" functional[38] $[\varepsilon(\omega)-1][\varepsilon(\omega)+2]^{-1}$ relaxes with the single relaxation time $3\tau_D/[\varepsilon(0)+2]$. By reference to Eqs. (5.6) and (5.7), however, it can be seen that the Kirkwood-Onsager functional obeys the relation

$$\frac{[\varepsilon(\omega)-1][2\varepsilon(\omega)+1]}{3\varepsilon(\omega)} = \frac{\varepsilon(0)-1}{3\varepsilon(0)}\left(\frac{1}{1-i\omega\tau_D/\varepsilon(0)} + \frac{2\varepsilon(0)}{1-i\omega\tau_D}\right) \qquad (5.12)$$

and that the characteristic relaxations of the longitudinal and transverse polarization both affect the frequency dependence of this functional. This was first shown by Berne[36] and Fulton.[14] We will discuss these results below.

Which of the functionals introduced above is used to interpret the properties of $\varepsilon(\omega)$ is a matter of choice, but, of course, the relationship to the fundamental quantity $\mathscr{R}(\omega)$ is different in each case. The Debye functional $\varepsilon(\omega)-1$ has most frequently been used to interpret experimental data, and we focus our attention on it. In the next two sections, we address the problem of relating the many-particle relaxation time $\mathscr{R}$ to the correlation times of single molecules and to extending the description of molecular motion beyond the diffusion limit.

Another way of looking at the various relationships between $\varepsilon(\omega)$ and the relaxation parameters follows from a macroscopic dielectric argument,[1,2] which may be used to relate the different functionals discussed above to the transform of the time-dependent normalized correlation function (Φ^{SH}) of the total moments $\mathbf{M}^{SH}$ of macroscopic samples of different shapes (SH),[23] that is,

$$f^{SH}[\varepsilon(\omega)] = 1 + i\omega\Phi^{SH}(\omega) \qquad (5.13)$$

The Debye functional is associated with a needle, the Glarum functional with a sphere in vacuo, and the Kirkwood-Onsager functional with the sphere embedded in an infinite sample of the same permittivity.[2] Titulaer and Deutch[9] investigated these relationships by a careful application of the linear response theory. Brot and coworkers[39a] have demonstrated in a computer simulation that the moment correlation functions of two samples of different shape yield consistent results for the permittivity if they are associated with the appropriate functionals. In our approach, we always evaluate the correlation functions for an infinite sample, but once we have evaluated $\varepsilon(\omega)$ in terms of these correlation functions, we can express any of the functions $f^{SH}(\varepsilon(\omega))$ in terms of those correlation functions.

Although the collective many-particle correlation functions of one sample may be related to those of another by this macroscopic argument, the fundamental problem of relating any of the collective correlation functions to the single-molecule motion is not solved. This is the problem that we take up in the next two sections, one of the results of which we have anticipated in Eq. (5.10). This equation and the above discussion enable us to conclude that, if the single-molecule correlation function relaxes as a single exponential, then so do the total moment $(\mathbf{M}^{SH})$ correlation functions of the needle and the sphere in vacuo, but that of the embedded sphere relaxes as two exponentials, which is equivalent to the readily proved statement that $\mathcal{R}_{SH}(\omega)$ for the embedded sphere is dependent upon long-range correlations. These results were obtained by a procedure closely related to ours by Sullivan and Deutch.[15]

Although the minimum number of relaxation times needed to describe the relaxation of the various functionals $f^{SH}(\varepsilon(\omega))$ discussed above can be obtained from the formal considerations in Section VI, it may be useful to present more intuitive arguments based upon the shapes of the samples associated with each functional. If we assume that the relaxation of the total moment $\mathbf{M}^{SH}$ of the sample is slow, and if there are no other slow variables that are strongly coupled to $\mathbf{M}^{SH}$, then we expect $\mathbf{M}^{SH}$ to decay as a single exponential, and for each additional relevant variable we find an additional exponential in the decay of $\mathbf{M}^{SH}$. For an infinite needle and for a sphere in vacuo, $\mathbf{M}^{SH}$ represents all the dipoles so that one relaxation time is expected in each case. For an embedded sphere, $\mathbf{M}^{SH}$ represents only the dipoles within the sphere, and these can interact with the dipoles outside the sphere; the latter must be collectively as "slow" a variable as $\mathbf{M}^{SH}$, so we expect the embedded sphere to relax with two relaxation times.

The arguments above can also be given in terms of projection operators. The transform of the normalized correlation function $\Phi^{SH}(\omega)$ corresponding to a particular $\mathbf{M}^{SH}$ can be expanded as indicated in Eqs. (5.3) and (5.4) with $\mathbf{M}^{SH}$ replacing M_a. Thus, the projections in the time evolution $(t_{M^{SH}})$ are orthogonal to $\mathbf{M}^{SH}$. If $\mathbf{M}^{SH}$ is the only slow variable, then all the slow motions are projected out of $\langle \mathbf{M}^{SH}(t_{M^{SH}})\mathbf{M}^{SH}(0)\rangle$, but if there are additional slow variables, then there remain some slow components in this correlation function. It follows that in those cases where $\mathbf{M}^{SH}$ is the only slow variable, the appropriate analogue of $T_a(k, \omega)$ in Eqs. (5.3) and (5.4) is independent of frequency, but if other slow variables exist, the frequency dependence of $T_a(k, \omega)$ must be retained. This is the case for the embedded sphere.

VI. RELATIONSHIP BETWEEN DIELECTRIC AND SINGLE-PARTICLE RELAXATION TIMES: ROTATIONAL DIFFUSION

The dielectric relaxation properties discussed in the last section are contained in the behavior of the response function $\Phi_{aa}(\mathbf{k}, \omega)$, which, as indicated in Eq. (5.2), is dependent upon the autocorrelation function of the *collective* property $M_a(\mathbf{k}, t)$. The corresponding short-range relaxation frequency $\mathcal{R}^{-1}(\omega)$, which is associated with $M_a(\mathbf{k}, t)$, is, as indicated in Eq. (5.8), dependent upon the autocorrelation function of $\dot{M}_a(\mathbf{k}, t)$ with time evolution projected along $M_a(\mathbf{k}, t)$. We wish to relate the collective, but intensive, transport property $\mathcal{R}(\omega)$ to the corresponding single-molecule relaxation properties. After doing this, we shall investigate dielectric relaxation and the corresponding molecular relaxation in the rotational diffusion limit.

In order to describe single-particle relaxation, consider the single-particle analogue $\phi_{aa}(\mathbf{k}, \omega)$ to $\Phi_{aa}(\mathbf{k}, \omega)$ in Eq. (5.2):

$$\phi_{aa}(\mathbf{k}, \omega) = \frac{\int_0^\infty dt\, e^{i\omega t}\langle \mu_a^1(t)\exp[i\mathbf{k}\cdot\mathbf{r}^1(t)]\, \mu_a^1 \exp[-i\mathbf{k}\cdot\mathbf{r}^1]\rangle}{\langle |\mu_a^1|^2\rangle} \qquad (6.1)$$

These quantities are all independent[39b] of a, and the dependence on $\mathbf{k}$, for low k, is negligible. In analogy to Eq. (5.3), we have

$$\phi_{aa}(\mathbf{k}, \omega) = \tau_s(\omega)[1 - i\omega\tau_s(\omega)]^{-1} \qquad (6.2)$$

where the *single-molecule* relaxation function $\tau_s(\omega)$ is independent of a and $\mathbf{k}$. We wish to relate $\tau_s(\omega)$ to the collective function $\mathcal{R}(\omega)$.

The function $\tau_s(\omega)$ is given by an expression analogous to that for $T_a^{-1}(\mathbf{k}, \omega)$ in Eq. (5.4):

$$\tau_s^{-1}(\omega) = \frac{\int_0^\infty dt\, e^{i\omega t} \langle \dot{\mu}_a^1 \big(t_{\{\mu_a\}}\big) \dot{\mu}_a^1 \rangle}{\langle |\mu_a^1|^2 \rangle} \tag{6.3}$$

where $t_{\{\mu_a\}}$ indicates that the time dependence is appropriately projected orthogonally to *all* the μ_a^j's, that is, to every molecular dipole moment in the system. In the projection scheme of Mori,[16] all "slow" variables must be projected out and, since every dipole moment in the system relaxes at a similar rate, each dipole is a "slow variable" and, as prescribed in Eq. (6.3), the projection operator must project time evolution orthogonally to *all* the moments. The rotational diffusion limit is appropriate when the relaxation rate of the molecular orientation (and hence the experimental frequency ω at which this relaxation is observed) is low compared to the relaxation rate of the molecular angular velocities and torques. It is then found that the relaxation of the dipole correlation function of a single molecule (in the presence of all the other molecules) is exponential, provided that the dipole lies along a principal axis of the diffusion tensor; that is, $\tau_s(\omega)$ as given by Eq. (6.3) is frequency-independent:

$$\tau_s(\omega) = \tau_s(0) \equiv \tau_s \tag{6.4}$$

In order to compare $\tau_s(\omega)$ with the intensive multiparticle function $\mathscr{R}(\omega)$, we rewrite Eq. (5.8) as

$$\mathscr{R}^{-1}(\omega) = (k_B T \varepsilon_0)^{-1} \rho \int_0^\infty dt\, e^{i\omega t} \langle \dot{\mu}_a^1 \big(t_{M_a}\big) \dot{\mu}_a^1 \rangle \dot{g}(\omega) \tag{6.5}$$

where ρ is the number density and the *dynamic coupling parameter* is defined as

$$\dot{g}(\omega) = 1 + N_0 \frac{\int_0^\infty dt\, e^{i\omega t} \langle \dot{\mu}_a^1 \big(t_{M_a}\big) \dot{\mu}_a^2 \rangle}{\int_0^\infty dt\, e^{i\omega t} \langle \dot{\mu}_a^1 \big(t_{M_a}\big) \dot{\mu}_a^1 \rangle} \tag{6.6}$$

The correlation functions in Eqs. (6.3) and (6.5) look very similar, but they have different projections; in Eq. (6.4), the projections, as we have argued, are along *all* the individual μ_a^j's, and in Eq. (6.5) they are along $M_a(\mathbf{k})$. However, a theorem known as the *corresponding macro-micro correlation the-*

orem $(CMMC)^{40,23}$ indicates that Eqs. (6.5) and (6.6) can be written equally well with t_{M_a} replaced by $t_{\{\mu_a\}}$. We thus have the following important connection between the collective $\mathcal{R}(\omega)$ and single-molecule $\tau_s(\omega)$ frequency-dependent correlation times:

$$\mathcal{R}(\omega) = \frac{3k_B T \varepsilon_0}{(\mu)^2 \rho}\left(\frac{\tau_s(\omega)}{\dot{g}(\omega)}\right) \tag{6.7}$$

where $\dot{g}(\omega)$ can be written as in Eq. (6.6) or as

$$\dot{g}(\omega) = 1 + N_0 \frac{\displaystyle\int_0^\infty dt\, e^{i\omega t}\langle \dot{\mu}_a^1\bigl(t_{\{\mu_a\}}\bigr)\dot{\mu}_a^2\rangle}{\displaystyle\int_0^\infty dt\, e^{i\omega t}\langle \dot{\mu}_a^1\bigl(t_{\{\mu_a\}}\bigr)\dot{\mu}_a^1\rangle} \tag{6.8}$$

and N_0 is the number of particles.

In the rotational diffusion limit specified by Eq. (6.4), the single-particle time $\tau_s(\omega)$ is independent of frequency.* From the corresponding macro-micro correlation theorem, it follows that $\mathcal{R}(\omega)$ and $\dot{g}(\omega)$ are also independent of ω in this limit: $\mathcal{R}(\omega) \to \mathcal{R}(0)$ and $\dot{g}(\omega) \to \dot{g}$. Furthermore, as indicated in Eq. (5.10), in this limit $\varepsilon(\omega)-1$ relaxes with a single relaxation time, the Debye time τ_D. We readily obtain the following useful relations for the diffusion limit:

$$\mathcal{R}(0) = \frac{3k_B T \varepsilon_0}{(\mu)^2 \rho}\left(\frac{\tau_s}{\dot{g}}\right) = \frac{3\varepsilon(0)\bigl(g^S/\dot{g}\bigr)\tau_s}{[\varepsilon(0)-1][2\varepsilon(0)+1]} \tag{6.9}$$

$$\tau_D = 3\varepsilon(0)\left(\frac{g^S}{\dot{g}}\right)\tau_s[2\varepsilon(0)+1]^{-1} \tag{6.10}$$

In the special case that $(g^S/\dot{g})=1$, Eq. (6.10) is known as the Glarum-Powles relation,[38,41,42] and it gives a useful interrelationship between the measured collective quantities τ_D and $\varepsilon(0)$ and the single-molecule property τ_s. However, g^S can differ appreciably from unity and although $\dot{g}$ is thought to be of the order of 1, it cannot be measured independently. Alternatively, Eq. (6.10) can be rewritten as

$$\tau_D = \frac{3\varepsilon_0 k_B T[\varepsilon(0)-1]}{(\mu)^2 \rho \dot{g}}\tau_s \tag{6.11}$$

*This is not true unless the dipole lies along a principal axis for rotational diffusion.

Actually, the single-particle projected time-dependent correlation functions appearing in Eqs. (6.3) and (6.5)–(6.8) should all have $\dot{\mu}_a^j$ replaced by $\dot{\mu}_a^j \exp(i\mathbf{k}\cdot\mathbf{r}^j)$, the projections indicated by $t_{\{\mu_a\}}$ should be along the $[\mu_a^j \exp(i\mathbf{k}\cdot\mathbf{r}^j)]$ rather than along all the $[\mu_a^j]$, and the projections indicated by t_{M_a} should be along $\sum \mu_a^j \exp(i\mathbf{k}\cdot\mathbf{r}^j)$ rather than along $\sum \mu_a^j$. Were it not for the fact that $\mathscr{R}(\omega)$ is independent of $\mathbf{k}$ at low k, and hence explicitly short-range, and that because of Eq. (6.7) the same must be true of $\dot{g}(\mathbf{k}, \omega)$, $\dot{g}(\mathbf{k}, \omega)$ could be of order N_0, cross correlations between molecular quantities could be of order unity, and the CMMC theorem would not be applicable. It is, therefore, the *short-range character of $\mathscr{R}(\omega)$ that enables us to relate $\varepsilon(\omega)$ to molecular quantities* in a meaningful way. Because of the short-range nature of $\mathscr{R}(\omega)$ at low k, we can therefore neglect all $\mathbf{k}$'s in Eqs. (6.3) and (6.5)–(6.8).

In Section VII, we discuss the dynamical problem in some detail, and in Sections VIII and IX, we return to a discussion of the dielectric constant $\varepsilon(0)$ for polarizable molecules.

VII. NONDIFFUSIONAL MOTION

A. Introduction

In the last two sections, we examined the dielectric relaxation of rigid, dipolar, nonpolarizable molecules that reorient diffusively; we found that $\varepsilon(\omega)$ -1 has a simple Lorentzian form [Eq. (5.10)]. This result is found experimentally to be valid at low frequencies for one-component liquids of small rigid molecules (see, for example, Refs. 43–46]; at higher frequencies, and in more complex liquids even at low frequencies, deviations from the simple Lorentzian result are observed. These deviations may be caused by both the relaxation of interaction-induced dipoles and the inadequacy of the simple diffusion model. In this section, we consider extensions of the theory needed to cope with the latter feature; some of the problems posed by the interaction-induced moments will be addressed in Sections IX and X.

Equations (5.6)–(5.9) are not only an exact representation for $\varepsilon(\omega)$, but in infinite systems they also establish the independence of $\mathscr{R}(\omega)$ from long-range correlations; $\mathscr{R}(\omega)$ is independent of $\mathbf{k}$ at low k and independent of the direction a over which the interactions are studied. It is therefore convenient to direct our study of the extensions beyond the diffusion limit to the quantity $\mathscr{R}(\omega)$. We do this by extending the general development of the Mori theory presented in Section VI and then formulating a semiquantitative theory (the "three-variable" theory) that can encompass numerous models for dielectric relaxation and can be used to correlate experimental data. The comparison of the formal developments in this section with experimental data is postponed until Section XI, following the generalization of this theory to include the effects of polarizability and interaction-induced dipoles. In Sections XI.A–C, we show how this theory can explain inertial oscillations in the longitudinal correlation functions (dipolarons) and the librational character of all orientational correlation functions at short times. In Section XI.D, we discuss the general problem of realistically modeling orientational correlation functions.

B. "Three-Variable" Theory

It has been shown elsewhere[47,23] that the "three-variable" or "third-order continued fraction" prescription for reorientational correlation functions is capable of describing many of the observed dynamical properties of simple liquids. This approach has been used extensively by Evans[46] and Gerschel[48] to interpret far-infrared absorption lineshapes. We shall develop this three-variable theory in a general form that is readily applicable to the various cases of interest to us. We let $\Lambda_1(t)$ be the "primary" dynamical variable, whose correlation function is of direct interest, such as $M_x(\mathbf{k})$ and $M_z(\mathbf{k})$. The other two "slow secondary" dynamical variables are chosen to be Λ_2,

$$\Lambda_2 = \dot{\Lambda}_1 \tag{7.1}$$

and Λ_3, which is closely related to $\ddot{\Lambda}_1$:

$$\Lambda_3 = \ddot{\Lambda}_1 + \Omega_R^2 \Lambda_1 \tag{7.2}$$

where

$$\Omega_R^2 = \frac{\langle |\dot{\Lambda}_1|^2 \rangle}{\langle |\Lambda_1|^2 \rangle} \tag{7.3}$$

With Λ_3 defined in this way, the Λ_i's are orthogonal, in the sense that

$$\langle \Lambda_i \Lambda_j^* \rangle = \delta_{ij} \langle |\Lambda_i|^2 \rangle \tag{7.4}$$

In the low-k cases of interest to us, Λ_1 is closely related to the sum of the components of the molecular dipoles μ^i along a laboratory axis, so that it is an orientational variable. Then, Λ_2 is an orientational flux, primarily determined by the molecular angular velocities Ω^i:

$$\dot{\mu}^i = \Omega^i \times \mu^i \tag{7.5}$$

Since

$$\ddot{\mu}^i = \dot{\Omega}^i \times \mu^i + \Omega^i \times (\Omega^i \times \mu^i) \tag{7.6}$$

Λ_3 is determined partially by the molecular torques $I\dot{\Omega}^i$, where I is the molecular moment of inertia about an axis perpendicular to the dipole axis. In a dense fluid, the torque terms [the first terms on the right of Eq. (7.6)] dominate the "kinetic terms" [the second terms on the right-hand sides of Eqs. (7.6) and (7.2)]. The three-variable theory allows for the possibility that the angular velocity and intermolecular torque, as well as the molecular orientation, relax on the "slow" observational time scale ω^{-1}. If Λ_2 and Λ_3 are indeed "slow," then the correlation functions of the orientational variables Λ_1 will relax nonexponentially, and $\varepsilon(\omega)-1$ will be non-Lorentzian. In our discussion, we consider only molecules for which the dipole lies along a symmetry axis.

The application of the Mori formalism[16] to the set of variables $\{\Lambda_1, \Lambda_2, \Lambda_3\}$ provides an expression for the spectral density or lineshape associated with this nondiffusional behavior. The predicted lineshape is given by

$$\Phi(\omega) = \int_0^\infty dt\, e^{i\omega t} \frac{\langle \Lambda_1(t)\Lambda_1^* \rangle}{\langle |\Lambda_1|^2 \rangle} = \left(-i\omega + \cfrac{\Omega_R^2}{-i\omega + \cfrac{\Omega_T^2}{-i\omega + \Gamma(\omega)}} \right)^{-1}$$

$$(7.7)$$

This expression involves the three parameters Ω_R, Ω_T, and $\Gamma(\omega)$: Ω_R was defined above and may be written, in the cases of interest to us, in terms of the "molecular free rotation frequency" ω_s,*

$$\omega_s = \left(\frac{2k_B T}{I} \right)^{1/2} \tag{7.8}$$

that is,

$$\Omega_R^2 = \frac{N_0 \left[(\mu)^2/3 \right] \omega_s^2}{\langle |\Lambda_1|^2 \rangle} \tag{7.9}$$

Ω_T^2 is related to the mean-squared "torque" by

$$\Omega_T^2 = \frac{\langle |\Lambda_3|^2 \rangle}{\langle |\Lambda_2|^2 \rangle} \tag{7.10}$$

*This expression is valid only if the dipole moment lies along a principal axis for the inertia tensor.

$\Gamma^{-1}(\omega)$ is a "torque" relaxation time given by

$$\Gamma(\omega) = \int_0^\infty dt\, e^{i\omega t} \langle \{\exp[i(1-P_3)\mathscr{L}t](1-P_3)\dot{\Lambda}_3\}$$
$$\times \{(1-P_3)\dot{\Lambda}_3\}^*\rangle \langle |\Lambda_3|^2\rangle^{-1}, \tag{7.11}$$

where P_3 is the projection operator, which, when acting on a quantity F, yields

$$P_3 F = \sum_{j=1}^3 \langle F\Lambda_j^*\rangle \langle |\Lambda_j|^2\rangle^{-1}\Lambda_j \tag{7.12}$$

We shall let $\Lambda_1 = M_x(\mathbf{k})$ or $M_z(\mathbf{k})$, and we shall denote the corresponding values of Ω_T and $\Gamma(\omega)$ by Ω_T^x, Ω_T^z and $\Gamma^x(\omega)$, $\Gamma^z(\omega)$, respectively. In this way, we can make use of Eq. (7.7) to obtain $\chi_{xx}^0(\omega)$ and $\chi_{zz}^0(\omega)$ through Eqs. (5.1) and (5.2); by comparing these results with Eqs. (5.6), (5.7), (4.5), (2.13), and (4.1), we obtain

$$\mathscr{R}_a(\omega) = \frac{3I\varepsilon_0}{2(\mu)^2\rho} \left\{ -i\omega + \frac{(\Omega_T^a)^2}{-i\omega + \Gamma^a(\omega)} \right\} \tag{7.13}$$

where $a = x$ or z. We know from Eq. (5.9) that for low k, $\mathscr{R}_a(\omega)$ must be independent of k and of a; consequently, Ω_T and Γ must be independent of k and a, which establishes their short-range character.

Equation (7.7) is an exact expression for $\Phi(\omega)$, but the approximate expression obtained by *assuming that* $\Gamma(\omega)$ *is independent of frequency* is much more useful. This assumption is justified provided that the set of variables $\{\Lambda_1, \Lambda_2, \Lambda_3\}$—the collective orientation, angular momentum, and intermolecular torques—form a "complete slow set" of dynamical variables. In Section XI.D, we discuss this assumption further, but here we assume that it is valid and that[*]

$$\Gamma(\omega) = \Gamma \tag{7.14}$$

The results obtained with this assumption constitute what we call the "three-variable theory." The diffusional theory corresponds to a first-order continued-fraction expansion, whereas the three-variable theory is equivalent to a third-order continued-fraction expansion.

---

[*]This condition can almost certainly not be true unless the dipole moment lies along a principal axis for the rotational motion in the liquid, as is the case for symmetric tops.

As a result of the above, we can now write the three-variable prescription for $\varepsilon(\omega)$ in a form that contains no implicit long-range correlations; to do this, we substitute Eq. (7.13) into Eq. (5.7):

$$\frac{\varepsilon(\omega)-1}{\varepsilon(0)-1} = \left(1 - i\omega\left[\frac{3Ig^S\varepsilon(0)}{2(k_BT)(2\varepsilon(0)+1)}\right]\left(-i\omega + \frac{\Omega_T^2}{-i\omega + \Gamma}\right)\right)^{-1}$$

$$(7.15)$$

Only Ω_T and Γ need be taken as adjustable parameters: $\varepsilon(0)$ can be determined from an experiment at $\omega = 0$; g^S can be determined by combining $\varepsilon(0)$ with the measured gas-phase dipole moment μ, as indicated in Eq. (4.1); I is a readily obtainable molecular quantity. In favorable cases, it may even be possible to evaluate the effective mean-square torque Ω_T^2 independently.

The figures in Section XI show typical Cole-Cole plots and far-infrared absorption spectra predicted by Eq. (7.15). We defer the discussion of data to Section XI because real molecules are polarizable and our discussion of rotating polarizable dipoles is not given until Section X. Many examples of calculations with Eq. (7.15) have been described by Evans et al.[46] and by Gerschel et al.[48]; in fitting the three-variable expansion to experimental data, these authors have often left the square-bracketed term in Eq. (7.15) as an adjustable parameter, but, as we have seen, this should not be necessary. Discussion of three-variable theory with various illustrative plots have also been given by Kivelson and Keyes[47] and specifically for dielectrics by Madden and Kivelson[23] and Guillot and Bratos[49].

C. An Extended CMMC Theorem

One often wishes to relate the behavior of the "three-variable" collective correlation function to the corresponding single-particle dipole correlation function. To do this, we must extend the diffusional CMMC theorem discussed in Section VI to the "three-variable" theory. A derivation of such an extended CMMC theorem was given in Ref. 23.

As discussed in Section VI, to describe the single-molecule correlation function,

$$\phi_{aa}(t) = \langle\left\{\mu_a^1(t)\exp\left[i\mathbf{k}\cdot\mathbf{r}^1(t)\right]\right\}\left\{\mu_a^1(0)\exp\left[i\mathbf{k}\cdot\mathbf{r}^1(0)\right]\right\}^*\rangle \quad (7.16)$$

the set of slow variables must include, at the very least, the dipole components of *all* the molecules in the sample. If we wish to construct a theory that is applicable when angular velocities and torques relax on the time scale of the orientations, we must use three sets of variables analogous to Λ_1, Λ_2, and Λ_3, the sets made up from $\mu_a^j \exp(i\mathbf{k} \cdot \mathbf{r}^j)$, $\dot{\mu}_a^j \exp(i\mathbf{k} \cdot \mathbf{r}^j)$, and η_a^i:

$$\eta_a^j = (\mu)^{-1} \exp(i\mathbf{k} \cdot \mathbf{r}^j) \left\{ \ddot{\mu}_a^j + \omega_s^2 \mu_a^j - \frac{f_a \omega_s^2}{1 + N_0 f_a} \left(\sum_{l \neq i} \mu_a^l - f_a \mu_a^j \right) \right\} \quad (7.17)$$

where each set j runs over all molecules in the sample, and

$$f_a = \frac{\langle \exp[i\mathbf{k} \cdot (\mathbf{r}^1 - \mathbf{r}^2)] \mu_a^1 \mu_a^2 \rangle}{\langle \mu_a^1 \mu_a^1 \rangle} \quad (7.18)$$

Without the terms containing f_a, η_a^j would have the same form as the collective Λ_3 variable introduced in Section VII.B, but even though f_a is of the order N_0^{-1}, the terms in η_a^j that contain f_a are needed to assure the orthogonality condition

$$\langle \eta_a^j \mu_a^j \rangle = 0 \quad (7.19)$$

to order f_a. Here, $I\eta_a^j$ is closely related to the intermolecular torque acting on a molecule, provided, as discussed near Eq. (7.6), the kinetic terms may be neglected. With this set of $3N_0$ "slow" variables, we can obtain the single-particle correlation functions $\phi(\omega)$ in Eq. (7.16), as well as the multi-particle correlation function $\Phi(\omega)$ in Eq. (7.7). Since both results are a consequence of the same "three-set-variable theory," the collective parameters can readily be related to the molecular ones.

With this formalism for the $3N_0$ variables, it can be shown that the single-particle correlation function $\phi(\omega)$ is identical in form to Eq. (7.7); details of this derivation may be found in Ref. 23. For the molecular case, the parameters Ω_R, Ω_T, and Γ that appear in Eq. (7.7) are independent of directional coordinate a. These parameters are:

Single: $\quad \Omega_R^2 \equiv \omega_s^2 = \dfrac{2k_B T}{I}$ $\qquad\qquad\qquad\qquad\qquad\qquad$ (7.20)

Single: $\quad \Omega_T^2 \equiv \omega_t^2 = \dfrac{\langle |\eta_a^i|^2 \rangle}{\omega_s^2}$ $\qquad\qquad\qquad\qquad\qquad$ (7.21)

Single:

$$\Gamma(\omega) \equiv \tau_t(\omega)^{-1}$$
$$= \int_0^\infty dt \, e^{i\omega t} \frac{\langle \{ \exp[i(1 - \mathsf{P}_{3N}) \mathscr{L} t](1 - \mathsf{P}_{3N}) \dot{\eta}_a^1 \} \{ (1 - \mathsf{P}_{3N}) \dot{\eta}_a^1 \} \rangle}{\langle |\eta_a^1|^2 \rangle}$$

$$(7.22)$$

where the operator $P_{[3N]}$ projects onto the $3N_0$ slow variables $\{\mu^i_a \exp(i\mathbf{k}\cdot\mathbf{r}^i)\}$, $\{\dot{\mu}^i_a \exp(i\mathbf{k}\cdot\mathbf{r}^i)\}$, and $\{\eta^i_a(\mathbf{k})\}$ [cf. Eq. (7.12)]. In dense fluids, τ_t can be identified with the torque relaxation time. In principle, ω_t can be evaluated from a computer simulation, leaving only τ_t as an adjustable parameter. Kivelson and Keyes[47] showed that, for different ranges of the parameters $\omega_s \tau_t$ and $\omega_t \tau_t$, very different reorientational motions could be described by the three-variable formula [Eq. (7.7) with $\Gamma(\omega) = \Gamma$], ranging from nearly free inertial motion to strongly torque-driven librations. If the single-particle function $\phi(t)$ is given by the general three-variable formula [Eq. (7.7)] with Ω_R, Ω_T, and Γ given by ω_s, ω_t, and τ_t^{-1}, respectively, then the collective correlation functions are also given by Eq. (7.7),[23] but with the collective parameters Ω_R, Ω_T, and Γ given as follows: For transverse modes,

$$(\Omega^x_R)^2 = \omega_s^2 \frac{2\varepsilon(0)+1}{3\varepsilon(0)g^S} \tag{7.23a}$$

and for longitudinal modes,

$$(\Omega^z_R)^2 = \omega_s^2 \frac{2\varepsilon(0)+1}{3g^S} \tag{7.23b}$$

with

$$\Omega^2_T = \omega_t^2 \ddot{g} \tag{7.24}$$

and

$$\Gamma = \frac{\tau_t^{-1}\ddot{g}}{\dddot{g}} \tag{7.25}$$

where

$$\ddot{g} = 1 + \frac{N_0 \langle \dot{\eta}^1_a \dot{\eta}^2_a \rangle}{\langle |\dot{\eta}^1_a|^2 \rangle} \tag{7.26}$$

and

$$\ddot{g}(\omega) = 1 + N_0 \int_0^\infty dt\, e^{i\omega t}$$
$$\times \frac{\langle \{\exp[i(1-P_{3N})\mathscr{L}t](1-P_{3N})\dot{\eta}^1_a\}\{(1-P_{3N})\dot{\eta}^2_a\}^*\rangle}{\tau_t^{-1}\langle |\dot{\eta}^1_a|^2 \rangle} \tag{7.27}$$

These correspondences between the collective and single-particle results constitute the *extended CMMC theorem*. The CMMC theorem holds provided the forms of the collective and single particle correlation functions obey the same one-, two-, or three-variable theory; if the collective variable is the moment of a sphere embedded in a medium of the same dielectric, then this is not the case because the corresponding $\mathscr{R}(\omega)$ for this collective variable is long range.

In a more complete treatment, the relaxation parameters $\tau_t(\omega)$ and $\ddot{g}(\omega)$ are frequency-dependent, but in the three-variable theory, this frequency dependence and that of $\Gamma(\omega)$ are neglected. Since $\mathscr{R}(\omega)$ is independent of $\mathbf{k}$ and a, it follows that in the three-variable theory, not only Ω_T and Γ, but also ω_t, τ_t, $\ddot{g}$, and $\ddot{g}$, are independent of $\mathbf{k}$ and a.

The single-particle orientational correlation time is defined as

$$\tau_s \equiv \phi_{xx}(\omega = 0) \tag{7.28}$$

and the Debye time is the corresponding dielectric correlation time, defined by

$$\tau_D \equiv \Phi(\omega = 0) \tag{7.29}$$

It follows that

$$\tau_s = \frac{\omega_t^2}{\omega_s^2} \tau_t \tag{7.30}$$

and that

$$\tau_D = \tau_s \frac{\ddot{g}^2}{\ddot{g}} \left(\frac{3\varepsilon(0)}{2\varepsilon(0)+1} \right) g^S \tag{7.31}$$

For strong, rapidly relaxing torques and moderate frequencies (i.e., $\omega \ll \Omega_T^2 \Gamma^{-1}$, $\omega \ll \Gamma^{-1}$), Eq. (7.7) reduces to the single Lorentzian form obtained in the rotational diffusion limit. Equation (7.31) is the *generalized Glarum-Powles formula* of Eq. (6.10), with

$$\dot{g} = \ddot{g}\,\ddot{g}^{-2} \tag{7.32}$$

A convenient form for the three-variable dielectric relaxation expression is

$$\frac{\varepsilon(\omega)-1}{\varepsilon(0)-1} \left[1 - \frac{\omega^2}{(\Omega_R^x)^2} - i\omega\tau_D \left(1 - \frac{i\omega}{\Gamma} \right)^{-1} \right]^{-1} \tag{7.33}$$

where Ω_R^x is defined in Eq. (7.23a), Γ in Eq. (7.25), and τ_D in Eq. (7.31). As explained above, Ω_R^x and $\varepsilon(0)$ can be determined independently from molecular data and $\varepsilon(0)$ from dielectric constant measurements, so that only the two adjustable parameters τ_D and Γ need be determined from the dielectric relaxation measurements. Here, τ_D can be determined from the low-frequency dielectric measurements and Γ from the high-frequency measurements; actually, since

$$\tau_D \Gamma = \frac{\Omega_T^2}{\left(\Omega_R^x\right)^2} \tag{7.34}$$

and Ω_T is an equilibrium quantity that may sometimes be obtained from computer calculations, τ_D may be the only adjustable parameter in the three-variable theory of dielectric relaxation. Equation (7.33) is similar to the expression used by Lobo et al.[50] in the absence of dielectric friction, their Eq. (2.18), and includes the term $\omega^2/(\Omega_R^x)^2$, which they associate with "inertial" contributions. While noting the inclusion of "inertial" contributions in the three-variable theory, one should stress that Eq. (7.33) does not reduce properly to the correct formula for the free molecule.

We leave the graphical study of the three-variable theory, the comparison of the theory with experiment, and the critique of the theory to Section XI, since a more realistic overview can be obtained with polarizable molecules (studied in Sections XIV and X) than with rigid dipoles.

VIII. POLARIZABLE NONPOLAR MOLECULES

A. General Development

In this section, we consider the static ($\omega = 0$) permittivity of molecules that do not have permanent dipoles; this permittivity is equivalent to the high-frequency (optical) permittivity of polar polarizable molecules because the permanent dipoles cannot respond to the high frequencies. We are therefore interested in $\varepsilon(\infty)$. As in the previous section, we wish again to base the theory on the response of the medium to the vacuum electric field $\mathbf{E}^0$ and to use a molecular rather than a continuum description to evaluate the quasi-susceptibility χ^0. The relevant molecular expression for χ^0 is given in Eq. (3.23) with $\mathbf{A}(\mathbf{k})$ defined in Eqs. (3.18) and (3.13). Here, $\underset{\sim}{\chi}^0$ can be rewritten as

$$\chi^0_{aa}(\infty) = \lim_{k \to 0} \frac{\alpha\rho}{\varepsilon_0} \left(1 + \frac{1}{\alpha} \left\langle \left(\sum_j \underset{\sim}{\alpha}^1 \cdot \mathsf{T}^{1j} \cdot \underset{\sim}{\alpha}^j \cdot \mathscr{T}^{j1} \cdot \underset{\sim}{\alpha}^1 \right)_{aa} \right\rangle \right.$$

$$\left. + \frac{1}{\alpha} \left\langle \left(\sum_j e^{i\mathbf{k} \cdot \mathbf{r}^{1j}} \underset{\sim}{\alpha}^1 \cdot \mathscr{T}^{1j} \cdot \underset{\sim}{\alpha}^j \right)_{aa} \right\rangle \right) \tag{8.1}$$

where $\mathscr{T}^{1j}$ is defined in Eq. (3.12), T^{1j} is the dipolar tensor defined in Eqs. (3.8) and (2.10), and α is the gas-phase isotropic polarizability:

$$\alpha = \tfrac{1}{3} \mathrm{Tr}\, \underset{\sim}{\alpha}^1 \tag{8.2}$$

The first average in Eq. (8.1) is independent of $\mathbf{k}$, and the quantity averaged has a dependence on $\mathbf{r}^{mn}$ that is short-range, that is, shorter-ranged than $(r^{mn})^{-3}$; it follows that this average is independent of the direction a indicated by the subscripts on χ^0_{aa}. The quantity in the second average in Eq. (8.1) has some contributions that vary as $(r^{mn})^{-3}$ and some that vary more rapidly; the former are proportional to the dipole operator T^{1j} and are $\mathbf{k}$-dependent, while the latter are independent of $\mathbf{k}$. By arguments totally analogous to those given in Section IV [cf. Appendix B near Eq. (B.12′)], we can therefore separate $\underset{\sim}{\chi}^0$ into long-range b^L and short-range b^S contributions:

$$\chi^0_{zz}(\infty) = \frac{\rho\alpha}{\varepsilon_0} [1 + b^S + 2b^L] \tag{8.3a}$$

$$\chi^0_{xx}(\infty) = \frac{\rho\alpha}{\varepsilon_0} [1 + b^S - b^L] \tag{8.3b}$$

where

$$b^S = \frac{1}{3} \mathrm{Tr} \left(\left\langle \frac{1}{\alpha} \sum_j \underset{\sim}{\alpha}^1 \cdot \mathsf{T}^{1j} \cdot \underset{\sim}{\alpha}^j \cdot \mathscr{T}^{j1} \cdot \underset{\sim}{\alpha}^1 \right\rangle + \mathbf{b}(\mathbf{k}) \right) \tag{8.4a}$$

$$b^L = \frac{2b_{zz} - b_{xx} - b_{yy}}{6} \tag{8.4b}$$

and

$$\mathbf{b} = \frac{1}{\alpha} \left\langle \sum_j \underset{\sim}{\alpha}^1 \cdot \mathscr{T}^{1j} \cdot \underset{\sim}{\alpha}^j \exp[i\mathbf{k} \cdot \mathbf{r}^{1j}] \right\rangle. \tag{8.4c}$$

As shown in Appendix B [Eq. (B.12′)], all terms that vary as r^{-3}, that is, all those terms that are proportional to T^{jl}, are included in b^L. Such terms are eliminated from b^S by the operation of taking the trace.

We can obtain useful expressions for $\varepsilon(\infty)$ by eliminating averages over long-range interactions; that is, we can eliminate b^L from the expressions. This can be done by taking the trace of $\underset{\sim}{\chi}^0$ and combining this result with Eqs. (2.13):

$$\frac{[\varepsilon(\infty)-1][2\varepsilon(\infty)+1]}{3\varepsilon(\infty)} = \frac{\rho\alpha}{\varepsilon_0}[1+b^S] \qquad (8.5)$$

Once again we have the Kirkwood functional, which depends only upon short-range interactions. [Compare with Eq. (4.1).]

Equation (8.5) is not the most convenient form in which to express the permittivity, both because b^S is rather large and because it can be evaluated for certain limiting cases. In particular, we can obtain an analytic expression for b^S as a function of $\rho\alpha$ for the case of a liquid composed of isotropically polarizable molecules in which the n-particle distribution function $g^{(n)}(1,2,\ldots)$ can be approximated by a product of pair distribution functions that are correct at long range but ignore the short-range liquid structure. More specifically, if we call the approximate form of the n-particle distribution function $\tilde{g}^{(n)}(1,2,\ldots)$, we have

$$\tilde{g}^{(n)}(1,2,\ldots,n) = \prod_{j<l}^{n} \tilde{g}^{(2)}(r^{jl}) \qquad (8.6)$$

where

$$\tilde{g}^{(2)}(r^{jl}) = \begin{cases} 1 & \text{for } r^{jl} \geq r_0 & (8.7a) \\ 0 & \text{for } r^{jl} < r_0 & (8.7b) \end{cases}$$

and we add the requirement that

$$\tfrac{4}{3}\pi r_0^3\rho = 1 \qquad (8.8)$$

If we label the value of b^S obtained by the use of $\tilde{g}^{(n)}(1,2,\ldots)$ as $\tilde{b}_S$, we can express b^S conveniently as

$$b^S = \tilde{b}_S + \delta b^0 \qquad (8.9)$$

where δb^0 is a residual quantity that should be small. In Appendix B, [Eq.

(B.12)], we find that

$$\tilde{b}_S = [(1-x)(1+2x)]^{-1}2x^2 \qquad (8.10)$$

where

$$x = \frac{\alpha\rho}{3\varepsilon_0} \qquad (8.11)$$

This is the result obtained by Felderhof.[10] *If we assume that δb is negligible,* then Eqs. (8.5), (8.9), and (8.10) lead to

$$\frac{\varepsilon(\infty)-1}{\varepsilon(\infty)+2} = \frac{\alpha\rho}{3\varepsilon_0} \qquad (8.12)$$

This is the *Clausius-Mossotti (Lorentz-Lorenz)* equation.[1-3] It is remarkably accurate for fluids composed of spherical molecules. We shall see why this is so. The Clausius-Mossotti equation is valid only for $\alpha\rho/\varepsilon_0 < 3$, and it predicts a phase transition at $\alpha\rho/\varepsilon_0 = 3$; this new phase is characterized by a cooperative buildup of induced moments. For rigid dipoles described by the Kirkwood relation, no such phase transition is observed, because the cooperative buildup of the permanent dipoles occurs only if accompanied by a long-range correlation of orientations.

Before discussing the effect of the correction δb^0 on the Clausius-Mossotti equation, we shall discuss the relationship between our approximate molecular derivation and the classical derivation of Lorentz and Onsager,[1,3] both of whom used continuum electrostatic theory. In both the Lorentz and the Onsager derivations of Eq. (8.12), it is assumed that

$$\mathbf{P} = \rho\alpha\cdot\mathbf{E}_{\text{local}} \qquad (8.13)$$

and that $\mathbf{E}_{\text{local}}$ is the field inside a small sphere about the molecule of interest; the material outside this sphere is treated as a uniform dielectric continuum. In the Lorentz model, the inner sphere is filled with dielectric material, and the field within this sphere is the Lorentz field $[\varepsilon(\infty)+2]\mathbf{E}/3$, which is the sum of the $\mathbf{E}^0$ field plus the field attributable to the external dielectric. It is assumed that the contribution to the local field at particle 1 arising from the dielectric material within the sphere vanishes so that $\mathbf{E}_{\text{local}}$ is the Lorentz field. In the Onsager model, the inner sphere is a cavity, and the local field is taken to be the sum of the cavity field $\mathbf{G}$, which consists of $\mathbf{E}^0$ and the field from the polarization of the surrounding dielectric, and the reaction field $\mathscr{E}_R$ arising from the polarization of the dielectric by the dipole

$\underset{\sim}{\alpha} \cdot (\mathbf{G} + \mathscr{E}^R)$ induced in a molecule that lies inside the cavity and has polariz-ability $\underset{\sim}{\alpha}$. The form of the polarization around the cavity is different from that around the Lorentz sphere; it is found that

$$\mathbf{G} = \frac{3\varepsilon(\infty)\mathbf{E}}{2\varepsilon(\infty)+1} \tag{8.14}$$

and

$$\mathscr{E}_R = \left[\left(\frac{2\varepsilon(\infty)+1}{2[\varepsilon(\infty)-1]} \right) \frac{r_0^3}{\alpha} - 1 \right]^{-1} \frac{3\varepsilon(\infty)\mathbf{E}}{2\varepsilon(\infty)+1} \tag{8.15}$$

where r_0 is the cavity radius. Combining Eqs. (8.13)–(8.15) with Eqs. (2.2) and (2.6), we obtain

$$\frac{[\varepsilon(\infty)-1][2\varepsilon(\infty)+1]}{3\varepsilon(\infty)} = \frac{\rho\alpha}{\varepsilon_0} \left[1 - \left(\frac{2\varepsilon(\infty)+1}{2[\varepsilon(\infty)-1]} \right)^{-1} \frac{\alpha}{r_0^3} \right]^{-1} \tag{8.16}$$

where the second factor has arisen from the reaction-field-induced polar-ization. The reaction field that arises following the introduction of the polarizable molecule into the cavity is the result of the modification of the polarization of the surrounding medium because of its interaction with the cavity contents. When r_0 is chosen to satisfy Eq. (8.8), the dielectric properties of the interior and exterior of the sphere are matched, and the net polarization of the surrounding medium reaches its Lorentz value; with this choice for r_0, Eq. (8.16) yields the Clausius-Mossotti expression for $\rho\alpha$. Our derivation of Eq. (8.12), or equivalently of Eq. (8.10), has run parallel to that of Onsager, except that we have performed a molecular (albeit for a simple model), rather than a continuum, electrostatic calculation of the fields acting on the molecules. Most other molecular approaches to the problem have in-corporated fluctuations into the Lorentz theory.[10,11]

Our analysis has been based, at the outset, on the superposition ap-proximation for the distribution function and upon the use of the approxi-mate distribution function $\tilde{g}^{(2)}$. Wertheim[7] and Logan[20] have shown how a rigorous theory of the permittivity of polarizable fluids, with $\mathbf{E}^0$ as the driv-ing field, may be developed in a way in which the superposition approxima-tion is avoided. The multiple scattering series is resummed before, rather than after, averages are taken. However, to obtain computationally tractable ex-pressions for the deviations from Clausius-Mossotti behavior, some ap-proximation scheme is necessary for the distribution function (or for the direct correlation function).[7,51] We have chosen to make the superposition approximation at the outset and then to consider correction terms. Our analysis provides a simple guide to the physics underlying the Clausius-Mossotti equation and to the corrections required to improve the results; these corrections are discussed below. In the next section, we extend this analysis to polar polarizable fluids.

In obtaining Eq. (8.12), we have not only assumed superposition for the n-particle distribution [Eq. (8.6)], but we have also assumed that the molecules are spherically symmetric and that certain short-range terms can be neglected in Eqs. (8.7). In order to extend our results, we must relax these conditions. When this is done, the quantity δb in Eq. (8.9) is no longer negligible, and the Clausius-Mossotti equation becomes

$$\frac{\varepsilon(\infty)-1}{\varepsilon(\infty)+2} = \frac{\alpha\rho}{3\varepsilon_0}(1+\delta b) \tag{8.17}$$

This is a convenient way of expressing the "exact" result, but it is δb^0 for which we have a *direct molecular expression* [viz. Eqs. (8.9), (8.10), and (8.4a)], and we need a procedure for evaluating δb from the value of δb^0 calculated from molecular theory. The relationship of δb to δb^0 is much the same as that of χ to χ^0. We can readily obtain an expression for δb in terms of δb^0 by comparing Eq. (8.17) with Eqs. (8.5) and (8.9)–(8.11). If δb^0 is sufficiently small for the relationship between δb and δb^0 to be linear, it can be seen that

$$\delta b = \delta b^0 \left\{ \frac{27\varepsilon(\infty)^2}{[\varepsilon(\infty)+2]^2[2\varepsilon(\infty)^2+1]} \right\} \tag{8.18}$$

Values of $\delta b/\delta b^0$ are given in Table I. Equivalently, if the $\varepsilon(\infty)$'s in this expression are evaluated by means of Eq. (8.12), a permissible procedure if only terms linear in δb^0 are retained, then

$$\delta b = \delta b^0 \frac{(2x+1)^2(1-x)^2}{3x^2+2x+1} \tag{8.19}$$

We see that, to lowest order in $\rho\alpha/3\varepsilon_0$, $\delta b \approx \delta b^0$.

B. Calculation of Correction Terms

We next focus on the actual evaluation of δb^0 by considering a number of the major corrections to the Clausius-Mossotti expression. Readers who are more interested in the general development might wish to avoid this section at a first reading. First, we shall consider some of the effects of molecu-

TABLE I

$\varepsilon(\infty)$	$\delta b/\delta b^0$
1.5	0.90
2	0.75
2.5	0.62
3	0.51
4	0.36
5	0.27

lar anisotropy on the dielectric permittivity. To do this, we introduce the gas-phase molecular anisotropic polarizability γ. We shall make use of the simple model introduced in Section VIII.A, in which the distribution is isotropic and is given by $\tilde{g}^{(n)}$ in Eqs. (8.6)–(8.8), but we shall include the anisotropic polarizablities. In the evaluation of b^S, specified in Eq. (8.4b), we shall treat γ/α as an expansion parameter; within our simple model the lowest-order contributing terms in γ/α are quadratic. There are two classes of $(\gamma/\alpha)^2$ terms contributing to b^S or, equivalently, to δb^0. In class (a), the anisotropic polarizability of molecule 1 contributes twice, and only the isotropic polarizabilities of the other molecules contribute in the chain of dipole–induced dipole interactions. Class (b) terms arise when only the isotropic polarizability of molecule 1 contributes but the anisotropic polarizability of another molecule in the chain appears twice. These terms are given pictorial representations in Appendix B, where it is shown that the terms in each class can be summed: For class (a), we show that

$$\delta b_a^0 = \frac{2}{9}\left(\frac{\gamma}{\alpha}\right)^2 \frac{2x^2}{(1-2x)(1+3x)} \tag{8.20}$$

and for class (b),

$$\delta b_b^0 = \frac{2}{9}\left(\frac{\gamma}{\alpha}\right)^2 (2x^2)^2(1-3x+\cdots) \tag{8.21}$$

The quantities δb_a^0 and δb_b^0 are molecular quantities, and the corresponding corrections δb_a and δb_b to the Clausius-Mossotti equation can be obtained from Eq. (8.18). Class (b) terms make a considerably smaller contribution to δb^0 than do class (a) terms; even for a large value of ε, such as the value of 2.5 for CS_2, $x \cong \frac{1}{3}$, which indicates that the class (b) term is about one-ninth that of the class (a) contribution. Furthermore, the first two terms in the series indicated in Eq. (8.21) almost cancel each other. Thus, we can usually neglect the class (b) contributions δb_b^0. For a highly anisotropically polarizable molecule such as CS_2, $\gamma/\alpha \approx 1$, and it follows that $\delta b_a^0 \approx \frac{4}{45}$ and $\delta b_a \approx +0.055$, a reasonably small correction term. Note that both the expression for δb in terms of δb^0 [Eq. (8.18)] and the expression for δb_a^0 [Eq. (8.20)] have been summed consistently to all powers in $\rho\alpha$.

The Clausius-Mossotti equation has been derived under the assumption that the structure of the fluid is given by the simple distribution function $\tilde{g}^{(n)}$ defined in Eq. (8.6). In addition, certain short-range terms were neglected; they arise during evaluation of averages in which the dipole in molecule 1 polarizes molecule 2 via a third molecule [cf. Eq. (B.5)]. We will now discuss the correction terms that arise when we include the nontrivial aspects of the local structure of the fluid; we shall express them as an expansion in α and γ/α:

$$\alpha\delta b_R^0 = \sum_{n=2}^{\infty} \alpha^n \sum_{m=0}^{n} b^{(n,m)} \left(\frac{\gamma}{\alpha}\right)^m \qquad (8.22)$$

To evaluate δb_R^0, we must evaluate b^S as specified in Eq. (8.4a), but with the average taken with the distribution functions $g^{(n)} - \tilde{g}^{(n)}$, where $g^{(n)}$ is the exact n-body distribution and $\tilde{g}^{(n)}$ is the approximate n-body distribution given in Eqs. (8.6)–(8.8).

The lead term in the expansion of δb_R in powers of the polarizability may be written

$$\sum_{m=0}^{2} b^{(2,m)} \left(\frac{\gamma}{\alpha}\right)^m = \frac{\rho}{3\alpha^2} \operatorname{Tr} \int d\mathbf{r}^{12} \, d\Omega^1 \, d\Omega^2 \, \boldsymbol{\alpha}^1 \cdot \mathbf{T}^{12} \cdot \boldsymbol{\alpha}^2 e^{i\mathbf{k}\cdot\mathbf{r}^{12}}$$
$$\times \left\{ g^{(2)}(\mathbf{r}^{12}, \Omega^1, \Omega^2) - \tilde{g}^{(2)}(r^{12}) \right\} \qquad (8.23)$$

where Ω^j represents the Eulerian angles describing the orientation of the jth molecule. The corrections contained in $b^{(2,m)}$ arise from the fact that it accounts for the short-range liquid structure that is neglected in $\tilde{g}^{(2)}(r^{12})$. Our expression for $\sum_m b^{(2,m)} (\gamma/\alpha)^m$ is identical to that obtained by Keyes and Ladanyi[12] by an extension of Felderhof's theory.[10] For spherical molecules, $g^{(2)}(\mathbf{r}^{12}, \Omega^1, \Omega^2) = g^{(2)}(r^{12})$, and the integrand in Eq. (8.23) has no dependence upon Ω^1 and Ω^2; the only dependence of the integrand upon the angular part of $\mathbf{r}^{12}$ enters through $\mathbf{T}^{12}\exp(i\mathbf{k}\cdot\mathbf{r}^{12})$, and when the trace is taken the integral vanishes. Hence, $b^{(2,m)}$ *vanishes for spherical molecules*. For nonspherical molecules, the orientation-dependent parts of $g^{(2)}(\mathbf{r}^{12}, \Omega^1, \Omega^2)$ make a nonvanishing contribution to $b^{(2,1)}$ and $b^{(2,2)}$. We will not evaluate these terms for nonspherical molecules, since this has been done by Keyes and Ladanyi[12]; they interpret these contributions in terms of the "shaped cavity" ideas of classical dielectric theory. The $\sum_m b^{(2,\,m)}(\gamma/\alpha)^m$ value obtained by Keyes and Ladanyi is comparable in magnitude, and opposite in sign, to the δb_a^0 value obtained above.

The first nonvanishing term in δb_R^0 for spherical molecules is $b^{(3,0)}$. It contains contributions from three-particle interactions, as well as others arising from two-particle interactions. For spherical molecules:

$$b^{(3,0)} = \tfrac{1}{3}\mathrm{Tr}\bigg\{ \rho \int d\mathbf{r}^{12}\, \mathsf{T}^{12}\cdot\mathsf{T}^{21}\big[\, g^{(2)}(r^{12}) - \tilde{g}^{(2)}(r^{12})\,\big]$$

$$+\, \rho^2 \int d\mathbf{r}^{12}\, d\mathbf{r}^{13}\, e^{i\mathbf{k}\cdot\mathbf{r}^{12}}\mathsf{T}^{13}\cdot\mathsf{T}^{32}$$

$$\times \big[\, g^{(3)}(r^{12}, r^{13}, r^{23}) - \tilde{g}^{(2)}(r^{12})\tilde{g}^{(2)}(r^{13})\tilde{g}^{(2)}(r^{23})\,\big]\bigg\} \qquad (8.24)$$

Notice that $g^{(3)}$, $g^{(2)}$, and $\tilde{g}^{(2)}$ are all density-dependent, the latter through the density dependence of r_0 as indicated in Eqs. (8.7) and (8.8). Therefore, as it stands, $b^{(3,0)}$ is *not* a term in a virial expansion; although this is not the form in which $b^{(3,0)}$ is usually expressed, it does show how $b^{(3,0)}$ arises only from the *nontrivial local liquid structure*, that part not contained in $\tilde{g}^{(3)}$. To obtain $b^{(3,0)}$ in more conventional form, we may rearrange Eq. (8.24) by using the results given in Appendix C, Eqs. (C.15)–(C.18), for the integrals over the products of T^{ij} tensors and $\tilde{g}^{(2)}$ factors. The three-body factor in Eq. (8.24) may be written as

$$\rho^2 \int d\mathbf{r}^{12}\, d\mathbf{r}^{13}\,\mathsf{T}^{13}\cdot\mathsf{T}^{32}e^{i\mathbf{k}\cdot\mathbf{r}^{12}}$$

$$\times \big[\, g^{(3)}(r^{12}, r^{13}, r^{23}) - \tilde{g}^{(2)}(r^{13})\tilde{g}^{(2)}(r^{23})\{1 + (\tilde{g}^{(2)}(r^{12}) - 1)\}\,\big]$$

$$= \rho^2 \int d\mathbf{r}^{12}\, d\mathbf{r}^{13}\,\mathsf{T}^{13}\cdot\mathsf{T}^{32}\big[\, g^{(3)}(r^{12}, r^{13}, r^{23}) - \tilde{g}^{(2)}(r^{13})\tilde{g}^{(2)}(r^{23})\,\big]$$

$$+ \frac{4\pi}{3}r_0^3\rho \left[\rho \int d\mathbf{r}^{12}\,\mathsf{T}^{13}\cdot\mathsf{T}^{31}\tilde{g}^{(2)}(r^{13})\right] \qquad (8.25)$$

In the first term on the right-hand side of Eq. (8.25), we have made use of the fact that $\tilde{g}^{(3)}(r^{12}, r^{13}, r^{23}) - \tilde{g}^{(2)}(r^{13})\tilde{g}^{(2)}(r^{23})$ is a short-range factor in order to eliminate $\exp(i\mathbf{k}\cdot\mathbf{r}^{12})$. In evaluating the second term in this equation, we have used Eq. (C.11) to evaluate the integral over $d\mathbf{r}^{12}$ and the fact that $[\tilde{g}^{(2)}(r^{13})]^2 = \tilde{g}^{(2)}(r^{13})$. Finally, combining Eqs. (8.24), (8.25), and (8.8), and noting that the particle labels are arbitrary, we obtain an expression for

$b^{(3,0)}$ in the conventional form[52,21]:

$$b^{(3,0)} = \tfrac{1}{3}\mathrm{Tr}\Big\{ \rho \int d\mathbf{r}^{12} \mathsf{T}^{12}\cdot\mathsf{T}^{21} g^{(2)}(r^{12})$$

$$+ \rho^2 \int d\mathbf{r}^{12}\, d\mathbf{r}^{13} \mathsf{T}^{13}\cdot\mathsf{T}^{32} \big\{ g^{(3)}(r^{12}, r^{13}, r^{23}) - \tilde{g}^{(2)}(r^{13})\tilde{g}^{(2)}(r^{32}) \big\} \Big\}$$

$$(8.26)$$

This expression is very similar, but not identical, to the usually quoted result of the Yvon-Kirkwood theory[52,21]; in the latter, the $\tilde{g}^{(2)}$ factors that appear in Eq. (8.26) are replaced by the full $g^{(2)}$. The Yvon-Kirkwood results are obtained in the renormalized theory of Wertheim,[7] in which the introduction of the fictitious distribution function $\tilde{g}^{(n)}$ is avoided. However, the numerical difference between Eq. (8.26) and Yvon-Kirkwood result is not likely to be large, since the requirement that $4\pi r_0^3 \rho/3 = 1$ guarantees that integrals over $r^{-3}\tilde{g}^{(2)}(r)$ closely approximate those over $r^{-3}g^{(2)}(r)$.

Even in the model based on the simplified distribution function $\tilde{g}^{(n)}$, short-range terms such as those appearing in Eqs. (B.5) and (C.18) affect the dielectric properties. These short-range terms were neglected in our derivation of $\tilde{b}^S$ and the Clausius-Mossotti equation and have not yet been mentioned in our discussion of correction terms. These terms enter through the three-molecule integrals involving $\tilde{g}^{(n)}$ for $n \geq 2$; therefore, they contribute corrections of the order α^3 to $\alpha \delta b_R^0$ and so affect $b^{(3,0)}$ but not $b^{(2,m)}$. Although these short-range corrections are likely to be small compared to those associated with the nonvanishing character of $g^{(n)} - \tilde{g}^{(n)}$, they *are significant for the model fluid*, the structure of which is actually given by $\tilde{g}^{(n)}$ with r_0 given by Eq. (8.8). This model fluid is of some interest as a reference fluid for computationally tractable theories of the dielectric properties of fluids of spherical molecules, such as those based on the mean spherical approximation.[51] For a spherical system, $b^{(2,0)}$ is zero, and for the model fluid the first term in Eq. (8.24) also vanishes. This means that the second virial coefficient, the correction to the Clausius-Mossotti formula that is linear in density, vanishes for the model fluid. For this "model fluid," the short-range contributions to $b^{(3,0)}$, which were omitted in our calculation of b^S and are not included in Eq. (8.26), are

$$b^{(3,0)} = \tfrac{1}{3}\rho^2 \mathrm{Tr}\int_0^{2r_0} \mathscr{S}(r^{12}) 4\pi(r^{12})^2\, dr^{12} \qquad (8.27)$$

where

$$\underset{\sim}{\mathscr{S}}(r^{12}) = \int d\mathbf{r}^3\, \tilde{g}^{(2)}(r^{13}) \mathsf{T}^{13}\cdot\mathsf{T}^{32} \tilde{g}^{(2)}(r^{32}) \qquad (8.28)$$

[see Eqs. (C.12) and (C.17)–(C.19)]. This short-range part of $\mathscr{S}(r^{12})$ is evaluated in Eqs. (C.16) and (C.18); if this value is substituted back into Eq. (8.27),

$$b^{(3,0)} = -\left(\frac{15}{16}\right)\rho^2\left(\frac{1}{3\varepsilon_0}\right)^2 \tag{8.29}$$

If we do not go to higher order in α, then it is consistent to set $\delta b^0 \equiv \delta b$, and we obtain

$$\frac{\varepsilon(\infty)-1}{\varepsilon(\infty)+2} = \frac{\alpha\rho}{3\varepsilon_0}\left[1 - \left(\frac{15}{16}\right)\left(\frac{\alpha\rho}{3\varepsilon_0}\right)^2\right] \tag{8.30}$$

Equation (8.30) is the result obtained from a graphical analysis of the dielectric constant of a dense fluid[51]; it is also the form predicted by the mean spherical approximation,[52] to this order in α. There has been discussion recently[53] of the fact that this expression does not contain a second dielectric virial coefficient, which is known to be nonvanishing for a true hard-sphere gas (for which r_0 is a fixed, density-independent parameter). However, our model reference fluid is not a true hard-sphere gas. For this reference fluid, we have exploited the cancellation of terms that appear at different orders in a virial expansion [as in passing from Eq. (8.24) to Eq. (8.26)]. Such cancellations are of central importance in understanding the success of the Clausius-Mossotti formula for a dense fluid. Conversely, our development makes it clear than an expression like Eq. (8.30) should not be used at low density, where $\tilde{g}^{(n)}$, with r_0 given in Eq. (8.8), is a very poor approximation to the true distribution function.

IX. POLARIZABLE POLAR MOLECULES: ZERO-FREQUENCY PERMITTIVITY

An analysis, in molecular terms, of the static electrical permittivity of an assembly of polar polarizable molecules proceeds along much the same lines as the analyses presented in the previous sections for the permittivity of rigid polar and polarizable nonpolar molecules. It is readily seen that for polarizable polar molecules $\varepsilon(\infty)$ is independent of the permanent dipole moments and is still given by Eq. (8.17). It then follows from Eqs. (2.13ab) and (3.21–3.23) that $\varepsilon(0)$ is given by the exact expression:

$$\frac{[\varepsilon(0)-1][2\varepsilon(0)+1]}{\varepsilon(0)} - \frac{[\varepsilon(\infty)-1][2\varepsilon(\infty)+1]}{\varepsilon(\infty)}$$

$$= \lim_{k \to 0} \mathrm{Tr}\langle \mathbf{M}(\mathbf{k})\mathbf{M}(\mathbf{k})^*\rangle \frac{\rho}{N_0 k_B T\varepsilon_0} \tag{9.1}$$

where the total dipole density $\mathbf{M}$ is defined by Eqs. (3.20) and (3.11). For nonpolarizable polar molecules, $\varepsilon(\infty) = 1$, and Eq. (9.1) reduces to Eq. (4.1); in particular, for this case the right-hand side of Eq. (9.1) is independent of the long-range correlation parameter g^L and dependent upon the short-range parameter g^S. However, for polarizable polar molecules, as we shall show, the right-hand side of Eq. (9.1) contains contributions that are dependent upon long-range correlations between permanent dipoles. An approximate expression for $\varepsilon(0)$ is that obtained from the Onsager model,[22] as extended by Fröhlich[35]:

$$\frac{[\varepsilon(0) - \varepsilon(\infty)][2\varepsilon(0) + \varepsilon(\infty)]}{\varepsilon(0)} = \left(\frac{\varepsilon(\infty) + 2}{3}\right)^2 \frac{(\mu)^2 g^S \rho}{\varepsilon_0 k_B T} \qquad (9.2)$$

where μ is the gas-phase dipole moment. We obtain this expression on the basis of a simple molecular model that is consistent with our discussion in the preceding sections. At the end of this section, we discuss the connection between our treatment and those of others, and the difficulties of extending the calculations beyond this simple model.

Since our ultimate goal is the study of the frequency-dependent permittivity of polarizable polar molecules, our approach to the static problem should be chosen so as to form a starting point for our later treatment of the dynamic problem. We first rewrite the expression in Eqs. (3.20) and (3.11) for the total instantaneous dipole density M_a as

$$M_a = \sum_j \hat{\mu}_a^j e^{i\mathbf{k}\cdot\mathbf{r}^j} \qquad (9.3)$$

with

$$\hat{\mu}^j = \left[1 + \underset{\sim}{\alpha}^j \cdot \mathsf{T}^{jm} \cdot \underset{\sim}{\alpha}^m \cdot \underset{\sim}{\mathscr{T}}^{mj} + \exp(i\mathbf{k}\cdot\mathbf{r}^{mj})\underset{\sim}{\alpha}^m \cdot \underset{\sim}{\mathscr{T}}^{mj}\right] \cdot \mu^j \qquad (9.4)$$

In this form, $\mathbf{M}$ is expressed as a sum of terms associated with the permanent dipole μ^j of each molecule, rather than as a sum of instantaneous molecular dipoles $\sum_j \mathbf{m}^j$, as indicated in Eq. (3.20). We next note that the time-dependence of the dipole moments arises from molecular reorientation and from fluctuations in the instantaneous molecular dipole moment resulting from variations in the interparticle separations and relative reorientations. In many cases, we expect the rotations of the permanent dipoles to be slow compared to the fluctuations in the induced dipoles, although these two phenomena may be correlated.[54-56] With this discussion in mind, we write the total instantaneous dipole density $M_a(\mathbf{k}, t)$ as

$$M_a(\mathbf{k}, t) = \mathscr{M}_a(\mathbf{k}, t) + \Delta M_a(\mathbf{k}, t) \qquad (9.5)$$

where $\mathscr{M}_a$ is the projection of $M_a(t)$ onto the total gas-phase moment $M_a^0(\mathbf{k}, t)$,

$$\mathbf{M}^0(\mathbf{k}) = \sum_j \mu^j \exp(i\mathbf{k}\cdot\mathbf{r}^j) \tag{9.6}$$

and $\Delta M_a(t)$ is the component of $M_a(t)$ orthogonal to $M_a^0(t)$. Then, $\mathscr{M}_a$ is given by

$$\mathscr{M}_a(\mathbf{k}, t) = \frac{\langle M_a(\mathbf{k}) M_a^0(-\mathbf{k})\rangle}{\langle |M_a^0(\mathbf{k})|^2\rangle} M_a^0(\mathbf{k}, t) \tag{9.7}$$

$\mathscr{M}$ contains not only the permanent gas-phase dipoles, but also an ensemble average of that part of the induced moment that is projected along the gas-phase dipole moment. Here, $\Delta\mathbf{M}$ is associated with fluctuation of the induced dipoles about the "effective" permanent moment $\mathscr{M}$ for a given set of orientational coordinates; the fluctuations occur due to changes in intermolecular coordinates. It is the time dependence of $\mathscr{M}(t)$ that we associate with molecular rotations, and it is $\Delta\mathbf{M}(t)$ that we believe fluctuates at much higher frequencies; we take up this argument in the next section. Here, we consider the static behavior of $\mathbf{M}$ by separating it into $\mathscr{M}$ and $\Delta\mathbf{M}$ components. For nonpolarizable dipolar molecules, $\Delta\mathbf{M} = 0$ and $\mathscr{M} = \mathbf{M}^0$.

To obtain $\mathscr{M}$, we must evaluate

$$\langle M_a(\mathbf{k}) M_a^0(-\mathbf{k})\rangle = \sum_{j,l} \langle \mu_a^j\mu_a^l\exp(i\mathbf{k}\cdot\mathbf{r}^{jl})\rangle + \sum_l \langle (\hat{\mu}^l - \mu^l)_a\mu_a^l\rangle$$
$$+ \sum_{j,l \neq j} \langle (\hat{\mu}^l - \mu^l)_a\mu_a^j\exp(i\mathbf{k}\cdot\mathbf{r}^{jl})\rangle \tag{9.8}$$

The first of these terms was evaluated in Section IV; the remaining averages have elements in common with those that we evaluated in both Sections IV and VIII, and the model we shall use for the liquid structure is an amalgam of those introduced previously. As always, we shall be interested in the low-k limit.

Consider the second term in Eq. (9.8); it gives that part of the dipole which is induced by a permanent molecular moment and is correlated with the orientation of the permanent dipole. This term is composed of two parts [cf. Eq. (9.4)]:

$$\langle (\hat{\mu}^1 - \mu^1)_a\mu_a^1\rangle = N_0 \lim_{k \to 0} \left\{ \left\langle \exp(i\mathbf{k}\cdot\mathbf{r}^{12})\left[\underset{\sim}{\alpha}^2\cdot\underset{\sim}{\mathscr{T}}^{21}\cdot(\mu^1\mu^1)\right]_{aa}\right\rangle \right.$$
$$\left. + \left\langle \left[\underset{\sim}{\alpha}^1\cdot\mathbf{T}^{12}\cdot\underset{\sim}{\alpha}^2\cdot\underset{\sim}{\mathscr{T}}^{21}\cdot(\mu^1\mu^1)\right]_{aa}\right\rangle \right\} \tag{9.9}$$

The first part gives the correlation between the permanent dipole of molecule 1 and the dipoles induced in all *other* molecules by molecule 1, and the second gives the correlation between the reaction field induced and the permanent dipole of molecule 1. The product $(\mu^1\mu^1)$ may be regarded as a symmetric dyad that depends upon the orientation of molecule 1 in exactly the same way as the polarizability tensor; the averages that occur in Eq. (9.9) are thus of exactly the same form as the averages in Eq. (8.1). As we showed in Section VIII, if Eq. (8.1) is evaluated with the approximate distribution function $\tilde{g}^{(n)}$ and if the anisotropy in the molecular polarizability is neglected, the Clausius-Mossottii formula for $\varepsilon(\infty)$ is obtained. Furthermore, we showed that the corrections that arose with a more accurate representation of the distribution function were short-range and small. Since the Clausius-Mossotti formula gives $\varepsilon(\infty)$ accurately for polar as well as nonpolar fluids, we anticipate that a good approximation to $\langle(\hat{\mu}^1 - \mu^1)_a \mu_a^1\rangle$ may be obtained in the same way. The result of this procedure is easily obtained from Eqs. (B.12) and (B.13):

$$\langle(\hat{\mu}^1 - \mu^1)_a \mu_a^1\rangle = \frac{\mu^2}{3}\left[\frac{-\nu_a x + 2x^2}{(1-x)(1+2x)}\right], \tag{9.10}$$

where $\nu_x = \nu_y = -1$, $\nu_z = 2$, and $x = \alpha\rho/3\varepsilon_0$.
   The third term in Eq. (9.8) is given by

$$\langle\exp(i\mathbf{k}\cdot\mathbf{r}^{12})(\hat{\mu}^2 - \mu^2)_a \mu_a^1\rangle = \lim_{k \to 0}\left\{\left\langle\sum_{j \neq 2}\exp(i\mathbf{k}\cdot\mathbf{r}^{j1})(\underset{\sim}{\alpha}^j\cdot\underset{\sim}{\mathscr{T}}^{j2}\cdot\mu^2)_a \mu_a^1\right\rangle\right.$$

$$\left. +\left\langle\sum_{j \neq 2}\exp(i\mathbf{k}\cdot\mathbf{r}^{21})(\underset{\sim}{\alpha}^2\cdot\underset{\sim}{\mathscr{T}}^{2j}\cdot\underset{\sim}{\alpha}^j\cdot\mathbf{T}^{j2}\cdot\mu^2)_a \mu_a^1\right\rangle\right\} \tag{9.11}$$

We wish to approximate this term in the same spirit, although it is clear that orientational correlation between the permanent dipoles must now be introduced. The simplest model for the liquid structure in which this might be achieved is one in which the n-body distribution function $g^{(n)}$ is written in the approximate form $\hat{g}^{(n)}$:

$$\hat{g}^{(n)}(12 \cdots n) = \tilde{g}^{(n)}(12 \cdots n)\left(1 + \sum_{i < j}^{n}\frac{h^{DD}(i, j)}{\tilde{g}^{(2)}(ij)}\right) \tag{9.12}$$

where $\tilde{g}^{(n)}$ is the approximate spherical reference system distribution function given in Eq. (8.6), and $h^{DD}(ij)$ is that part of the pair distribution function (of the polar polarizable fluid) which accounts for the correlation between the molecular dipole moments. $h^{DD}(ij)$ contains the terms in g^{110} and g^{112} from the invariant expansion of Eq. (4.4). The function $\hat{g}^{(n)}$ may be viewed as a superposition approximation to $g^{(n)}$:

$$\hat{g}^{(n)}(12\cdots n) \cong \prod_{i<j} \left[\tilde{g}^{(2)}(ij) + h^{DD}(ij) \right] \tag{9.13}$$

in which only the lowest order terms in h^{DD} are kept. If, in addition, the molecules are treated as isotropically polarizable, and those configurations in which molecule 1 appears in the DID chains of Eq. (9.11) are neglected, then $\mathscr{T}^{12}$ may be replaced by its uniform medium value, and we find

$$\langle (\hat{\mu}^2 - \mu^2)_a \mu_a^1 \exp(i\mathbf{k}\cdot\mathbf{r}^{12}) \rangle$$
$$= \left[(1-x)(1+2x) \right]^{-1} \left\{ -\nu_a x + 2x^2 \right\} \langle \mu_a^1 \mu_a^2 \exp(i\mathbf{k}\cdot\mathbf{r}^{12}) \rangle \tag{9.14}$$

If we now combine Eqs. (4.5), (9.10), and (9.14), we obtain

$$\lim_{k\to 0} \langle M_a(\mathbf{k}) M_a^0(-\mathbf{k}) \rangle = N(\tfrac{1}{3})(\mu)^2 \left(g^S + \nu_a g^L \right)$$
$$- N(\tfrac{1}{3}) \left(g^S + \nu_a g^L \right)(\mu)^2 x f(x) \nu_a$$
$$+ N(\tfrac{1}{3})(\mu)^2 \left(g^S + \nu_a g^L \right) 2x^2 f(x) \tag{9.15}$$

where $x = \alpha\rho/3\varepsilon_0$, g^S and g^L are, respectively, the short- and long-range factors defined in Section IV, and

$$f(x) = (1-x)^{-1}(1+2x)^{-1} \tag{9.16}$$

If we denote the value of $\mathscr{M}$ obtained for this model by $\hat{\mathscr{M}}$, it readily follows from Eqs. (9.7), (4.5), and (9.15) that

$$\hat{\mathscr{M}}_a(\mathbf{k}) = \sum_j \bar{\mu}_a^j \exp(i\mathbf{k}\cdot\mathbf{r}^j) \tag{9.17}$$

where $\bar{\mu}_a^j$ is given by

$$\bar{\mu}_a^j = \mu_a^j \left[1 - \nu_a x f(x) + 2x^2 f(x) \right] \tag{9.18}$$

Since from Eq. (9.10) we have

$$\frac{\langle \hat{\mu}_a^j \mu_a^j \rangle}{\mu^2/3} \mu_a^j = \mu_a^j \left(1 + \frac{\langle (\hat{\mu}^j - \mu^j)_a \mu_a^j \rangle}{\mu^2/3} \right)$$
$$\equiv \bar{\mu}_a^j \tag{9.19}$$

the quantity $\bar{\mu}^j$ may be recognized as the *effective permanent dipole* of molecule j, that is, as the mean value of $\hat{\mu}^j$ for a given orientation of the permanent dipole of molecule j averaged over all other degrees of freedom. In our model for the liquid structure, the net effect of the projection specified in Eq. (9.7) is, therefore, a change in the magnitude of the molecular moment from its gas-phase value to an effective mean value comprising the gas-phase moment plus the moments it induces. This net effect is calculated as though all other molecules in the fluid occupied their equilibrium positions. The transverse and longitudinal effective moments, $\bar{\mu}_x^j$ and $\bar{\mu}_z^j$, are unequal because of the dependence of the induced moment of a molecule on the orientation of permanent dipoles at long range. In our treatment, it is consistent to replace $x = \alpha\rho/3\varepsilon_0$ by $[\varepsilon(\infty)-1][\varepsilon(\infty)+2]^{-1}$, the Clausius-Mossotti result. Introducing this substitution into Eq. (9.18), we find

$$\bar{\mu}_a^j = \mu_a^j \frac{[\varepsilon(\infty)+2]/3}{[\varepsilon(\infty)]^{\delta_{az}}} \qquad (9.20)$$

where δ_{az} is one if $a = z$, and zero otherwise.

The quasi-susceptibility $\chi^0(0)$, which is related to $\varepsilon(0)$ by Eqs. (2.13a,b), is given by

$$\chi_{aa}^0(0) - \chi_{aa}^0(\infty) = \lim_{k \to 0} (k_B T N_0 \varepsilon_0)^{-1} \rho \langle |M_a(\mathbf{k})|^2 \rangle \qquad (9.21)$$

Making use of Eq. (9.3) and the orthogonality of $\mathcal{M}_a$ and ΔM_a, we find

$$\langle |M_a(k)|^2 \rangle = \langle |\mathcal{M}_a(\mathbf{k})|^2 \rangle + \langle |\Delta M_a(\mathbf{k})|^2 \rangle \qquad (9.22)$$

The first term of Eq. (9.22) is associated with the intensity of the reorientational part of the spectrum of M_a, whereas the second term, which [from Eqs. (9.3), (9.5), and (9.6)] can be written as

$$\langle |\Delta M_a(\mathbf{k})|^2 \rangle = \left\langle \left| \sum_j (\hat{\mu}_j - \bar{\mu}_j) e^{i\mathbf{k}\cdot\mathbf{r}^j} \right|^2 \right\rangle \qquad (9.23)$$

is therefore the intensity of the "neighbor-fluctuation" or "collision-induced" part of the spectrum of M_a. For our structural model, in which the induced dipole terms are treated in the "uniform medium approximation," such neighbor-fluctuation terms do not occur. Thus, within our model, Eq. (9.21) becomes

$$\chi_{aa}^0(0) - \chi_{aa}^0(\infty) = \lim_{k \to 0} (k_B T \varepsilon_0 N_0)^{-1} \rho \langle |\hat{\mathcal{M}}_a(\mathbf{k})|^2 \rangle \qquad (9.24)$$

Equation (9.24) can be combined with Eqs. (9.20), (9.17), and (9.7) to yield

$$\chi_{zz}^0(0) - \chi_{zz}^0(\infty) = \frac{(\mu)^2}{3k_BT\varepsilon_0}(g^S + 2g^L)\left(\frac{\varepsilon(\infty)+2}{3\varepsilon(\infty)}\right)^2 \qquad (9.25a)$$

and

$$\chi_{xx}^0(0) - \chi_{xx}^0(\infty) = \frac{(\mu)^2}{3k_BT\varepsilon_0}(g^S - g^L)\left(\frac{\varepsilon(\infty)+2}{3}\right)^2 \qquad (9.25b)$$

It is clear that simply taking the trace of χ^0, which gives the Kirkwood functional on the left-hand side of Eq. (9.1), will not remove all the long-range dipolar correlation parameters g^L from the right-hand side of that equation. The difference between this and the rigid-dipole case arises from the interference between the long-range parts of the induced moments and the long-range correlation between permanent dipoles. However, if Eq. (9.25a) is multiplied by $[\varepsilon(\infty)]^2$ and added to twice Eq. (9.25b), one obtains a formula for $\varepsilon(0)$ that is independent of the long-range term g^L; this results in the Onsager-Fröhlich formula in Eq. (9.2).

As in the case of the Clausius-Mossotti formula, we have shown how the Fröhlich formula may be recovered from a very simple representation of the fluid structure. This representation includes only pair dipolar orientational correlations between molecules, and the positional correlations are described through $\tilde{g}^{(n)}$. The model is thus one of permanent dipoles floating in a fluid of polarizable molecules, the fluid itself being described by the "uniform-medium" approximation introduced in Appendix B.

In attempting to assess the validity of this representation, it is useful to distinguish between its effect on the two averages given by Eqs. (9.9) and (9.11). The effect of the approximations on $\langle(\hat{\mu}^1 - \mu^1)_a\mu_a^1\rangle$ is well understood because of its close relationship to Eq. (8.1), which we know leads to the Clausius-Mossotti formula under the same approximations and correctly includes all long-range terms (as shown in Appendix B). It is interesting to note that in a dilute solution of polar molecules in a nonpolar solvent, $\langle(\hat{\mu}^1 - \mu^1)_a\mu_a^1\rangle$ dominates the difference between $\mathcal{M}_a$ and M_a^0; furthermore, in vibrational infrared spectra of centrosymmetric molecules, this is the only term that appears (due to the lack of coherence between the vibrational phases of different molecules). For both these cases, then, we expect the simple model introduced above to give a good representation of the effects of the induced dipole.

The effect of the approximations on the average in Eq. (9.11) is far more difficult to assess. If, within the limitations of the superposition approximation, one were to retain [as indicated in Eq. (9.13)] higher-order terms in h^{DD} than are included in $\hat{g}^{(n)}$, some improvement might be expected. There is, however, a fundamental drawback to this procedure: terms such as $h^{DD}(1j)h^{DD}(j2)$ make long-range contributions in r^{12} to the first term on the right-hand side of Eq. (9.11). This suggests that our neglect of the three-body dipole-dipole correlations results in the discarding of long-range dipolar terms and, as we have seen earlier, unless all long-range dipole-dipole correlations are handled exactly in a theory of the permittivity, the theory will lead to inconsistencies with Maxwell's equations. We conclude that the neglect of higher-order dipolar correlation may be serious, but it is unlikely that dealing with such terms through the superposition approximation is sufficient; yet this is the only approach that is currently open in a theory that is based upon the distribution function. A dielectric theory of the polar polarizable fluid that avoids this limitation has been developed by Wertheim,[57a] and by Høye and Stell[57b].

For the reasons discussed above, we shall not attempt to extend the Fröhlich expression. However, in view of the difficulties encountered in establishing this expression, it is of interest to review other treatments of the polar polarizable fluid. We have used a molecular approach in which a simplified distribution function was assumed. Onsager's calculation[22] was carried out for an isotropically polarizable dipolar molecule inside a spherical cavity in a uniform medium of dielectric constant $\varepsilon(0)$; consequently, short-range orientational correlations between molecules, including those that cause g^S to differ from unity, are neglected in his model. If, in our treatment, we omitted the $g^{110}(ij)$ term from $h^{DD}(ij)$ in Eq. (9.12), we would have $g^S = 1$ and would obtain the Onsager result. Indeed, for a real molecular system, there seems to be no reason to retain g^{110} while neglecting other short-range parts of $g^{(2)}(ij)$. On the other hand, examination of Eqs. (9.25a, b) shows that it is essential to include the $g^{112}(ij)$ term in our model distribution function; otherwise, the values of $\varepsilon(0)$ determined from Eqs. (9.25a) and (9.25b) would be inconsistent.

The Kirkwood[24] approach for treating the g^S correlations in fluids consisting of rigid dipoles is to calculate the electrostatic response of the exterior matter to the moment of an embedded sphere with dipoles in a fixed configuration. As Buckingham[32] pointed out, this procedure cannot be generalized to treat polarizable polar molecules, since the moment of the embedded sphere is determined not only by the configuration of its permanent dipoles but also by its moments induced by the exterior matter. Thus, we cannot select a given fixed configuration and average over the configurations of the exterior molecules. Fröhlich[35] adopted another approach: he calculated the "internal" moment of a polarizable dipole placed in a cavity carved out of a medium of dielectric $\varepsilon(\infty)$. The "internal" moment is the sum of the gas-phase moment and the moment induced by the reaction field to this permanent dipole. Fröhlich found that the internal moment was $\mu^i[\varepsilon(\infty)+2]/3$, and he then followed the Kirkwood procedure,[24] replacing the permanent moment by the "internal" moment to obtain Eq. (9.2). There are clear physical similarities between our molecular model and Fröhlich's continuum calculation.

Since the left-hand side of Eq. (9.1) is the correct dielectric functional for a sphere embedded in a sample of the same material, we find, in the manner of the discussion following Eq. (5.13), that the factor $\text{Tr}\langle|M_\alpha(k)|^2\rangle$ on the right-hand side of Eq. (9.1) can be replaced by $\langle \mathbf{M}^{SP}\cdot\mathbf{M}^{SP}\rangle$, where $\mathbf{M}^{SP}$ is the total moment of an embedded sphere. However, the left-hand side of Eq. (9.2), the Fröhlich equation, is not the same functional as that in Eq. (9.1); neither is the right-hand side equal to $\text{Tr}\langle|M_\alpha(k)|^2\rangle$ or equivalently to $\langle \mathbf{M}^{SP}\cdot\mathbf{M}^{SP}\rangle$. In his criticism of the Fröhlich theory, Wertheim, who has recently constructed an exact theory of the polarization of a medium of polarizable dipoles,[57a] incorrectly identifies $g^S\mu^2\{[\varepsilon(\infty)+2]/3\}^2$, which appears on the right-hand side of Eq. (9.2), as the mean-square moment of a sphere embedded in a sample of the same dielectric material. On the basis of this identification, Wertheim concludes that Eq. (9.2) is inconsistent, since the functional on the left is not correct for such a sphere. By combining Eqs. (9.25a, b), (9.1), and (2.13a, b), we can, in fact, calculate $\text{Tr}\langle|M_\alpha(k)|^2\rangle$, which is equal to $\langle \mathbf{M}^{SP}\cdot\mathbf{M}^{SP}\rangle$, and show that it is not the right-hand side of Eq. (9.2), but

$$\langle \mathbf{M}^{SP}\cdot\mathbf{M}^{SP}\rangle = \left(\frac{\varepsilon(\infty)+2}{3\varepsilon(\infty)}\right)^2 \frac{(\mu)^2}{3k_BT\varepsilon_0}\left\{g^S\left[1+2\varepsilon^2(\infty)\right]-2g^L\left[\varepsilon^2(\infty)-1\right]\right\}$$

$$(9.26)$$

We see that $\langle \mathbf{M}^{SP}\cdot\mathbf{M}^{SP}\rangle$ involves both short-range (g^S) and long-range (g^L) correlations between permanent dipoles. This can be understood by noting that although the $\mathbf{M}^{SP}$'s in this average include only those molecular dipoles that lie within the sphere, the values of the induced parts of these dipoles are determined in part by permanent dipoles throughout the sample; that is,

$$\langle \mathbf{M}^{SP}\cdot\mathbf{M}^{SP}\rangle = \left\langle \sum_{ij}\mathbf{m}^i\cdot\mathbf{m}^j\right\rangle \qquad (9.27)$$

where the sums run over molecules inside the sphere, but $\mathbf{m}^j$ may depend upon the configuration of permanent dipoles outside the sphere. Felderhof[58] has argued that the Fröhlich formula is related to the mean-squared moment of a spherical sample of dipoles embedded in an infinite continuum of dielectric constant $\varepsilon(\infty)$.

We can use our development of the theory to obtain explicit expressions for the long-range interaction. In a manner analogous to that used in Section IV to obtain an explicit relation for g^L for nonpolarizable dipoles, we can make use of Eqs. (9.25a, b), (2.13), and (9.1) to obtain a similar expression for polarizable dipoles:

$$g^L = -\frac{\varepsilon(0)\rho\left\{g^S\mu[\varepsilon(\infty)+2]/3\right\}^2}{[2\varepsilon(0)+\varepsilon(\infty)]^2 k_B T\varepsilon_0} \tag{9.28}$$

We also see that the long-range term in the distribution function has the same form for polarizable molecules as for rigid dipoles [viz. Eq. (4.18)], except that the apparent dipole $\tilde{\mu}^1$ is given by

$$\tilde{\mu}^1 = \mu^1\left(\frac{\varepsilon(\infty)+2}{3}\right)g^S\frac{3\varepsilon(0)}{2\varepsilon(0)+\varepsilon(\infty)} \tag{9.29}$$

rather than by Eq. (4.19).

X. POLARIZABLE POLAR MOLECULES: FREQUENCY-DEPENDENT PERMITTIVITY

The dynamical problem for polarizable polar molecules reduces to that for nonpolarizable polar molecules plus additional high-frequency contributions. To understand this, we first recall that in Eq. (9.3) we set $M_a(t)$ equal to $\mathcal{M}_a(t)+\Delta M_a(t)$, where $\mathcal{M}_a(t)$ is the projection of $M_a(t)$ onto the collective gas-phase dipole moment $M_a^0(0)$. It follows that

$$\int dt\, e^{i\omega t}\langle M_a(t)M_a(0)\rangle = \frac{1}{k_B T V}\int_0^\infty dt\, e^{i\omega t}\{\langle\mathcal{M}_a(t)\mathcal{M}_a(0)\rangle$$
$$+\langle\Delta M_a(t)\Delta M_a(0)\rangle+\langle\mathcal{M}_a(t)\Delta M_a(0)\rangle$$
$$+\langle\Delta M_a(t)\mathcal{M}_a(0)\rangle\} \tag{10.1}$$

Next we recall that $\mathcal{M}_a(t)$, as indicated in Eq. (9.7), is proportional to $M_a^0(t)$. In the uniform medium approximation that we used to obtain both the Clausius-Mossotti and the Fröhlich equations, $\mathcal{M}_a(t)=\hat{\mathcal{M}}_a(t)$ as given in Eqs. (9.17) and (9.20) or, equivalently,

$$\hat{\mathcal{M}}_x(t) = M_x^0(t)\frac{\varepsilon(\infty)+2}{3} \tag{10.2a}$$

$$\hat{\mathcal{M}}_z(t) \cong M_z^0(t)\frac{\varepsilon(\infty)+2}{3\varepsilon(\infty)} \tag{10.2b}$$

The first of the four average terms in Eq. (10.1) can be handled in the same way that the rigid nonpolarizable dipoles were treated in Sections IV and V. For the molecular model, which makes use of the *approximate distribution function* in Eq. (9.12), the model that gives rise to the Fröhlich result in Eq. (9.2), we have seen that

$$\langle \Delta M_a(0) \, \Delta M_a(0) \rangle = 0 \tag{10.3}$$

and

$$\langle \hat{\mathcal{M}}_a(0) \, \Delta M_a(0) \rangle = 0 \tag{10.4}$$

In addition, we expect the relaxation time associated with $\Delta \mathbf{M}(t)$ to be very different from that associated with $\hat{\mathcal{M}}(t)$, which suggests that correlations such as those appearing in the third and fourth terms in Eq. (10.1) are probably very small. Thus, it is likely that, to the extent to which our simple molecular model holds, only the first term in Eq. (10.1) need be retained. From Eqs. (5.1), (3.21)–(3.23), (2.13b), and (10.1), we find that in the low-k limit

$$\frac{\varepsilon(\omega) - \varepsilon(\infty)}{\varepsilon(0) - \varepsilon(\infty)} = 1 + i\omega \Phi_{xx}(\omega) \tag{10.5a}$$

for transverse modes, and

$$\frac{\varepsilon(\omega) - \varepsilon(\infty)}{\varepsilon(0) - \varepsilon(\infty)} \left[\frac{\varepsilon(0)}{\varepsilon(\omega)} \right] = 1 + i\omega \Phi_{zz}(\omega) \tag{10.5b}$$

for longitudinal modes. The function $\Phi_{aa}(\omega)$ can be evaluated by means of Eq. (7.7):

$$1 + i\omega \Phi_{aa}(\omega) = \left[1 - \frac{i\omega}{(\Omega_R^a)^2} \left(-i\omega + \frac{\Omega_T^2}{-i\omega + \Gamma(\omega)} \right) \right]^{-1} \tag{10.6}$$

Here Ω_T^2 is given by Eq. (7.24). In the three-variable theory, $\Gamma(\omega)$ is independent of ω and is given in Eq. (7.25). The quantity $(\Omega_R^a)^2$ can be calculated with the aid of Eqs. (7.9), (2.13a, b), (3.22), (9.25), and (9.2):

$$(\Omega_R^x)^2 = \frac{\omega_s^2}{\varepsilon(0) g^S} \left(\frac{2\varepsilon(0) + \varepsilon(\infty)}{3} \right) \tag{10.7a}$$

$$(\Omega_R^z)^2 = \frac{\varepsilon(0)(\Omega_R^x)^2}{\varepsilon(\infty)} \tag{10.7b}$$

From Eqs. (7.34), (7.24), (7.25), (10.7a), and (7.30), we find that

$$\tau_D = \varepsilon(0) g^S \left(\frac{3}{2\varepsilon(0) + \varepsilon(\infty)} \right) \frac{(\ddot{g})^2}{\ddot{g}} \tau_s \qquad (10.8a)$$

or

$$\tau_D = \frac{\varepsilon(0) g^S}{\omega_s^2} \left(\frac{3}{2\varepsilon(0) + \varepsilon(\infty)} \right) \frac{\Omega_T^2}{\Gamma} \qquad (10.8b)$$

Equation (10.8a) is an extension to polarizable dipoles of the Glarum-Powles relation in Eq. (6.10).

The results in Eqs. (10.5a), (10.6), and (10.7a) can be re-expressed as

$$\frac{\varepsilon(\omega) - \varepsilon(\infty)}{\varepsilon(0) - \varepsilon(\infty)} = \frac{1}{1 - i\omega\tau_D(\omega)} \qquad (10.9)$$

where in analogy to Eq. (5.11) we have

$$\tau_D(\omega) = \mathscr{R}(\omega)[\varepsilon(0) - \varepsilon(\infty)] \qquad (10.10)$$

and in analogy to Eq. (5.8) we have

$$\mathscr{R}(\omega)^{-1} = (k_B T N_0 \varepsilon_0)^{-1} \rho \lim_{k \to 0} \int_0^\infty dt\, e^{i\omega t} \langle \dot{\mathcal{M}}_a(k, t_{\mathcal{M}_a}) \dot{\mathcal{M}}_a(k, 0)^* \rangle \qquad (10.11)$$

The subscript on t in Eq. (10.11) indicates projection along $\mathcal{M}_a(k)$. For longitudinal relaxation, the equation corresponding to Eq. (10.9) is

$$\left(\frac{\varepsilon(\omega) - \varepsilon(\infty)}{\varepsilon(0) - \varepsilon(\infty)} \right) \frac{\varepsilon(0)}{\varepsilon(\omega)} = \frac{1}{1 - i\omega\tau_D(\omega)\varepsilon(\infty)/\varepsilon(0)} \qquad (10.12)$$

which is analogous to Eq. (5.6). For rigid dipoles, as described by Eqs. (5.6) and (5.7), the ratio of the longitudinal to transverse relaxation times is $\varepsilon(0)^{-1}$, whereas for polarizable dipoles, as described by Eqs. (10.9) and (10.12), the ratio is $\varepsilon(\infty)/\varepsilon(0)$.

In dealing with the dynamical problem of a real molecular liquid, to which the Fröhlich limit that we have discussed above is only a first approxima-tion, several difficulties arise. If a "time-scale separation"[54-56] exists be-tween the reorientational relaxation of $\mathcal{M}_a$ and the relaxation of the near neighbor intermolecular fluctuations (cf. ΔM_a), then the spectrum of the correlation function of the total moment will be reasonably well represented by the sum of a reorientational spectrum, which will dominate the total spectrum at low frequency, and a spectrum due to near-neighbor inter-molecular collision-induced fluctuations, which will dominate at high frequency.[55,56] If the molecular fluctuations are included in this way, then the results in this section are no longer correct, because $\varepsilon(0) - \varepsilon(\infty)$ is no longer a measure merely of $\langle |\mathcal{M}_x|^2 \rangle$, but a measure of $[\langle |\mathcal{M}_x|^2 \rangle + \langle |\Delta M_x|^2 \rangle]$. If there is a time-scale separation, then the analysis is greatly simplified, because the effect on the spectrum of the cross-correlations be-tween $\mathcal{M}$ and ΔM is small, because separate expressions for the intensities of the reorientational and collision-induced spectrum may be written down, and because one can readily modify the results above by the introduction of $\varepsilon(\omega_0)$, where ω_0^{-1} is much shorter than the decay time characteristic of $\mathcal{M}(t)$ and much longer than the decay time characteristic of $\Delta M(t)$. In this case, $\varepsilon(\omega_0)$ replaces $\varepsilon(\infty)$ in Eqs. (10.5a,b), Eq. (10.7a) becomes

$$(\Omega_R^x)^2 = \frac{\omega_s^2 \left[2\varepsilon(0)\varepsilon(\omega_0) + \varepsilon^2(\infty) \right]}{g^S 3\varepsilon(0)\varepsilon(\infty)} \tag{10.13}$$

Eq. (10.7b) becomes

$$(\Omega_R^z)^2 = \frac{\omega_s^2}{g^S} \frac{\left[2\varepsilon(0)\varepsilon(\omega_0) + \varepsilon^2(\infty) \right]\varepsilon(\omega_0)}{3\varepsilon^3(\infty)} \tag{10.14}$$

Eq. (10.8a) becomes

$$\tau_D = \frac{g^S 3\varepsilon(0)\varepsilon(\infty)}{\left[2\varepsilon(0)\varepsilon(\omega_0) + \varepsilon^3(\infty) \right]} \left(\frac{\ddot{g}^2}{\ddot{g}} \tau_s \right) \tag{10.15}$$

and the left-hand side of Eq. (9.2) becomes

$$\left[\varepsilon(0) - \varepsilon(\omega_0) \right] \left[2\varepsilon(0)\varepsilon(\omega_0) + \varepsilon(\infty)^2 \right] \left[\varepsilon(0)\varepsilon(\infty) \right]^{-1}$$

However, even if the time-scale separation holds, the presence of a collision-induced "background" arising from $\langle \Delta M_a(t) \Delta M_a(0) \rangle$ limits the information concerning short-time reorientational dynamics that may be obtained from the observed spectrum, because the amplitude of the reorienta-tional spectrum at the high frequencies corresponding to short times may be comparable to that of the collision-induced background, and it is difficult to separate the two effects unambiguously. We have discussed the reorientational spectrum at length, but relatively little is known about the collision-induced spectrum for polar molecules, although for nonpolar molecules it has been well studied. (See, e.g. Ref. 59.) The reorientational intensity may be interpreted in terms of an "effective" permanent dipole moment for the molecules in the material in much the same way as light-scattering intensities have been calculated in terms of effective molecular polarizabilities.[12,56] The existence of a time-scale separation in the analogous light-scattering problem has been studied by computer simulation[55,56]; this work suggests that, while the reorientational and collision-induced processes occur at the same rates in nonviscous fluids of small molecules such as CO or N_2,[56] time-scale separation is feasible for CS_2, where the reorientation is appreciably slower than in collision-induced processes,[55] especially close to the triple point. (See Note Added in Proof)

If the time-scale separation is valid and the collision-induced effects are separable, one must still account for the intensity of the reorientational part of the spectrum. As we discussed in Section IX, it is a formidable problem to go beyond the Fröhlich limit in obtaining these results. Computationally feasible procedures might be developed based upon Wertheim's theory.[57a]

If the gas-phase dipole is not along a molecular symmetry axis, the anisotropic rotational motion will complicate the results. Furthermore, the intermolecular contributions to the dipole cause the axis of the effective dipole to be sensitive to the molecular environment.

XI.   APPLICATIONS OF DYNAMICAL THEORY

In this section, we continue our study of the dynamics of polarizable polar molecules and their effect upon the permittivity. The simple Debye theory with its single relaxation channel is clearly not adequate to describe the full frequency range of dielectric response, and so we investigate the consequences of the more flexible three-variable theory. In Section XI.A, we indicate how both librational and inertial oscillations appear in the three-variable theory, the latter being known as *dipolarons*.[50] In XI.B, we seek to identify librational oscillations in the experimental data, whereas in XI.C we do the same for dipolarons. In XI.D, we critique the three-variable theory.

A.   Librations and Dipolarons

To study the nature of the oscillatory characteristics of the reorientational motion, we note that the "three-variable" theory described earlier can be reformulated in terms of three coupled linear-transport equations:

$$\frac{\partial}{\partial t}\begin{bmatrix} \langle \Lambda_1(t)\Lambda_1^* \rangle \\ \langle \Lambda_2(t)\Lambda_1^* \rangle \\ \langle \Lambda_3(t)\Lambda_1^* \rangle \end{bmatrix} = \begin{bmatrix} 0 & 1 & 0 \\ -\Omega_R^2 & 0 & 1 \\ 0 & -\Omega_T^2 & -\Gamma \end{bmatrix}\begin{bmatrix} \langle \Lambda_1(t)\Lambda_1^* \rangle \\ \langle \Lambda_2(t)\Lambda_1^* \rangle \\ \langle \Lambda_3(t)\Lambda_1^* \rangle \end{bmatrix} \quad (11.1)$$

For guidance we recall from section VII the significance of the parameters which appear in this equation. Consider first the application to single particle motion. Ω_R is then the free rotational frequency (ω_s), the reciprocal of the time required for the orientational correlation function to change sign in the absence of torques. Such "inertial" oscillations occur only if this time is short compared to the time over which the torques have a significant effect on the orientational motion. For the dense fluids of interest to us, the frequencies $\Omega_T = \omega_t$ and $\Gamma = \tau_t^{-1}$ are properties of the intermolecular torques[64,55]. To explain the significance of Ω_T and Γ it is convenient to imagine that a molecule is at a site, and if it were to sit at that site for a sufficiently long time, it would undergo a librational oscillation with mean frequency Ω_T. Γ is the rate at which molecules "jump" between sites, so that if Γ is very large compared to Ω_T, the librational character will be lost. Librational and inertial oscillations have been predicted by means of the three variable theory[47], and they may be distinguished by whether the oscillatory frequency is Ω_T or Ω_R. Similar considerations apply to the collective cases where the primary variables are M_x and M_z. Indeed, as we saw in section VII, Ω_T and Γ differ from the corresponding single particle properties only by the short range parameters $\ddot{g}$ and $\bar{g}$, which we expect to be close to unity, so that the Λ_3 variable is the same in all three cases. Furthermore, since Ω_R^2 is similar to ω_s, (Eq. 7.23a), the inertial effects are similar in the transverse and single particle relaxation functions. However, Ω_R^z is larger than ω_s by a factor of $(\varepsilon(0))^{1/2}$, so the inertial frequency for the longitudinal motion may be much larger than for the transverse or single particle case; it is as if at large permittivity the longitudinal variable had a reduced moment of inertia.

The solutions of Eqs. (11.1) yield

$$\langle \Lambda_1(t)\Lambda_1^* \rangle = \sum_{j=1}^{3} a_j \exp(-\lambda_j t) \quad (11.2)$$

where the frequencies λ_j are the eigenvalues of the transport matrix in Eq. (11.1). Where all three λ_j's are positive numbers, $\langle \Lambda_1(t)\Lambda_1^* \rangle$ is purely dissipative, but under some conditions, two of the λ_j's are complex and correspond to oscillatory behavior. The condition under which the eigenvalues for the cubic secular equation corresponding to Eq. (11.1) become complex is

$$\left[1 - 3\left(\Omega_R^2 + \Omega_T^2\right)\Gamma^{-2}\right]^3 \leq \left[1 - \tfrac{9}{2}\left(\Omega_T^2 - 2\Omega_R^2\right)\Gamma^{-2}\right]^2 \qquad (11.3)$$

so that this is the condition for the existence of oscillatory behavior. If

$$\frac{\Omega_R^2}{\Gamma^2} \gtrsim 0.035 \qquad (11.4)$$

then the condition in Eq. (11.3) is automatically satisfied. For $\Omega_R^2/\Gamma^2 \lesssim$ 0.035, there is a range of values of Ω_T^2/Γ^2 for which condition (11.3) is not satisfied. (See Table II.) For these small values of Ω_R^2/Γ^2, below 0.035, the oscillations are called *librational* if Ω_T^2/Γ^2 is greater than the minimum values indicated in Table II, and they are called *inertial* if Ω_T^2/Γ^2 is smaller than the maximum values indicated in Table II. Of particular interest is the situation where

$$\Omega_R^2 \ll \Gamma^2 \qquad (11.5)$$

for which we find that librations occur if

$$\frac{\Omega_T^2}{\Gamma^2} > 0.25 \qquad (11.6)$$

and inertial oscillations occur if, approximately,

$$\frac{\Omega_T^4}{\Gamma^4} \lesssim 4\frac{\Omega_R^2}{\Gamma^2}. \qquad (11.7)$$

TABLE II

Ω_R^2/Γ^2	Inertial Oscillations $(\Omega_T^2/\Gamma^2)_{\max}$	$(\Omega_T^2/\Gamma^2)^2$	Librational Oscillations $(\Omega_T^2/\Gamma^2)_{\min}$
0.0001	0.020	0.0001	0.250
0.01	0.180	0.008	0.250
0.025	0.260	0.017	0.279
0.030	0.277	0.019	0.285
0.035	0.292	0.021	0.293

For single-particle correlations, Eq. (11.6), the condition for librational oscillations, can be converted with the aid of Eqs. (7.20)–(7.22) and (7.30) to

$$\omega_s^2 \tau_s^2 \geq \frac{\langle |\eta_x^1|^2 \rangle}{4\omega_s^4} = \frac{\omega_t^2}{4\omega_s^2} \tag{11.6a}$$

where $\langle |\eta_x^1|^2 \rangle \gg \omega_s^4$. With the aid of the same relations, the condition for inertial oscillations becomes

$$\omega_s^2 \tau_s^2 \leq 4 \tag{11.7a}$$

Equation (11.6a) indicates that for very slow reorientations, that is, in highly viscous systems where molecules are subject to strong torques, the molecules are trapped for a long period of time in a torsional well, and they librate coherently with an oscillatory frequency that is quite close to ω_t. For single molecules Eq. (11.7a) cannot be readily satisfied, since even for nearly free rotations we expect $\omega_s^2 \tau_s^2 > 1$; therefore, we do not expect "inertial oscillations" in the single-particle correlation functions for liquids.

For the collective motions encountered in dielectric relaxation, the results are slightly different from those for individual particles. For both the longitudinal ($\Lambda_1 = M_z$) and transverse ($\Lambda_1 = M_x$) motions, the condition in Eq. (11.6) can be rewritten, with the aid of Eqs. (7.24), (7.25), and (7.30), as

$$\omega_s^2 \tau_s^2 \geq \frac{(\ddot{g})^2}{\ddddot{g}} \left(\frac{\langle |\eta^1|^2 \rangle}{4\omega_s^4} \right) = \frac{(\ddot{g})^2}{\ddddot{g}} \left(\frac{\omega_t^2}{4\omega_s^2} \right) \tag{11.6b}$$

This is virtually the same condition as for the single-particle correlations [Eq. (11.6a)], since $\ddot{g}^2 / \ddddot{g}$ is probably of order unity. Thus, librational oscillations appear under about the same conditions for both single and collective correlation functions. These librations are discussed more fully in Section XII. For inertial oscillations, however, the situation is different. For transverse collective motions, we make use of Eq. (7.23a), in addition to Eqs. (7.24) and (7.25), to rewrite the condition for inertial oscillations in Eq. (11.7) as

$$\omega_s^2 \tau_s^2 \leq \frac{4[2\varepsilon(0) + \varepsilon(\infty)]}{3g^S \varepsilon(0)} \left(\frac{(\ddot{g})^2}{(\ddddot{g})^4} \right) \tag{11.7b}$$

which is even more difficult to satisfy than the corresponding single-particle condition in Eq. (11.7a). Therefore, *inertial oscillations are not expected in the transverse modes*. In order to study the longitudinal modes, we insert the value of $(\Omega_R^z)^2$, as given in Eqs. (10.7a,b), into the condition in Eq. (11.7). Then,

$$\omega_s^2 \tau_s^2 \leq \frac{4[2\varepsilon(0)+\varepsilon(\infty)]}{3g^S \varepsilon(\infty)} \left(\frac{(\ddot{g})^2}{(\ddot{g})^4} \right) \qquad (11.7c)$$

or, if combined with Eq. (7.31),

$$\varepsilon(0)\varepsilon(\infty) \geq \frac{\omega_s^2 \tau_s \tau_D}{4} \left(\frac{(\ddot{g})^2}{\ddot{g}} \right) \qquad (11.7c')$$

Equation (11.7c') is useful because τ_s and τ_D should be relatively independent of $\varepsilon(0)$, $\ddot{g}$ and $\ddot{g}$ should be approximately 1, and g^S does not appear explicitly. Thus, at large $\varepsilon(0)$, we expect Eq. (11.7c') to be satisfied; *Eq. (11.7c') is the condition for the existence of longitudinal "inertial" oscillations or dipolarons.*

Under the conditions of Eq. (11.7c'), the complex solutions $\lambda_{\pm}$ for the eigenvalues of the transport matrix in Eq. (11.1) can be approximated by

$$\lambda_{\pm} = \frac{\omega_s^2 \tau_s}{2} \pm i\omega_d \qquad (11.8)$$

where ω_d is the *dipolaron frequency*

$$\omega_d = \Omega_R^z = \frac{\omega_s \left(2\varepsilon(0)+\varepsilon(\infty)/3g^S\right)^{1/2}}{\varepsilon(\infty)} \qquad (11.9)$$

This frequency differs from that given by Lobo et al.[50] (their Eq. 1.1); if we set $\varepsilon(\infty) = 1$ and eliminate μ^2 by means of the Kirkwood relation in Eq. (4.1), we find that their dipolaron frequency is $3\{3[\varepsilon(0)-1]g^S/2\varepsilon(0)\}^{1/2}$ times our result in Eq. (11.9). However, our result agrees with the $k \rightarrow 0$ limit formula given by Pollock and Alder.[60]

A physical picture of dipolarons can be readily devised. As explained in section IV, at high $\varepsilon(0)$ the molecular dipole projections along **k** almost cancel so that M_z is very small, whereas the perpendicular dipolar projections reinforce so that $\langle |M_x|^2 \gg N\mu^2/3$; as a consequence, as explained in section V, a very small degree of molecular reorientation can cause the sign of M_z to change but not that of M_x nor of the molecular dipole. Thus M_z reorientations occur in very short times compared to molecular and M_x reorientations. At very short times the molecular rotations are inertial, that is, $d\langle \mu_a^j(t)\mu_a^j(0)\rangle/dt$ is proportional to $(2k_B T/I)$, and it is the behavior of this molecular rotation at short times and through small angles that determines the flipping or oscillation of M_z. Thus the oscillations of M_z are dependent upon I and are inertial.

In our discussion of dipolarons, we have assumed that Eq. (11.5) is satisfied, that is, the torques relax in less than a free rotational period. Even though this condition is usually satisfied for molecular and M_x rotations, it may well be violated for M_z oscillations. If Eq. (11.5) is indeed violated for M_z, then, as discussed, M_z oscillates under all conditions, but the smaller the value of $\omega_s^2 \tau_s \tau_t$, the more inertial the motion, and the larger the value of $\omega_s^2 \tau_s \tau_t$, the more librational the motion, with a reasonable dividing point of $\omega_s^2 \tau_s \tau_t \simeq 0.29$, as indicated in Table II.

B. Experimental Data and Model Calculations

In this section, we carry out numerical calculations and examine experimental data on $CHCl_3$, which we take as representative of a typical organic liquid of low polarity. We compare the data with calculations made with the three-variable theory, the theory in which $\Gamma(\omega)$ is independent of ω. Although in Section XI.C we modify them somewhat, we start out with parameters that correspond to the experimentally determined quantities for chloroform ($CHCl_3$) at 295 K[43]:

$$\varepsilon(0) = 4.71 \tag{11.10a}$$

$$\varepsilon(\infty) = 2.13 \tag{11.10b}$$

$$I = 25.4 \times 10^{-39} \text{ g} \cdot \text{cm}^2 \tag{11.10c}$$

$$\tau_D = 6.36 \text{ ps} \tag{11.10d}$$

$$g^S = 1 \tag{11.10e}$$

In addition, we set

$$\ddot{g} = \bar{g} = 1 \tag{11.10f}$$

The "rotational frequencies" of interest are ω_s, Ω_R^x, and Ω_R^z, which are given by Eqs. (7.8) and (10.7a, b), at $T = 300$ K:

$$\omega_s^2 = 3.24 \text{ (ps)}^{-2} \tag{11.11a}$$

$$(\Omega_R^x)^2 = 2.64 \text{ (ps)}^{-2} \tag{11.11b}$$

$$(\Omega_R^z)^2 = 5.85 \text{ (ps)}^{-2} \tag{11.11c}$$

With the aid of Eq. (10.8), we can then calculate τ_s,

$$\tau_s = 5.20 \text{ ps} \tag{11.12}$$

and we can readily determine that

$$\omega_s^2 \tau_s^2 = 87 \tag{11.13a}$$

$$\left(\Omega_R^x\right)^2 \tau_D^2 \approx 100 \tag{11.13b}$$

We apply the three-variable theory within the confines of the Fröhlich model as described in Section X. Thus, $\varepsilon(\omega)$ is given by Eq. (10.5a), together with Eqs. (10.6), (10.7a), and (10.8b). We have still to evaluate the parameters Ω_T and Γ, but from Eq. (10.8b) we see that only one of these can be adjusted independently; we take the torquelike frequency Ω_T as the independent variable. Selected values of Ω_T^2 and the corresponding values of Γ, obtained by means of Eq. (10.8b), are given in Table III.

We have made use of the three-variable theory, together with the values of Ω_T shown in Table III, to calculate the quantities plotted in Figs. 1–4. From Table III, we see that for all the choices of Ω_T the inequalities corresponding to the librational regime are satisfied; hence, none of the parameters chosen yield dipolarons.

In Fig. 1, we have plotted $\varepsilon''(\omega)$, the imaginary part of $\varepsilon(\omega)$, versus the real part, $\varepsilon'(\omega)$. These Cole-Cole plots are given for curves A, B, and C; a curve representing the experimental data[43] for $CHCl_3$ up to 140 cm^{-1} is also included. At frequencies higher than 1 GHz, the measured quantities from which $\varepsilon(\omega)$ is determined are the absorption coefficient $\alpha(\omega)$ and the refractive index $n(\omega)$, which are related to $\varepsilon'(\omega)$ and $\varepsilon''(\omega)$ by the relations

$$n = \left(\frac{1}{2}\varepsilon'\right)^{1/2}\left\{\left[1+\left(\frac{\varepsilon''}{\varepsilon'}\right)^2\right]^{1/2}+1\right\}^{1/2} \tag{11.14}$$

$$\alpha = (2\omega/c)\left(\frac{1}{2}\varepsilon'\right)^{1/2}\left\{\left[1+\left(\frac{\varepsilon''}{\varepsilon'}\right)^2\right]^{1/2}-1\right\}^{1/2} \tag{11.15}$$

TABLE III[a]

Curve	Ω_T^2 (ps)$^{-2}$	Γ (ps)$^{-2}$	$\Gamma\tau_D$	$\dfrac{\left(\Omega_R^x\right)^2}{\Gamma^2}$	$\dfrac{\left(\Omega_R^z\right)^2}{\Gamma^2}$	$\dfrac{\Omega_T^2}{\Gamma^2}$
A	200	12	76	0.018	0.04	1.39
B	50	3	19	0.29	0.64	5.56
C	20	1.2	7.6	1.5	4.0	13.9

[a]See Figs. 1–4.

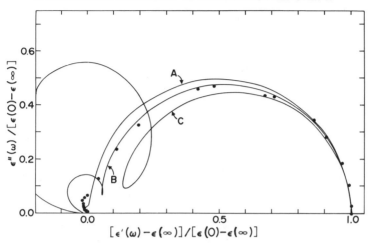

Fig. 1.   Cole-Cole plots (continuous curves) calculated from the three-variable theory, with the parameters given in Table III, are compared with experimental data[43] on $CHCl_3$ at 295 K (circles).

The experimental and calculated curves for $\alpha(\omega)$ and $n(\omega)$ are given in Figs. 2 and 3.

The Cole-Cole plot of case A in Fig. 1 is very similar to a Debye semi-circle except for a small "high-frequency knob" that can be associated with librations. As depicted in Fig. 1, the experimental data for $CHCl_3$ gives rise to a Cole-Cole plot much like that of case B, but the agreement is not quantitative at high frequencies. For case C, the Cole-Cole plot displays very striking high-frequency librational behavior. Although at low and moderate frequency a choice of $\Omega_T^2 \approx 50$ (ps)$^{-2}$, as in case B, yields quite good agreement with $CHCl_3$ data, at high frequency the three-variable theory predicts much greater librational character than is observed.

The far-infrared absorption coefficient $\alpha(\omega)$ predicted by the Debye theory has a plateau (the "Debye plateau"), which extends out to infinite ω:

$$\alpha(\omega)_{\text{plateau}} \rightarrow \frac{\varepsilon(0) - \varepsilon(\infty)}{\varepsilon^{1/2}(\infty)\tau_D} \qquad (11.16)$$

This plateau is indicated by the dotted line in Fig. 2. In the three-variable theory, at high frequency $\alpha(\omega)$ decays, in agreement with experiment; this decay is due to the inclusion of inertial effects. In both the three-variable calculation and the experimental data, the absorption rises above the plateau

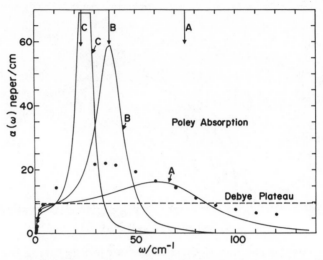

Fig. 2. Far-infrared absorption spectra generated as described in the caption to Fig. 1. The circles are the experimental data[43] on the $CHCl_3$ at 295 K. The horizontal dotted line shows the Debye plateau appropriate to the calculated curves. The vertical arrows show the frequencies Ω_T, which may be seen to correspond to the frequency of the librational feature for each curve.

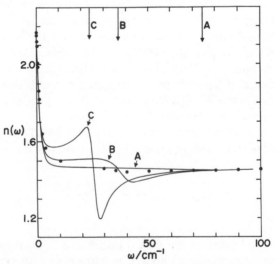

Fig. 3. Refractive index curves corresponding to the data given in Figs. 1 and 2. The vertical arrows show the frequencies Ω_T, and the circles indicate the experimental data[43] on $CHCl_3$ at 295 K.

at intermediate frequencies; this phenomenon is known as the *Poley absorption*. The maximum appears at a frequency close to Ω_T (the vertical arrows in the figure) and is due to librational oscillations. The refractive index $n(\omega)$ is plotted in Fig. 3; the Debye theory yields a monotonically decaying function indistinguishable from curve A on the scale of the figure, and librational absorption gives rise to the dispersion peaks observed for curves B and C.

It is clear from Figs. 1–3 that for a fluid such as $CHCl_3$, the three-variable theory can qualitatively reproduce the differences between the experimental data and the predictions of the simple Debye equation. Since we have fixed τ_D by the low-frequency experimental behavior of $\varepsilon(\omega)$, the three-variable theory should be examined at intermediate and high frequencies. The free parameter Ω_T determines the librational frequency, and we can see from the positions of the librational peaks in Figs. 2 and 3 that the choice of Ω_T used in curve B gives a predicted peak position in closest agreement with ex-

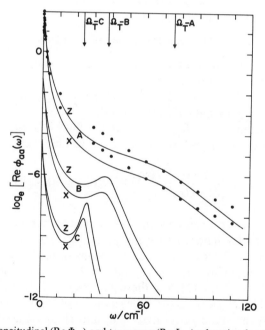

Fig. 4. The longitudinal (Re Φ_{zz}) and transverse (Re Φ_{xx}) relaxation functions corresponding to the calculated curves of Figs. 1–3. The experimental points for $CHCl_3$ at 295 K (circles) are obtained from the experimental permittivity data[43] by means of Eq. (9.5). For clarity, curves B and C are vertically offset by -3 and -6, respectively. The vertical arrows show the frequencies Ω_T.

periment. Curve B also gives a satisfactory representation of the experimental Cole-Cole plot at intermediate frequencies (see Fig. 1). However, in all three figures it is seen that the librational feature predicted with the curve B parameters is too narrow, so that the theory becomes unreliable at frequencies around Ω_T. This reflects a basic limitation of the three-variable theory, which is discussed in Section XI.D; this limitation is a consequence of the simple exponential form for the torque relaxation that is implicit in the theory.

For completeness, we show in Fig. 4 the real parts of the relaxation functions $\Phi_{xx}(\omega)$ and $\Phi_{zz}(\omega)$ corresponding to the parameters of curves A, B, and C and to the experimental data of $CHCl_3$. The relaxation functions are obtained from the permittivity through Eqs. (10.5a, b). This representation of the reorientational spectrum is the first-rank analogue to the spectrum that is observed directly in a light-scattering experiment. As we have noted above, all of curves A, B, and C are within the librational regime, so that none of the qualitative differences between $\Phi_{xx}(\omega)$ and $\Phi_{zz}(\omega)$ that characterize dipolaron behavior are expected or observed. Again, the underdamped nature of the librations predicted by the three-variable theory, relative to the behavior of the experimental data, is apparent.

C.   On the Observation of Dipolarons

As we have shown in Section XI.A, the dipolaron oscillation occurs in the longitudinal relaxation function $\Phi_{zz}(\omega)$ and not in the transverse $\Phi_{xx}(\omega)$ or molecular function $\phi(\omega)$. It may be possible to observe dipolarons directly by measuring the longitudinal susceptibility χ_{zz}^0 in an oblique incidence reflectivity measurement.[61] Alternatively, if the fluid is such that it supports dipolarons, a dipolaron peak will be detected in the function

$$\frac{[\varepsilon(\omega) - \varepsilon(0)]\,\varepsilon(\infty)}{[\varepsilon(0) - \varepsilon(\infty)]\,\varepsilon(\omega)} = i\omega \lim_{k \to 0} \Phi_{zz}(\mathbf{k}, \omega) \qquad (11.17)$$

which may be formed from coventionally obtained permittivities. To demonstrate what kind of fluid will have this property, we present Cole-Cole plots and absorption coefficients calculated by means of the three-variable theory with parameters ranging from those in Table III to those subject to the inequalities (11.7a, b, c) and (11.7c′); these parameters are listed in Table IV.

Equation (11.7c′) shows that the observation of dipolarons is favored by high values of the static permittivity $\varepsilon(0)$ and short reduced reorientation times $\omega_s \tau_s$. The parameters of curve A of Table III form a useful reference point, as they give rise to near-Debye-like dielectric behavior. The curve A parameters do not satisfy the inequality in Eq. (11.7c′) and, as shown in Fig. 5, the corresponding longitudinal and transverse relaxation functions have

TABLE IV[a,b]

Curve	$\varepsilon(0)$	τ_D (ps)	τ_s (ps)	Ω_T^2 (ps)$^{-2}$	Γ (ps)$^{-1}$	$\Gamma\tau_D$	$\dfrac{(\Omega_R^x)^2}{\Gamma^2}$	$\dfrac{(\Omega_R^z)^2}{\Gamma^2}$
A	4.71	6.36	5.20	200	12	76	0.018	0.04
D	24.71	6.36	13.27	200	14	89	0.011	0.13
E	24.71	2.0	4.17	200	44	89	0.0012	0.013

[a]See Figs. 5–7.
[b]$\varepsilon(\infty) = 2.13$; $I = 25.4 \times 10^{-39}$ g·cm^2; $g^S = 1$.

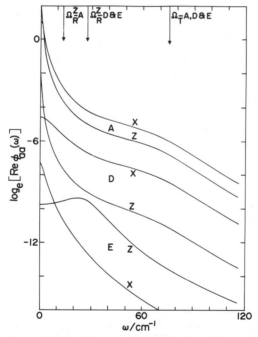

Fig. 5. The longitudinal and transverse relaxation functions calculated for the three-variable theory with the parameters given in Table IV. As discussed in the text, the parameters of curve E are within the dipolaron regime. The vertical arrows show the frequencies Ω_T and Ω_R^z, which are the predicted librational and dipolaron frequencies in the three-variable theory. For clarity, curves D and E are offset by -3 and -6, respectively.

similar shapes with a librational peak at about the Ω_T value of 75 cm^{-1}. If we arbitrarily increase $\varepsilon(0)$ by 20 but keep all other parameters the same as in curve A, then, as shown by curve D in Fig. 5, the difference between $\Phi_{xx}(\omega)$ and $\Phi_{zz}(\omega)$ increases markedly. However, even the curve D parameters (given in Table IV) do not satisfy the inequality (11.7c') and therefore do not exhibit dipolarons; the only shifted-absorption peak is the librational one at approximately 75 cm^{-1}. Curve D indicates that the librational character of $\Phi_{zz}(\omega)$ is enhanced by an increase in $\varepsilon(0)$; similar effects have been found in a computer simulation of water.[62]

For dipolarons to exist in fluids with molecular reorientations similar to those of $CHCl_3$—that is, for the inequality (10.7c') to be satisfied with the reorientation times used to generate curves A and D—an unphysically large value for the permittivity $\varepsilon(0)$ would be required. The inequality can, however, be satisfied with physically realizable values of $\varepsilon(0)$ if the reorientation is more inertial, for example, if τ_D is reduced to 2.0 ps as in the curve E of Fig. 5 and Table IV. For curve E, the values of the permittivity and of the product $\tau_D\omega_S$ are similar to those of CH_3CN at room temperature. The longitudinal relaxation function $\Phi_{zz}(\omega)$ now shows a clear dipolaron peak that is not present in $\Phi_{xx}(\omega)$ and that occurs at the predicted dipolaron frequency Ω_R^z (see the vertical arrow in Fig. 5).

As illustrated in Figs. 6 and 7, the absorption coefficient and the permittivity for case E exhibit the characteristic dipolaron signature. In con-

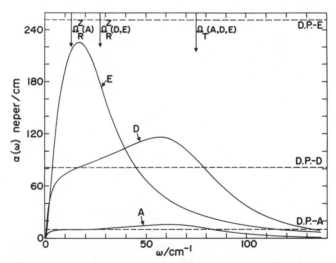

Fig. 6. The far-infrared absorption spectrum calculated from the three-variable theory with the parameters of Table IV. Curve E exhibits dipolaron behavior. The horizontal (dashed) lines show the location of the Debye plateau for the different curves.

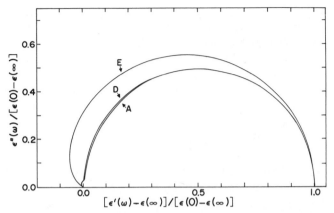

Fig. 7. The Cole-Cole plots calculated from the three-variable theory with the parameters of Table IV. Note that curve E, which exhibits dipolaron behavior, lies outside the Debye semicircle at all frequencies.

TABLE V
Molecular Dynamics Data on Point Dipoles[60][a]

μ	$(1)^b$ $\varepsilon(0)$	$(2)^c$ g^S	$(3)^d$ $\dfrac{\tau_D}{\tau_s \dot{g}}$	$(4)^d$ $\dfrac{4[2\varepsilon(0)+1]}{3g^S}$	$(5)^e$ $\omega_s^2 \tau_s^2$	$(6)^f$ $\dfrac{\omega_d^M}{\omega_s}$	$(7)^g$ $\dfrac{\omega_d}{\omega_s}$
1.8	1.8	1.1	1.3	5.7	1.0	1.5	1.2
3.5	5.8	1.4	1.9	12.0	1.0	2.0	1.7
5.3	21	2.4	3.5	23.4	1.7	2.8	2.4
6.2	38	3.3	4.9	31.2	2.6	3.3	2.8
6.7	59	4.4	6.5			3.6	3.0

$^a\mu^* = \mu(\varepsilon_{LJ}/\sigma^3)^{1/2}$; $\omega_s = (2k_BT/I)^{1/2} = 10.4(\varepsilon_{LJ}/m\sigma^2)^{1/2}$; $I = 0.025m\sigma^2$; $k_BT = 1.35\varepsilon_{LJ}$; $\rho = 0.8$, where ε_{LJ} and σ are Lennard-Jones potential parameters and m is the molecular mass.

^{b}Taken from Ref. 60.

^{c}Evaluated by means of Eq. (4.1) and data in column 1.

^{d}Evaluated by means of Eq. (6.10) and data in columns 1 and 2.

^{e}Estimate from graphs of Pollock and Alder.[60] Poor (and probably low) estimates show that they depend upon short-time decays, while τ_s depends in large part upon long-time behavior.

$^f\omega_d^M/[\varepsilon_{LJ}/m\sigma^2]^{1/2}$ are the molecular dynamics results of Ref. 60.

^{g}Calculated from Eq. (11.9) and data in columns 1 and 2.

trast to the behavior of the absorption coefficient for fluids in the librational regime (as represented by curves $A-D$), the absorption coefficient predicted by the curve E parameters does not rise above the Debye plateau; there is no Poley absorption. As shown in Fig. 6 the absorption maximum for curve E is close to the dipolaron frequency (indicated by the vertical arrow) and, as seen in Fig. 7, the Cole-Cole plot for curve E lies *outside* the Debye semicircle. The behavior of liquids described by parameters similar to those of curve E is characteristic of inertial reorientational motion. Such behavior is observed experimentally for globular polar molecules such as $(CH_3)_3CCl$ (see Fig. 2, page 13 of Ref. 46).

The longitudinal relaxation function may be readily studied for a model fluid in a computer simulation. Pollock and Alder[60] have carried out molecular dynamics calculations on point dipolar particles; they found longitudinal collective dipolarons but damped single-particle and transverse collective modes. Their data are presented in Table V. Since $\bar{g}$ and $\ddot{g}$ are assumed to be close to unity, comparison of columns 4 and 5 in the table indicates that the dipolaron condition in Eq. (11.7c') is probably satisfied, particularly at high $\varepsilon(0)$; however, the determination of τ_s is not very accurate. Comparison of columns 6 and 7 indicates that the dipolaron frequency predicted by Eq. (11.9) is reasonably accurate; Pollock and Alder have attributed the discrepancies to k-dependent contributions, since their calculations were made at quite high values of k.

D. Are Three Variables Enough?

In the preceding sections, we have developed a description of the frequency dependence of the permittivity, a description based upon the Mori generalized Langevin equation. We have carried out simultaneously an analogous calculation for the reorientation of the individual dipoles and have compared the two analyses to obtain a molecular interpretation of the parameters that appear in the theory of dielectric relaxation. The merit of this approach is that the correspondence between the collective and single-particle properties can be drawn with careful regard to the long-range dipole-dipole correlations. In the past, this correspondence has often been made semi-intuitively; for example, the correlation function of the total moment of a sphere embedded in an infinite medium has been associated with the single molecular dipole correlation function, a procedure that leads in the rotational diffusion limit to a single Lorentzian form rather than to a two-Lorentzian form for the frequency-dependent Kirkwood equation [cf. Eq. (5.12)]. To obtain manageable expressions, however, we have had to limit our description of dielectric relaxation to the three-variable theory of reorientation. It is therefore appropriate to consider here the limitations of this theory and to indicate where further generalizations are desirable.

The three-variable prescription clearly cannot describe the frequency dependence of the permittivity under those circumstances where a number of "primary variables"[54] contribute to the permittivity. Examples of this kind are fluids with more than one polar component, fluids composed of flexible molecules, and liquid crystals where the "primary variables" include, respectively, the dipole density of each component, the dipole density of each conformer, and the cartesian components of the dipole density along and perpendicular to the director[63]. Here, we are not concerned with these complex fluids; our concern is that, even for simple one-component fluids of rigid molecules, the three-variable theory of molecular reorientation may not be adequate. This section is devoted to a clarification of this point.

Numerous other theories of single-molecule reorientation have been developed, many of them described in reviews by Steele.[64] In contrast to our parallel use of Mori theory for both molecular and collective relaxation, for most of the other theories it is not easy to relate the molecular relaxation to the dielectric relaxation, since there is usually no analogous way to carry out the calculations for both the collective and molecular properties.

For a one-component system, the third-order expansion in Eq. (7.7) is exact, provided that $\Gamma(\omega)$ is frequency-dependent. However, to discuss the general case where $\Gamma(\omega)$ is indeed frequency-dependent, it is often more useful to start with $\Phi(\omega)$ in a somewhat different form, the form obtained by a second-order memory function expansion. This gives

$$\Phi(\omega) = \left[-i\omega + \hat{K}_1(\omega) \right]^{-1} \tag{11.18}$$

$$\hat{K}_1 = \frac{\Omega_R^2}{-i\omega + \hat{K}_2(\omega)} \tag{11.19}$$

where $\hat{K}_1$ and $\hat{K}_2$ are the Laplace transforms of the first and second memory functions, $K_1(t)$ and $K_2(t)$. The second-order expansion yields

$$\hat{K}_2(\omega) = \int_0^\infty dt\, e^{i\omega t} \frac{\langle \{\exp[i(1-\mathsf{P}_2)\mathscr{L}t]\ddot{\Lambda}_1\}\{(1-\mathsf{P}_2)\ddot{\Lambda}_1\}\rangle}{\langle |\ddot{\Lambda}_1|^2 \rangle} \tag{11.20}$$

where P_2 is the projection operator that projects into Λ_1 and $\dot{\Lambda}_1$. The quantity $\hat{K}_2(\omega)$ is closely connected with the autocorrelation function of the torques $\ddot{\Lambda}_1$, that is, with the relaxation of the torques. An extremely useful form for the permittivity of the dipoles can be obtained from Eqs. (11.18)–(11.20), (5.1), and (2.13b):

$$\frac{\varepsilon(\omega)-1}{\varepsilon(0)-1} = \left(1 - \frac{i\omega \hat{K}_2(\omega)}{(\Omega_R^x)^2} - \frac{\omega^2}{(\Omega_R^x)^2} \right)^{-1} \tag{11.21}$$

[The a independence of $\hat{K}_2(\omega)$ is established through Eq. (7.13).] Since Ω_R^x is an easily obtained equilibrium property [see Eq. (7.23a)], *this expression therefore emphasizes the fact that the interesting dynamical behavior of $\varepsilon(\omega)$ is directly dependent upon the relaxation of the torques as contained in $K_2(\omega)$.* This expression should serve as a convenient starting point for theories of $\varepsilon(\omega)$.

In the third-order expansion of Eq. (7.7),

$$\hat{K}_2(\omega) = \frac{\Omega_T^2}{-i\omega + \Gamma(\omega)} \tag{11.22}$$

Since this development is exact, any inadequacy of the three-variable theory must be due to a breakdown of the assumption that $\Gamma(\omega)$ is frequency-independent, that is, that $\hat{K}_2(\omega)$ is a simple Lorentzian or, equivalently, that the autocorrelation function for the effective torques is exponential in time. This simple three-variable result is adequate at frequencies $\omega \ll \Gamma$, since then $\hat{K}_2(\omega) \approx \hat{K}_2(0)$. However, circumstances arise where the exponential memory function does not appear to be sufficiently realistic.

Recently, McConnell[65] has thoroughly studied the inertial corrections to rotational diffusion. In his Langevin model, a stochastic prescription is used for the torques, but inertial terms are treated fully. In discussing permittivity, he assumes that the relaxation function of the dipole density (Φ_{xx}) may be replaced by the single-molecule correlation ϕ. Under this approximation, Eq. (11.21) becomes

$$\frac{\varepsilon(\omega)-1}{\varepsilon(0)-1} = \left(1 - i\omega \frac{\hat{K}_2^s(\omega)}{\omega_s^2} - \frac{\omega^2}{\omega_s^2}\right)^{-1} \tag{11.23}$$

where $\hat{K}_2^s$ is the second memory function for the single-molecule motion. McConnell shows that in the case of high rotational friction $(\tau_s^{-2}/\omega_s^2) \ll 1$), the Langevin model gives a formula of the Rocard type for the Debye functional,

$$\frac{\varepsilon(\omega)-1}{\varepsilon(0)-1} = \left[(1 - i\omega\tau_s)\left(1 - \frac{i\omega}{\omega_s^2\tau_s}\right)\right]^{-1} \tag{11.24}$$

for a symmetric-top molecule. Equations (11.23) and (11.24) are identical if $\hat{K}_2^s(\omega) = \hat{K}_2^s(0)$ and

$$\hat{K}_2^s(0) = \tau_s\omega_s^2 \tag{11.25}$$

But this is exactly the value of $K^s(0)$ in the three-variable theory, as can be seen by combining Eqs. (7.21), (7.22), (7.30), and (11.22). The Langevin theory may therefore be seen to neglect all the frequency dependence of the second memory function, that is, to treat the torque relaxation as instantaneous. In the high-friction limit, then, we see that the description of inertial effects given by the Mori theory (when the single-particle simplification is made) is the same as that given by the Langevin theory. Since Eq. (11.21) correctly includes many-particle correlations, whereas Eq. (11.23) does not, we are led to suggest that an improved Rocard formula might be [cf. Eq. (7.23a)]

$$\frac{\varepsilon(\omega)-1}{\varepsilon(0)-1} = \left[(1 - i\omega\tau_D)\left(1 - \frac{i\omega 3\varepsilon(0)g^S}{\tau_D\omega_s^2[2\varepsilon(0)+\varepsilon(\infty)]}\right)\right]^{-1} \quad (11.26)$$

We will now look briefly at two circumstances of direct relevance to dielectric studies in which the simple forms for $\hat{K}_2(t)$ in the three-variable and Langevin theories prove inadequate. The first is the phenomenon of dielectric friction, and the second concerns the behavior of orientational correlation functions obtained at very high frequency in far-infrared spectroscopy.

Dielectric friction arises because in polar fluids the intermolecular torques have dipole-dipole contributions that are expected to relax slowly, on the same time scale as the permittivity itself. Consequently, even at the lowest frequencies at which the orientational correlation function is studied, $\hat{K}_2(\omega)$ may be appreciably frequency-dependent. Numerous authors[66,67] have calculated the dielectric friction from a model of a dipole rotating in a cavity in a continuum characterized by a frequency-dependent permittivity. Since the polarization of the medium does not respond instantaneously to changes in the dipole orientation, the reaction field $\mathscr{E}_R$ lags behind the dipole direction, so that there is a torque $\mu \times \mathscr{E}_R$ exerted on the dipole whose size increases with the rate of dipole rotation. This torque is identified with the dielectric friction.

Recently, we have developed[68] a molecular theory of this phenomenon within a Mori framework very similar to the one used here. The theory calculates the effect of dielectric friction on the single-particle correlation function $\phi(\omega)$; however, the generalization to the collective function $\Phi_{aa}(\omega)$ is conceptually straightforward. For the single particle, the torquelike variable $\ddot{\mu}_a^j$ is divided ino two parts, one due to short-range interactions and the other to dipole-dipole terms, and the two are assumed to be uncorrelated. Thus, $\hat{K}_2(\omega)$ consists of a short-range term that is assumed to be relaxing rapidly and a dipole-dipole term whose relaxation is calculated explicitly. In appropriately restricted models of this relaxation, the results of the continuum theories are recovered. However, because the theory is molecular, numerous relaxation channels, which are not readily discernible in the continuum analysis, have been found. These channels involve longitudinal and transverse relaxation of the polarization of the surrounding medium via both rotational and translational relaxation, as well as rotation and translation of the probe molecule.[68]

The development of far-infrared[69] and time-domain[70] spectroscopy, which probe dielectric relaxation in the high-frequency (5–200 cm^{-1}) domain, have provided a stimulus for the study of the short-time behavior of orientational correlation functions. As we have seen, at those high frequencies one does not measure the real $\varepsilon'(\omega)$ and imaginary $\varepsilon''(\omega)$ parts of the permittivity, but the absorption coefficient $\alpha(\omega)$ and the refractive index $n(\omega)$. These are related to the permittivity through the relations in Eqs. (11.14) and (11.15). As is well known,[23,64,46] with appropriately chosen parameters the three-variable expression Eq. (7.7), with $\Gamma(\omega) = \Gamma$, will give the first six moments of the frequency spectrum correctly; furthermore, at low frequency this expression reduces to a Lorentzian. Consequently, the expression that appears in Eq. (7.7), with $\Gamma(\omega) = \Gamma$, is sufficiently flexible to fit an observed spectrum over a *limited* frequency range. The prime test of the three-variable theory is to see whether consistent values of the parameters τ_D and Ω_T^2 can describe the dielectric relaxation over a wide frequency range. If this test is passed, a more exacting demand is that the values of these parameters correspond to the values determined independently by experiment or theory. Librational features, such as those predicted by the three-variable theory, are observed in the far-infrared absorption spectra of polar fluids, but the parameter τ_D determined by fitting Eq. (7.7) with $\Gamma(\omega) = \Gamma$ to the experimental data is not consistent with that obtained from low-frequency experiments.[71] This indicates that the *three-variable theory is not an adequate description of dielectric phenomena over the entire observed frequency range.*

The significance of the high-frequency results is somewhat obscured by the fact that the interaction-induced moments ΔM_a can make a substantial contribution to the far-infrared absorption of real liquids. These interaction-induced moments arise because the molecules are polarizable and because the dipole moment can be altered upon intermolecular contact; they may relax on a time scale comparable to that characteristic of the relaxation time for intermolecular torques. Therefore, it is *not clear whether the observed high-frequency spectra arise from nondiffusional molecular reorientation, from the relaxation of interaction-induced moments, or from a combination of both.*

The same uncertainties are not present in computer simulation studies of molecular reorientation. The liquids studied to date by this method have been primarily nonpolar, but the conclusions drawn from them should also be appropriate to the polar case. Lynden-Bell and McDonald[72] studied the single-molecule reorientation of a tetrahedral molecule at low (critical-point) and high (triple-point) densities. They used the correlation function given by the three-variable theory, determined ω_s^2 and ω_t^2, and varied the value of τ_s to obtain a best fit to the simulated correlation function. They found that the fit was very good at low density, but at high density the fitted function was far too oscillatory at short and intermediate times. Similar conclusions were reached by Tildesley and Madden,[73] who studied CS_2 along the normal SVP curve below the boiling point. They compared the memory function $K_1^s(t)$ obtained from the simulation with that calculated from Eqs. (11.19) and (11.22), using the value of ω_t^2 obtained from the computer experiment and a value of τ_s obtained from Eqs. (7.28) and (7.30). The calculated $K_1^s(t)$ showed an oscillatory feature with the same frequency as that appearing in the memory function obtained by computer simulation, but it was much less damped. Consequently, we conclude that the calculated reorientation function is too oscillatory at short times. It would therefore appear that the three-variable theory *can* account correctly for the frequency of the librational features, such as those observed in far-infrared spectroscopy, but that it underestimates their damping. The problem can be traced[73] to the fact that, at the times at which the oscillations occur in $K_1^s(t)$, the torque correlation function itself is still highly oscillatory. Consequently, the exponential form for $K_1^s(t)$ is inappropriate. The conclusion of both simulation studies is that the *time scale separation between the decay of the torque and angular velocity becomes poorer as the fluid becomes increasingly dense*, and as a result the three-slow-variable description becomes increasingly defective at short times. The oscillations in $K_1^s(t)$ and in $K_2^s(t)$ are not in phase[73]; consequently, the libration is not harmonic in character. Singer et al.[74] have studied the nature of the oscillations that occur in the torque and angular-velocity correlation functions of model diatomics.

When a fluid, albeit one composed of rigid polarizable molecules, becomes extremely viscous, it would appear that even the expanded set of "circumstances" we describe above is incomplete. "Anomalous" dielectric relaxation, which is often observed in such fluids, must be described by functions such as the Cole-Davidson skewed arc, the Cole-Cole circular arc, or the Havaril-Nagama function.[75] To generate such functions for our theory requires a torque-torque autocorrelation function that relaxes with a continuous distribution of exponentials; this is discussed elsewhere.[76]

We conclude that the deficiencies of presently available theories of molecular reorientation (and consequently of dielectric relaxation) in dense liquids arise from our inability to describe the dynamics of the intermolecular torques and not from inadequate descriptions of the inertial motion. As Singer et al.[74] discuss, the nonharmonic character of the torques precludes the success of theories of the "itinerant oscillator" type.

XII.   VV LIGHT SCATTERING

In our previous considerations, we studied infinite samples in which the permittivity is uniform in time and space. In light scattering, one detects light scattered off dielectric inhomogeneities, and in this section we extend our studies of dielectric material to the mean-square fluctuations responsible for light scattering. We restrict our analysis to small fluctuations of thermodynamic quantities that retain the local isotropic character of the permittivity; such fluctuations give rise to polarized, isotropic VV scattering.[77] In order to study depolarized light scattering, we would have to describe anisotropic fluctuations in the permittivity, and we have eschewed such considerations in this article. We have also not considered fluctuations that can be described only in molecular terms, such as dipole–induced-dipole fluctuations.

Isotropic fluctuations in permittivity can be related, at least in part, to fluctuations in thermodynamic variables such as density ρ and temperature T. Since we are interested in the way light is scattered by dielectric fluctuations, our interest should focus on spatial variations over distances that are at least as long as the wavelength of the light. Over such distances, the relaxation of fluctuations in the thermodynamic variables is hydrodynamic and slow; the relevant distances q^{-1} and corresponding times are very large compared to molecular distances r_c and times. (Here, $\mathbf{q}$ is the scattering vector.) Thus, if the fluctuation takes place over a length q^{-1}, we can choose a macroscopic distance L over which the thermodynamic properties of the fluid are uniform, provided that

$$L \ll q^{-1} \tag{12.1a}$$

If L is sufficiently large compared to the microscopic length r_c, that is,

$$r_c \ll L \tag{12.1b}$$

within a cube of side L about some general point $\mathbf{r}$, then the microscopic state of the fluid will be the equilibrium state prescribed by the local values of the thermodynamic variables. Because of the inequality in Eq. (12.1b), the response of the material inside the cube can be described by dielectric theory [cf. the discussion following Eq. (2.6)]. The response will be that of a uniform medium at equilibrium, and the dielectric permittivity will be appropriate to the local values of the thermodynamic state variables at $\mathbf{r}$.

In view of the discussions above, we can express the local polarization $\mathbf{P}(\mathbf{r}, t)$ in a small range of order L about the point $\mathbf{r}$, as

$$\mathbf{P}(\mathbf{r}, t) = \int d\mathbf{r}' \int_{-\infty}^{t} dt' \underset{\sim}{\chi}(\mathbf{r}, t; \mathbf{r}-\mathbf{r}', t - t') \cdot \mathbf{E}(\mathbf{r}', t) \tag{12.2}$$

where $\chi(\mathbf{r}, t; \mathbf{r} - \mathbf{r}', t - t')$ is a nonlocal susceptibility that vanishes if

$$|\mathbf{r} - \mathbf{r}'| > r_c$$

and χ varies with the absolute value of $\mathbf{r}$ and t through the spatial and temporal variation of the local thermodynamic state variables.

Equation (12.2) may be rewritten as

$$\mathbf{P}(\mathbf{r}, t) = \int d\mathbf{k} \int d\omega \, \chi(\mathbf{r}, t; \mathbf{k}, \omega) \cdot \mathbf{E}(\mathbf{k}, \omega) \exp[-i(\mathbf{k} \cdot \mathbf{r} + \omega t)] \quad (12.3)$$

where $\mathbf{E}(\mathbf{k}, \omega)$ is defined in Eq. (2.3b), and

$$\chi(\mathbf{r}, t; \mathbf{k}, \omega) = \int d\mathbf{r}'' \int_0^\infty dt'' \exp[i(\mathbf{k} \cdot \mathbf{r} + \omega t'')] \chi(\mathbf{r}, t; \mathbf{r}'', t'') \quad (12.4)$$

The wavelength $2\pi/k$ of the optical electric field is such that

$$\frac{2\pi}{k} \gg r_c$$

so that we may replace $\chi(\mathbf{r}, t; k, \omega)$ by its low-k limit, $\chi(\mathbf{r}, t; \omega)$. Furthermore, the optical frequencies are much greater than the rate of molecular reorientation so that [cf. Eq. (3.23)] we may replace $\chi(\mathbf{r}, t; \omega)$ by its high-frequency limit $\chi(\mathbf{r}, t; \infty)$. As the fluctuations in the thermodynamic variables are small, we may expand χ about its value for the system χ_{unif}. Combining these results, we have for the susceptibility fluctuations $\delta\chi$:

$$\chi(\mathbf{r}, t; \mathbf{k}, \omega) = \mathbf{I}\chi_{\text{unif}}(\omega = \infty) + \delta\chi(\mathbf{r}, t; \infty) \quad (12.5)$$

We shall consider light scattering where the incident radiation field $\mathbf{E}^i(\mathbf{k}_i, \omega_i)$ is monochromatic with angular frequency ω_i and propagation vector $\mathbf{k}_i$ polarized in the direction of the unit vector $\hat{\mathbf{n}}_i$:

$$\mathbf{E}^i = \hat{\mathbf{n}}_i E_0^i \delta(\mathbf{k}_i - \mathbf{k}) \delta(\omega_i - \omega) \quad (12.6)$$

The scattered radiation has frequency ω_f, propagation vector $\mathbf{k}_f$, and polarization along unit vector $\hat{\mathbf{n}}_f$. If the scattered spectrum $I_{if}(\mathbf{q}, \omega_f, R)$ is observed at a great distance R from the sample, the classical expression for the spectrum is[77]

$$I_{if}(\mathbf{q}, \omega_f, R) = \frac{|E_0^i|^2 k_f^4}{16\pi^2 R^2 \varepsilon_0^2} \frac{1}{2\pi} \int_{-\infty}^\infty dt \exp[i(\omega_f - \omega_i)t]$$

$$\times \langle \delta\hat{\chi}_{if}(\mathbf{q}, 0; \infty)^* \delta\hat{\chi}_{if}(\mathbf{q}, t; \infty) \rangle \quad (12.7)$$

where the scattering vector $\mathbf{q}$ is defined as

$$\mathbf{q} = \mathbf{k}_i - \mathbf{k}_f \tag{12.8}$$

and

$$\delta\hat{\chi}_{if}(\mathbf{q}, t; \infty) = \int d\mathbf{r}\, e^{i\mathbf{q}\cdot\mathbf{r}} \hat{\mathbf{n}}_f \cdot \delta\chi(\mathbf{r}, t; \infty) \cdot \hat{\mathbf{n}}_i \tag{12.9}$$

The dependence of $\chi(\mathbf{r}, t; \infty)$ on $\mathbf{r}$ and t arises from the dependence of χ on the value of the thermodynamic state variables ξ; that is, $\chi(\mathbf{r}, t; \infty) = \chi(\xi(\mathbf{r}, t); \infty)$, where $\xi = \{\xi_1, \xi_2, \ldots\}$ gives a complete prescription of the thermodynamic state. ξ^0 gives the equilibrium state of the whole system so that

$$\chi_{unif}(\infty) = \chi(\xi^0; \infty) \tag{12.10}$$

and

$$\delta\chi(\xi(\mathbf{r}, t); \infty) = \sum_j \left(\frac{\partial\chi(\xi; \infty)}{\partial\xi}\right)_{\xi^0} \delta\xi_j(\mathbf{r}, t) \tag{12.11}$$

where

$$\delta\xi(\mathbf{r}, t) = \xi(\mathbf{r}, t) - \xi^0 \tag{12.12}$$

If we now Fourier-transform Eq. (12.11), we obtain [cf. Eq. (12.9)]:

$$\delta\chi(\mathbf{q}, t; \infty) = \left(\frac{\partial\chi(\xi; \infty)}{\partial\xi_j}\right)_{\xi^0} \delta\xi_j(\mathbf{q}, t) \tag{12.13}$$

The correlation function that determines the light-scattering spectrum (Eq. 12.7) is thus obtained by first evaluating χ ($\xi; \omega = \infty$) at a different set of values of the constraints ξ, so that the derivative in Eq. (11.13) may be determined (see below). The time-dependent correlation functions of the hydrodynamic variables $\langle\delta\xi_j(\mathbf{q}, t)\delta\xi_l(\mathbf{q})\rangle$ may be found from hydrodynamic considerations.[77]

In order to obtain molecular expressions for $\delta\chi$, we refer to the convenient molecular expression for ε that we obtained in Section VIII, the extended Clausius-Mossotti relation

$$\frac{\varepsilon(\infty) - 1}{\varepsilon(\infty) + 2} = \frac{\alpha\rho}{3\varepsilon_0}(1 + \delta b) \tag{12.14}$$

where δb can be described in terms of averages over short-range molecular quantities. We will assume at first that $\delta b = 0$, that is, that the Clausius-Mossotti equation holds exactly. Since $\partial \varepsilon / \partial \xi = \partial \chi / \partial \xi$, it can then be seen readily that

$$\left(\frac{\partial \chi}{\partial \xi} \right)_0 = \left[\frac{\varepsilon + 2}{3} \right]^2 \frac{\alpha}{\varepsilon_0} \left(\frac{\partial \rho}{\partial \xi_j} \right)_0 \tag{12.15}$$

where $\varepsilon = \varepsilon(\infty)$ and where $(\partial \chi / \partial \xi_j)_0$ and $(\partial \rho / \partial \xi_j)_0$ are evaluated at equilibrium, at $\xi_j = \xi_j^0$. Since the number density itself is a well-characterized constraint in an ensemble average, we choose $\xi_1 = \delta \rho$. If we then combine Eqs. (12.15) and (12.11), we obtain

$$\langle \delta \chi (\mathbf{q}, t) \partial \chi (\mathbf{q}, 0)^* \rangle = \left(\frac{\varepsilon + 2}{3} \right)^4 \frac{\alpha^2}{\varepsilon_0^2} \langle \delta \rho (\mathbf{q}, t) \delta \rho^* (\mathbf{q}, 0) \rangle \tag{12.16}$$

This is a well-known result,[77] which states that the mean-square isotropic dielectric fluctuation is proportional to the mean-square density fluctuation; the isotropic polarizability α is that of the gas-phase molecules, but the *local field effects* are accounted for by the Lorentz factor $[(\varepsilon + 2)/3]^2$.

Next, we shall assume that the factor δb in Eq. (11.14) is small but not negligible; only terms linear in δb will be retained. In Section VIII, we found that it was the small quantity δb^0, rather than δb, that could be directly related to molecular quantities, and the linearized relationship between δb and δb^0 is given in Eq. (8.18). It follows from Eqs. (12.15) and (8.17) that, through terms linear in δb^0,

$$\left(\frac{\partial \varepsilon}{\partial \xi_j} \right)_0 = \left(\frac{\varepsilon + 2}{3} \right)^2 \left(\frac{\alpha}{\varepsilon_0} \right) \left\{ \left(\frac{\partial \rho}{\partial \xi_j} \right)_0 \left(1 + \frac{\delta b^0 9 \varepsilon (-4\varepsilon^4 + 10\varepsilon^3 + 7\varepsilon - 4)}{(2\varepsilon^2 + 1)^2 (\varepsilon + 2)^2} \right. \right.$$
$$\left. \left. + \rho \frac{\partial b^S}{\partial \xi_j} \left(\frac{27\varepsilon^2}{(\varepsilon + 2)^2 (2\varepsilon^2 + 1)} \right) \right) \right\} \tag{12.17}$$

Once again it seems reasonable to choose the density ρ as one of the variables, $\delta \xi_1 = \delta \rho$, but other variables might now also be important. Thus,

$$\delta \chi (\mathbf{q}, t) = \left(\frac{\varepsilon + 2}{3} \right)^2 \left(\frac{\alpha}{\varepsilon_0} \right) \delta \rho (q, t) \left(1 + \frac{\delta b^0 9 \varepsilon (-4\varepsilon^4 + 10\varepsilon^3 + 7\varepsilon - 4)}{(\varepsilon + 2)^2 (2\varepsilon^2 + 1)^2} \right.$$
$$\left. + \left(\frac{\partial b^S}{\partial \rho} \right)_0 \rho \frac{27\varepsilon^2}{(\varepsilon + 2)^2 (2\varepsilon^2 + 1)} \right)$$
$$+ \left(\frac{3\varepsilon^2}{2\varepsilon^2 + 1} \right) \left(\frac{\alpha \rho}{\varepsilon_0} \right) \sum_{j \neq 1} \left(\frac{\partial b^S}{\partial \xi_j} \right) \delta \xi_j (\mathbf{q}, t) \tag{12.18}$$

.

where b^S is specified by Eq. (8.9). Small terms linear in δb^0 or $\partial b^S/\partial \rho$ have been neglected in the $\delta\rho$ term but have been retained in the $\delta\xi_j$ terms, where they play a major role. The dependence of $\delta\chi$ upon variables other than the density are likely to be small, since $(\partial b^S/\partial \xi_j)_0$ is usually small. As an example, one might choose ξ_2 equal to the temperature fluctuation δT so that at constant density $(\delta\rho = 0)$ we have

$$\delta\chi(\mathbf{q}, t) = \frac{3\varepsilon^2}{2\varepsilon^2 + 1}\left(\frac{\alpha}{\varepsilon_0}\right)\rho\left(\frac{\partial b^S}{\partial T}\right)_0 \delta T(\mathbf{q}, t) \qquad (12.19)$$

Although we do not know much about $(\partial b^S/\partial T)_0$ except that it is a short-range molecular quantity, the factor $3\varepsilon^2/(2\varepsilon + 1)$ increases as ε increases, as one goes from gas to liquid, but increases less rapidly than does the factor $[(\varepsilon + 2)/3]^2$, which appears in the expression for density fluctuation.

XIII. SUMMARY

We have developed a consistent molecular theory of dielectrics in which static and dynamic phenomena for rigid dipoles, polarizable dipoles, and nonpolar materials have been analyzed on a common framework. Because of the long-range dipolar interactions, the n-body distribution functions $g^{(n)}(1, 2, \ldots)$ and, in particular, the two-body function $g^{(2)}(1, 2)$, depend upon sample shape. This means that although $\varepsilon(\omega)$ is independent of sample shape, the formal molecular expressions for $\varepsilon(\omega)$ are different for differently shaped samples. Throughout, we have used an infinite dielectric sample because the absence of boundary effects makes the interpretation of $\varepsilon(\omega)$ in molecular terms most straightforward.

The first problem in developing a molecular theory of $\varepsilon(\omega)$ is to obtain a formal relationship between the molecular quantities and $\varepsilon(\omega)$ or $\chi(\omega)$. The polarization of the sample is equal to the susceptibility χ times the Maxwell field $\mathbf{E}$, but it can be written equally well as a quasi-susceptibility χ^0 times the vacuum field $\mathbf{E}^0$. This latter approach is useful both because the $\mathbf{E}^0$ field is a true external perturbation that allows the straightforward use of linear response theory and because χ^0 can be expressed readily in terms of correlation functions of molecular quantities. The relationships between χ_{aa}^0 and $\varepsilon(\omega)$ are given in Eqs. (2.13a, b), and the relationships between χ_{aa}^0 and the molecular quantities are given in Eq. (3.21). The susceptibility χ_{aa}^0 is directly related to the dipolar correlation functions, but this correlation function is a collective quantity involving sums of molecular quantities.

The relationship between χ^0 and the equilibrium fluid structure is discussed in Sections IV, VIII, and IX. For rigid nonpolarizable molecules, we obtain the Kirkwood relationship between the mean-square moment of a small sphere in the sample and $\varepsilon(0)$. We have shown that with reasonable, simple models for the distribution functions $g^{(n)}$, one can obtain the Clausius-Mossotti equation [Eq. (8.12)] for polarizable nonpolar molecules and the Fröhlich equation [Eq. (9.2)] for polarizable polar molecules. In particular, to obtain the Clausius-Mossotti equation, we have set $g^{(n)} = \tilde{g}^{(n)}$, where $\tilde{g}^{(n)}$ is a superposition of two-body correlations that are uniform with a hard-core lower cutoff. [See Eqs. (8.6)–(8.8).] In deriving the Fröhlich equation, we let $g^{(n)} = \hat{g}^{(n)}$, where $\hat{g}^{(n)}$ is equal to $\tilde{g}^{(n)}$ times a pair-distribution correction that accounts for the correlation between dipoles [Eq. (9.13)]. The Clausius-Mossotti equation relates $\varepsilon(0)$ to the properties of single nonpolar molecules, while the Fröhlich equation connects $\varepsilon(0)$ to the properties of a "local" group of polar polarizable molecules. We have also shown, in principle in most cases and actually in some, how improvements in the structural model, that is, improvements in the form of $g^{(n)}$, will lead to corrections to the Clausius-Mossotti and Fröhlich equations. Some long-range dipole-dipole correlations have been omitted in the Fröhlich equation, and this certainly reduces the usefulness of the expression.

Although the direct correlation function has proved to be a powerful tool in studying the dielectric constant $\varepsilon(0)$ of polarizable molecules, we have developed our formalism (Sections IV, VIII, and IX) in terms of the distribution functions $g^{(n)}$. We have done so because we believe that people have more intuitive understanding of $g^{(n)}$ and because we found it a convenient property to use in unifying our studies of dielectrics; however, it appears that, in obtaining accurate results by computational methods, the direct correlation function approach is the preferred method.[5,7] In discussing relaxation phenomena, we have made an attempt to separate the explicitly long-range contributions to $\varepsilon(\omega)$ from the explicitly short-range ones, because the latter can be compared more readily to molecular effects.

By means of analogous Mori continued-fraction expansions for single-molecule and collective properties, we can obtain interrelationships between the collective and single-particle relaxation parameters. We have called the results of this analysis the "corresponding micro-macro correlation," CMMC. This is a nontrivial result, since the two expansions involve quite different projection operators. One of the results obtained by this procedure is a relationship between single-molecule reorientation times τ_s and the Debye dielectric relaxation time τ_D [Eqs. (6.10) and (7.31)]; this is a generalized form of the Glarum-Powles relation.

The molecular approach to the dynamics of the collective correlation function resolves one of the long-standing controversies of dielectric theory concerning which functional of the dielectric permittivity is most closely related to the single-particle correlation function. Even though the functional $\{[\varepsilon(\omega)-1][2\varepsilon(\omega)+1]/\varepsilon(\omega)\}$ is the quasi-susceptibility $\chi^0(\omega)$ for a small embedded sphere (which means that only dipoles in the small sphere need be included in its evaluation), the relaxation dynamics of the small sphere are strongly affected by interactions with particles in the surrounding medium. This results in a relaxation process involving two [Eq. (5.12)], rather than one, relaxation times for the small sphere, even in the case of rotational diffusion, where the molecular rotations relax with a single decay time.

A particularly useful expression for $\varepsilon(\omega)$ is given in Eq. (11.21). The ω^2/Ω_R^2 contribution is an inertial term that is readily evaluated but is significant only at very high frequencies. The $\hat{K}_2(\omega)$ term is very closely associated with the correlation function for intermolecular torques [Eq. (11.20)], and it is the dynamical behavior of the torques that one must describe in order to understand the frequency dependence of $\varepsilon(\omega)$. The shift of focus from the stochastic process of reorientation to the study of torques is, we believe, both theoretically and physically sound. In Sections VI, VII, and XI, we have explored a few models for $\hat{K}_2(\omega)$. If $\hat{K}_2(\omega) \cong \hat{K}_2(0)$ and $\omega^2/\Omega_R^2 \ll 1$, one recaptures the Debye expression of Eq. (5.10); the rotational motion is then said to be diffusional, and there is only one adjustable parameter. If $\hat{K}_2(\omega)$ is represented by a single complex Lorentzian [Eq. (11.22), $\Gamma(\omega) = \Gamma(0)$], we have the "three-variable theory," which, with two adjustable parameters, can explain at least qualitatively many features of dielectric behavior over a wide range of frequencies. In particular, the three-variable theory encompasses both librational oscillations [Eq. (11.6)] and longitudinal inertial oscillations (dipolarons) [Eq. (11.7)]. In detail, however, the three-variable theory does not seem to give a quantitative description of the molecular rotations contributing to dielectric relaxation, and more detailed studies of the torquelike memory function $\hat{K}_2(\omega)$ are needed, even for simple fluids.

For simple fluids, the frequency dependence of the torque correlation function is detectable at high frequency only and may be regarded as important in fine-tuning the description of the rotational dynamics. For many fluids of experimental interest, however, the character of the torque relaxation is reflected at all frequencies. In particular, at high viscosity, the strong coupling with translational motions and the hindered nature of the relative reorientation of neighboring molecules appears to alter completely the dynamic behavior of the intermolecular torques; this "anomalous" relaxation, which gives rise to $\varepsilon(\omega)$'s describable by such functions as Cole-Davidson skewed arcs and Cole-Cole circular arcs, is under study, as are the analogous effects in oriented liquids. An understanding of the intermolecular torques is also crucial to the study of two-component fluids. For such fluids, one might be tempted to divide the moment operator M_α into two components, M_α^1 and M_α^2, corresponding to the two components, but one would then have to calculate the static averages $\langle|M_\alpha^1|^2\rangle$, $\langle|M_\alpha^2|^2\rangle$, and $\langle M_\alpha^1(M_\alpha^2)\rangle$, none of which can be related directly to $\varepsilon(0)$, whereas the total quantity $\langle|M_\alpha|^2\rangle$ can be expressed in terms of $\varepsilon(0)$. (See Table VI.) It would thus seem preferable to retain the total dipole M_α as a variable and to relegate to the torque autocorrelation function $\hat{K}_2(\omega)$ the division of the problem into components for each species; this is the physically meaningful place to discuss the interactions among molecules, including those between different species, while taking advantage of the short-range nature of $\hat{K}_2$.

The dynamics of $\varepsilon(\omega)$ for polarizable polar molecules is a very complex problem, since it can involve the interplay of rotations with the intermolecular fluctuations that modulate the dipole moment. However, within the scope of the model that leads to the Fröhlich equation in the static case, cross-coupling of the rotations to these other motions should vanish. In this case, one can obtain expressions for the rotational contributions to $\varepsilon(\omega)$ [Eqs.

TABLE VI

Selected Results for Infinite Samples (Three-Variable Theory, Fröhlich Model):
$\Delta M_\alpha = 0,\ \Gamma(\omega) = \Gamma(0)$

	Single particle	Transverse mode, x	Longitudinal mode, z
$3\langle|M_\alpha|^2\rangle/\mu^2$	1	$\dfrac{3Ng^S\varepsilon(0)}{2\varepsilon(0)+\varepsilon(\infty)}$	$\dfrac{3Ng^S}{[2\varepsilon(0)+\varepsilon(\infty)]\varepsilon(\infty)}$
$\langle|M_\alpha|^2\rangle/k_BTV\varepsilon_0$		$\varepsilon(0)-\varepsilon(\infty)$	$[\varepsilon(0)-\varepsilon(\infty)]/\varepsilon(0)\varepsilon(\infty)$
Ω_R^2/ω_s^2 [a]	1	$\dfrac{2\varepsilon(0)+\varepsilon(\infty)}{3g^S\varepsilon(0)}$	$\dfrac{2\varepsilon(0)+\varepsilon(\infty)}{3g^S\varepsilon(\infty)}$
$\Phi_{\alpha\alpha}(0)$ [b]	τ_s	τ_D	$\tau_D\varepsilon(\infty)/\varepsilon(0)$
Ω_T^2/ω_t^2	1	$\ddot{g}$	$\ddot{g}$
Γ^{-1}/τ_t [c]	1	$\ddot{g}/\bar{g}$	$\ddot{g}/\bar{g}$
Dipolarons [d]			
$\frac{1}{4}\omega_s^2\tau_s^2 <$	1^e	$\left(\dfrac{2\varepsilon(0)+\varepsilon(\infty)}{3g^S\varepsilon(0)}\right)^e$	$\dfrac{2\varepsilon(0)+\varepsilon(\infty)}{3g^S\varepsilon(\infty)}$
Librations [d]			
$\dfrac{4\omega_s^4\tau_s^2}{\omega_t^2} >$	1	1	1

[a] $\omega_s^2 = 2k_BT/I$.

[b] $\tau_D = \tau_s\left(\dfrac{3\varepsilon(0)}{2\varepsilon(0)+\varepsilon(\infty)}\right)g^S\left(\dfrac{\ddot{g}^2}{\bar{g}}\right)$.

[c] $\tau_s = (\omega_t^2/\omega_s^2)\tau_t$.

[d] $\ddot{g}/\bar{g}^2 = 1$ and $\omega_s^2 \ll \omega_t^2$.

[e] Condition cannot be satisfied.

TABLE VII

Summary of Important Results (Three-Variable Theory, Fröhlich Model):
$$\Delta M_\alpha = 0; \ \Gamma(\omega) = \Gamma(0)$$

$$\frac{\varepsilon(\omega) - \varepsilon(\infty)}{\varepsilon(0) - \varepsilon(\infty)} = \left(1 - \frac{i\omega K_2(\omega)}{(\Omega_R^x)^2} - \frac{\omega^2}{(\Omega_R^x)^2}\right)^{-1}$$

$$K_2(\omega) = \Omega_T^2 [-i\omega + \Gamma(\omega)]^{-1}$$

$$\frac{[\varepsilon(0) - \varepsilon(\infty)][2\varepsilon(0) + \varepsilon(\infty)]}{\varepsilon(0)} = \left(\frac{\varepsilon(\infty) + 2}{3}\right)^2 \frac{(\mu)^2 g^S \rho}{\varepsilon_0 k_B T}$$

(10.5)–(10.8)], expressions very similar to those applicable for rigid dipoles but with effective moments altered by polarizability effects. A summary of important results is given in Tables VI and VII. Although the Fröhlich model, and consequently some of the expressions dependent upon it, may be inadequate because of the neglect of some long-range dipole-dipole correlations, the separation of rotations of an effective dipole from other motions and the neglect of cross-correlations between these processes may be justified on more general grounds.

APPENDIX A

In this appendix, we evaluate the $\langle |M_\alpha|^2 \rangle$'s in Eqs. (4.5a, b). We require an expression for

$$N_0 \langle C_{1,m}(\Omega^1) C_{1,m'}^*(\Omega^2) \exp(i\mathbf{k} \cdot \mathbf{r}^{12}) \rangle$$

$$= \frac{\rho}{\Omega^2} \int d\mathbf{r}^{12} \, d\Omega^1 \, d\Omega^2 \, g(\mathbf{r}^{12}, \Omega^1, \Omega^2) C_{1,m}(\Omega^1) C_{1,m'}^*(\Omega^2) \exp(i\mathbf{k} \cdot \mathbf{r}^{12})$$

$$\tag{A.1}$$

where $\Omega = 4\pi$ for a linear molecule and $8\pi^2$ for a symmetric top. [See Eqs. (4.3a, b, c).]

The expansion of the distribution function in spherical harmonics [Eq. (4.4)] may be substituted into this expression; the integrations over Ω^1 and Ω^2 can be performed with the aid of the orthonormality properties of the spherical harmonics[25]:

$$N_0 \langle C_{1,m}(\Omega^1) C_{1,m'}^*(\Omega^2) \exp(i\mathbf{k} \cdot \mathbf{r}^{12}) \rangle$$

$$= \sum_{l'l''} \frac{\rho}{9} \begin{pmatrix} 1 & 1 & l'' \\ -m & m' & m'' \end{pmatrix} (2l+1) \times \int d\mathbf{r}^{12} g^{11l''}(r^{12})$$

$$\times \left[C_{l''m''}(\hat{\mathbf{r}}^{12}) C_{l\mu}^*(\hat{\mathbf{r}}^{12}) j_l(kr^{12}) C_{l\mu}(\hat{\mathbf{k}}) \right] (i)^{l-l''} (-1)^m \tag{A.2}$$

where $\hat{\mathbf{k}}$ and $\hat{\mathbf{r}}^{12}$ represent the orientational angles associated with $\mathbf{k}$ and $\mathbf{r}^{12}$, respectively, and $g^{11l'''}$ is specified in Eq. (4.4). The contents of the square brackets arise from the Rayleigh expansion of $\exp(i\mathbf{k}\cdot\mathbf{r}^{12})$.[25] Performing the integration over $\hat{\mathbf{r}}^{12}$, we obtain

$$N_0 \langle C_{1,m}(\Omega^1) C^*_{1,m'}(\Omega^2) \exp(i\mathbf{k}\cdot\mathbf{r}^{12}) \rangle$$

$$= \frac{4\pi\rho}{9} \begin{pmatrix} 1 & 1 & l \\ -m & m' & \mu \end{pmatrix} (-1)^m C_{l\mu}(\hat{\mathbf{k}}) \int dr\, r^2 g^{11l}(r)\, j_l(kr)$$

$$\text{(A.3)}$$

For $\hat{\mathbf{k}}$ parallel to z, we have $C_{l\mu}(\hat{\mathbf{k}}) = \delta_{\mu,0}$, so that

$$3N \langle C_{1,m}(\Omega^1) C^*_{1,m'}(\Omega^2) \exp(i\mathbf{k}\cdot\mathbf{r}^{12}) \rangle = \left(g^S - 1 + \gamma_m g^L \right) \delta_{m,m'} \quad \text{(A.4)}$$

where g^S and g^L are given by Eqs. (4.6) and (4.7), $\gamma_0 = 2$, and $\gamma_{\pm 1} = -1$. This gives the result in Eqs. (4.5a, b).

APPENDIX B

In this appendix, we carry out the manipulations required to obtain the Clausius-Mossotti formula, Eq. (8.12) and to evaluate some of the correction terms [Eqs. (8.5), (8.20), (8.21)]. We begin by obtaining an expression for the value of $\overline{\underset{\sim}{\alpha}^1 \cdot \mathscr{T}^{12} \cdot \underset{\sim}{\alpha}^2}$, where $\mathscr{T}^{12}$ is given by Eq. (3.12), and the bar denotes an average weighted by the probability of finding molecules 1 and 2 separated by $\mathbf{r}^{12}$, over the positions and orientations of all the molecules in the sample except molecules 1 and 2. The first few terms are given by

$$\overline{\underset{\sim}{\alpha}^1 \cdot \mathscr{T}^{12} \cdot \underset{\sim}{\alpha}^2} = \left[g^{(2)}(1,2) \right] \left\{ \underset{\sim}{\alpha}^1 \cdot T^{12} \cdot \underset{\sim}{\alpha}^2 + \underset{\sim}{\alpha}^1 \cdot T^{12} \cdot \underset{\sim}{\alpha}^2 \cdot T^{21} \cdot \underset{\sim}{\alpha}^1 \cdot T^{12} \cdot \underset{\sim}{\alpha}^2 + \cdots \right\}$$

$$+ \frac{\rho}{\Omega} \int dr^3\, d\Omega^3 \left[g^{(3)}(1,2,3) \right]$$

$$\times \left\{ \underset{\sim}{\alpha}^1 \cdot T^{13} \cdot \underset{\sim}{\alpha}^3 \cdot T^{32} \cdot \underset{\sim}{\alpha}^2 + \underset{\sim}{\alpha}^1 \cdot T^{13} \cdot \underset{\sim}{\alpha}^3 \cdot T^{31} \cdot \underset{\sim}{\alpha}^1 \cdot T^{12} \cdot \underset{\sim}{\alpha}^2 \right.$$

$$\left. + \underset{\sim}{\alpha}^1 \cdot T^{12} \cdot \underset{\sim}{\alpha}^2 \cdot T^{23} \cdot \underset{\sim}{\alpha}^3 \cdot T^{32} \cdot \underset{\sim}{\alpha}^2 + \cdots \right\}$$

$$+ \left(\frac{\rho}{\Omega} \right)^2 \int dr^3\, d\Omega^3\, dr^4\, d\Omega^4 \left[g^{(4)}(1,2,3,4) \right]$$

$$\times \left\{ \underset{\sim}{\alpha}^1 \cdot T^{13} \cdot \underset{\sim}{\alpha}^3 \cdot T^{34} \cdot \underset{\sim}{\alpha}^4 \cdot T^{42} \cdot \underset{\sim}{\alpha}^2 + \cdots \right\} + \cdots \quad \text{(B.1)}$$

where $\Omega = 4\pi$ for a linear molecule and $8\pi^2$ for a symmetric-top molecule.

This expression includes all terms up to those involving three dipole tensors. Rather than write out these integrals at every stage in the argument, it is convenient to represent them diagrammatically. In the first stage, the stage in which the Clausius-Mossotti relationship is obtained, we represent the particle distribution functions by the simple form $\tilde{g}^{(n)}$ given in Eqs. (8.6), (8.7a, b). The corresponding diagram consists of dots (•, ○, and ⊗) that represent molecular polarizabilities and lines that represent dipole tensors; a scalar product is indicated by junction of a dot and a line. With each pair of dots (labeled i and j) in a diagram, we associate a factor $\tilde{g}(ij)$; thus,

$$(\overset{1 \quad 2}{\bullet\!\!-\!\!\bullet})_{ab} = (\underset{\sim}{\alpha}^1 \cdot \underset{\sim}{T}^{12} \cdot \underset{\sim}{\alpha}^2)_{ab}\, \tilde{g}(12) \qquad (B.2)$$

When two dots touch, they belong to the same molecule. When an *even* number of lines enter and leave a particle (indicated by one or more contiguous dots), integration over the position and orientation of that particle is implied. Finally, each diagram is multiplied by $(\rho/\Omega)^{n-2}$, where n is the number of particle labels in the diagram. In this way, Eq. (B.1) may be rewritten

$$\tilde{g}(12)\underset{\sim}{\alpha}^1 \cdot \mathscr{T}^{12} \cdot \underset{\sim}{\alpha}^2 = \overset{1 \quad 2}{\bullet\!\!-\!\!\bullet} + \overset{1 \quad 2}{\rlap{\;\;\bullet}\bullet\!\!=\!\!\bullet} + \cdots\cdots$$

$$+ \overset{1 \quad 2 \quad 3}{\bullet\!\!-\!\!\bullet\!\!-\!\!\bullet} + \underset{1 \quad 2}{\overset{3}{\bullet}\!\!-\!\!\bullet} + \underset{1 \quad 2}{\bullet\!\!-\!\!\overset{3}{\bullet}} + \cdots$$

$$+ \overset{1 \quad 2 \quad 3 \quad 4}{\bullet\!\!-\!\!\bullet\!\!-\!\!\bullet\!\!-\!\!\bullet} + \cdots\cdots \qquad (B.1')$$

The diagrams provide a convenient and suggestive picture of the multiple scattering series.

For an axially symmetric molecule, the molecular polarizability may be expanded into a scalar and a symmetric traceless part:

$$\alpha^i_{\alpha\beta} = \alpha\delta_{\alpha\beta} + \gamma\left[\hat{e}_\alpha\hat{e}_\beta - \frac{\delta_{\alpha\beta}}{3}\right] \qquad (B.3)$$

where $\hat{e}$ is a unit vector along the molecular axis, or

$$\overset{i}{-\!\!\bullet\!\!-} = \overset{i}{-\!\!\circ\!\!-} + \overset{i}{-\!\!\otimes\!\!-} \qquad (B.3')$$

The complicated diagrams may be broken down into simple ones. To see this, note that $\tilde{g}^{(n)}$ does not depend on the orientations of the molecules, so that integration over the orientation of i yields

$$\underset{\bullet}{\overset{i}{\rule{1.5em}{0.4pt}}} = \Omega \underset{\circ}{\overset{i}{\rule{1.5em}{0.4pt}}} \tag{B.4a}$$

$$\underset{\bullet\bullet}{\overset{i}{\rule{1.5em}{0.4pt}}} = \Omega \underset{\circ}{\overset{i}{\rule{1.5em}{0.4pt}}} \alpha \left\{ 1 + \frac{2}{9}\left(\frac{\gamma}{\alpha}\right)^2 \right\} \tag{B.4b}$$

$$\underset{\bullet\bullet\bullet}{\overset{i}{\rule{1.5em}{0.4pt}}} = \Omega \underset{\circ}{\overset{i}{\rule{1.5em}{0.4pt}}} \alpha \left\{ 1 + \frac{2}{3}\left(\frac{\gamma}{\alpha}\right)^2 + \frac{2}{27}\left(\frac{\gamma}{\alpha}\right)^2 \right\} \tag{B.4c}$$

We will also need the value of the integral over the spatial coordinates of the ith dot ($\circ$) (viz. Appendix C); these can be illustrated by

$$\overset{1 \quad i \quad 2}{\bullet\!-\!\circ\!-\!\bullet} = -\left(\frac{\alpha\rho}{3\epsilon_0}\right)\overset{1 \quad 2}{\bullet\!-\!\bullet} + \text{short}-\text{range terms} \tag{B.5}$$

and

$$\overset{i}{\underset{1}{\rule{1.5em}{0.4pt}\,\Omega\,\rule{1.5em}{0.4pt}}} = \alpha\rho\,(6r_0^{\,3}\,\pi\epsilon_0^{\,2})^{-1}\,\overset{}{\underset{1}{\bullet\!\!\bullet}} \tag{B.6}$$

The short-range terms will be discussed below; they appear for $r^{12} < 2r_0$. If r_0 is chosen to satisfy Eq. (8.8), the algebraic factor in Eq. (B.6) becomes $\frac{2}{9}(\alpha\rho^2/\epsilon_0^2)$.

The diagrams are evaluated by the following procedure, which will be illustrated with the three-bond diagrams (denoted by $\overset{(3)}{\bullet\!\!\curvearrowright\!\!\bullet}$) in Eq. (B.1):

1. Integrate over the orientation of all *single* dots (except 1 and 2):

$$\overset{(3)}{\underset{1 \qquad 2}{\bullet\!\!\curvearrowright\!\!\bullet}} = \underset{1 \quad 2 \quad 3 \quad 4}{\bullet\!-\!\circ\!-\!\circ\!-\!\bullet}\,(\Omega)^2 + \underset{1 \quad 2}{\overset{}{\bullet\!\!-\!\!\bullet}}\,(\Omega)$$

$$+\ \underset{1 \quad 2}{\bullet\!-\!\!\bullet\bullet}\,(\Omega) + \underset{1 \quad 2}{\overset{}{\bullet\!\!\equiv\!\!\bullet}} \tag{B.7'}$$

2. Starting with the straightest chain, integrate over the positions of the white dots; Eqs. (B.5) and (B.6) give the values obtained by integrating over all space, but remember that, because of the $\tilde{g}(ij)$ factors, a

dot cannot occupy the space associated with another dot. Thus, those configurations for which overlap occurs must be removed from the integral before Eqs. (B.5) and (B.6) may be used. This may be accomplished by use of the integral in Eq. (C.20). Thus, for example,

$$
\underset{1\ \ 3\ \ 4\ \ 2}{\bullet\!-\!\circ\!-\!\circ\!-\!\bullet} \ \xrightarrow{\ \underset{\sim}{dr^{14}}\ } \ -(3\epsilon_0)^{-1}\ \alpha\rho\ \underset{1\ \ 3\ \ 2}{\bullet\!-\!\circ\!-\!\bullet} \ -\ \underset{1\ \ \ 2}{\overset{3}{\bullet\!-\!\bullet}}\ \frac{4\pi}{3}\ \rho r_0{}^3 \tag{B.7''}
$$

where the short-range terms from Eq. (B.5) have been neglected. When we make use of Eq. (8.8), the second diagram on the right-hand side partially cancels one of the other diagrams in Eq. (B.7'):

$$
\underset{1\ \ 2}{\overset{3}{\bullet\!-\!\bullet}} \ -\ \underset{1\ \ 2}{\overset{3}{\bullet\!-\!\bullet}} \ =\ \underset{1\ \ 2}{\overset{3}{\bullet\!-\!\bullet}} \tag{B.7'''}
$$

Evaluating all the diagrams of Eq. (B.7) in this way and taking advantage of the cancellations, we obtain for the value of the diagrams in Eq. (B.7):

$$
\underset{1\ \ \ \ 2}{\overset{(3)}{\bullet\!\wedge\!\!\wedge\!\!\wedge\!\bullet}} = \left(\frac{\alpha\rho}{\epsilon_0}\right)^2 \underset{1\ \ 2}{\bullet\!-\!\circ} + \frac{2}{9}\left(\frac{\alpha\rho^2}{\epsilon_0{}^2}\right)\left(\underset{1\ \ \ \ 2}{\otimes\!-\!\bullet} + \underset{1\ \ \ \ 2}{\bullet\!-\!\otimes\!\bullet}\right) + \underset{1\ \ \ \ 2}{\otimes\!=\!\otimes} \tag{B.8}
$$

Similarly, the terms in $\tilde{g}(12)\overline{\underset{\sim}{\alpha}^1\!\cdot\!\underset{\sim}{\mathscr{T}}^{12}\!\cdot\!\underset{\sim}{\alpha}^2}$ involving two dipole tensors become

$$
\underset{1\ \ \ \ 2}{\overset{(2)}{\bullet\!\wedge\!\!\wedge\!\bullet}} = -\left(\frac{\alpha\rho}{3\epsilon_0}\right)\underset{}{\bullet\!-\!\bullet} \tag{B.9}
$$

and those involving four tensors become

$$
\underset{1\ \ \ \ 2}{\overset{(4)}{\bullet\!\wedge\!\!\wedge\!\bullet}} = -\left(\frac{\alpha\rho}{3\epsilon_0}\right)\underset{1\ \ \ \ 2}{\overset{(3)}{\bullet\!\wedge\!\!\wedge\!\bullet}} - \frac{2}{9}\left(\frac{\gamma}{\alpha}\right)^2\underset{1\ \ \ 2}{\bullet\!-\!\bullet} - 2\left(\frac{\alpha\rho}{3\epsilon_0}\right)^3 \alpha^{-1}\left(\underset{1\ \ \ \ 2}{\otimes\!-\!\bullet} + \underset{1\ \ \ \ 2}{\bullet\!-\!\otimes\!\bullet}\right)
$$

$$
+ \underset{1\ \ 3\ \ 2}{\otimes\!\!=\!\!\otimes\!-\!\bullet} + \underset{2\ \ \ \ 1}{\bullet\!-\!\otimes\!\!=\!\!\otimes} - \left(\frac{2\alpha\rho}{3\epsilon_0}\right)\underset{}{\otimes\!\!\bowtie\!\!\otimes} \tag{B.10}
$$

Two or more lines between two particles indicate higher-order dipole–induced-dipole (DID) interactions, the order being given by the number of such lines. An interesting feature of these results is that the third-order pair terms [such as the last diagram of Eq. (B.8) and the last three terms in Eq. (B.10)] vanish for molecules with purely isotropic polarizabilities.

If we neglect the polarizability anisostropy, then Eqs. (B.7)–(B.10) give

$$\underset{1 \quad 2}{\text{⬤〰〰⬤}} \equiv \sum_n \overset{(n)}{\underset{1 \quad 2}{\text{⬤〰〰⬤}}} = (1 - x + 3x^2 - 5x^3 + \cdots) \underset{1 \quad 2}{\text{○—○}} \tag{B.11}$$

where $x = \alpha\rho/3\varepsilon_0$. These are just the first four terms obtained by Felderhof[10] in the series expansion of the dipole propagator in a uniform medium, a series which sums to

$$\underset{1 \quad 2}{\text{⬤〰〰⬤}} = [(1 - x)(1 + 2x)]^{-1} \underset{}{\text{○—○}} \tag{B.12}$$

and is the result given in Eq. (8.10). We cannot claim to have derived Eq. (B.12), since we have not obtained an expression for the general term in our series; nevertheless, it is possible (but tedious) to show that a given term calculated as above agrees with the value obtained by expanding Eq. (B.12). Our result agrees with the continuum result because, by neglecting the short-range terms from Eq. (B.5), we are ignoring the finite particle size in a certain class of excluded volume effects. The effect of the missing terms is discussed in Section VIII.B.

Equation (B.12), rewritten in the notation of the main text, is simply

$$\alpha^2 \overline{\mathscr{T}_{aa}^{12}} = [(1 - x)(1 + 2x)]^{-1} \alpha^2 T_{aa}^{12} \tag{B.12'}$$

This equation gives all the long-range contributions to the tensor **b** [Eq. (8.4c)] for isotropically polarizable particles. It follows from Eq. (C.7) that the long-range part of **b** is a traceless tensor and consequently does not contribute to b^S [see Eq. (8.4a)].

We will now obtain the terms that appear in $\chi^0_{\alpha\alpha}(\mathbf{k})$ [Eq. (8.1)] when we retain the polarizability anisotropy to order $(\gamma/\alpha)^2$ while retaining the approximation $g^{(n)} \simeq \tilde{g}^{(n)}$; that is, we derive Eqs. (8.20) and (8.21). In the diagrammatic notation, we may write Eq. (8.1) as

$$\chi^0_{\alpha\alpha} = \frac{\alpha\rho}{\epsilon_0}\left[1 + \frac{\rho}{\alpha}\int d\underset{\sim}{r}^{12} \, d\Omega_1 \, d\Omega_2 \left(\underset{1 \quad 2}{\text{⬤〰〰⬤}} \, e^{i\underset{\sim}{k} \cdot \underset{\sim}{r}^{12}} + \underset{1}{\text{⬤⬤}} \right)_{\alpha\alpha} \right] \tag{B.13}$$

The term $[\text{⬤〰〰⬤} \exp i\underset{\sim}{k} \cdot \underset{\sim}{r}^{12}]$ on the right-hand side gives a vanishing contribution to the trace of χ^0, and this situation is unaltered by introduction of an anisotropic polarizability as long as $g^{(n)} = \tilde{g}^{(n)}$. Within this approximate scheme, that is, $g^{(n)} = \tilde{g}^{(n)}$, we need examine only the last term in Eq. (B.13). As a consequence of Eq. (B.4), the only diagrams affected are those that contain two or more dots associated with the same particle. When Eqs. (B.8)–(B.10) are substituted into ∫, two classes of terms are found to give contributions in $O(\gamma/\alpha)^2$: Class (a) are those terms of which

are the first few members. The first few terms of this series are found to be

$$\text{Class}\,(a) = \frac{2}{9}\left(\frac{\gamma}{\alpha}\right)^2 2x^2(1 - x + 7x^2 - 13x^3 + \cdots) \qquad (B.14)$$

We propose that this series be resumed as indicated in Eq. (8.20). Class (b) are those terms in which the anisotropy contributes to the renormalization of the dipole propagators in the reaction field; for example,

Equations (B.8)–(B.10) suffice to evaluate the first few members of this class, and these terms are given in Eq. (8.21). As discussed in the main text, these terms show that the class (b) contribution is small; for that reason, we will not evaluate the higher-order terms.

APPENDIX C

In this appendix, we obtain the values, which appear in the text, of a number of integrals involving products of dipole tensors. The most systematic way to handle these integrals is to express the dipole tensors in spherical form and then to use the properties of the spherical harmonics. The cartesian-spherical (C-S) transformation coefficients $(12, m|\alpha\beta)$ introduced by Stone[78] provide a compact way of handling the transformation. The dipole tensor elements, which are defined by Eqs. (2.10) and (3.8a,b), may be written as

$$T_{\alpha\beta}(\mathbf{r}^{12}) = \frac{6^{1/2}C_{2m}(\hat{\mathbf{r}}^{12})}{4\pi\varepsilon_0(r^{12})^3}(12, m|\alpha\beta) \qquad (C.1)$$

where C_{2m} is a spherical harmonic.[25]

First, since it enters in Eq. (2.12), we require the Fourier transform $\hat{T}_{\alpha\beta}$ of the weighted quantity $T_{\alpha\beta}\tilde{g}(12)$:

$$\hat{T}_{\alpha\beta}(\mathbf{k}) = \int d\mathbf{r}\, \tilde{g}(r) T_{\alpha\beta}(\mathbf{r}) e^{i\mathbf{k}\cdot\mathbf{r}} = \frac{6^{1/2}}{4\pi\varepsilon_0}(12, m|\alpha\beta)\int_{r_0}^{\infty} dr\, r^{-1}$$

$$\times \sum_l (2l+1) i^l j_l(kr) \int d\hat{\mathbf{r}}\, C_{l\mu}(\hat{\mathbf{k}}) C_{l\mu}^*(\hat{\mathbf{r}}) C_{2m}(\hat{\mathbf{r}}) \qquad (C.2)$$

where, in the second equality, we have used the Rayleigh expansion[25] of $e^{i\mathbf{k}\cdot\mathbf{r}}$. Making use of the orthogonality of the spherical harmonics,[25] we obtain

$$\hat{T}_{\alpha\beta}(\mathbf{k}) = -\varepsilon_0^{-1} 6^{1/2}(12, m|\alpha\beta) C_{2m}(\hat{\mathbf{k}})\int_{r_0}^{\infty} dr\, r^{-1} j_2(kr) \qquad (C.3)$$

The radial integral is readily performed by making use of the expression[79]

$$j_n(z) = z^n\left(-\frac{1}{z}\frac{d}{dz}\right)^n \frac{\sin z}{z} \qquad (C.4)$$

and the resulting expression for $\hat{T}$ is

$$\hat{T}_{\alpha\beta}(\mathbf{k}) = -6^{1/2}\varepsilon_0^{-1}(12, m|\alpha\beta) C_{2m}(\hat{\mathbf{k}})\frac{j_1(kr_0)}{kr_0} \qquad (C.5)$$

For $\mathbf{k}$ parallel to the z direction, $C_{2m}(\hat{\mathbf{k}}) = \delta_{m0}$. In the $k \to 0$ limit,

$$\lim_{k \to 0} \hat{T}_{\alpha\beta}(\hat{\mathbf{k}}) = -\frac{1}{3\varepsilon_0} 6^{1/2}(12, 0|\alpha\beta) \qquad (C.6)$$

Notice that

$$(12,0|zz) = -2(12,0|xx) = -2(12,0|yy) \qquad (C.7)$$

so that any tensor proportional to $\hat{T}$ is traceless.

This integral may be used to evaluate not only the transform of $T\tilde{g}$, but also the transform $T(\mathbf{k})$ of the full static dipole-dipole propagator T given by Eq. (2.10), with $\omega/c \to 0$,

$$T_{\alpha\beta}(\mathbf{k}) \equiv \int d\mathbf{r}\, e^{i\mathbf{k}\cdot\mathbf{r}} F(r,0) = \lim_{\cdot r_0 \to 0} \hat{T}_{\alpha\beta}(\mathbf{k}) - \frac{1}{3\varepsilon_0}\delta_{\alpha\beta} \qquad (C.8)$$

From Eq. (C.5), we find

$$\lim_{r_0 \to 0} \hat{T}_{\alpha\beta}(\mathbf{k}) = -\frac{6^{1/2}}{3\varepsilon_0} C_{2m}(\mathbf{k})(12, m|\alpha\beta) = -\frac{1}{\varepsilon_0}\left(\hat{k}_\alpha \hat{k}_\beta - \frac{\delta_{\alpha\beta}}{3}\right) \qquad (C.9)$$

Consequently, we obtain Eq. (2.12).

The second integral that we require is $\int_{V_0} d\mathbf{r}\, T_{\alpha\beta}(\mathbf{r}' - \mathbf{r})$, where the integral is over the volume V_0, a sphere of radius r_0 about the origin, and $\mathbf{r}'$ lies outside the sphere. This integral is needed in the reduction of Eq. (8.24) to Eq. (8.25). Here, we make use of the properties of the irregular solid harmonic,

$$I_{lm}(\mathbf{r}) = r^{-(l+1)}C_{lm}(\hat{r}) \tag{C.10}$$

under a change of origin.[80] We have

$$\int_{V_0} d\mathbf{r}\, T_{\alpha\beta}(\mathbf{r}' - \mathbf{r}) = -\frac{6^{1/2}}{4\pi\varepsilon_0}(12, m|\alpha\beta)\int_{V_0} d\mathbf{r}\, I_{2m}(\mathbf{r}' - \mathbf{r})$$

$$= -\frac{6^{1/2}}{4\pi\varepsilon_0}(12, m|\alpha\beta)\sum_{s=0}^{\infty}(-1)^{2+m}\left(\frac{(2s+5)!}{4!2s!}\right)^{1/2}$$

$$\times\begin{pmatrix} s+2 & s & 2 \\ m-t & t & -m \end{pmatrix}I_{2+s,m+t}(\mathbf{r}')\int_0^{r_0} dr\, r^{2+s}\int d\hat{r}\, C_{st}(\hat{r})$$

$$= r_0^3 T_{\alpha\beta}(\mathbf{r}')4\pi/3 \tag{C.11}$$

This result is used in the discussion below Eq. (8.25).

For the understanding of Eqs. (4.9), (8.25), and (8.28)–(8.30), we also require the convolution

$$\mathscr{I}_{\alpha\beta}(\mathbf{r}^{12}) = \int d\mathbf{r}^3\, \tilde{g}^{(2)}(\mathbf{r}^{13})T_{\alpha\gamma}(\mathbf{r}^{13})T_{\gamma\beta}(\mathbf{r}^{32})\tilde{g}^{(2)}(\mathbf{r}^{32})$$

$$= \varepsilon_0^{-2}6(12, m|\alpha\gamma)(12, m'|\gamma\beta)(2\pi)^{-3}$$

$$\times \int d\mathbf{k}\, e^{-i\mathbf{k}\cdot\mathbf{r}^{12}}C_{2,m}(\hat{k})C_{2,m'}(\hat{k})\left(\frac{j_1(kr_0)}{kr_0}\right)^2 \tag{C.12}$$

where in the second equality we have made use of the convolution theorem. We now contract the product of spherical harmonics[25] and perform the angular integral over $\hat{k}$, making use of the Rayleigh expansion[25]

$$\mathscr{I}_{\alpha\beta}(\mathbf{r}^{12}) = (12, m|\alpha\gamma)(12, m'|\gamma\beta)\frac{3}{4\pi^2\varepsilon_0^2}$$

$$\times \sum_L (2L+1)(-1)^M\begin{pmatrix} 2 & 2 & L \\ m & m' & -M \end{pmatrix}\begin{pmatrix} 2 & 2 & L \\ 0 & 0 & 0 \end{pmatrix}$$

$$\times C_{LM}(\hat{r}^{12})i^L\int dk\, k^2\left(\frac{j_1(kr_0)}{kr_0}\right)^2 j_L(kr^{12}) \tag{C.13}$$

The techniques developed by Stone[78] enable us to combine $3-j$ symbols and C-S coefficients; if this is done, then

$$\mathscr{I}_{\alpha\beta}(\mathbf{r}^{12}) = 2^4 \cdot 3 \cdot 5 i^L (2L+1)^{1/2} \begin{pmatrix} 2 & 2 & L \\ 0 & 0 & 0 \end{pmatrix} (4\pi\varepsilon_0)^{-2} W(1212; 1L) C_{LM}(\hat{\mathbf{r}}^{12})$$

$$\times \pi^{3/2} 2^{-3/2} (r^{12})^{-1/2} r_0^{-3} \int dk\, k^{-3/2} J_{L+1/2}(kr^{12}) \left[J_{3/2}(kr_0) \right]^2$$

$$(C.14)$$

In this step, we have transformed the integral dk into a standard form in terms of the normal Bessel functions $J_n(x)$, which may, in large part, be found in standard texts.[79,81] For the special case of $1 \equiv 2$, $r^{12} = 0$ and we must have $L = 0$; we then find

$$\mathscr{I}_{\alpha\beta}(0) = \frac{\delta_{\alpha\beta}}{6 r_0^3 \pi \varepsilon_0^2} \qquad \text{for } 1 \equiv 2, \, r^{12} = 0 \qquad (C.15)$$

By the use of standard integrals,[81] we can show that

$$\mathscr{I}_{\alpha\beta}(\mathbf{r}^{12}) = -(3\varepsilon_0)^{-1} T_{\alpha\beta}^{12} \qquad (r^{12} > 2r_0) \qquad (C.16)$$

and

$$\mathscr{I}_{\alpha\beta}(\mathbf{r}^{12}) = -T_{\alpha\beta}^{12} \frac{(1 - r^2 16 r_0^{-2}) r^4}{r_0^4 16 \varepsilon_0} + u(r^{12}) \qquad (r^{12} < 2r_0) \quad (C.17)$$

Equation (C.16) is relevant to the discussion below Eq. (4.9). In Eq. (C.14), $u(r^{12})$ is a nonstandard integral:

$$u(r^{12}) = \frac{2^4}{r_0^3} \int dx\, x^2 \left[\frac{j_1(x)}{x} \right]^2 j_0 \left(\frac{x r^{12}}{r_0} \right) \qquad (r^{12} < 2r_0) \qquad (C.18)$$

For our development, we only require the integral

$$\int dr\, u(r) r^2 = -\frac{5}{16 \varepsilon_0^2} \qquad (C.19)$$

which has been evaluated by other authors.[21] Equations (C.15)–(C.19) are useful in reducing Eq. (8.24) to (8.25); in addition, Eqs. (C.17)–(C.19) are specifically important in the discussion of Eqs. (8.28)–(8.30).

In the analysis of the multiple scattering series that appears in Appendix B, particularly in the discussion of Eq. (B.7′), we encounter integrals such as

$$\mathcal{H}_{\alpha\beta}(\mathbf{r}^{13},\mathbf{r}^{12}) = \int d\mathbf{r}^{14} \left[T^{34}(\mathbf{r}^{14} - \mathbf{r}^{13}) \tilde{g}(\mathbf{r}^{14} - \mathbf{r}^{13}) \right.$$

$$\left. \cdot T^{24}(\mathbf{r}^{12} - \mathbf{r}^{14}) \tilde{g}(\mathbf{r}^{12} - \mathbf{r}^{14}) \right]_{\alpha\beta}$$

$$\times \tilde{g}(\mathbf{r}^{13}) \tilde{g}(\mathbf{r}^{14}) \tilde{g}(\mathbf{r}^{12}) \qquad (C.20)$$

in which $\mathbf{r}^{12}$ and $\mathbf{r}^{13}$ are held fixed. The factor $\tilde{g}(\mathbf{r}^{14})$ prevents particle 4 from entering the space occupied by particle 1. This integral may be evaluated by an extension of the techniques discussed above. By setting $\tilde{g} = 1 + [\tilde{g} - 1]$, we immediately obtain

$$\mathcal{H}_{\alpha\beta}(\mathbf{r}^{13},\mathbf{r}^{12}) = \tilde{g}(r^{13}) \tilde{g}(r^{12}) \mathcal{I}_{\alpha\beta}(\mathbf{r}^{12} - \mathbf{r}^{13}) - Q_{\alpha\beta}(\mathbf{r}^{13},\mathbf{r}^{12}) \quad (C.21)$$

where $\mathcal{I}$ is given above and $\mathbf{Q}$ is given by

$$Q_{\alpha\beta}(\mathbf{r}^{13},\mathbf{r}^{12}) = \frac{6}{(4\pi\varepsilon_0)^2} \tilde{g}(r^{13}) \tilde{g}(r^{12}) \int_{V_0} d\mathbf{r}^{14}$$

$$\times (12, m|\alpha\gamma)(12, m'|\gamma\beta) I_{2,m}(\mathbf{r}^{14} - \mathbf{r}^{13})$$

$$\times \tilde{g}(\mathbf{r}^{14} - \mathbf{r}^{13}) I_{2,m'}(\mathbf{r}^{12} - \mathbf{r}^{14}) \tilde{g}(\mathbf{r}^{12} - \mathbf{r}^{14}) \quad (C.22)$$

where V_0 means that the domain of integration is a sphere of radius r_0 about the origin and where we have made use of Eqs. (C.10) and (C.1). The factors $\tilde{g}(\mathbf{r}^{13})$ and $\tilde{g}(\mathbf{r}^{12})$ and the limits on the integral imply that $r^{14} < r^{13}$ and $r^{14} < r^{12}$, and that to a very good approximation we may set $\tilde{g}(\mathbf{r}^{14} - \mathbf{r}^{13})$ and $\tilde{g}(\mathbf{r}^{12} - \mathbf{r}^{14})$ to unity. We may then shift the origins of the solid harmonics as in Eq. (C.12); this leads to

$$Q_{\alpha\beta}(\mathbf{r}^{12},\mathbf{r}^{13}) = \frac{6}{(4\pi\varepsilon_0)^2} (12, m|\alpha\gamma)(12, m'|\gamma\beta)$$

$$\times \sum_{s=0} \sum_{s'=0} (-1)^{2+m} \left[\frac{(2s+5)!}{4!2s!} \right]^{1/2}$$

$$\times \left[\frac{(2s'+5)!}{4!2s'!} \right]^{1/2} \begin{pmatrix} s+2 & s & 2 \\ m-t & t & -m \end{pmatrix} \begin{pmatrix} s'+2 & s' & 2 \\ m'-t & t' & -m \end{pmatrix}$$

$$\times \tilde{g}(r^{13}) I_{2+s, m+t}(-\mathbf{r}^{13}) I_{2+s', m'+t'}(\mathbf{r}^{12}) \tilde{g}(r^{12})$$

$$\times \left\{ \delta_{ss'} \delta_{tt'} \frac{r_0^{2s+3}}{2s+3} \left(\frac{4\pi}{2s+1} \right) \right\} \qquad (C.23)$$

where the factor in curly brackets is the value of the integral over the two regular harmonics. The values of $s \neq 0$ in this sum may often be neglected, since:

1. The radial factors may be written

$$r_0^{-3} \tilde{g}(r^{13}) \left[\frac{r^{13}}{r_0} \right]^{s+3} \tilde{g}(r^{12}) \left[\frac{r^{12}}{r_0} \right]^{s+3}$$

   where the factors in square brackets are necessarily less than unity.
2. In the analysis of Appendix B, Q creates a graph in which the vector $\mathbf{r}^{13}$ appears in a loop. An angular integral will then always ensure that only the $s = 0$ terms from Q will survive. For our purposes, we may therefore neglect the $s \neq 0$ terms and obtain

$$Q_{\alpha\beta}(\mathbf{r}^{12}, \mathbf{r}^{13}) = -\frac{4\pi r_0^3}{3} T_{2,m}(\mathbf{r}^{13}) T_{2,m'}(\mathbf{r}^{12}) \delta_{mm'} \qquad (C.24)$$

APPENDIX D

In our development, we have replaced the retarded dipole field propagation F specified in Eq. (2.10) by its static limit. The justification for this was mentioned briefly in Section II and has been discussed by Fulton[18a] and by Titulaer and Deutch.[18b] As a consequence of this replacement, our theory does not describe the propagation of electromagnetic fields through the medium and does not yield the relation between $\mathbf{k}$ and ω. A complete theory that would accomplish this would be tantamount to a molecular derivation of Maxwell's equations and of the constitutive equations within the material. Such a theory, however, is not our goal; we seek merely a molecular expression for $\varepsilon(\omega)$. We are content to ascribe to Maxwell's equations the role of determining the properties of the propagating waves and to regard $\varepsilon(\omega)$ as a measurable property of the material. The intensive character of $\varepsilon(\omega)$ implies that it can be determined only by intermolecular correlations over ranges for which the retardation part of the propagator is irrelevant. Fulton[18a] has shown how the retardation terms lead to additional $\mathbf{k}$- and ω-dependent contributions in the relationship between χ and χ^0, as well as in the expression for $\langle M_\alpha(\mathbf{k}, t) M_\alpha(-\mathbf{k}) \rangle$; the resulting expression for χ, however, remains unaltered by the retardation terms. Similar conclusions were reached by Titulaer and Deutch.[18b]

NOTE ADDED IN PROOF

The time scale separation and the validity of the Fröhlich formula have recently been investigated in a computer simulation of the dielectric properties of CH_3CN; see D. M. F. Edwards and P. A. Madden, *Mol. Phys.* **51**, xxx (1984).

REFERENCES

1. C. J. F. Böttcher, *Theory of Electric Polarisation*, Vol. 1, Elsevier, New York, 1973; C. J. F. Böttcher and P. Bordewijk, *Theory of Electric Polarisation*, Vol. 2, Elsevier, New York, 1978.

2. C. Brot, in *Dielectric and Related Processes*, Vol. 2, M. Davies, Ed., a Specialist Periodical Report, Chemical Society, London, 1975, p. 1.

3. A. D. Buckingham, in *M.T.P. Review of Physical Chemistry*, Vol. 2, A. D. Buckingham, Ed., Butterworths, London, 1972, p. 241.

4. G. Nienhuis and J. M. Deutch, *J. Chem. Phys.* **55**, 4213 (1971).

5. G. S. Stell, G. Patey, and J. Høye, *Adv. Chem. Phys.* **48**, 183 (1981).

6. J. M. Deutch, *Ann. Rev. Phys. Chem.* **24**, 301 (1973).

7. M. S. Wertheim, *Ann. Rev. Phys. Chem.* **30**, 471 (1979).

8. R. Kubo, *J. Phys. Soc. Japan* **12**, 570 (1957).

9. U. M. Titulaer and J. M. Deutch, *J. Chem. Phys.* **60**, 1502 (1974).

10. B. U. Felderhof, *Physica* **76**, 486 (1974).

11. P. Mazur, *Adv. Chem. Phys.* **1**, 309 (1956).

12. T. F. Keyes and B. M. Ladanyi, *Mol. Phys.* **33**, 1063, 1099 (1977).

13. D. E. Sullivan and J. M. Deutch, *J. Chem. Phys.* **64**, 3870 (1976); R. L. Fulton, *J. Chem. Phys.* **62**, 3676 (1975).

14. R. L. Fulton, *Mol. Phys.* **29**, 405 (1975).

15. D. E. Sullivan and J. M. Deutch, *J. Chem. Phys.* **62**, 2130 (1975); D. E. Sullivan, Ph.D. thesis, MIT, 1976.

16. H. Mori, *Prog. Theor. Phys.* **33**, 423 (1965); **34**, 399 (1965).

17. D. Jackson, *Classical Electrodynamics*, Wiley, New York, 1962; see also Refs. 20 and 4 for a discussion of the behavior at the origin.

18. (a) R. L. Fulton, *J. Chem. Phys.* **63**, 77 (1975); (b) U. M. Titulaer and J. M. Deutch, *J. Chem. Phys.* **60**, 2703 (1974).

19. A. D. Buckingham, *Adv. Chem. Phys.* **12**, 107 (1967).

20. D. E. Logan, *Mol. Phys.* **44**, 1271 (1981); **46**, 271, 1155 (1982); S. Kielich, in *Dielectric and Related Processes*, Vol. 1, M. Davies, Ed., a Specialist Periodical Report, Chemical Society, London, 1972, p. 192.

21. D. E. Logan and P. A. Madden, *Mol. Phys.* **46**, 715, 1195 (1982), and references therein.

22. L. Onsager, *J.A.C.S.* **58**, 1486 (1936).

23. D. Kivelson and P. A. Madden, *Mol. Phys.* **30**, 1749 (1975).

24. J. G. Kirkwood, *J. Chem. Phys.* **7**, 911 (1939).

25. D. M. Brink and G. R. Satchler, *Angular Momentum*, Clarendon Press, Oxford, 1975.

26. A. J. Stone, *Mol. Phys.* **36**, 241 (1978). The expression we use is for linear molecules; however, it will be sufficient to discuss dipolar correlations for symmetric tops.

27. L. Blum and A. J. Torruella, *J. Chem. Phys.* **56**, 303 (1972).

28. J. S. Høye and G. Stell, *J. Chem. Phys.* **61**, 562 (1974).

29. The following contain applications, with references to the theory: (a) E. L. Pollock and B. J. Alder, *Physica* **102A**, 2 (1980); (b) G. N. Patey, D. Levesque, and J. J. Weis, *Mol. Phys.* **45**, 733 (1982); (c) D. J. Adams, *ibid.* **40**, 1261 (1980); **42**, 907 (1981).

30. G. N. Patey, D. Levesque, and J. J. Weis, *Mol. Phys.* **38**, 1635 (1979) and references therein; see also Ref. 5.

31. (a) D. Chandler, *Ann. Rev. Phys. Chem.* **29**, 441 (1978); and in *The Liquid State of Matter*, E. N. Montroll and J. L. Lebowitz, Eds., North-Holland, Amsterdam, 1982). (b) D. Chandler, *Faraday Disc. Chem. Soc.* **66**, 71 (1978); D. E. Sullivan and C. G. Gray, *Mol. Phys.* **42**, 443 (1981); P. T. Cummings and G. Stell, *ibid.*, **46**, 383 (1982). (c) D. Chandler, C. G. Joslin, and J. M. Deutch, *ibid.* **47**, 871 (1982).

32. A. D. Buckingham, *Proc. Roy. Soc. London* **A238**, 235 (1956).

33. (a) O. Steinhauser, *Mol. Phys.* **46**, 827 (1982); (b) J. Böhm, K. R. McDonald, and P. A. Madden, *Mol. Phys.* **49**, 347 (1983).

34. G. Nienhuis and J. M. Deutch, *J. Chem. Phys.* **56**, 235, 5511 (1972).

35. H. Fröhlich, *Theory of Dielectrics*, Oxford Univ. Press, London, 1958.

36. B. J. Berne, *J. Chem. Phys.* **62**, 1154 (1975).

37. P. Debye, *Polar Molecules*, Dover, New York, 1954.

38. (a) S. H. Glarum, *J. Chem. Phys.* **33**, 1371 (1960); (b) R. H. Cole, *J. Chem. Phys.* **42**, 637 (1965).

39. (a) C. Brot, G. Bossis, and C. Hesse-Bezot, *Mol. Phys.* **40**, 1053 (1980). (b) S. Bratos and G. Targus, *Phys. Rev.* **A24**, 1591 (1981).

40. (a) T. F. Keyes and D. Kivelson, *J. Chem. Phys.* **56**, 1057 (1971). (b) B. Guillot and S. Bratos, *Mol. Phys.* **37**, 991 (1979).

41. J. G. Powles, *J. Chem. Phys.* **21**, 633 (1953).

42. J. M. Deutch, *Faraday Symp. Chem. Soc. London* **11**, 26 (1977).

43. J. Goulon, J. L. Rivail, J. W. Fleming, J. Chamberlain, and G. W. Chantry, *Chem. Phys. Lett.* **18**, 211 (1972).

44. N. Hill, W. E. Vaughan, A. H. Proce, and M. Davies, *Dielectric Properties and Molecular Behavior*, Van Nostrand, New York, 1969.

45. G. Bossis, *Mol. Phys.* **31**, 1897 (1976).

46. (a) M. W. Evans, in *Dielectric and Related Processes*, Vol. 3, M. Davies, Ed., a Specialist Periodical Report, Chemical Society, London, 1977. (b) M. Evans, G. Evans, and R. Davis, *Adv. Chem. Phys.* **44**, 255 (1980).

47. D. Kivelson and T. Keyes, *J. Chem. Phys.* **57**, 4599 (1972).

48. (a) A. Gerschel, *Faraday Symp. Chem. Soc. London* **11**, 115 (1977); (b) A. Gerschel, I. Dimicoli, J. Jaffre, and A. Riou, *Mol. Phys.* **32**, 679 (1976).

49. B. Guillot and S. Bratos, *Phys. Rev.* **A16**, 424 (1977). S. Bratos and B. Guillot, *J. Molec. Structure*, **84**, 195 (1982).

50. R. Lobo, J. E. Robinson, and S. Rodriquez, *J. Chem. Phys.* **59**, 5992 (1973).

51. M. S. Wertheim, *Mol. Phys.* **25**, 211 (1973).

52. H. W. Graben, G. S. Rushbrooke, and G. Stell, *Mol. Phys.* **30**, 373 (1975); G. Stell and G. S. Rushbrooke, *Chem. Phys. Lett.* **24**, 531 (1974).

53. A. D. Buckingham and C. G. Joslin, *Mol. Phys.* **40**, 1513 (1980).

54. T. F. Keyes, D. Kivelson, and J. P. McTague, *J. Chem. Phys.* **55**, 4096 (1971).

55. P. A. Madden, *Mol. Phys.* **49**, 193 (1983); see also P. A. Madden and D. J. Tildesley, *Phil. Trans. Roy. Soc. Lond.* **A293**, 419 (1979).

56. D. Frenkel and J. P. McTague, *J. Chem. Phys.* **72**, 2801 (1980).

57. (a) M. S. Wertheim, *Mol. Phys.* **36**, 1217 (1978). (b) J. S. Høye and G. Stell, *J. Chem. Phys.* **73**, 461 (1980); **72**, 1597 (1980).

58. B. U. Felderhof, *J. Phys.*, *C* **12**, 2423 (1979).

59. T. I. Cox and P. A. Madden, *Mol. Phys.* **43**, 287 (1981).

60. E. L. Pollock and B. J. Alder, *Phys. Rev. Lett.* **46**, 950 (1981).

61. G. Ascarelli, *Chem. Phys. Lett.* **39**, 23 (1976); see also M. R. Philpott, *Ann. Rev. Phys. Chem.* **31**, 97 (1981), for a discussion of techniques used to observe the optical frequency longitudinal susceptibility.

62. R. W. Impey, P. A. Madden, and I. R. McDonald, *Mol. Phys.* **46**, 513 (1982).

63. D. M. F. Edwards and P. A. Madden, *Mol. Phys.* **48**, 471 (1983).

64. W. A. Steele, in *Intermolecular Spectroscopy and Dynamical Properties of Dense Systems*, (Proc. Int. School Physics E. Fermi *LXXV*), J. Van Kranendonk, Ed., North-Holland, Amsterdam, 1980, p. 325; W. A. Steele, *Adv. Chem. Phys.* **34**, 000 (1976).

65. J. McConnell, *Rotational Brownian Motion and Dielectric Theory*, Academic Press, New York, 1980.

66. T.-W. Nee and R. Zwanzig, *J. Chem. Phys.* **52**, 6353 (1970).

67. J. B. Hubbard and P. G. Wolynes, *J. Chem. Phys.* **69**, 998 (1978).

68. P. A. Madden and D. Kivelson, *J. Phys. Chem.* **86**, 4244 (1982).

69. J. Chamberlain, *Infrared Phys.* **11**, 25 (1971); see also Ref. 46 and J. Chamberlain in *Vibrational Spectra and Structure*, vol. 8, J. Durig, Ed., Elsevier, New York, 1978.

70. R. H. Cole, *Ann. Rev. Phys. Chem.* **28**, 283 (1977).

71. J. Yarwood and M. W. Evans, private communication.

72. R. M. Lynden-Bell and I. R. McDonald, *Mol. Phys.* **43**, 1429 (1981); *Chem. Phys. Lett.*

73. D. J. Tildesley and P. A. Madden, *Mol. Phys.* **49**, 193 (1983).

74. E. Detyna, K. Singer, J. V. L. Singer, and A. Taylor, *Mol. Phys.* **41**, 31 (1980); **37**, 1239 (1979).

75. See, e.g., G. Williams, *Chem. Rev.* **72**, 55 (1972), and references therein.

76. D. Kivelson and R. McPhail, *J. Chem. Phys.* (in press).

77. B. J. Berne and R. Pecora, *Dynamic Light Scattering* Wiley, New York, 1976.

78. A. J. Stone, *J. Phys. A* **9**, 485 (1976).

79. M. Abramovitz and I. A. Stegun, *Handbook of Mathematical Functions*, Dover, New York, 1965.

80. R. J. A. Tough and A. J. Stone, *J. Phys. A* **10**, 1261 (1977).

81. I. S. Gradsteyn and I. M. Ryzhik, *Tables of Integrals, Series and Products*, 4th ed., Academic Press, New York, 1980.

AUTHOR INDEX

Numbers in parentheses are reference numbers and indicate that the author's work is referred to although his name is not mentioned in the text. Numbers in *italics* show the pages on which the complete references are listed.

Abraham, F. F., 187, 187(106), *249*
Abraham-Schrauner, B., 165(64), *248*
Abramovitz, M., 559(79), *566*
Abramowitz, M., 293(20), *409*
Adams, D. J., 483(29), *565*
Adams, G. E., 168(66), *248*
Adelman, S. A., 237(197), 239(197), *252*
Aizenman, M., 121, 121(107), 125, 125(107), *139*
Alastuey, A., 191(116), 234(180), *250, 251*
Alder, B. J., 483(29), 528, 528(60), *565, 566*
Alexanian, M., 99, 99(75), 103, 103(75), 104, 104(75), *138*
Alfrey, T., 231(150), *251*
Ali-Zade, P. G., 172(75), *249*
Alms, G. R., 423(14), 440(14), 441(14), 444(14), 448(14), 454(56), *464, 465*
Altenberger, A. R., 157(238), *253*
Alterman, Z., 87, 87(67), *138*
Andersen, H. C., 193(119), 241(231), 242(231), *250, 253*
Applequist, J., 449(38,39), *464, 465*
Ascarelli, G., 534(61), *566*
Aval, G. M., 450(42), *465*

Bacquet, R., 232(156), 246(156), *251*
Badiali, J. P., 236(191,192), *252*
Bajpai, R. K., 131, 131(129), *139*
Bak, T. A., 121, 121(107), 125, 125(107), *139*
Barber, M., 234(176), 246(176), *251*
Barber, M. N., 237(208), *252*
Barker, J. A., 187, 187(106), *249*
Barlow, C. A., Jr., 143(7), 152(7), 168(7), 236(7), 237(7), *247*
Barnsley, M., 88(69), 96(73), 99(76), 103, 103(77), 104, 104(73), *138*

Barnsley, M. F., 108, 108(84), *138*
Barrett, J. J., 450(42), *465*
Barrow, J. D., 126, 126(109), *139*
Basu, S., 450(41), *465*
Battaglia, M. A., 440, 440(30), *464*
Bauer, D. R., 454(55), 455(55), *465*
Baus, M., 151(24), 160(24), 161(52), 234(179), *247, 248, 251*
Bell, G., 161(239), 176(80), 177, 177(80), 180(80), *249, 253*
Bell, G. M., 154(39,40), 170(39,40), 174(81), 176, 176(39,40,89), 180(39,40, 89), 207(89), *248, 249*
Benoit, H., 264, 264(7), 265(7), *408*
Ben Reuven, A., 413(1), 420(1), 422(11,12), *463, 464*
Berg, D. W., 231(150), *251*
Berne, B. J., 413(1), 420(1), *463*, 487(36), 488, 488(36), 544(77), 545(77), 546(77), *565, 566*
Bernstein, S., 107(83), *138*
Beysens, D., 421(7), *463*
Bhatnagar, P. L., 107(80), *138*
Bhuiyan, L. B., 146(19), 177(90,91,92), 180(90,91,92,97,98), 181(90,92,98), 182(90,91,92), 183(91), 184(90,91,92,98), 196(92), 207(90,91,92,98), 220(92), 225(98), 227(92), 229(141), *247, 250*
Birge, R. R., 452(46,48), 453(48,49), *465*
Blatz, P. L., 115, 115(101), *139*
Blum, L., 157(44), 160(48,49), 161(51,55), 177, 177(95), 187(107), 189(95,107), 191(109), 196(109), 206, 206(130), 229(141), 237(198), 238(214), 239(198,214,226,227), *248, 249, 250, 252, 253*, 480(27), *565*
Bluman, G. W., 28, 28(31), *137*

567

SUBJECT INDEX